Meteorology: Weather, Climate and the Environment

Meteorology: Weather, Climate and the Environment

Editor: Ben Perry

R CALLISTO REFERENCE

www.callistoreference.com

Callisto Reference,
118-35 Queens Blvd., Suite 400,
Forest Hills, NY 11375, USA

Visit us on the World Wide Web at:
www.callistoreference.com

ISBN: 978-1-64116-567-9 (Hardback)

Cataloging-in-Publication Data

Meteorology : weather, climate and the environment / edited by Ben Perry.
    p. cm.
Includes bibliographical references and index.
ISBN 978-1-64116-567-9
1. Meteorology. 2. Weather. 3. Climate. 4. Environmental protection. I. Perry, Ben.
QC861.3 .M48 2022
551.5--dc23

Table of Contents

Permissions

List of Contributors

Index

Preface

Meteorology is the scientific study of atmospheric phenomena, occurring in the lower stratosphere and troposphere. It also encompasses the study of weather. An important area of meteorological studies is the observation, understanding and forecasting of weather. Some of the focus areas of meteorology are climate modeling, climate change, remote sensing and air quality. Atmospheric parameters are measured using radar, weather balloons, satellites and buoys. They are relayed to weather centers to predict the future state of the atmosphere. Meteorological studies operate on four meteorological scales- microscale, mesoscale, global and synoptic scales. This book unravels the recent studies in the field of meteorology. Also included in this book is a detailed explanation of the various weather, climate and environment studies that are under the purview of meteorology. The readers would also gain knowledge that would broaden their perspective about this field.

Various studies have approached the subject by analyzing it with a single perspective, but the present book provides diverse methodologies and techniques to address this field. This book contains theories and applications needed for understanding the subject from different perspectives. The aim is to keep the readers informed about the progresses in the field; therefore, the contributions were carefully examined to compile novel researches by specialists from across the globe.

Indeed, the job of the editor is the most crucial and challenging in compiling all chapters into a single book. In the end, I would extend my sincere thanks to the chapter authors for their profound work. I am also thankful for the support provided by my family and colleagues during the compilation of this book.

Editor

Sensitivity Study on the Influence of Parameterization Schemes in WRF_ARW Model on Short- and Medium-Range Precipitation Forecasts in the Central Andes of Peru

Aldo S. Moya-Álvarez ⓘ,[1] **Daniel Martínez-Castro,**[1,2] **José L. Flores,**[1] **and Yamina Silva**[1]

[1]*Instituto Geofísico del Perú, Lima, Peru*
[2]*Instituto de Meteorología de Cuba, La Habana, Cuba*

Correspondence should be addressed to Aldo S. Moya-Alvarez; amoya@igp.gob.pe

Academic Editor: Mario M. Miglietta

A sensitivity study of the performance of the Weather Research and Forecasting regional model (WRF, version 3.7) to the use of different microphysics, cumulus, and boundary layer parameterizations for short- and medium-term precipitation forecast is conducted in the Central Andes of Peru. Lin-Purdué, Thompson, and Morrison microphysics schemes were tested, as well as the Grell–Freitas, Grell 3d, and Betts–Miller–Janjic cumulus parameterizations. The tested boundary layer schemes were the Yonsei University and Mellor–Yamada–Janjic. A control configuration was defined, using the Thompson, Grell–Freitas, and Yonsei University schemes, and a set of numerical experiments is made, using different combinations of parameterizations. Data from 19 local meteorological stations and regional and global gridded were used for verification. It was concluded that all the configurations overestimate precipitation, but the one using the Morrison microphysical scheme had the best performance, based on the indicators of bias (*B*) and root mean square error (RMSE). It is recommended not to use the Betts–Miller–Janjic scheme in this region for low resolution domains. Categorical forecast verification of the occurrence of rainfall as a binary variable showed detection rates higher than 85%. According to this criterion, the best performing configuration was the combination of Betts–Miller–Janjic and Morrison. Spatial verification showed that, even if all the configurations overestimated precipitation in some degree, spatial patterns of rainfall match the TRMM and PISCO rainfall data. Morrison's microphysics scheme shows the best results, and consequently, this configuration is recommended for short- and medium-term rainfall forecasting tasks in the Central Andes of Peru and particularly in the Mantaro basin. The results of a special sensitivity experiment showed that the activation or not of cumulus parametrization for the domain of 3 km resolution is not relevant for the precipitation forecast in the study region.

1. Introduction

Mesoscale meteorological models are a powerful tool, both for operational simulation and for atmospheric investigations [1]. In this sense, the Weather Research and Forecasting (WRF) model [2] is currently one of the most used in the world for these purposes, since it supports the use of very high-resolution grids for domains in any region of the planet and allows changes in its physics schemes configuration to tune it for regional conditions without having to compile it every time. Its development began in the

second half of the 1990s as a collaborative partnership, mainly between the National Center for Atmospheric Research (NCAR), the National Oceanic and Atmospheric Administration (NOAA), the Air Force Time Agency (AFWA), the Naval Research Laboratory, the University of Oklahoma, and the Federal Aviation Administration (FAA), all of the United States of America.

In the present work, a research is carried out to evaluate how the WRF model simulates rainfall in the Peruvian Central Andes. For validation, observational data from 19 stations in the region are used, as well as gridded observational data. Most of the rainfall in Peru is concentrated in the period between the months of September and April [3], defining a marked seasonality, with a dry season between May and August [4–6]. In this region, rainfall plays an important economic role, since 71% of the arable land in the basin depends on it for crops [7].

Regional weather numerical models, including WRF, have been previously used to carry out studies on different mountainous regions, including part of the Andes [8–11]. However, in Peruvian Central Andes, numerical modeling studies are scarce. In a recent paper [12], the influence of orography on the diurnal cycle of rainfall in the Central Andes has been investigated using WRF. However, there are no studies evaluating the sensitivity of the model to changes in its configuration; for example, the use of certain parameterization schemes of physical processes or the characteristics of the domains, which is of great interest due to the orographic complexity of the region, may be relevant to the forecasting skill of the model for operational applications. In the present investigation, WRF sensitivity to changes in its cumulus parameterization, microphysics, and boundary layer schemes is evaluated, relative to the results of its short- and medium-term precipitation forecast in the study region.

In the specialized bibliography, as well as in the user manual, there are dissimilar parametrization schemes of the different physical processes that WRF solves, among them those of boundary layer, clusters, and microphysics, which would be the objects of verification of this work. Thus, among others, for the boundary layer, there are the Mellor–Yamada–Janjic scheme (MYJ) [13] and the Yonsei University scheme (YSU) [14]. For convection, the parameterizations of Betts–Miller–Janjic (BMJ) [15], Grell 3D (GRELL3), and Grell–Freitas [16], were used, and for microphysics, the parameterizations of Thompson [17], Morrison [18], and Lin-Purdué [19] were used.

High-resolution numerical modeling has specific issues to deal with, as the choice to include or not cumulus parameterization schemes (CPS), and explicit microphysics. The WRF user manual (v. 3.7) suggests the explicit solution of convection for grids with these resolutions; however, in the scientific literature, there does not seem to be a total consensus on this aspect. For example, in [20], it is proposed that an explicit treatment of convection for a resolution, for example, of 4 km, shows a more accurate description of the physics of convective systems but not necessarily provides a better forecast. Thus, in [21], although it is stated that "Idealized modeling demonstrates that, at a fine resolution such as 2 or 4 km, classic super-cellular convective features including a reflectivity hook, midlevel rotation, and storm splitting can be resolved explicitly, and CPS is not needed," but it is concluded that it is not appropriate to assume that a simulation that uses a grid of small spacing (<5 km) will not need a CPS. This will depend on the synoptic forcing and the time of the year.

Consequently, and bearing in mind that the objective of this research is to verify the sensitivity of WRF to the use of different parametrization schemes for punctual precipitation forecast in this complex region, it was decided to include the

convection parameterization schemes also for the domain of 3 km resolution. However, a special numerical experiment was made, using the best performing configuration, deactivating the cumulus parameterization for the domain of 3 km resolution to investigate the sensitivity of the forecast to this procedure.

2. Data and Methodology

2.1. Initial and Border Conditions. The experiments were carried out using model WRF v. 3.7 (Weather Research and Forecasting, version 3.7). The initial and boundary conditions were taken from the "Global Operational Analysis" of the National Center of Environmental Prediction (NCEP), final analysis FNL (https://rda.ucar.edu/datasets/ds083.2/), every 6 hours, with horizontal resolution of 1° × 1°. The analysis contains surface data and 26 higher air levels and includes the following surface variables: sea level pressure, higher air variables (geopotential height, air temperature, and zonal and meridional components of the wind and vertical air velocity), sea surface temperature, and soil parameters. Even though there are currently Global Forecast System (GFS) outputs with resolution of 0.5° and 0.25° available, in the present research, border conditions of FNL with coarser resolution were used, with the objective that the physical WRF parameterizations determine the behavior of the forecast in the largest possible measure. On the contrary, for a sensitivity study, analysis data are preferred over prognostic output, which would have introduced implicit additional forecast uncertainty.

2.2. Topography Data. The topographic data Global 30 Arc-Second Elevation (GTOPO30) of the Geological Survey of the United States (USGS), which is a default option in the model, were replaced by the digital elevation model of the Shuttle Radar Topography Mission (SRTM; https://dds.cr.usgs.gov/srtm/version2_1/) [22, 23], with a resolution of 90 m, which for South America has an average horizontal error of 9.0 m and an average absolute vertical error of 6.2 m and improves about 10 times (both in spatial resolution and in vertical accuracy) the continental digital elevation model GTOPO30 [24].

2.3. Model Configuration. The simulations are performed for three domains (Figure 1), whose characteristics are specified in Table 1. The unidirectional nesting technique has been applied in the simulations, running all three domains at the same time. The model is initialized at 12 UTC in all cases.

The model was configured with 28 vertical levels, up to approximately 16,000 meters, for which the variable "e_vert" was used in the configuration of the "namelist.input." In this case, 57% of the levels were framed below 6000 meters.

A "control" simulation was defined, and the cumulus, boundary layer, and microphysical schemes were subsequently changed in the rest of the numerical experiments (only one at a time).

To carry out the sensitivity study, we have previously considered previous work results in similar conditions,

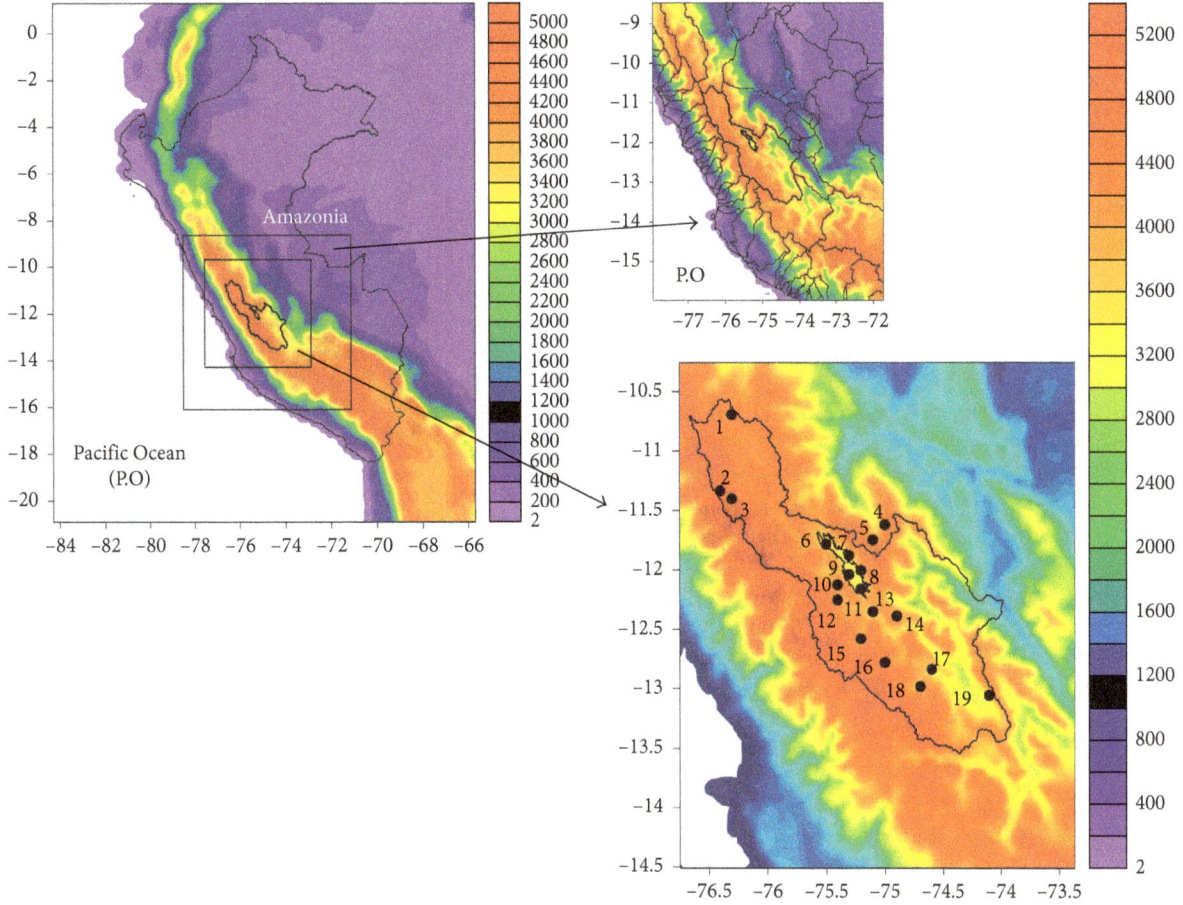

FIGURE 1: Domains used in the simulations and distribution of the stations considered for the verification of the results.

TABLE 1: Main characteristics of domains, initial, and boundary conditions.

Characteristics	Domain 1	Domain 2	Domain 3
Central point	Latitude: 10°S Longitude: 75°S	Latitude: 12.25819°S Longitude: 74.8356°S	Latitude: 12.39203°S Longitude: 75.0548°S
Horizontal grid step	18 km	6 km	3 km
Dimension (XYZ)	$115 \times 140 \times 28$	$115 \times 142 \times 28$	$125 \times 161 \times 28$
Time step	90 s	36 s	18 s
Initial and contour conditions	FNL $1.0^0 \times 1.0^0$	Simulation of domain 1	Simulation of domain 2

TABLE 2: Parameterization schemes used in the experiments.

Parameters	Experiments					
	CTR	C_BMJ	C_GRELL3	MP_MR	MP_LP	BL_MYJ
Microphysics	Thomson	Thomson	Thomson	Morrison	Lin et al.	Thomson
Cumulus	Grell–Freitas	Betts–Miller–Janjic	Grell 3D	Grell–Freitas	Grell–Freitas	Grell–Freitas
Boundary layer	Yonsei university	Yonsei university	Yonsei university	Yonsei university	Yonsei university	Mellor–Yamada–Janjic
Radiation LW	RRTMG	RRTMG	RRTMG	RRTMG	RRTMG	RRTMG
Radiation SW	RRTMG	RRTMG	RRTMG	RRTMG	RRTMG	RRTMG

focusing on orographic complexity and/or tropical condition. Thus, the parametrization schemes used in the control simulation (CTR in Table 2) and in the rest of experiments were mostly based on the suggestions of the WRF v. 3.7 "User Guide" and in previous results of sensitivity studies and applications of the model in the Andes [12] and in

domains with complex topography in tropical and subtropical regions, such as the Himalayas [25, 26], the south of India [27], and the Caribbean [28].

For the control experiment, the following parameterization schemes were selected: for microphysics, the Thompson scheme, the soil model is the so-called Unified

Noah Land Surface Model [29]; for convection, the Grell–Freitas parameterization, it explicitly represents updraft and downdraft and includes the "detrainment" of cloud and ice.

For the boundary layer, the YSU scheme was used, and for the surface layer, the MM5 similarity scheme was used [30–34].

The radiation model was the RRTMG (rapid radiative transfer model for general circulations models) [35], which is a relatively recent version of the RRTM (rapid radiative transfer model) [36], with a better representation of cloudiness, unresolved by the mesh of the model. It applies the k-correlated method to implement an algorithm that is characterized by its speed [37].

Six configurations of the model were tested, which have been summarized in Table 2: the control configuration (CTR); C_BMJ and C_GRELL3, with Betts–Miller–Janjic and Grell 3D as cumulus parameterizations; MP_MR and MP_LP, where Morrison and Lin-Purdué parameterizations have been used cumulus; and BL_MYJ, where the Mellor–Yamada–Janjic boundary layer scheme was applied. Notice that the acronyms of the configurations are composed by the type of the scheme that substitutes the one in the control configuration, CTR (C, convection; MP, microphysics; BL, boundary layer), and two or three letters referring to the applied scheme.

2.4. Study Period and Verification of Results.

The simulations were carried out with a forecast horizon of up to 10 days, for which 9 tens (periods of ten consecutive days) of the months of December, January, and February of the years 2007, 2009, 2010, 2011, and 2012 were selected. All the selected dates belong to the rainy period of the year. The particular tens were chosen among the rainiest for each month. Figure 2 shows the average of 24-hour rainfall for each forecast period for the 19 weather stations considered for punctual verification during the 9 selected tens.

The stations considered belong to the observation network of the National Meteorology and Hydrology Service of Peru (SENAMHI), whose data are subject to verification.

The tens considered were the following:

(1) First ten of January 2007

(2) Second ten of January 2007

(3) Second ten of February 2007

(4) Third ten of December 2007

(5) Second ten of February 2009

(6) First and ten of February 2010

(7) Third ten of February 2010

(8) Third ten of January 2011

(9) First ten of February 2012.

The coordinates and altitude of the stations considered for the study are shown in Table 3.

The verification of the results is carried out in three ways: punctual verification the forecast of 24-hour precipitation values, using numerical descriptive statistical measures, the second, punctual verification of the occurrence of

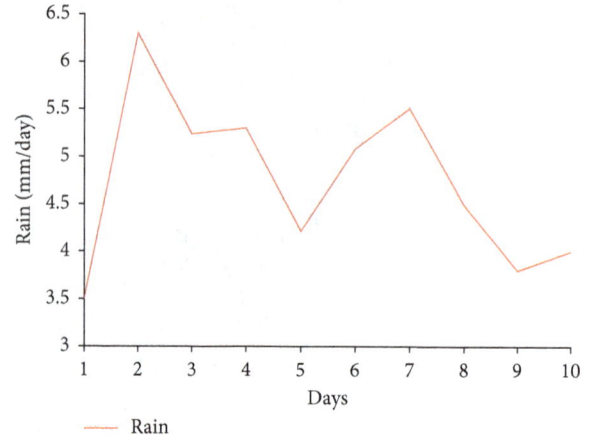

FIGURE 2: Mean of accumulated 24-hour precipitation for each forecast period in the 19 meteorological stations considered during the 9 selected tens.

precipitation as a categorical binary variable, and the third, verification of the ability of the model to reproduce patterns of spatial distribution of precipitation. Model output was interpolated using the Cressman method [38] to the stations grid points and compared to the stations grid point with the "in situ" measured data on the basis of the following statistics: bias (B), root mean squared error (RMSE), and mean absolute error (MAE) are calculated as follows:

$$B_k = \sum_{i=1}^{N_s} \sum_{j=1}^{N_p} \frac{\text{Pf}_{ij} - \text{Po}_{ij}}{N},$$

$$\text{RMSE}_k = \sqrt{\sum_{i=1}^{N_s} \sum_{j=1}^{N_p} \left(\frac{\text{Pf}_{ij} - \text{Po}_{ij}}{N}\right)^2}, \qquad (1)$$

$$\text{MAE}_k = \sum_{i=1}^{N_s} \sum_{j=1}^{N_p} \left|\frac{\text{Pf}_{ij} - \text{Po}_{ij}}{N}\right|,$$

where B_k, RMSE_k, and MAE_k are the bias, root mean square error, and absolute mean error for the k forecast term ($k = 1, 10$), respectively. Pf_{ij} and Po_{ij} are the simulated (forecast) and observed precipitation values, respectively, for the ith station and the jth period of ten days ($i = 1, 19$; $j = 1, 9$).

"B" estimates the difference of the mean value of the numerical forecast relative to the observations; "RMSE" is a measure of the magnitude of the mean error of the simulation with respect to the observations, while "MAE" constitutes the mean absolute deviation of the simulated variable with respect to the observations.

The binary categorical verification was carried out with the help of the so-called "contingency table," used in [39], but in our case, the fuzzy verification technique was not used, but direct point-to-point verification was used, which is based on the point-to-point comparison of the predicted and observed fields and the calculation of several indices, such as the probability of detection (POD), the proportion of

TABLE 3: Stations used to validate the simulations.

Number	Name	Abbreviation	Longitude	Latitude	Altitude (m)	Location
1	Cerro de Pasco	CER	−76.3	−10.693	4260	North
2	Yantac	YAN	−76.4	−11.333	4600	North
3	Marcapomacocha	MAR	−76.3	−11.4	4413	North
4	Ricran	RIC	−75	−11.619	3500	Center
5	Comas	COM	−75.1	−11.748	3300	Center
6	Jauja	JAU	−75.5	−11.784	3322	Center
7	Ingenio	ING	−75.3	−11.879	3450	Center
8	Santa Ana	STA	−75.2	−12.004	3295	Center
9	Huayao	HYO	−75.3	−12.038	3308	Center
10	S. J. de Jarpa	JAR	−75.4	−12.125	3726	Center
11	Vieques	VIQ	−75.2	−12.16	3186	Center
12	Laive	LAI	−75.4	−12.252	3990	Center
13	Pilchaca	PIL	−75.1	−12.35	3570	Center
14	Pampas	PAM	−74.9	−12.388	3260	Center
15	Huancapi	HCP	−75.2	−12.58	3800	South
16	Huancavelica	HCV	−75	−12.78	3676	South
17	Acobamba	ACO	−74.6	−12.838	3236	South
18	Lircay	LIR	−74.7	−12.982	3150	South
19	La Quinua	QUI	−74.1	−13.055	3260	South

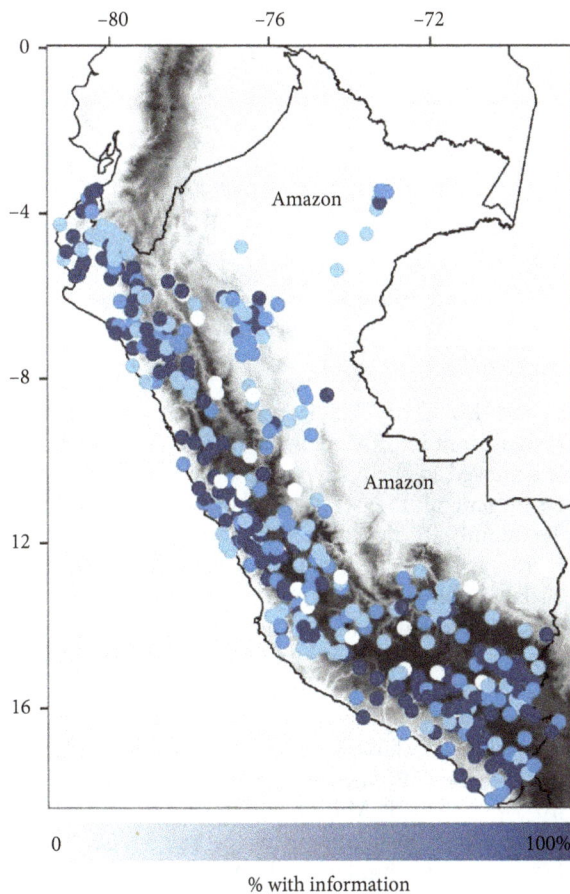

FIGURE 3: Network of stations used for the elaboration of PISCO (taken from [42]), with percentages of data per station, considering the period 1981–2016.

false alarms (FAR), the bias (B), and the "negative correct" (NC). As a criterion of the precipitation event, a 0.5 mm threshold was used.

In this case, a modification to the ACC (accuracy) index is proposed, and the Weighted Effectiveness Index (WEI) is introduced (2), which increases the relative weight of the misses, based on the criterion that a precipitation event that has occurred without being predicted, will have a greater negative impact on society and the economy than a false alarm:

$$\text{WEI} = \frac{\text{hits} + \text{negative corrects}}{\text{hits} + \text{negative corrects} + \text{falses alarms} + 1.25 * \text{misses}},$$

(2)

where hits = the event was forecast and it happened; false alarms = the event was forecast and it did not happen; negative corrects = the event was not forecast and it did not happen; and misses = the event was not forecast and it happened.

The spatial verification was carried out using the gridded data of the product 3B42 of the "Tropical Rainfall Measuring Mission" (TRMM 3B42), which is composed of a set of global "multisatellite" precipitation analysis data with a horizontal resolution of 0.25° and time resolution of 3 h, available between 50°S and 50°N [40, 41] and of the "Peruvian Interpolate data of the SENAMHI's Climato-logical and Hydrological Observations" (PISCO) [42]. In this sense, it is important to point out that PISCO data have little information in the area of the Peruvian Amazonia (Figure 3), which could influence the generated precipitation patterns [42]. To obtain the spatial distribution of bias, the TRMM grid was adjusted (by interpolation) to the output grid of WRF and subsequently the difference of the predicted grid and that of TRMM was calculated.

3. Results and Discussion

3.1. Results of Punctual Verification: 24-Hour Cumulative Precipitation. Table 4 shows the results of the simulations in each domain in terms of forecast quality measures.

TABLE 4: Statistics of the point verification of precipitation forecast for the 3 simulated domains and 6 combinations of parametrizations, including the control experiment, average of all the stations considered.

WRF_DOM	Statistics	Parametrizations					
		CTR	C_BMJ	C_GRELL3	MP_MR	MP_LP	BL_MYJ
1: 18 km	B (mm/day)	7.35	12.08	**6.63**	6.82	7.97	8.73
	RMSE (mm/day)	13.59	17.08	**12.41**	12.49	14.27	14.03
	MAE (mm/day)	10.17	13.84	**9.37**	9.63	10.62	11.18
2: 6 km	B (mm/day)	6.01	6.71	5.78	**4.66**	6.06	6.19
	RMSE (mm/day)	12.16	11.67	11.76	**10.62**	12.35	12.28
	MAE (mm/day)	9.09	9.23	8.92	**8.07**	9.18	9.31
3: 3 km	B (mm/day)	5.31	5.71	5.46	**3.68**	4.87	5.98
	RMSE (mm/day)	11.69	10.21	11.3	**10.02**	11.68	12.17
	MAE (mm/day)	9.7	8.08	8.52	**7.32**	8.45	9.19

The smaller errors are given in bold.

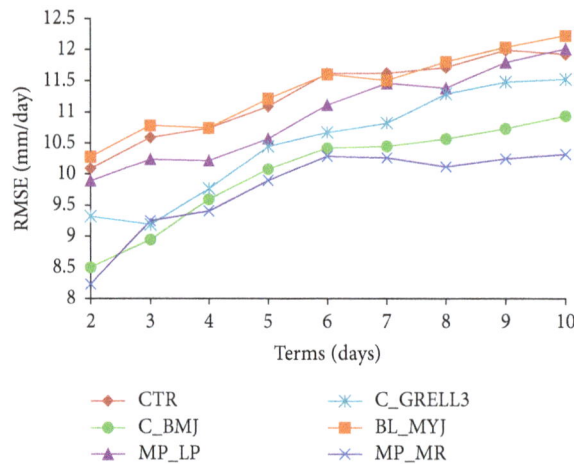

FIGURE 4: Evolution of the "RMSE" average of domains 2 and 3 of all stations with each configuration scheme used, for the 10 days of simulation horizon.

It can be seen that the model overestimated precipitation in all domains regardless of the configuration. This is consistent with [12], who found that WRF shows a significant positive bias over the Tropical Andes, when comparing the results of the diurnal cycle produced by the model with reference to the TRMM 3B42 product.

For the 18 km domain, the best results were obtained with the C_GRELL3 scheme (6.63 mm/day), which at the same time showed an "RMSE" of 12.41 mm/day and an "MAE" of 9.37 mm/day. The poorer result was obtained when the configuration C_BMJ was applied, which showed a significantly higher "B" than the rest of the experiments (12.08 mm/day).

For the 6 km domain, the MP_MR configuration gave the best results, with a bias of 4.66, "RMSE" of 10.62, and "MAE" of 8.07 mm/day. For this domain, all the configurations showed a remarkable improvement of their indicators in relation to the domain of 18 km. In the particular case of C_BMJ, it showed bias of 6.71 and "RMSE" of 11.67, very similar to the rest of the configurations, so it can be partially concluded that this configuration can be used for high-resolution domains, for example, 6 km, but not for low-resolution domains, this is 18 km in this case.

For the 3 km domain, the MP_MR configuration also showed the best indicators with bias of 3.68, "RMSE" of

10.02, and "MAE" of 7.32. The verification indicators did not show significant improvement for this domain in relation to those achieved in the 6 km domain.

So in general, the MP_MR configuration gave the best results, mainly for the finer domains.

Figure 4 shows the evolution of the average "RMSE" for the 10 days of the simulation horizon, between domains 2 and 3 with each configuration used. It is noted that the "RMSE" increases as the simulation period increases. The configuration C_BMJ and MP_MR shows relatively low RMSE values in all forecast days, growing slowly from day 1 to day 5, but from the sixth day of forecast on, C_BMJ continues increasing the error, while MP_MR curve shows almost no slope until day 10. Here, the MP_MR scheme showed the best behavior regarding RMSE during the whole period.

Figure 5 shows similar results as Figure 4, but for the B indicator. As can be seen, the lowest "B" was shown by the MP_MR configuration, producing the less rainfall overestimation. Figures 4 and 5 indicate that the MP_MR was not only the configuration showing the best results in a general way but also had the most stable behavior throughout the forecast horizon. Based on the point verification, it is further shown that all the tested configurations overestimate precipitation in the study area.

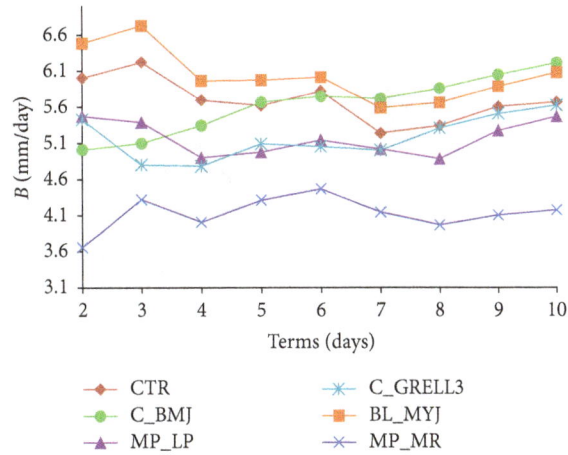

FIGURE 5: Evolution of the average "B" of domains 2 and 3 with each configuration, for the 10 days of the simulation horizon.

TABLE 5: Statistics of the point verification for the domains 2, 3, and 6 combinations of parametrizations, including the control experiment, average of all the stations considered.

WRF_DOM	Statistics	Parametrizations					
		CTR	C_BMJ	C_GRELL3	MP_MR	MP_LP	BL_MYJ
2: 6 km	POD (%)	81.54	**94.39**	90.61	94.04	92.28	92.72
	FAR (%)	28.40	28.55	29.00	29.10	**28.48**	28.77
	NC (%)	38.80	61.90	57.71	**65.31**	62.87	62.78
	B	**1.14**	1.32	1.28	1.33	1.29	1.30
	WEI (%)	66.33	**74.57**	72.36	74.3	73.90	73.84
3: 3 km	POD (%)	89.90	**93.07**	90.96	88.95	90.61	90.88
	FAR (%)	26.70	29.08	28.63	**26.63**	26.69	27.30
	NC (%)	**62.80**	62.74	59.61	61.23	58.20	59.38
	B	1.23	1.31	1.27	**1.21**	1.24	1.25
	WEI (%)	**74.18**	73.75	72.96	73.67	73.98	73.80

The smaller errors are given in bold.

TABLE 6: Statistics of the point quantitative verification for every ten, simulated with CPS and without CPS in domain 3.

CPS use	Statistics	Tens								
		1	2	3	4	5	6	7	8	9
CPS	B (mm/day)	**6.04**	**4.55**	1.76	2.64	**5.636**	5.97	**3.69**	−1.06	**3.97**
	RMSE (mm/day)	**9.66**	**9.20**	7.72	7.95	**11.01**	10.75	10.09	11.04	**12.81**
	MAE (mm/day)	**7.34**	**6.53**	4.09	5.28	**8.73**	7.52	**7.71**	9.14	**9.54**
Not CPS	B (mm/day)	6.92	5.15	**1.52**	**2.04**	5.637	**5.57**	3.95	**−0.97**	4.55
	RMSE (mm/day)	11.09	9.93	**6.71**	**7.28**	11.30	**10.42**	**9.80**	**10.58**	13.26
	MAE (mm/day)	8.46	7.22	**3.94**	**4.95**	8.97	**7.19**	7.73	**8.58**	9.95

The smaller errors are given in bold.

3.2. Results of Punctual Verification: The Occurrence of Precipitation as a Categorical Binary Variable. Table 5 shows the statistics of categorical binary verification for the different configurations. In a general sense, it can be noticed that the levels of detection of the different configurations are greater than 85%. However, the B was greater than 1, indicating that the model overestimates the presence of precipitations in the region. In the table, it is observed that, for the 6 km domain, the highest POD index (94.39%) was obtained with the C_BMJ configuration, while the highest "negative correct" of 65.31% was obtained with the MP_MR

configuration. The highest WEI was that of the C_BMJ scheme. The false alarms and B indices show close values for all the configurations.

For the 3 km domain, also the C_BMJ configuration showed the best POD (93.07%), with an NC of 62.74; however, the highest NC (62.8%), and WEI (74.18) were shown by the control configuration. In general, for the two domains, the CTR, C_BMJ, and MP_MR configurations were the most effective.

Additional verifications were performed, only for the stations located in the northern, central, and southern

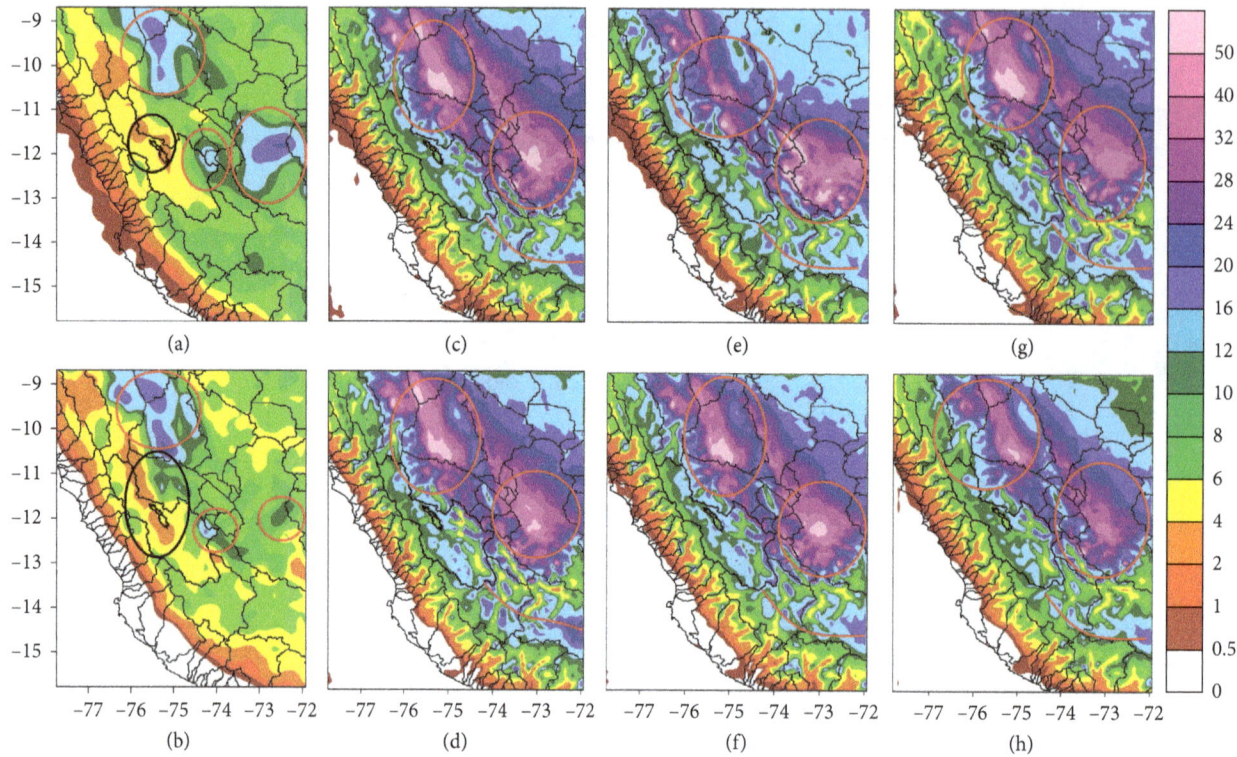

FIGURE 6: Spatial distribution of precipitation (mm) for the 6 km domain, average of all periods studied: (a) TRMM, (b) PISCO, (c) CTR, (d) MP_LP, (e) C_BMJ, (f) BL_MYJ, (g) C_GRELL3, and (h) MP_MR.

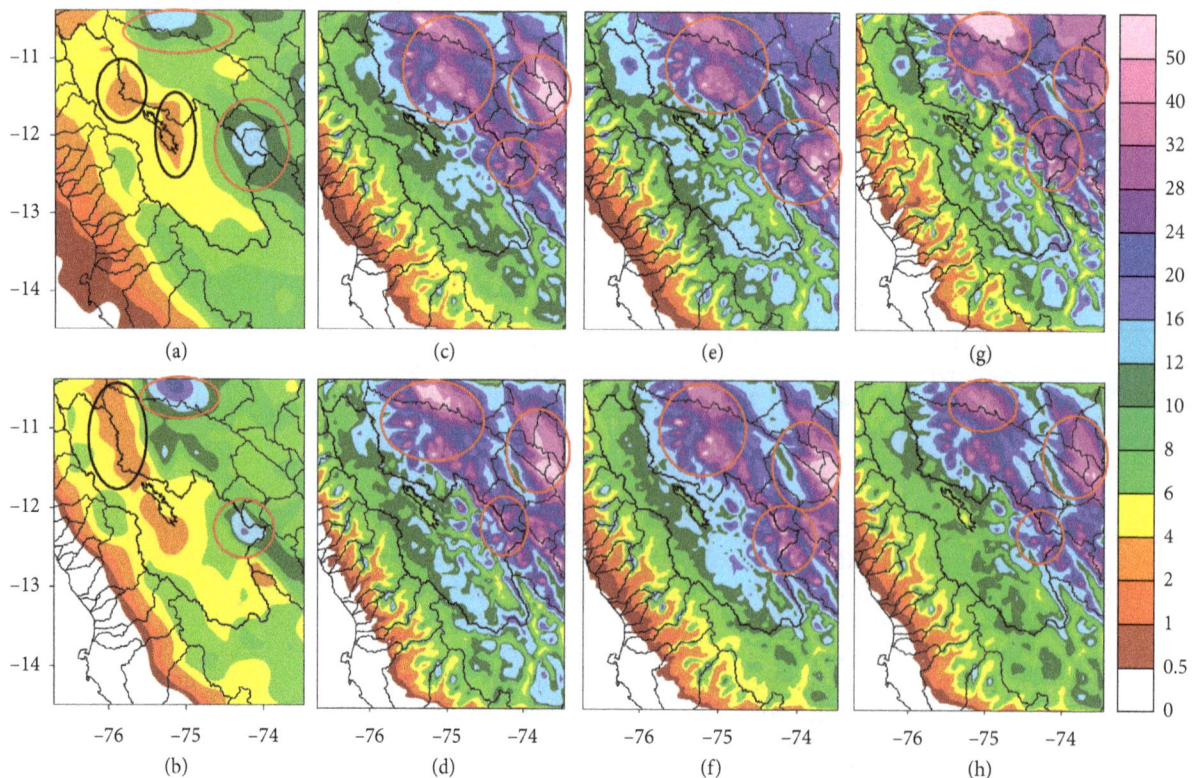

FIGURE 7: Spatial distribution of precipitation (mm) for the 3 km domain, average of all periods studied: (a) TRMM, (b) PISCO, (c) CTR, (d) MP_LP, (e) C_BMJ, (f) BL_MYJ, (g) C_GRELL3, and (h) MP_MR.

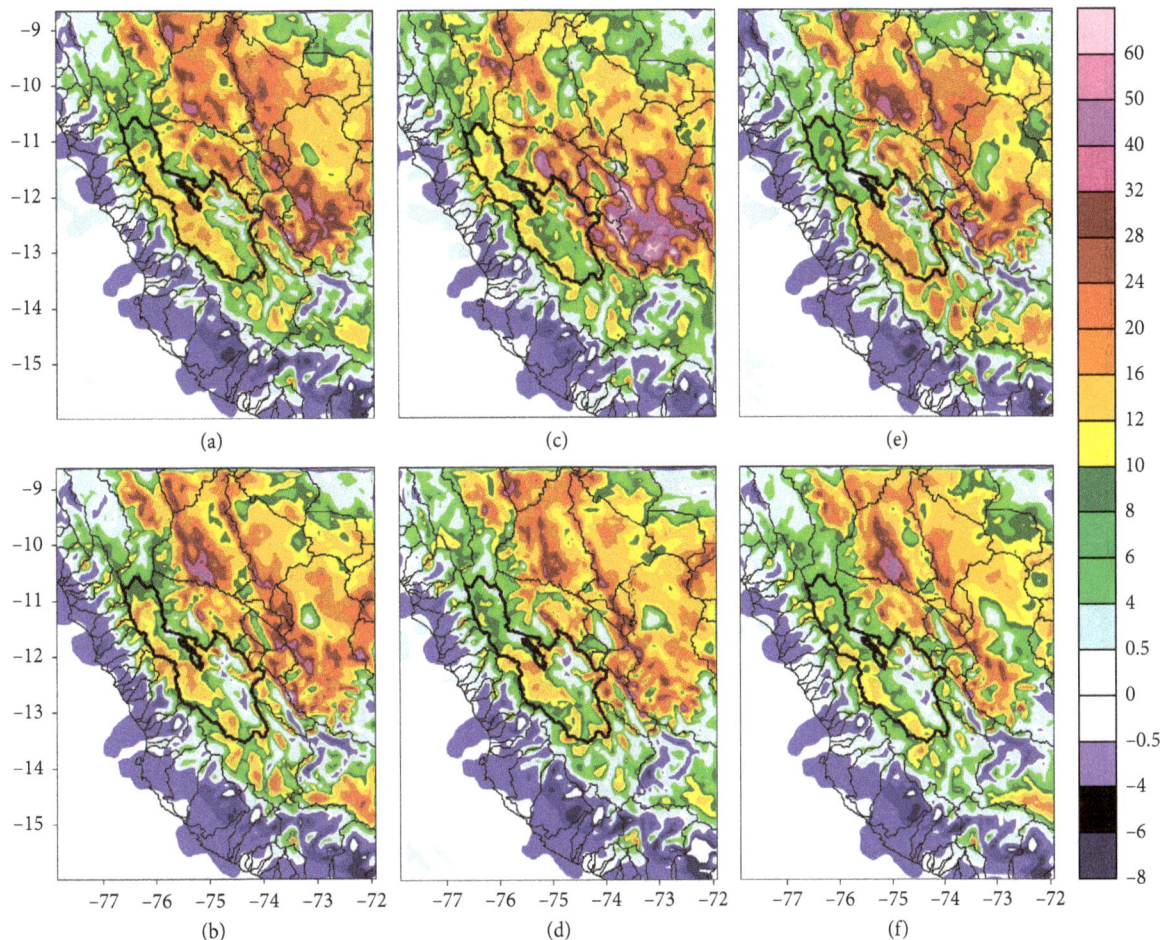

FIGURE 8: Spatial distribution of bias B (mm/day) for the 6 km domain, corresponding to the second day of forecast: (a) CTR, (b) MP_LP, (c) C_BMJ, (d) BL_MYJ, (e) C_GRELL3, and (f) MP_MR.

regions of the basin, both quantitatively and qualitatively. However, the results obtained were similar to those shown considering all the basin stations, and we do not show them in this paper.

As we had mentioned in the introduction, from the results of the experiments with the different parameterizations, a new experiment was carried out using the MP_MR configuration (which showed the best B and RMSE indicators for the entire forecast period) with the objective to verify the need or not to use CPS for the 3 km domain in the study region. In this sense, the results are shown in Table 6. In the table, the number of the tens corresponds to the order indicated in the Data and Methodology, and the best results for each ten are indicated in bold type. As can be seen, when looking at all the tens, the best results were obtained in some cases when CPS was used, and in others, when the cumulus scheme was deactivated. So, this result confirms the idea that it is not appropriate to assume that, for domains of less than 5 km, the parameterization of clusters must be deactivated, as was concluded in [21].

3.3. Results of the Verification of the Spatial Precipitation Field Forecast. Figure 6 shows the average spatial distribution of

the precipitation from the TRMM and PISCO databases and the model output for the different configurations used for the studied periods, corresponding to the 6 km domain. Regarding the data of TRMM and PISCO (Figures 6(a) and 6(b)), it can be seen that both databases show very similar patterns of precipitation. In this sense, the greater difference is that TRMM shows a maximum of rainfall around 12°S and 73°W, which is less notorious in PISCO, due to the fact that in that region there are scarce meteorological stations. Two maximum precipitation regions have been highlighted in red circles in Figures 6 and 7 in the zone of Amazonia, to the east of the mountain range. Another precipitation maximum of smaller area (also surrounded by a red circle in the figures) is located at the southeast of the Mantaro Basin, which contour line is represented by a thick black line in the figure.

An elongated rainy area which starts from the south of the basin and extends towards the southeast has been indicated with a thick red curve in each figure. Another significant feature is the relative spatial rainfall minimum shown by PISCO along part of the eastern boundary of the basin, mostly in its northern half, indicated by black circles in Figures 6(a) and 6(b). In TRMM, this zone is less extended to the north.

The maps in Figures 6(c)–6(h) show the output of WRF for the different configurations used in the experiments.

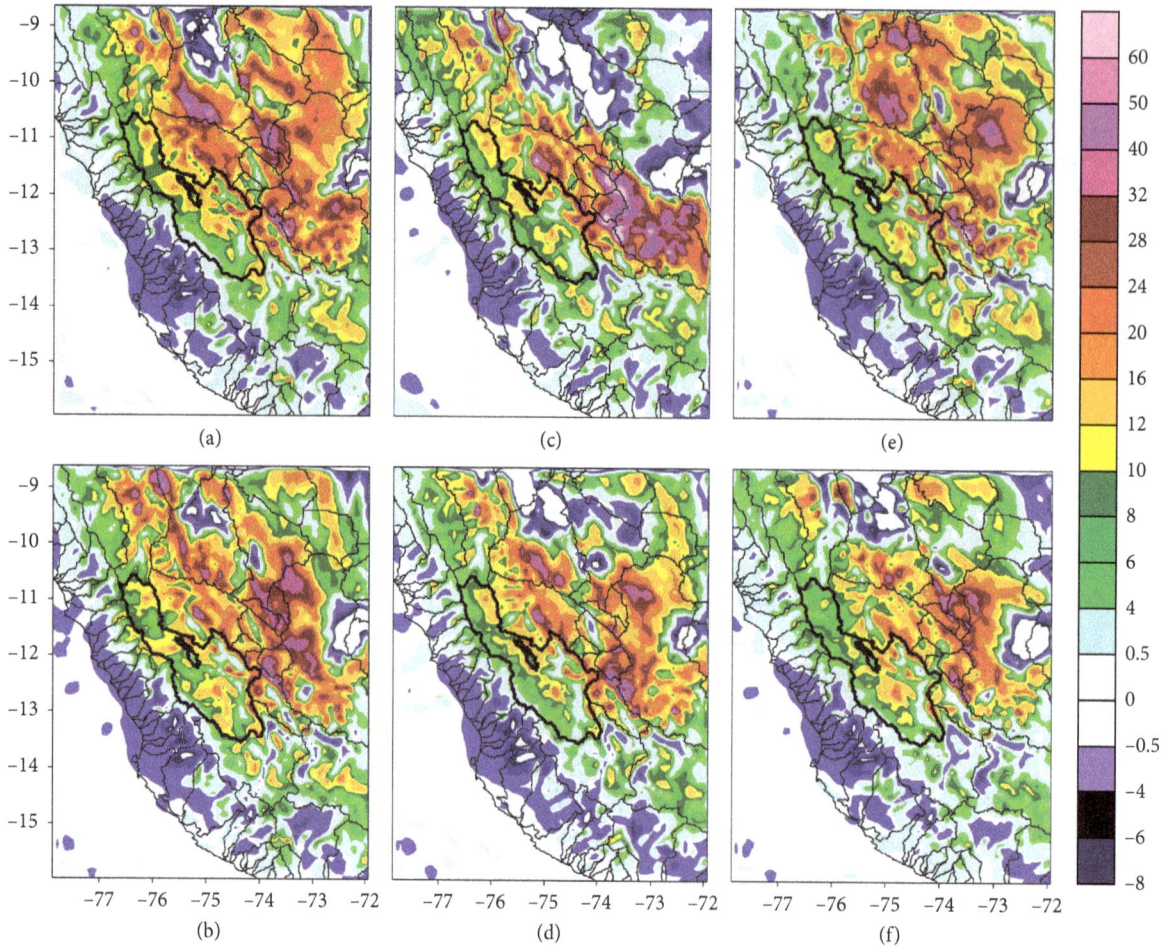

FIGURE 9: Spatial distribution of the *B* (mm/day) for the 6 km domain, corresponding to the sixth day of forecast: (a) CTR, (b) MP_LP, (c) C_BMJ, (d) BL_MYJ, (e) C_GRELL3, and (f) MP_MR.

It can be noticed that, in general, all configurations reproduced correctly the precipitation patterns shown by PISCO and TRMM but overestimating the reference accumulations, shown in Figures 6(a) and 6(b), mainly on the eastern slope of the mountain range (hot spots). However, it must be mentioned that it has been found [43] that in this sector of the Peruvian mountain range, TRMM underestimates precipitation by at least 40%, so that the overestimation of WRF relative to the "real precipitation value" in that case could be much lower. The configuration which best reproduced the maximum in 12°S–72°W, relative to TRMM was C_BMJ.

It must be pointed out that none of the configurations reproduced the less rainy strip shown by PISCO and, to a lesser extent, also by TRMM, along the eastern limit of the basin in its northern half (black circles in Figures 6(a) and 6(b)). In fact, WRF produces more precipitation in the eastern sector of the basin than in the west, which matches the observations only for the southern half of the basin.

Figure 7 shows the zoomed-in view of TRMM and PISCO precipitation patterns and WRF output for the 3 km domain. In TRMM and PISCO (Figures 7(a) and 7(b)), local maxima and minimum of precipitation are indicated as in Figure 6, using red circles for maxima and with a black circle for the minimum of precipitation in the eastern limit of the

northern half of the basin, practically covering the valley (which is represented by a small closed black contour near the western boundary of the basin). In a general sense, it is observed that the basin in its southern half is rainier towards the east, while in the northern half it is rainier towards the west, which can be seen more clearly in the PISCO data. It can also be noticed, more clearly in TRMM, that the south of the valley is drier than the north.

As shown in Figures 7(c)–7(h), the model reproduces the precipitation nuclei to the east of the basin, although in a general sense it overestimates accumulated rainfall. It can be seen that toward the southern half of the basin, the MP_MR configuration reproduced more clearly the fact that the eastern sector is rainier than the western. The CTR configuration also reproduces this pattern but much more extended to the west. The rest of the configurations were less precise in this regard. In general, the MP_MR configuration was the one reproducing best the spatial distribution of rainfall in the south of the basin. For the northern half of the basin, all the configurations reflected the eastern sector as rainier than the western, which does not reproduce the patterns in TRMM and PISCO.

From the above considerations, it can be concluded that for the 6 km domain, the precipitation pattern shown by

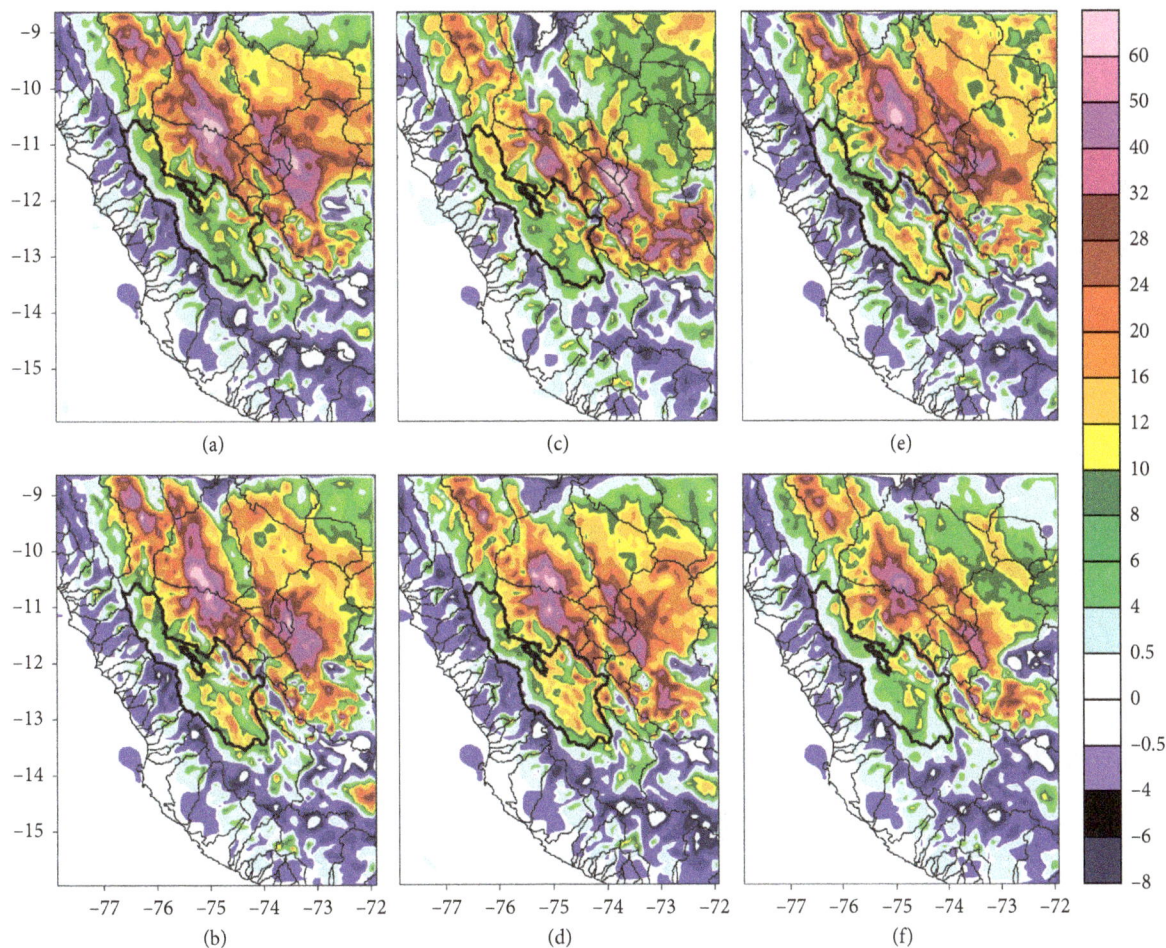

FIGURE 10: Spatial distribution of the B (mm/day) for the 6 km domain, corresponding to the tenth day of forecast: (a) CTR, (b) MP_LP, (c) C_BMJ, (d) BL_MYJ, (e) C_GRELL3, and (f) MP_MR.

C_BMJ matches better the PISCO and TRMM patterns, although the MP_MR configuration seems to overestimate less than the rest. For domain 3, all the configurations reflected the maximum precipitation to the east of the basin, although the MP_MR configuration better reflected the spatial distribution of rainfall in the southern half of the basin. Towards the northern half of the basin, none of the configurations reproduces well the observed patterns of TRMM and PISCO.

In order to quantify the spatial distribution of the bias of the model, the bias (B) maps of both domains were made for different forecast horizons. However, in this case, bias will be calculated relative to TRMM data, considering that PISCO reflects a significant deficit of rainfall to the southeast of the basin. Figure 8 shows the spatial B distribution for the second day of forecast of the 6 km domain, which confirms that in general all the configurations overestimate the precipitation; however, it is noticed that it underestimates both for the western sector of the mountain range and for the coast, although this sector of the Andes is much less rainy than its eastern sector. In this case, it can be seen that the MP_MR configuration produces the least overestimation, as we had observed in the general picture.

Figures 9 and 10 show the results for the sixth and tenth days of forecast, where it is visually noticed that the MP_MR configuration has the lowest bias. In this case, it is significant that, for the sixth day the configuration, C_BMJ shows a noticeable improvement in relation to the second, specifically in the Amazonia zone, with insignificant overestimations.

Figures 11–13 are analogous to Figures 8–10, but for the 3 km domain. As for the 6 km domain, in this case, the model overestimates precipitation, and even in the interior of the basin, some of the configurations overestimate more than in the 6 km domain.

For this domain, in all cases, it is also noticed that MP_MR showed the best results, but with the particularity that outside the basin, B increased with forecast horizon, while the behavior inside the basin was more stable.

Thus, from a spatial point of view, the model clearly shows an overestimation in the Mantaro basin and, in general, in the entire eastern sector of the ridge, while in its western sector, underestimation of precipitation was observed. From the spatial point of view, it is confirmed that MP_MR was the configuration with the lowest overestimation, behaving in a stable way from the second to the tenth forecast day, with a moderate increase in RMSE as the forecast horizon increased.

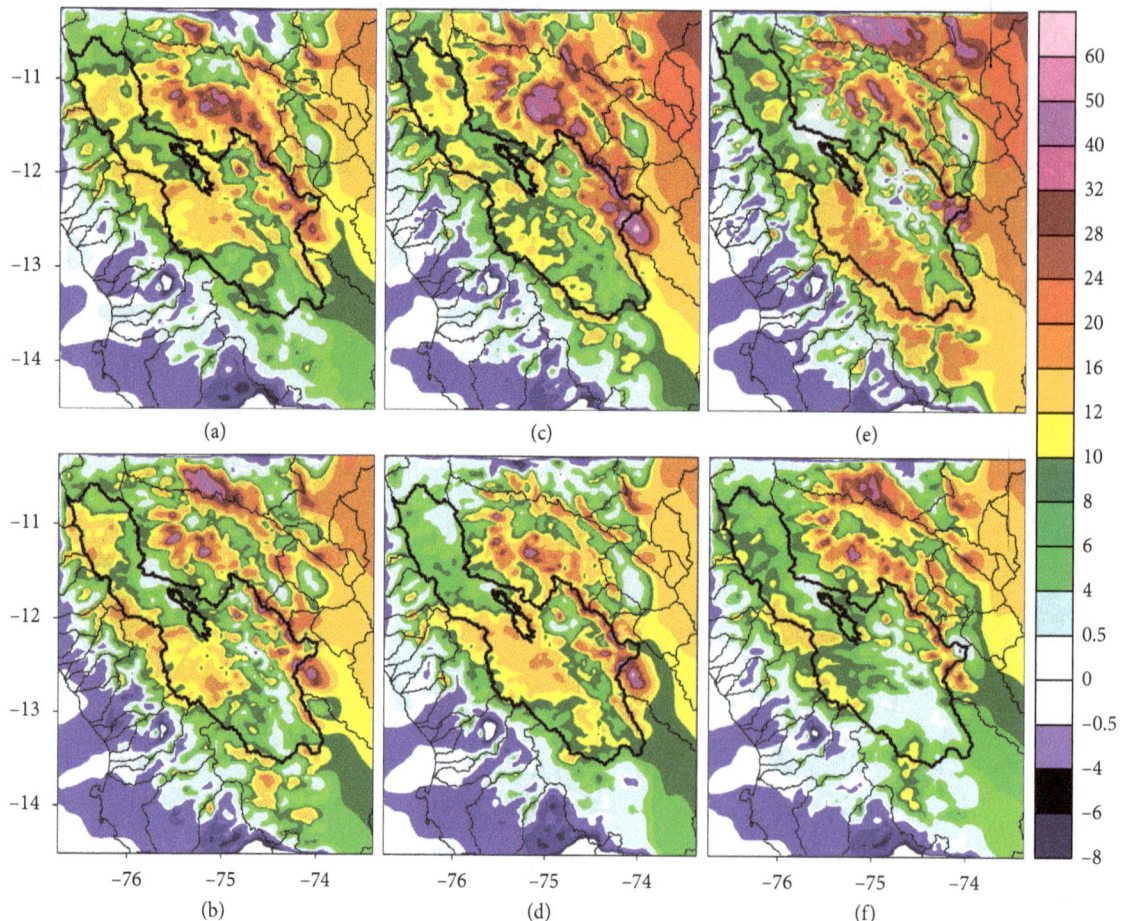

FIGURE 11: Spatial distribution of B (mm/day) for the 3 km domain, corresponding to the second day of forecast: (a) CTR, (b) MP_LP, (c) C_BMJ, (d) BL_MYJ, (e) C_GRELL3, and (f) MP_MR.

From the above arguments, it can be concluded that the sensitivity to the microphysical scheme was higher than to the convective scheme, which indicates the importance of the microphysics parameterization in WRF for precipitation forecasting tasks in the study region. As an example, Figure 14 shows the vertical profile of rain water mixing ratio (Qrain) produced by the 3 microphysical schemes used for the Huayao station and, in parentheses, the predicted rainfall in each case, averaged for all tens used in study. It can be noticed the correspondence between "Qrain" and predicted precipitation.

Similar to the results obtained in this research, in [25], a sensitivity study was made of simulated monsoon precipitation to the cloud microphysics schemes in WRF for the summer period in the valley of Langtang, Himalaya. It was obtained that, in a general sense, the model underestimated the accumulated precipitation in 10 days, and also the Morrison microphysical scheme showed the best results.

Our results for the Andes are consistent with the finding of [26], where it was concluded that, in general, the double-moment Morrison's scheme represents more correctly the microphysics processes in the Himalaya, possibly because of the better representation of the warm and cold processes related with the formation and evolution of the parameters of particle size distributions, in comparison with one-moment parameterizations.

4. Summary and Conclusions

Simulations were developed with the regional model WRF, for domains of 18, 6, and 3 km of spatial resolution, with the main objective of determining the ability of several configurations of the model to forecast the field of precipitation in the short and medium terms in the complex orography conditions of the Central Andes of Peru. The initial and boundary conditions were generated by the NCEP "Global Operational Analysis," final analysis FNL, every 6 hours, with horizontal resolution of $1° \times 1°$. The behavior of the numerical forecasts for the first 10 days of simulation was analyzed, concluding that the mean square error for the higher resolution domains increases with the simulation term, in general until the tenth day. The lowest mean square error was obtained for the MP_MR configuration (Morrison for microphysics and Grell–Freitas for convection) during the whole period, remaining with little variation after the sixth forecast day. This configuration also provided the lowest positive bias, although in all cases a clear overestimation of rainfall was observed.

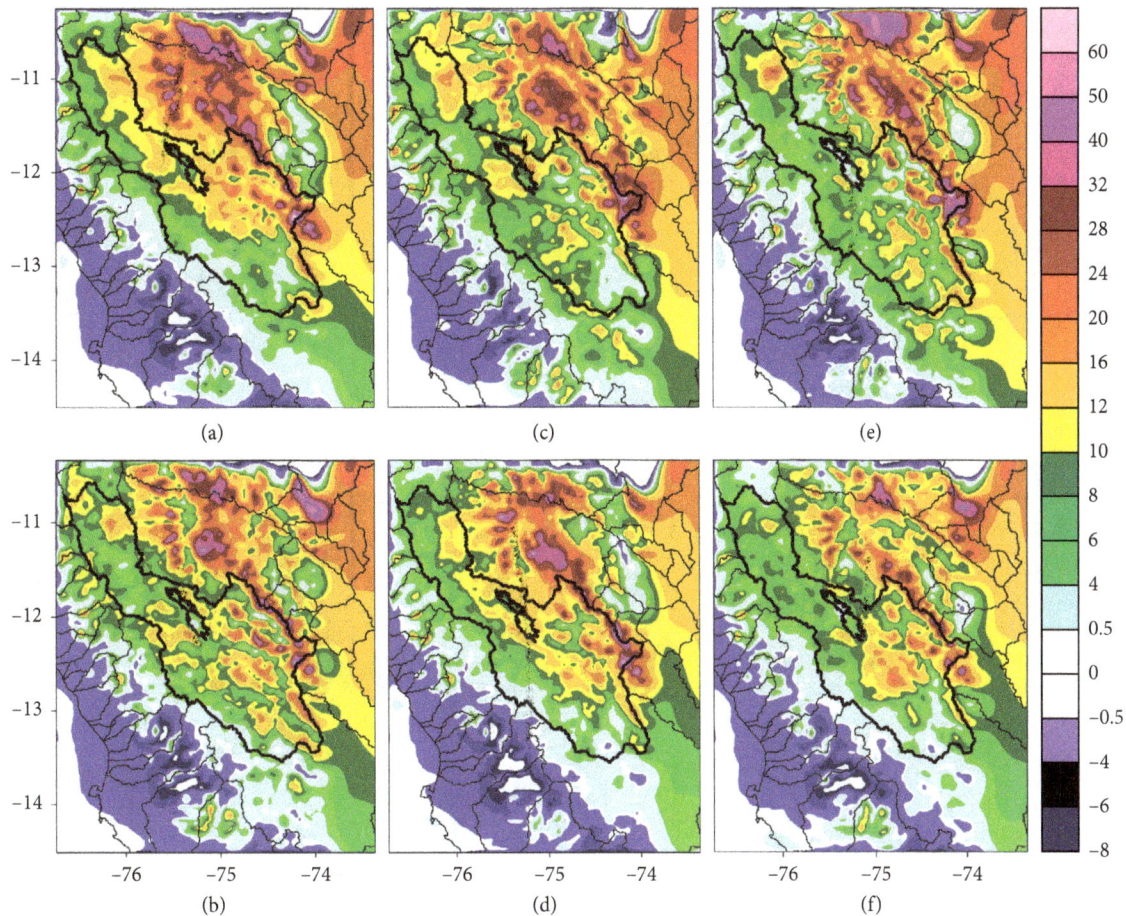

FIGURE 12: Spatial distribution of the B (mm/day) for the 3 km domain, corresponding to the sixth day of forecast: (a) CTR, (b) MP_LP, (c) C_BMJ, (d) BL_MYJ, (e) C_GRELL3, and (f) MP_MR.

The point verification showed that, in general, all the tested configurations overestimate rainfall in the region. In this sense, the MP_MR configuration performed better, according to the bias and RMSE indicators, which alludes not only to the entire forecast period but also to all the forecast terms. It is important to notice that from the sixth day on, the RMSE for MP_MR did not increase more, which does not apply to the rest of the schemes used. Another important result in this case is that the C_BMJ configuration (Thompson microphysics and Betts–Miller–Janjic convection) showed poor results for the 18 km domain, so it is recommended not to use it in this region for low-resolution domains.

The categorical binary verification showed detection rates of precipitation above 85% in all cases. The B was all positive, which indicates that the model overestimated the presence of rainfall in the region. In general, the configurations showing the best indicators were C_BMJ and MP_MR.

The spatial verification showed that all the schemes produced rainfall distribution patterns quite similar to those of TRMM and PISCO, although in all cases with positive bias in precipitation accumulations. In this sense, the configuration that less overestimated precipitation was MP_MR. It also reproduced better precipitation pattern in the southern half of the basin. Regarding the northern half of the basin, all the configurations showed the eastern sector as rainier than the west, which does not agree with TRMM and PISCO.

The bias spatial distribution generally showed positive values in the basin and in general throughout the eastern sector of the Andes and the Amazonia, while towards the west of the Andes and the coast, it showed values close to zero or negative. The configuration with lower bias was MP_MR, which also showed the lowest values of RMSE, in both cases for the two domains analyzed. The C_BMJ configuration, although with higher bias and RMSE, showed improvement for the greater terms, which is an indicator to be taken into account for its use in longer-term forecast.

Based on the results, it is concluded that the MP_MR and C_BMJ configurations were the ones with the best results in a general sense. Consequently, the MP_MR configuration is recommended for short- and medium-term rainfall forecasting tasks in the Central Andes of Peru and particularly in the Mantaro basin. The Betts–Miller–Janjic cumulus parameterization will be further investigated by the authors for the region in combination with other physical schemes to explore its possible application potential in medium range precipitation forecasting, using multiconfiguration ensembles. It was also concluded that the activation or not of cumulus parametrization for the domain of 3 km resolution is not relevant for the precipitation forecast in the study region.

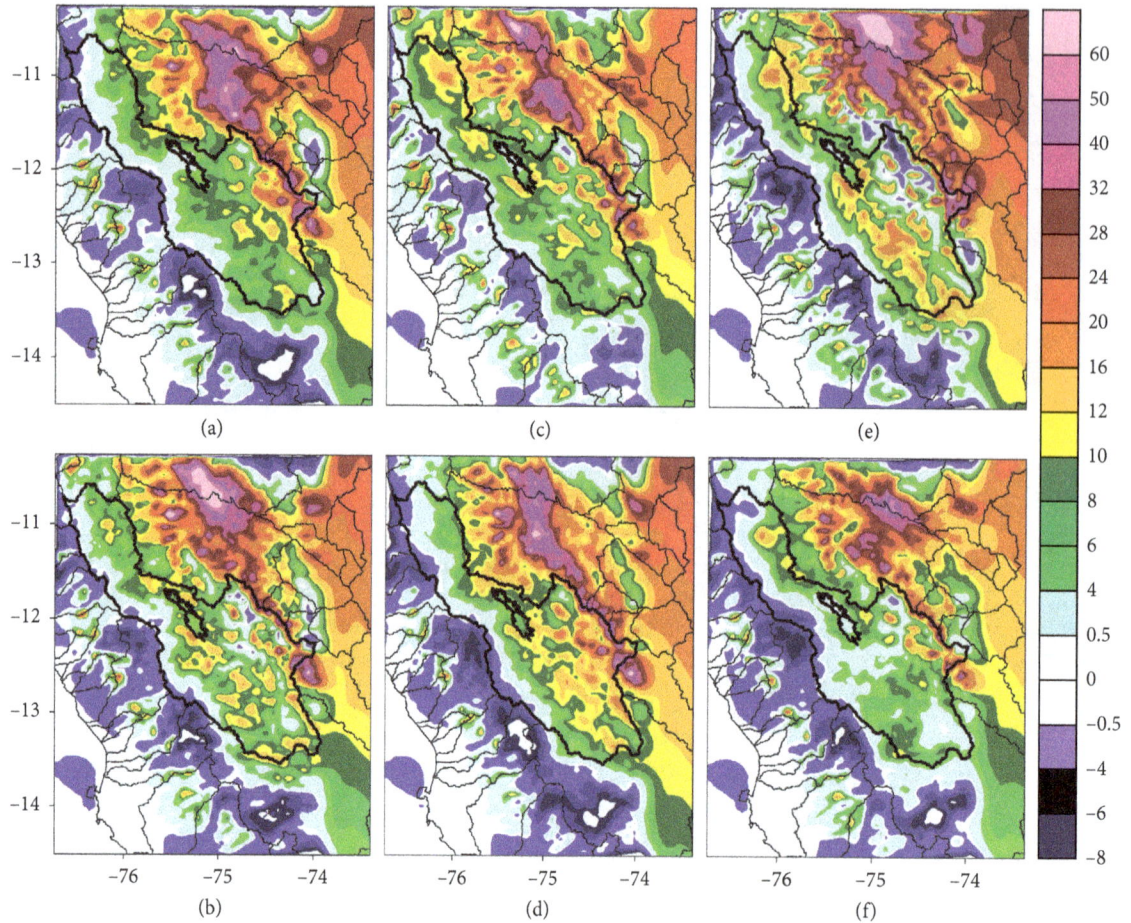

FIGURE 13: Spatial distribution of the B (mm/day) for the 3 km domain, corresponding to the tenth day of forecast: (a) CTR, (b) MP_LP, (c) C_BMJ, (d) BL_MYJ, (e) C_GRELL3, and (f) MP_MR.

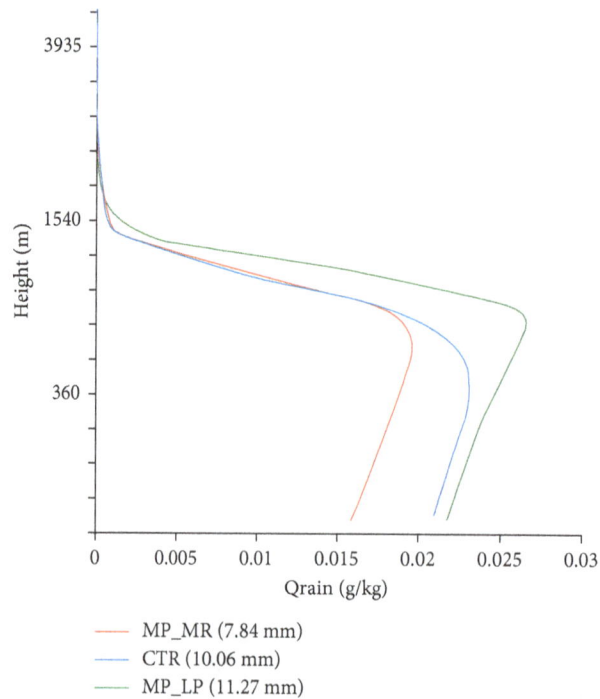

FIGURE 14: Vertical profile of "rain water mixing ratio (Qrain)" produced by the 3 microphysical schemes used for the Huayao station and, in parentheses, the predicted rainfall in each case, average of 10 study cases taken randomly.

Conflicts of Interest

The authors declare that there are no conflicts of interest regarding the publication of this paper.

Acknowledgments

The present study comes under the project "MAGNET-IGP: Strengthening the research line in physics and microphysics of the atmosphere (Agreement no. 010-2017-FONDECYT)." The authors would like to thank the CONCYTEC, Peru, for financial support and Inter-American Institute for Cooperation on Agriculture (IICA) for administrative support. This work was done using computational resources, HPC-Linux Cluster, from Laboratorio de Dinámica de Fluidos Geofísicos Computacionales at Instituto Geofísico del Perú (Grants 101-2014-FONDECYT, SPIRALES2012 IRD-IGP, Manglares IGP-IDRC, and PP068 program). The authors also thank NASA for TRMM precipitation data, NCEP for FNL analysis data, and SENAMHI for observational precipitation data.

References

[1] Y. H. Kuo, J Bresch, M. D. Cheng et al., "Summary of a mini workshop on cumulus parameterization for mesoscale models," *Bulletin of the American Meteorological Society*, vol. 78, no. 3, pp. 475–491, 1997.

[2] W. Skamarock, J. Klemp, J. Dudhia et al., *A Description of the Advanced Research WRF Version 3*, NCAR Technical Note, NCAR/TN–468+STR, National Center for Atmospheric Research (NCAR), Mesoscale and Microscale Meteorology Division, Boulder, CO, USA, 2008.

[3] Y. Silva, K. Takahashi, and R. Chávez, "Dry and wet rainy seasons in the Mantaro river basin (Central Peruvian Andes)," *Advances in Geosciences*, vol. 14, pp. 261–264, 2008.

[4] P. Aceituno, "On the functioning of the southern oscillation in the South American, sector. Part II. Upper-air circulation," *Journal of Climate*, vol. 2, no. 4, pp. 341–355, 1989.

[5] M. Vuille, G. Kaser, and I. Juen, "Glacier mass balance variability in the Cordillera Blanca, Peru and its relationship with climate and the large-scale circulation," *Global and Planetary Change*, vol. 62, no. 1-2, pp. 14–28, 2008.

[6] R. Garreaud, "The Andes climate and weather," *Advances in Geosciences*, vol. 22, pp. 3–11, 2009.

[7] A. G. Martínez, E. Núñez, Y. Silva et al., "Vulnerability and adaptation to climate change in the Peruvian Central Andes: results of a pilot study," in *Proceedings of the International Conference on Southern Hemisphere Meteorology and Oceanography (ICSHMO)*, pp. 297–305, Foz do Iguaçu, PR, Brazil, April 2006.

[8] B. S. Barrett, R. Garreaud, and M. Falvey, "Effect of the Andes cordillera on precipitation from a midlatitude cold front," *Monthly Weather Review*, vol. 137, no. 9, pp. 3092–3109, 2009.

[9] M. Viale and F. A. Norte, "Strong cross-barrier flow under stable conditions producing intense winter orographic precipitation: a case study over the subtropical central Andes," *Weather and Forecasting*, vol. 24, no. 4, pp. 1009–1031, 2009.

[10] P. A. Jiménez, J. Dudhia, J. F. González-Rouco et al., "An evaluation of WRF's ability to reproduce the surface wind over complex terrain based on typical circulation patterns," *Journal of Geophysical Research: Atmospheres*, vol. 118, no. 14, pp. 7651–7669, 2013.

[11] T. M. Weckwerth, I. J. Bennett, L. Jay Miller et al., "An observational and modeling study of the processes leading to deep, moist convection in complex terrain," *Monthly Weather Review*, vol. 142, no. 8, pp. 2687–2708, 2014.

[12] C. Junquas, K. Takahashi, T. Condom et al., "Understanding the influence of orography on the precipitation diurnal cycle and the associated atmospheric processes in the central Andes," *Climate Dynamics*, vol. 50, no. 11-12, pp. 3995–4017, 2017.

[13] Z. I. Janjic, "Nonsingular implementation of the Mellor-Yamada level 2.5 scheme in the NCEP meso model," *NCEP Office Note*, no. 437, p. 61, 2002.

[14] S.-Y. Hong, Y. Noh, and J. Dudhia, "A new vertical diffusion package with an explicit treatment of entrainment processes," *Monthly Weather Review*, vol. 134, no. 9, pp. 2318–2341, 2006.

[15] Z. I. Janjic, "The step-mountain eta coordinate model: further developments of the convection, viscous sublayer, and turbulence closure scheme," *Monthly Weather Review*, vol. 122, no. 5, pp. 927–945, 1994.

[16] G. A. Grell and S. R. Freitas, "A scale and aerosol aware stochastic convective parameterization for weather and air quality modeling," *Atmospheric Chemistry and Physics*, vol. 14, no. 10, pp. 5233–5250, 2014.

[17] G. Thompson, P. R. Field, R. F. Rasmussen, and W. D. Hall, "Explicit forecasts of winter precipitation using an improved bulk microphysics scheme. Part II: implementation of a new snow parameterization," *Monthly Weather Review*, vol. 136, no. 12, pp. 5095–5115, 2008.

[18] H. Morrison, G. Thompson, and V. Tatarskii, "Impact of cloud microphysics on the development of trailing stratiform precipitation in a simulated squall line: comparison of one- and two-moment schemes," *Monthly Weather Review*, vol. 137, no. 3, pp. 991–1007, 2009.

[19] Y. L. Lin, R. D. Farley, and H. D. Orville, "Bulk parametrization of the snow field in a cloud model," *Journal of Climate and Applied Meteorology*, vol. 22, no. 6, pp. 1065–1092, 1983.

[20] J. Done, C. A. Davis, and M. Weisman, "The next generation of NWP: explicit forecasts of convection using the weather research and forecasting (WRF) model," *Atmospheric Science Letters*, vol. 5, no. 6, pp. 110–117, 2004.

[21] E. K. Gilliland and C. M. Rowe, "A comparison of cumulus parameterization schemes in the WRF model," in *Proceedings of the 87th AMS Annual Meeting and 21st Conference on Hydrology*, San Antonia, TX, USA, January 2007.

[22] T. G. Farr, P. A. Rosen, E. Caro et al., "The shuttle radar topography mission," *Reviews of Geophysics*, vol. 45, no. 2, 2007.

[23] E. Rodriguez, C. S. Morris, and J. E. Belz, "A global assessment of the SRTM performance," *Photogrammetric Engineering and Remote Sensing*, vol. 72, no. 3, pp. 249–260, 2006.

[24] D. B. Gesch, K. L. Verdin, and S. K. Greenlee, "New land surface digital elevation model covers the earth," *Eos, Transactions American Geophysical Union*, vol. 80, no. 6, pp. 69-70, 1999.

[25] A. Orr, C. Listowski, M. Cottet et al., "Sensitivity of simulated summer monsoonal precipitation in Langtang Valley, Himalaya, to cloud microphysics schemes in WRF," *Journal of Geophysical Research: Atmospheres*, vol. 122, no. 12, pp. 6298–6318, 2017.

[26] R. K. Shrestha, P. J. Connolly, and M. W. Gallagher, "Sensitivity of WRF cloud microphysics to simulations of a con-

vective storm over the Nepal Himalayas," *The Open Atmospheric Science Journal*, vol. 11, no. 1, pp. 29–43, 2017.

[27] M. Rajeevan, A. Kesarkar, S. B. Thampi, T. N. Rao, B. Radhakrishna, and M. Rajasekhar, "Sensitivity of WRF cloud microphysics to simulations of a severe thunderstorm event over Southeast India," *Annales Geophysicae*, vol. 28, no. 2, pp. 603–619, 2010.

[28] Y. G. Mayor and M. D. S. Mesquita, "Numerical simulations of the 1 May 2012 deep convection event over Cuba: sensitivity to cumulus and microphysical schemes in a high-resolution model," *Advances in Meteorology*, vol. 2015, Article ID 973151, 16 pages, 2015.

[29] M Tewari, F. Chen, W. Wang et al., "Implementation and verification of the unified NOAH land surface model in the WRF model," in *Proceedings of the 20th Conference on Weather Analysis and Forecasting/16th Conference on Numerical Weather Prediction*, pp. 11–15, Seattle, WA, USA, January 2004.

[30] C. A. Paulson, "The mathematical representation of wind speed and temperature profiles in the unstable atmospheric surface layer," *Journal of Applied Meteorology*, vol. 9, no. 6, pp. 857–861, 1970.

[31] A. J. Dyer and B. B. Hicks, "Flux–gradient relationships in the constant flux layer," *Quarterly Journal of the Royal Meteorological Society*, vol. 96, no. 410, pp. 715–721, 1970.

[32] E. K. Webb, "Profile relationships: the log-linear range, and extension to strong stability," *Quarterly Journal of the Royal Meteorological Society*, vol. 96, no. 407, pp. 67–90, 1970.

[33] D. Zhang and R. A. Anthes, "A high-resolution model of the planetary boundary layer—sensitivity tests and comparisons with SESAME-79 data," *Journal of Applied Meteorology*, vol. 21, no. 11, pp. 1594–1609, 1982.

[34] A. C. M. Beljaars, "The parameterization of surface fluxes in large-scale models under free convection," *Quarterly Journal of the Royal Meteorological Society*, vol. 121, no. 522, pp. 255–270, 1994.

[35] M. J. Iacono, J. S. Delamere, E. J. Mlawer et al., "Radiative forcing by long-lived greenhouse gases: calculations with the AER radiative transfer models," *Journal of Geophysical Research*, vol. 113, p. D13, 2008.

[36] E. J. Mlawer, S. J. Taubman, P. D. Brown, M. J. Iacono, and S. A. Clough, "Radiative transfer for inhomogeneous atmospheres: RRTM, a validated correlated-k model for the longwave," *Journal of Geophysical Research: Atmospheres*, vol. 102, no. 14, pp. 16663–16682, 1997.

[37] R. M. Goody and Y. L. Yung, *Atmospheric Radiation: Theoretical Basis*, Oxford University Press, Oxford, UK, 1995, https://www.bookdepository.com/Atmospheric-Radiation-Theoretical-Basis-R-M-Goody/9780195102918.

[38] G. P. Cressman, "An operational objective analysis system," *Monthly Weather Review*, vol. 87, no. 10, pp. 367–374, 1959.

[39] E. E. Ebert, "Fuzzy verification of high-resolution gridded forecasts: a review and proposed framework," *Meteorological Applications*, vol. 15, no. 1, pp. 51–64, 2008.

[40] C. Kummerow, J. Simpson, O. Thiele et al., "The status of the tropical rainfall measuring mission (TRMM) after two years in orbit," *Journal of Applied Meteorology*, vol. 39, no. 12, pp. 1965–1982, 2000.

[41] G. J. Huffman, D. T. Bolvin, E. J. Nelkin et al., "The TRMM multisatellite precipitation analysis (TMPA): quasi-global, multiyear, combined-sensor precipitation estimates at fine scales," *Journal of Hydrometeorology*, vol. 8, no. 1, pp. 38–55, 2007.

[42] C. Aybar, W. Lavado-Casimiro, A. Huerta et al., *Uso del Producto Grillado "PISCO" de precipitación en Estudios, Investigaciones y Sistemas Operacionales de Monitoreo y Pronóstico Hidrometeorológico, Nota Técnica 001 SENAMHI-DHI-2017*, Senamhi, Lima, Peru, 2017, ftp://ftp.senamhi.gob.pe/PISCO_v2.0/PISCO-Prec-v2.0.pdf.

[43] S. P. Chavez and K. Takahashi, "Orographic rainfall hot spots in the Andes-Amazon transition according to the TRMM precipitation radar and in situ data," *Journal of Geophysical Research: Atmospheres*, vol. 122, no. 11, pp. 5870–5882, 2017.

Investigation of Vorticity during Prevalent Winter Precipitation in Iran

Iman Rousta [ID],[1,2] Farshad Javadizadeh [ID],[3] Fatemeh Dargahian,[4] Haraldur Ólafsson,[5] Amin Shiri-Karimvandi,[6] Sayed Hossein Vahedinejad,[7] Mehdi Doostkamian,[6] Edgar Ricardo Monroy Vargas,[8] and Anayat Asadolahi[6]

[1]Department of Geography, Yazd University, Yazd 8915818411, Iran
[2]Senior Researcher, Institute for Atmospheric Sciences, University of Iceland and Icelandic Meteororological Office (IMO), Bustadavegur 7, IS-108 Reykjavik, Iceland
[3]Department of Environment, Collage of Natural Resource, Bandar Abbas Branch, Islamic Azad University, Bandar Abbas, Iran
[4]Desert Research Division, Research Institute of Forests and Rangelands, Agricultural Research Education and Extension Organization (AREEO), Tehran, Iran
[5]Department of Physics, University of Iceland, Institute for Atmospheric Sciences and Icelandic Meteororological Office (IMO), Bustadavegur 7, IS-108 ReykjaviK, Iceland
[6]Department of Geography, University of Zanjan, Zanjan 3879145371, Iran
[7]Department of Geography, University of Kharazmi, Tehran 1491115719, Iran
[8]Department of Civil Engineering, Universidad Catolica de Colombia, Bogotá, Colombia

Correspondence should be addressed to Iman Rousta; irousta@yazd.ac.ir and Farshad Javadizadeh; javadizadeh2020@yahoo.com

Academic Editor: Andrew D. Jensen

In this study, precipitation data for 483 synoptic stations, and the U&V component of wind and HGT data for 4 atmospheric levels were respectively obtained from IRIMO and NCEP/NCAR databases (1961–2013). The precipitation threshold of 1 mm and a minimum prevalence of 50% were the criteria based on which the prevalent precipitation of Iran was identified. Then, vorticity of days corresponding to prevalent winter precipitation was calculated and, by performing cluster analysis, the representative days of vorticity were specified. The results showed that prevalent winter precipitation vorticity in Iran is related to the vorticity patterns of low pressure of Mediterranean-low pressure of Persian Gulf dual-core, low pressure closed of central Iran-high pressure of East Europe, Ural low pressure-Middle East High pressure, Saudi Arabia low pressure-Europe high pressure, and high-pressure belt of Siberia-low pressure of central Iran. At the same time, the most intense vorticity occurred when the climate of Iran was influenced by a massive belt pattern of Siberian high pressure-low pressure of central Iran. However, at the time of prevalent winter precipitation in Iran, an intense vorticity is drawn with the direction of Northeast and Northwest from the center of Iraq to the south of Iran.

1. Introduction

Changes in extreme weather and climate events have significant impacts and are among the most serious challenges to society in coping with a changing climate [1–9]. The most important factor in the formation and guidance of atmospheric systems is vorticity process [10–12]. In the middle latitudes, at the synoptic scale, the important dynamic properties are those which are related to rotating particles in the air [13]. Dessouky

and Jenkinson [14]and Jenkinson and Collison [15]are two examples of studies in different parts of the world focusing on the role of vorticity conditions and amount at different atmosphere levels in precipitation. Based on the general format of weather types identified by Lamb, Dessouky and Jenkinson investigated vorticity and the direction of flows in pressure systems producing severe storms in the UK, and systems producing drought and wet years in Egypt [14]. Jenkinson's method has been tested by many researchers in different areas of the world. They showed that using this method can make it possible to identify weather types and quantitatively calculate their intensity and weakness. Since the convergence leads to upside movements, some studies have investigated vorticity advection, divergence, and vertical motions combined with high levels of atmosphere jet stream as an evidence for the development of surface low pressure [16–19]. In another study, Nakamura showed that high levels of jet stream simultaneously occur with divergence and relative positive vorticity advection [20]. Different case studies show that the presence and intensity of relative positive vorticity advection with the vertical arrangement of the wind caused by the jet stream changes provide favorable conditions for increasing vertical movements and uplink, and creating low surface pressure [21–23]. Vincent studied the development of cyclones in the South Pacific convergence zone using vorticity [24]. In another study, Wang examined the relative vorticity of ocean winds and its impact on the development of tropical cyclones in the South China Sea. He concluded that the winter tropical cyclone genesis in the South China Sea happens due to vertical shear of the horizontal winds and low-level atmospheric vorticity [25]. In other studies around the world, Alpert et al. investigated the horizontal distribution and the vertical profile of the relative vorticity over the Mediterranean region over a period of 5 years [26]. Bartzokas and Metaxas estimated the seasonal values of the geostrophic relative vorticity at four grid points in the Mediterranean by using pressure data for the period 1873–1988 [27]. Ruiz and Vargas studied the 500 hPa vorticity distribution over Argentina and its association with large-scale precipitation on a climatological basis [28]. Xoplaki et al. studied the wet season Mediterranean precipitation variability, and showed that since the mid-nineteenth century, precipitation steadily increased with a maximum in the 1960s and decreased since then [29]. There are a few climatological studies of vorticity in Iran. Golmohammadian and Pishvaie is one of the research projects concentrating on the relationship between vorticity and other synoptic indicators in Iran. This study used the model of area circulation to create monthly vorticity indicators in south of Iran for eight points on two surfaces of land and 500 hPa. The results showed that there is a trough in the East Mediterranean during all months of the year, and vorticity indicators with temperature are better than precipitation in analyzing the climatic responses to the selected station of Shiraz [30]. Alijani and Zahehi analyzed Azerbaijan precipitation to statistically and synoptically determine the types of air masses affecting this area. For this purpose, the daily precipitation data of the Tabriz station for the period of 1961–1995 and also pressure data at 12 am at ground level and 500 hPa were used. Consequently, 11 types of air masses in the precipitation of Azerbaijan were identified, with only 3 of them being high pressure [31]. In another study, Meshkati and Moradi examined the pressure trough of the Red Sea from the dynamic viewpoint, revealing that if the advection of relative positive vorticity takes place in the east of the Mediterranean or north of the Red Sea, the pressure trough of the Red Sea moves to the east of the

Mediterranean Sea and influences the west and northwest of Iran. On the other hand, if the advection of relative positive vorticity occurs in the northeast of the Red Sea, small low-pressure cells are separated from the trough of pressure on the Red Sea and move towards the northeast and affect the west, southwest, and south of Iran [32]. In contrast, Mofidi et al. investigating fall precipitation in the northern coast of Iran, showed that a high-pressure center on the west of the Caspian Sea and negative vorticity on the sea in lower levels of the atmosphere combined with strong currents and prevailing in the north-south direction are the main factors causing heavy and extreme fall precipitation in all the synoptic patterns of the Caspian [33].

The objective of this study is to investigate the vorticity over the prevalent winter precipitation in Iran, for a period of 53 years, along with its seasonal variations. This study, thus, seeks to understand from a climatological point of view the dynamic background of the troposphere over the regions that have an effect on the occurrence of prevalent winter precipitation in the study area. Furthermore, it is intended to gain a better insight into the dynamic mechanisms responsible for this kind of precipitation in Iran. Since prevalent and extreme precipitation is a dangerous phenomenon that will have environmental damages, especially in areas with small amounts of annual precipitation (e.g. Iran), it is critical to study the dynamic features that can help to better identify such precipitation. According to studies conducted by researchers inside and outside Iran, it is crucial to review the vorticity status of the atmosphere in prevalent winter precipitation in Iran. This can lead to an understanding of the relationship between atmospheric rotations and the surface environment of land in order to recognize different states of vorticity and their impact on convergence and divergence at different levels of the atmosphere, leading to the identification of prevalent precipitation in Iran. Therefore, in this study, we have tried to study the vorticity patterns leading to prevalent winter precipitation in Iran.

2. Materials and Methods

This study sought to examine the vorticity status of the atmosphere during prevalent winter precipitation in Iran. To this end, winter precipitation data of 483 stations were gathered from the Meteorological Organization of Iran (IRIMO). The spatial distribution of stations is shown in Figure 1. After sorting, a database with the size of 7187×4383 was formed for a period of 53 years. Then, prevalent winter precipitation data were extracted from precipitation data with a minimum amount of a millimeter.

After the formation of database, three criteria were taken into account to determine the days with prevalent precipitation:

(1) A minimum precipitation of one mm during the specific day

(2) A precipitation lasting for at least two consecutive days

(3) A precipitation covering at least 50% of the area (spatial continuity condition)

(a)

(b)

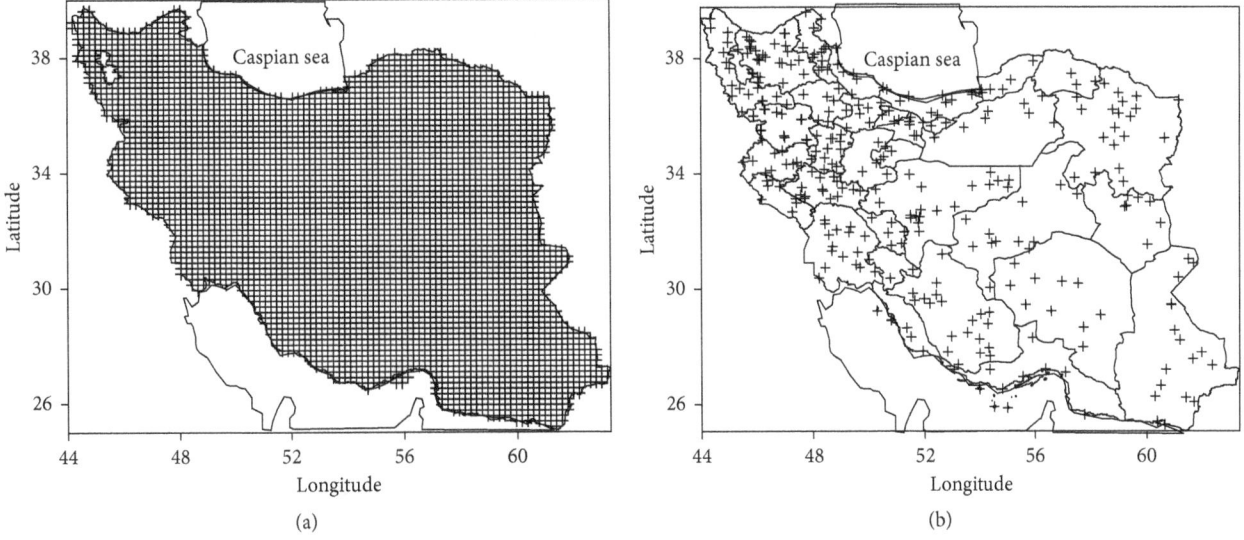

FIGURE 1: Regular grid of the precipitation database (a) and spatial distribution of studied stations (b).

By setting the conditions, just prevalent precipitation were selected for each of the cells studied and the relative concept of prevalent precipitation was observed for different regions of the country. The second criterion was having the precipitation for at least 2 consecutive days. Using this criterion, we were able to make a distinction between precipitation with a systematic (synoptic) origin and local precipitation that occurred due to convection or topography. After extracting the prevalent winter precipitation, pressure and U&V wind component data corresponding to the prevalent rainy days were extracted from NCEP/NCAR database with a spatial resolution of 2.5 × 2.5 degree [34]. Finally, using the programming in GrADS software [35], vorticities of 1000, 850, 700, and 500 hPa for the days with prevalent precipitation were calculated

The vorticity is a vector quantity defined as the curl (cross-product) of the velocity vector. The relative vorticity is given as follows [36]:

$$\vec{U} = \nabla \times \vec{V} = \left(\frac{\partial w}{\partial y} - \frac{\partial v}{\partial z}\right)\hat{i} + \left(\frac{\partial u}{\partial z} - \frac{\partial w}{\partial x}\right)\hat{j} + \left(\frac{\partial v}{\partial x} - \frac{\partial u}{\partial y}\right)\hat{k}.$$

(1)

In Cartesian coordinates, a large fraction of the rotating fluid systems with which we are interested exhibit rotation in the horizontal plane (i.e., midlatitude cyclones, hurricanes, and tornadoes). Consequently, dynamic meteorology is most often, though not exclusively, interested in the vertical component of the relative vorticity. It is generally expressed as follows [36]:

$$\zeta = \hat{k}.\vec{u} = \hat{k}.\nabla * \vec{V} = \frac{\partial v}{\partial x} - \frac{\partial u}{\partial y}.$$

(2)

In the northern hemisphere, a positive vorticity indicates a cyclonic motion and a negative vorticity demonstrates an anticyclonic motion. Anticlockwise and clockwise movements are respectively called cyclonic and anticyclonic movement [32,37–40].

Subsequently, cluster analysis of 1000 hPa vorticity was used in order to identify vorticity patterns of prevalent winter precipitation in Iran. In the next step, the result of this cluster analysis was obtained with the aim of classifying the atmospheric vorticity data and detecting representative days. Cluster analysis is a method where the variables are classified based on the characteristics desired in certain groups. The aim of cluster analysis is to find out the real groups of people and reduce the volume of data. In other words, the aim is to identify a smaller number of groups so that similar data are grouped together in a way that within-group variation is minimized and between-group variation is maximized. In this method, data are grouped based on the distance or similarity between them. There are several ways to measure the distance between the data. One of the most popular methods is the Euclidean distance [7]. Lund correlation method was used in order to choose the representative days of groups derived from classifying data on atmospheric vorticity. Thus, a representative day of a particular group is the one with the highest similarity to the maximum number of days in the group. The correlation coefficient indicates the degree of similarity of patterns of two maps. But a certain correlation coefficient threshold must be determined. The correlation coefficient value in such cases typically varied between 0.5 and 0.7 [41]. In the current study, the cut-off point for identifying the representative days was a correlation coefficient of 0.5. Thus, the day with the highest number of correlation coefficient values greater than 0.5 with other days of the same group was regarded as the representative day.

3. Results and Discussion

Five patterns were identified as a result of the implementation of cluster analysis on 1000 hPa data with prevalent winter precipitation. The results are displayed in Figure 2 and Table 1.

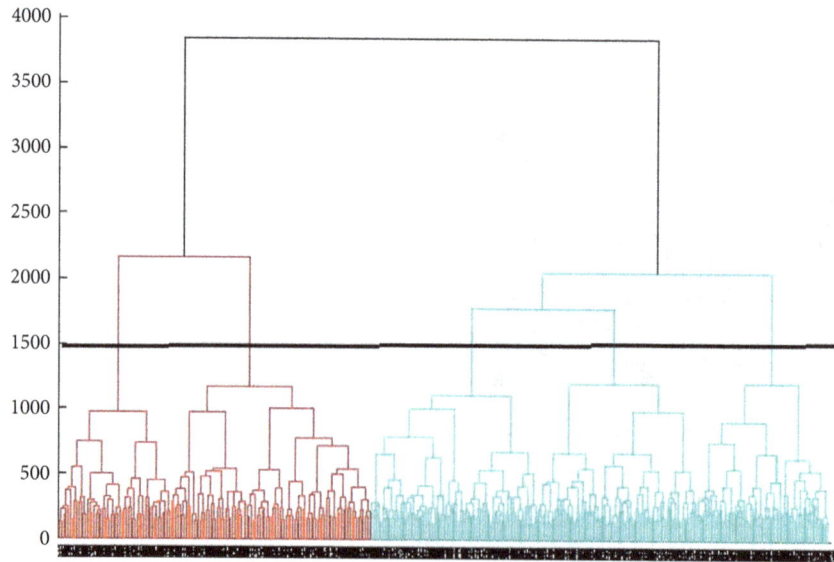

FIGURE 2: Dendrogram of cluster analysis on the Earth's surface pressure data.

TABLE 1: Some of the characteristics of identified patterns of prevalent winter precipitation in Iran.

1000 hPa vorticity patterns	Prevalence percentage	Overall precipitation	Frequency
Mediterranean low pressure-low-pressure dual-core Persian Gulf	50.132	3603	77
Iran closed low pressure-high pressure of east Europe	81.035	5824	73
Low pressure of Ural-high pressure of the Middle East	56.992	4096	60
European high pressure-low pressure of Saudi Arabia	58.314	4191	52
High-pressure belt of Siberia-low pressure of central Iran	55.322	3976	93

It is observed that the longest winter precipitation pattern in Iran (with a frequency of 93 days) occurred when the Siberian high-pressure belt-central Iran low pressure prevailed over Iran. On the other hand, the largest amount of overall precipitation in Iran happened when the low pressure of central Iran-high pressure of East Europe dominated the country (Table 1). However, at the time of the precipitation pattern of low pressure of the Mediterranean-low pressure of the Persian Gulf dual-core, the overall amount of precipitation reached 3603 mm.

In what follows, we provide a detailed description of the experimental results, present our interpretation, and draw a number of experimental conclusions.

3.1. The First Pattern: Low Pressure of Persian Gulf Dual-Core-Low Pressure of East Mediterranean. As illustrated in

Figure 3(a), the pressure and vorticity is 1000 hPa on Iran, and because of the exposure to dual-core low pressure in the Persian Gulf, the wind velocity is lower in the pole part; finally, in the eastern part of the low-pressure system, mass and density increase, and by increasing pressure, convergence will occur. While in its western part, due to the reduction of input mass from the north and increasing output mass from the southern part in the middle band, the mass and density are reduced and thus divergence will occur. However, the divergence of airflow in the low-pressure west of the Persian Gulf caused decreased airflow on the surface and vorticity advection of $-2.5 \times 10^{-5}_{s^{-1}}$ (m/s) on Iran's southern half. However, Lashkari showed that in a low-pressure flow, depending on the density of air and the input flow of air to its east, convergence occurs, and in its western part, divergence occurs due to the high speed of airflow [42–45]. Such a situation is clearly shown in Figure 3(a). On the other hand, the exposure of the west and northwest of the country to the East Mediterranean low pressure (wave front) caused a vertical rise and the convergence of airflow from a lower width at 1000 hPa (Figure 3(a)). So, the position of the low-pressure systems is critical because the systems usually have the strongest winds and the greatest waves [46]. With this interpretation, at the level of 850 hPa (Figure 3(b)) on Iran, positive vorticity advection of $-2.5 \times 10^{-5}_{s^{-1}}$ can be seen, which represents the planetary flows and vertical movements in the balance. As observed, vorticity minimum area on the north of the Black Sea can be detected coinciding with the full-height center of 1540 geopotential m that flows downward in a clockwise manner onto the northwest regions of the country.

Considering the amount of vorticity at 850 hPa, positive vorticity advection causes the flow of convergence on the lower level and intensification of flow of divergence in the upper level of the atmosphere [47]. On the other hand, at the levels of 700 and 500 hPa (Figures 3(c) and 3(d)), in accordance with the model 850 hPa, low-height center of 2960

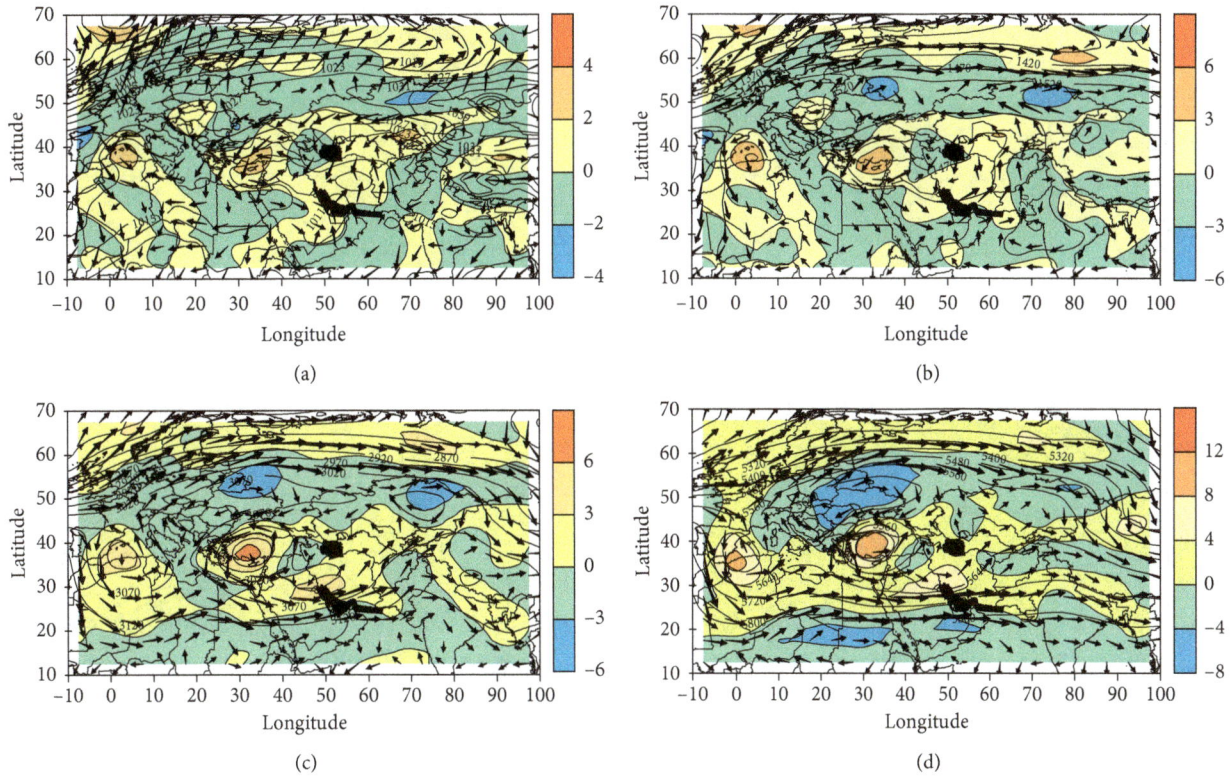

FIGURE 3: (a) 1000 hPa height (contours), vorticity ($5 \times 10^{-5}_{S^{-1}}$) (shaded), and wind flow (m/s) (vectors). (b) 850 hPa height, vorticity ($5 \times 10^{-5}_{S^{-1}}$), and wind flow (m/s). (c) 700 hPa height, vorticity ($5 \times 10^{-5}_{S^{-1}}$), and wind flow (m/s). (d) 500 hPa height, vorticity ($5 \times 10^{-5}_{S^{-1}}$), and wind flow (m/s).

geopotential meters is placed on the south of Turkey and East Mediterranean. The altitude trough is seen on the southwest and the Persian Gulf as positive vorticity advection on Iran causes a rise in the undercurrent convergence and vertical velocity of the air at the upper levels. At the 500 hPa level in Figure 3(d) on the southwest of the country, the altitude trough with positive vorticity advection of $8 \times 10^{-5}_{S^{-1}}$ shows convergence of rising flows of lower levels and upper level divergence. As a result of these conditions, the ground low pressure is constantly strengthened and deepened especially if it is accompanied with adequate moisture. After forming along the southwestern-northeast, the flows of the trough move to higher latitudes [48] and are converted to a dynamic mode and cause prevalent precipitation on Iran. In this pattern, the cyclone located on Iran is a dynamic system that synchronizes with the synoptic system of the middle levels of the atmosphere. This synoptic arrangement on Iran, with the supply of moisture from the Mediterranean and the Persian Gulf, has had a particular dynamic situation in the incidence of prevalent precipitation.

3.2. The Second Pattern: Central Iran Closed Low Pressure-High Pressure of Europe.
In this model, according to Figure 4(a) on a map of 1000 hPa level, a low-pressure area with centrality of 1010 hPa and 1000 hPa as an inverse pressure trough is completely placed on the central plateau

of Iran. The stretch of the south flows towards the low-pressure north on Iran and causes the reduction of airflow and airflow divergence on Iran and negative vorticity advection of $-3 \times 10^{-5}_{S^{-1}}$. On the other hand, on Central Europe, the high-pressure trough of 1022 hPa passes over the Black Sea and the Mediterranean and moves towards the lower latitudes in the central region of the Red Sea. This pressure arrangement is associated with the transmission of high-latitude cold and wet weather. In the upper level of 850 hPa (Figure 4(b)), we see low-height center of the Urals that is placed at a high level and west of the low-pressure center of land surface.

Moving up from the ground level, the low-height center of the upper level tends to move toward the northwest. In this level, convergence of degradation flows is observed behind the low-height trough where the negative vorticity advection of $-3 \times 10^{-5}_{S^{-1}}$ is placed to the north of the Red Sea and East Mediterranean. In contrast, the southern and eastern half of the country is placed on the elevation range of the trough and positive vorticity advection on the axis of the trough derived from the Urals low-height center on the west and southwest of $3 \times 10^{-5}_{S^{-1}}$, causing the highest amount of convergence at 850 hPa. On the other hand, studying the maps of the higher levels of 700 and 500 hPa shows that, similar to 850 hPa of height, a trough with centrality of 2980 geopotential meter is placed on the border to the west of the country and the Persian Gulf (Figures 4(c) and 4(d)). Positive vorticity advection of $5 \times 10^{-5}_{S^{-1}}$ per second coincides

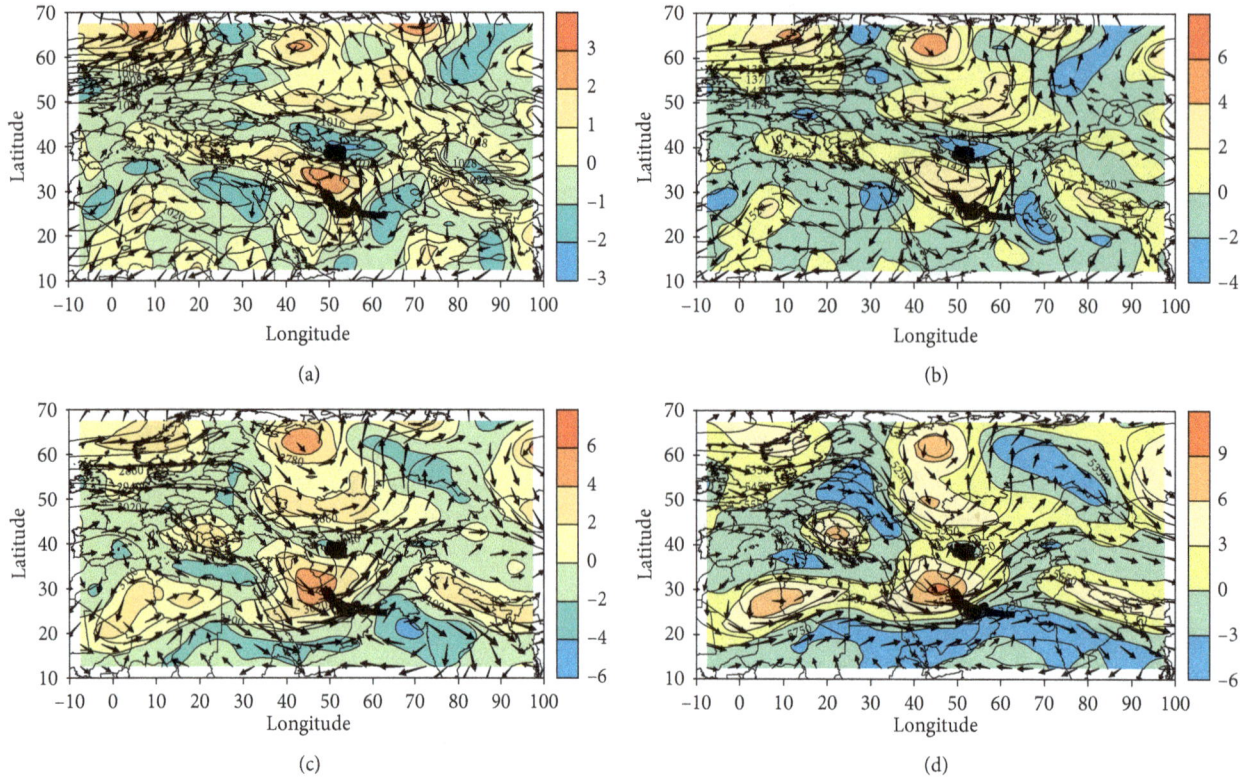

FIGURE 4: (a) 1000 hPa height (contours), vorticity ($5 \times 10_{S^{-1}}^{-5}$) (shaded), and wind flow (m/s) (vectors). (b) 850 hPa height, vorticity ($5 \times 10_{S^{-1}}^{-5}$), and wind flow (m/s). (c) 700 hPa height, vorticity ($5 \times 10_{S^{-1}}^{-5}$), and wind flow (m/s). (d) 500 hPa height, vorticity ($5 \times 10_{S^{-1}}^{-5}$), and wind flow (m/s).

with the axis of the trough. The convergence of subsurface flows and maximum upside cyclonic movements in 700 hPa all over of Iran were observed at 500 hPa level as its two lower levels. The axis of low height trough with centrality 5440 geopotential meter and positive vorticity advection with amount of $7 \times 10_{S^{-1}}^{-5}$ coincide with the low height axis is placed on Iraq and Saudi Arabia. Convergence and descending of northern cold flows in the western slope are clearly observed. It is observed that, in the higher layers of the atmosphere, just above the surface low-pressure center on Iran, the weather is divergent and air climbing from the lower layer is extended and exits from the rising air. The low-height center of the levels of 850, 700, and 500 hPa are placed under the area of upper level divergence. As a result, the deep divergence in the upper layers caused the forced air vertically in lower layers. As a result, low-pressure system of ground is strengthened and positive vorticity advection and convergence of airflow occur in lower levels and continue up to 500 hPa in Iran in lower levels and continues up to 500 hPa on Iran. In this pattern, there was a short wave at the middle level, and positive vorticity at the lower level of the atmosphere. These dynamic systems have provided ascending conditions. Eventually, with the provision of moisture by the Oman Sea and the Persian Gulf, the dynamic system has caused prevalent precipitation in Iran.

3.3. The Third Pattern: Low Pressure of the Urals-High Pressure of the Middle East. According to Figure 5(a), at the 1000 hPa level, a very strong low-pressure center of 990 hPa with several closed curves is placed on the west of Russia widely. Two pressure tongues are derived from it; one is with pressure of 1010 hPa in a southwesterly direction drawn through the Black Sea to the Mediterranean. The other tongue is in the southeast, crossing the Caspian to central areas of Iran (1014 hPa). This situation has led to taking sufficient moisture by passing through the Caspian Sea and transferring it to the central regions of Iran. As a result of such action, widespread precipitation has occurred especially in areas to the southeast of the country. In addition, on the northeast of Africa, Egypt, Syria, Iraq, and Saudi Arabia, a wide area of relatively high pressure of 1022 hPa is observed that in accordance with the low-pressure tongue of the Urals causes positive vorticity advection of $2 \times 10_{S^{-1}}^{-5}$ on the west and northwest of Iran.

At the level of 850 hPa in Figure 5(b), which matches the pressure pattern of 1000 hPa, the Ural low-pressure center is placed on west of Russia with centrality of 1200 geopotential meter. An elevation trough is observed in East Europe and the Mediterranean and on Iran passing from the Caspian Sea to the Persian Gulf. The cold weather of Scandinavia and

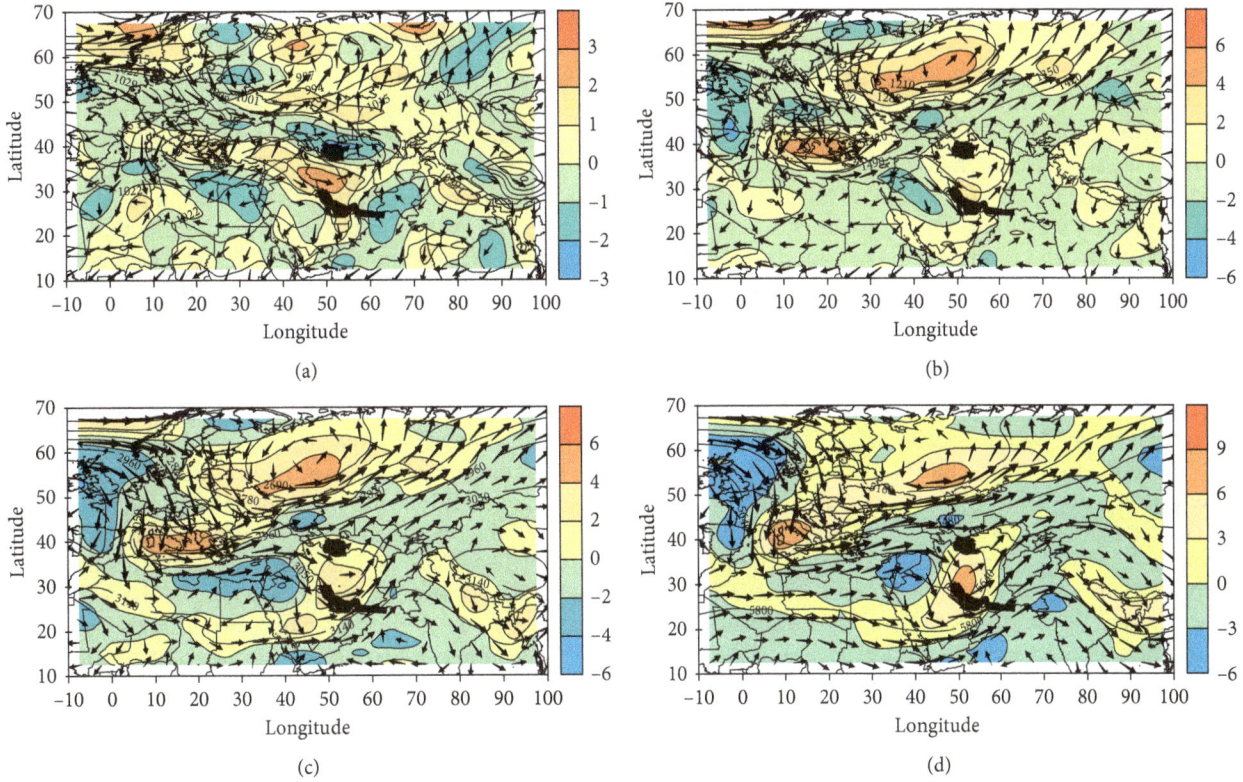

FIGURE 5: (a) 1000 hPa height (contours), vorticity $(5 \times 10_{S^{-1}}^{-5})$ (shaded), and wind flow (m/s) (vectors). (b) 850 hPa height, vorticity $(5 \times 10_{S^{-1}}^{-5})$, and wind flow (m/s). (c) 700 hPa height (hPa), vorticity $(5 \times 10_{S^{-1}}^{-5})$, and wind flow (m/s). (d) 500 hPa height, vorticity $(5 \times 10_{S^{-1}}^{-5})$, and wind flow (m/s).

Siberia transfers to lower latitudes by the low height centrality and its convergence with wet flows of Mediterranean and the Black Sea from the trough of the Mediterranean and its transfer to Iran by a relatively deep trough, based on the center of the country. In its movement from stack to the trough, the air has downward movement and with convergence, so the trough is the center of positive vorticity advection with amount of $2 \times 10_{S^{-1}}^{-5}$ and the maximum of convergence located on the trough axis [5, 47]. At the levels of 700 and 500 hPa, Figures 5(c) and 5(d), an elevation model of 850 hPa can also be clearly observed. Negative vorticity advection of $-4.5 \times 10_{S^{-1}}^{-5}$ is placed on East Mediterranean and the east of Iraq and Jordan. It causes reduction and air convergence at the trough axis and rising flows in front of the trough at this level. On the other hand, vorticity advection at the 700 hPa level of $3 \times 10_{S^{-1}}^{-5}$ on the trough axis is centered on Iran and its eastern slope of elevation. Convergence of cyclonic flows and rising air masses in these areas, especially the eastern half of the country, can be observed. As it can be observed, on the upper level (i.e., 500 hPa), the trough axis based on Iran is drawn with slight change to the below layers on the southwest areas of the Persian Gulf and lower latitudes to Saudi Arabia. The deepening trough above leads to durable stability and convergence of the flows related to Mediterranean, northern parts of the Red Sea, the Persian Gulf, and the Sea of Oman. As in front of the trough (its east range), positive vorticity advection of $6 \times 10_{S^{-1}}^{-5}$ leads to the severe divergence of

uppermost level, convergence of undercurrent, and thus, the formation of heavy precipitation, especially on the eastern half and south of the country. In fact, the areas are placed in front of the warm front. This convergence in the upper layers leads to the accumulation of air, which is dropped over the high-pressure of its bottom surface. The weather descended is replaced by diverged air at ground level, and even if the thickness of the layer of convergence at higher levels on the thickness of the low-level divergence is more, or in other words, the trough level is closer to the Earth's surface, the high pressure of Middle East is amplified. As a result, this pressure arrangement at different atmospheric levels strengthens the pressure trough on Iran and expands the front part of trough and maximum wind vorticity at the levels of 850, 700, and 500 hPa, and the upper-level divergence on the eastern half of Iran and required dynamic conditions for prevalent precipitation on Iran. In this pattern, a relatively deep trough in the middle level of the atmosphere has been formed on Iran. The establishment of the above situation created and strengthened the Ural low pressure and the high pressure of the east and west of Iran and subsequently made the northern-southern flows intense at the surface and the lower levels of the atmosphere. Therefore, the above two systems flew the cold and moist air from the Caspian Sea, Black Sea, and Mediterranean Sea to Iran. When this cold air crosses the warmer seas, it causes prevalent precipitation in Iran.

3.4. The Fourth Pattern: Low-Pressure of Saudi Arabia and High-Pressure Europe. According to Figure 6(a), at the level of 1000 hPa, a very broad area of high pressure 1030 hPa can be seen on Europe from parts of Scandinavia to the Mediterranean in the south and the east of the borders of Russia. The high-pressure tongue enters the country from the northwest and the Caspian Sea. The establishment of this high-pressure center dramatically increases the circulation in the lower levels of the atmosphere over the entire Caspian Sea region. As it moves into northern parts of the Caspian Sea, the cyclonic circulation increases, and the maximum negative vorticity is observed at the northern end of the Caspian Sea. On the other hand, a broad low-pressure center of 1010 hPa is entirely placed on Saudi Arabia, Iraq, and the Persian Gulf. The high pressure of Europe has created a strong gradient on the Black Sea, Turkey, Iraq, and especially the northern half of the country, and positive vorticity advection of $2.5 \times 10^{-5}_{S^{-1}}$ winds on the country's northwest. The pressure arrangement leads to warm air advection of lower latitudes in the northern half of the country and diffusion of very cold weather in northern Europe to the region. At the 850 hPa level in Figure 6(b), the high height center of 1580 geopotential meters is located perfectly on Europe and its tongue of 1520 geopotential meters on the northern strip. On the other hand, low height center with the amount of 1400 geopotential meters is placed on the north of Kazakhstan with its troughs from the northeast influencing the borders of the country. This height situation led to a convergence of very cold flows of northern Europe and Siberia and its precipitation on the northern regions of the country. On the other hand, the very strong height trough with several contours at 1440 geopotential meters is placed on Iraq and Saudi Arabia. The situation follows the advection of warm and moist currents from the Mediterranean and the Red seas to the Persian Gulf on Iran. Hence, that vorticity advection in the eastern tongue of a low height of $3 \times 10^{-5}_{S^{-1}}$ corresponding to intense flows of north leads to increase of mass and density and increasing convergence pressure in the level above the country.

At the 850 hPa level, the high height of 1580 geopotential meters (Figure 6(b)) is entirely located on Europe, and a tongue of it drawn on the northern strip of Iran (1520 geopotential meters). The deployment of this high pressure leads to a widespread and continuous north-northeast flow in the lower levels on the northern half of Iran. On the other hand, a low-height center of 1,400 geopotential meters is located in northern Kazakhstan, with its tongue infiltrating from the northeast of the borders of Iran. This height features led to the convergence of the very cold currents of northern Europe and Siberia and its downfall over the northern regions of Iran. Besides, a very strong trough is located in Iraq and Saudi Arabia and this situation has led to the convergence of hot and humid currents from the Mediterranean Sea, Red Sea, and Persian Gulf on Iran ($3 \times 10^{-5}_{S^{-1}}$). In these conditions, the vorticity convergence of the eastern tongue of low height has been adapted to the northward currents on Iran. As

a result, all of these situations lead to increase in mass and density of airflow on the country, which has resulted in an increase in pressure and convergence at the 850 hPa level.

At the 700 hPa level in Figure 6(c), the height arrangement of the lower level is affected by the south elevation trough, whose center of 3000 geopotential meters is placed on southern Turkey, Iraq, Syria, and East Mediterranean. Transfer of moisture from the Mediterranean Sea, the Red Sea, and the Persian Gulf, especially positive vorticity advection of $2.5 \times 10^{-5}_{S^{-1}}$, can be observed on the eastern slopes of the trough. At the 500 hPa level in Figure 6(d), the lower level of the axis of the southern trough, perfectly on the Red Sea and the middle of it, is drawn. The positive vorticity advection area corresponding to the trough with amount of $7 \times 10^{S^{-1}}_{-5}$ intensifies the divergence in upper levels and convergence in the lower levels. Saudi Arabia has low-pressure condition, and there are enhanced uplink moves in Iran and the Middle East by intensity of divergence in higher levels. Due to the establishment of a relatively deep trough at the midlevels between the Caspian Sea and the Aral Lake, the vorticity on the northern regions, especially the eastern parts of the Caspian Sea, are mostly positive, and in the surface map, a low pressure or cyclonic circulation dominated on the Aral Lake. In contrast, at the surface, a dynamic high pressure has been established on the whole region of the southern part of the Caspian Sea and the regions between the Caspian Sea and the Black Sea [49]. These conditions and the location of the low pressure of Saudi Arabia on the southern half of Iran have led to a fairly severe vorticity convergence at the lower and middle levels of the Iran's atmosphere. These features provided the climbing conditions, especially in the northern half of the country, and have caused the widespread precipitation in Iran.

In the precipitation pattern, with the outbreak of widespread precipitation on Iran leading to the release of latent heat of vaporization of rain, it added to the intensity of flotation and uplink movements. By intensifying uplink movements and expansion of air, the vorticity of the system is further increased, and by intensifying the vorticity of low height centers on the Middle East, it added to the severity of divergence in the upper levels of the atmosphere. In this pattern, at the middle level of the atmosphere, the front part of the trough with direction of northeast-southwest is located on Iran. The southward expansion of this trough has strengthened the surface low pressure. Therefore, with the availability of ascending conditions and with the humidity advection from the Mediterranean Sea and the Persian Gulf water resources caused the prevalent precipitation of Iran.

3.5. The Fifth Pattern: High Pressure Belt of Siberia-Central Iran Low Pressure. According to Figure 7(a), the low pressure of central Iran with pressure of the central core of 1010 hPa is placed in Iran in the northeast direction, whose south tongue with north side caused the rising of wet flows from the Sea of Oman and the Persian Gulf on Iran. Negative vorticity advection on Iran of $-1.5 \times 10^{-5}_{S^{-1}}$ can be seen due to the placement of the western part of low pressure on the country and reducing input airflow from north and

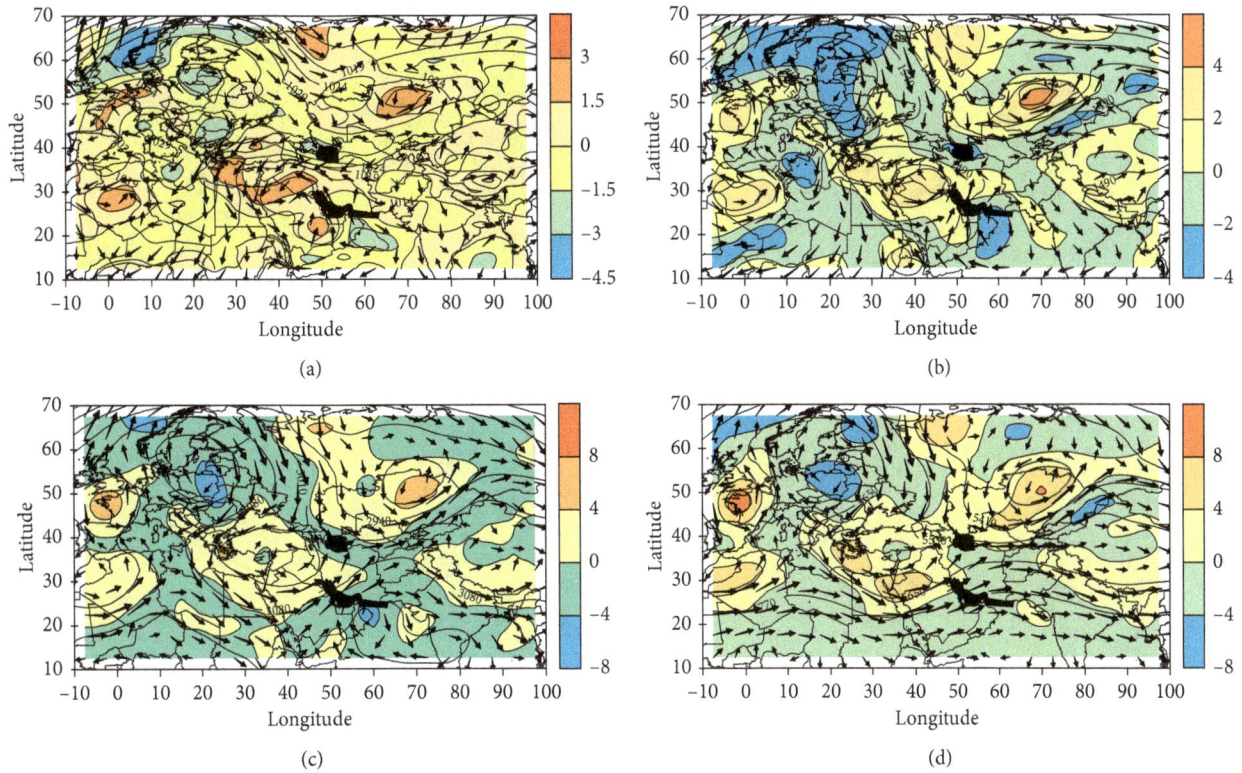

FIGURE 6: (a) 1000 hPa height (contours), vorticity ($5 \times 10^{-5}_{S^{-1}}$) (shaded), and wind flow (m/s) (vectors). (b) 850 hPa height, vorticity ($5 \times 10^{-5}_{S^{-1}}$), and wind flow (m/s). (c) 700 hPa height, vorticity ($5 \times 10^{-5}_{S^{-1}}$), and wind flow (m/s). (d) 500 hPa height, vorticity ($5 \times 10^{-5}_{S^{-1}}$), and wind flow (m/s).

increasing output mass in the south part of low pressure, reducing mass and the density of pressure, leading to divergence. On the other hand, the Siberian high pressure is reinforced and its range expanded to the northern half of the country causing the loss of cold air, decreasing airflow into the country, and exacerbating the divergence on the surface of the ground. Pressure arrangement of two low-pressure centers in the Persian Gulf and Siberian high pressure caused a sharp pressure and exacerbating instability on the northeastern areas of the country. At 850 hPa in Figure 7(b) as well as sea level, the low-height belt of central Iran with a central height of 1360 and 1400 geopotential meters is placed on Iran and the northeast. Positive vorticity advection on Iran of $3 \times 10^{-5}_{S^{-1}}$, matching the central core of low height, can be seen on Iran. However, this value in northwest reaches $2 \times 10^{-5}_{S^{-1}}$. On the map, at the pressure of 700 and 500 hPa in Figures 7(c) and 7(d), the axis of the trough is placed in the East and Southeast of Iran, and vorticity advection with the amount of $5 \times 10^{-5}_{S^{-1}}$ can be seen in this area. Considering that the placing of low pressure of central Iran in below levels caused the rise and convergence at lower levels of the atmosphere on the eastern and southeastern part and on the northwest of Iran at 700 hPa level, a low height center with center based at 2960 geopotential meters causes rising and convergence in the lower layers of the atmosphere. Positive vorticity advection resonance indicates rising and positive convergence at the levels of 700 and 500 hPa. In this pattern, in the higher layers, air density is

reduced, and reducing density increases vertical airflow in the lower layers of the atmosphere. In such a case, the flow of cold weather cools down due to adiabatic flow and causes the column of airflow on low pressure of central Iran to be colder than the surrounding environment, and the thickness of the column of airflow is reduced. Therefore, due to the decrease in the density of the air column at the levels of 850 and 700 hPa, curves at lower altitude are placed at surrounding area and can be seen on maps of 850, 700, and 500 hPa as low height cores (Figures 7(b)–7(d)). According to the process of turning at low pressure on the surface of the Earth (counterclockwise rotation), the flow of cold air is placed behind the low pressure [48, 50]. In other words, the western half of the low pressure of central Iran is colder than the eastern half, so the low-pressure of central Iran is stronger and axis of low pressure with increasing height at the levels of 850, 700, and 500 hPa inclined to the colder weather in the west, and the low-pressure system on Iran extended to southeast-northwest. This pressure arrangement at different atmospheric levels and placement of trough axis on Iran causes the intensity of vorticity advection and vertical rising of air at levels of 1000, 850, 700, and 500 hPa and divergence of air in the upper levels of the atmosphere. In this pattern, the trough axis positioning on the southern half of the country and low pressure on the surface has strengthened the upstream flow and exacerbated the instability in the region. These dynamic conditions have resulted in humidity advection from the Persian Gulf and

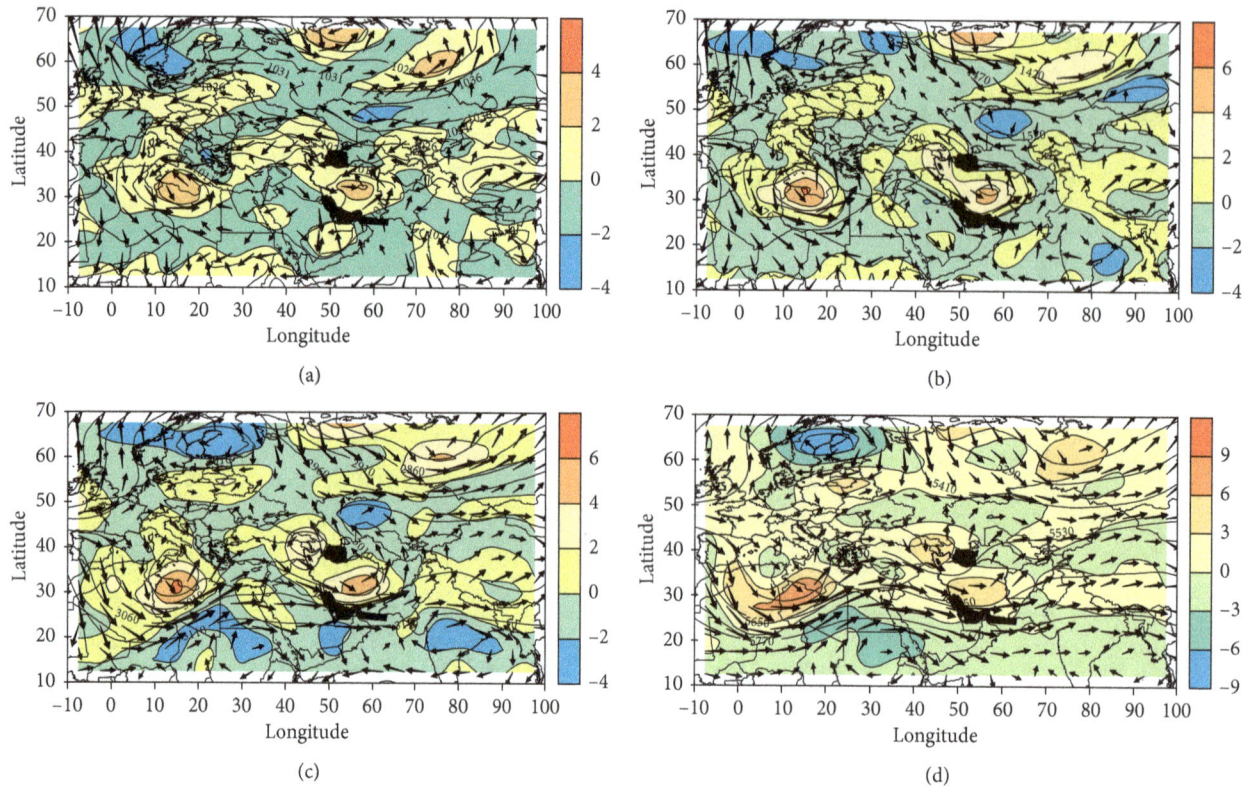

FIGURE 7: (a) 1000 hPa height (contours), vorticity ($5 \times 10^{-5}_{S^{-1}}$) (shaded), and wind flow (m/s) (vectors). (b) 850 hPa height, vorticity ($5 \times 10^{-5}_{S^{-1}}$), and wind flow (m/s). (c) 700 hPa height, vorticity ($5 \times 10^{-5}_{S^{-1}}$), and wind flow (m/s). (d) 500 hPa height, vorticity ($5 \times 10^{-5}_{S^{-1}}$), and wind flow (m/s).

the Oman Sea to Iran and the formation of prevalent precipitation.

Low-pressure systems have played a major role in the occurrence of winter-prevalent precipitation in Iran in all patterns. By creating instability in the atmosphere, these systems have led to the formation of baroclinic states, especially in the lower level of atmosphere. In the common mechanism of the occurrence of prevalent precipitation in all patterns was the role of midlatitude's trough (such as the Mediterranean trough) in strengthening low-pressure systems of lower levels of atmosphere, as well as the water resources such as the Persian Gulf, the Oman Sea, the Mediterranean Sea, the Caspian Sea, and the Black Sea in the provision of required humidity for that dynamic systems.

4. Conclusion

In this study, the status of vorticity of the atmosphere of pressure centers on Iran and the mechanisms of climate changes on widespread winter precipitation in Iran were investigated. For this purpose, the data of prevalent winter precipitation in Iran were identified on the basis of a millimeter precipitation threshold. After extracting prevalent winter precipitation in Iran, the U&V wind component and pressure data corresponding to the widespread rainy days were extracted from the databases of NCEP/NCAR and vorticities of these days were calculated. Then, by the implementation of cluster analysis, representative days for

each group were specified. The results of this study showed that the atmospheric vorticity status during prevalent winter precipitation in Iran is influenced by the interaction of low-pressure patterns of the Persian Gulf double core-low pressure of the East Mediterranean, central Iran closed low pressure-high pressure of Europe, low pressure of the Urals-high-pressure of the Middle East, low pressure of the Urals-high-pressure of Europe, and the high-pressure belt of Siberia-low pressure of central Iran.

In addition, the placement of the low pressure of dual-core of the Persian Gulf-East Mediterranean low pressure on Iran at levels of 850, 700, and 500 hPa of maximum positive vorticity caused rising and convergence of airflow at higher levels and the bottom surfaces. As a result of such conditions, the low pressure of land surface with moisture injected from the Persian Gulf and the Oman Sea and trough on Iran in high levels have strengthened its atmosphere. Positive vorticity advection on Iran is continued up to 500 hPa. In this regard, Alijani et al. investigated Iran's low-pressure role in the intensification of positive vorticity. The results of this study showed that this low pressure was the main cause of moisture transmission and the occurrence of precipitation in the first decade of July 1994 in the southeast of Iran [51]. The formation of this low pressure on the northern side of the Persian Gulf, while increasing the cyclonic motion in the southern part of the country, by creating suitable southern streams in the southeast of Iran and transferring the moisture content of the Oman Sea in a thin layer to the studied area caused the extreme precipitation in south-eastern Iran.

However, at the time of central Iran closed low pressure-high pressure of Europe, the center of low pressure on Iran caused convergence and rising airflow to higher latitudes. The placement of trough axis at levels of 850, 700, and 500 hPa caused increased positive vorticity advection on Iran and rising air in the lower atmosphere layers and divergence in the upper levels of the atmosphere. Therefore, the positive vorticity maximum is continued up to 500 hPa. In fact, the high pressure on the Middle East and the influence of the low-pressure tongue of the Urals on Iran and closing of the surface of trough to the Earth's surface caused the high pressure of Middle East to strengthen.

As a result of this pressure arrangement at different levels of the atmosphere, the pressure trough on Iran is strengthened. These conditions exacerbated dynamic conditions for the occurrence of widespread precipitation on Iran. On the other hand, the influence of the low-pressure tongue in Saudi Arabia on Iran at the ground level caused positive vorticity advection and influenced north flowing towards Iran. The low pressure of Saudi on Earth strengthened itself by the exposure to the wave of west wind at higher levels and exacerbated the divergence, leading to strengthened rising movements on Iran. Latent heat of vaporization of precipitation led to strengthening the uplink flow on Iran with these conditions increasing by strengthening the uplink flows and expansion and vorticity of the system, causing divergence in the upper levels of atmosphere. Vorticity state of the atmosphere during the formation of the central Iran low pressure made the flow of cold weather in the western half of low pressure placed on Iran colder than its eastern half. This strengthens the central Iran low pressure and causes the inclination of the low-pressure axis at levels of 850, 700, and 500 hPa in the western and northwestern areas in Iran. Pressure arrangement in different atmospheric levels and placing trough axis on Iran led to intensifying the positive vorticity advection on Iran. In the studies of Golmohammadian and Pishvaei about the production of daily rotation index and its effect on the temperature and precipitation of the northeastern part of Iran, it has been concluded that during the warm seasons, a ridge pattern has an absolute sovereignty in this region, which represents the emergence of a tropical belt on the area. As a result, Mashhad has a warm and dry climate in the warm seasons, while this cannot be seen in the cold seasons [52]. The average monthly vorticity indicator in the coldest half of the year has the highest value, indicating the frequency of cyclonic systems. In most cases, increasing the vorticity rate in one region leads to a decrease in temperature and an increase in precipitation. The major difference between the third and fourth patterns with other patterns is that the low-pressure systems of these patterns have an origin from the outside of the Iran. But in the first, second, and fifth patterns, the core of the low-pressure systems has been formed inside Iran.

Conflicts of Interest

The authors declare that they have no conflicts of interest.

Acknowledgments

This work was supported by Vedurfelagid, Rannis, and Rannsoknastofa I Vedurfraedi. Iman Rousta is deeply grateful to his supervisor (Haraldur Olafsson, Professor of Atmospheric Sciences, Department of Physics, University of Iceland, Institute for Atmospheric Sciences and Icelandic Meteorological Office), for his great support, kind guidance, and encouragement.

References

[1] I. Rousta, M. Nasserzadeh, M. Jalali et al., "Decadal spatial-temporal variations in the spatial pattern of anomalies of extreme precipitation thresholds (case study: northwest Iran)," *Atmosphere*, vol. 8, no. 12, p. 135, 2017.

[2] C. Data, *Guidelines on Analysis of Extremes in a Changing Climate in Support of Informed Decisions for Adaptation*, World Meteorological Organization, Geneva, Switzerland, 2009.

[3] A. Taimor, A. Qhasem, and I. Rousta, "Analyzing of 500 hpa atmospheric patterns in the incidence of pervasive and sectional rainfall in Iran," *Planning and arrangement of space*, vol. 16, no. 4, pp. 1–24, 2012.

[4] M. Soltani, I. Rousta, and S. S. M. Taheri, "Using Mann–Kendall and time series techniques for statistical analysis of long-term precipitation in gorgan weather station," *World Applied Sciences Journal*, vol. 28, no. 7, pp. 902–908, 2013.

[5] I. Rousta, M. Doostkamian, E. Haghighi, H. R. G. Malamiri, and P. Yarahmadi, "Analysis of spatial autocorrelation patterns of heavy and super-heavy rainfall in Iran," *Advances in Atmospheric Sciences*, vol. 34, no. 9, pp. 1069–1081, 2017.

[6] I. Rousta, M. Soltani, W. Zhou, and H. H. N. Cheung, "Analysis of extreme precipitation events over central plateau of Iran," *American Journal of Climate Change*, vol. 5, no. 3, p. 297, 2016.

[7] M. Soltani, P. Laux, H. Kunstmann et al., "Assessment of climate variations in temperature and precipitation extreme events over Iran," *Theoretical and Applied Climatology*, vol. 126, no. 3, pp. 775–795, 2016.

[8] I. Rousta, M. Doostkamian, A. Taherian, E. Haghighi, H. G. Malamiri, and H. Ólafsson, "Investigation of the spatio-temporal variations in atmosphere thickness pattern of Iran and the middle east with special focus on precipitation in Iran," *Climate*, vol. 5, no. 4, p. 82, 2017.

[9] I. Rousta, M. Doostkamian, E. Haghighi, and B. Mirzakhani, "Statistical-synoptic analysis of the atmosphere thickness pattern of Iran's pervasive frosts," *Climate*, vol. 4, no. 3, p. 41, 2016.

[10] J. R. Harman, *Synoptic Climatology of the Westerlies: Process and Patterns*, Association of American Geographers, Washington, DC, USA, 1991.

[11] L. A. Mofor and C. Lu, "Generalized moist potential vorticity and its application in the analysis of atmospheric flows," *Progress in Natural Science*, vol. 19, no. 3, pp. 285–289, 2009.

[12] N. R. Council, *The Atmospheric Sciences: Entering the Twenty-First Century*, The National Academies Press, Washington, DC, USA, 1998.

[13] M. Soltani, P. Zawar-Reza, F. Khoshakhlagh, and I. Rousta, "Mid-latitude cyclones climatology over Caspian Sea Southern Coasts–North of Iran," in *Proceedings of 21st Conference on Applied Climatology*, pp. 1–7, American Meteorological Society (AMS), London, UK, 2014, https://ams.

confex.com/ams/21Applied17SMOI/webprogram/Paper246601.html.

[14] T. Dessouky and A. Jenkinson, "An objective daily catalogue of surface pressure, flow, and vorticity indices for Egypt and it's use in monthly rainfall forecasting," *Meteorological Research Bulleting, Egypt*, vol. 11, pp. 1–25, 1975.

[15] A. Jenkinson and F. Collison, "An initial climatology of gales over the North Sea," *Synoptic Climatology Branch Memorandum*, vol. 62, p. 18, 1977.

[16] D. Conway and P. Jones, "The use of weather types and air flow indices for GCM downscaling," *Journal of Hydrology*, vol. 212, pp. 348–361, 1998.

[17] P. Jones, M. Hulme, and K. Briffa, "A comparison of Lamb circulation types with an objective classification scheme," *International Journal of Climatology*, vol. 13, no. 6, pp. 655–663, 1993.

[18] R. M. Trigo and C. C. DaCAMARA, "Circulation weather types and their influence on the precipitation regime in Portugal," *International Journal of Climatology*, vol. 20, no. 13, pp. 1559–1581, 2000.

[19] I. Phillips and G. McGregor, "The relationship between synoptic scale airflow direction and daily rainfall: a methodology applied to Devon and Cornwall, South West England," *Theoretical and Applied Climatology*, vol. 69, no. 3, pp. 179–198, 2001.

[20] H. Nakamura, "Horizontal divergence associated with zonally isolated jet streams," *Journal of the atmospheric sciences*, vol. 50, no. 14, pp. 2310–2313, 1993.

[21] M. R. Sinclair, "A diagnostic study of the extratropical precipitation resulting from Tropical Cyclone Bola," *Monthly Weather Review*, vol. 121, no. 10, pp. 2690–2707, 1993.

[22] C. Mattocks and R. Bleck, "Jet streak dynamics and geostrophic adjustment processes during the initial stages of lee cyclogenesis," *Monthly Weather Review*, vol. 114, no. 11, pp. 2033–2056, 1986.

[23] R. A. Maddox and C. A. Doswell, "An examination of jet stream configurations, 500 mb vorticity advection and low-level thermal advection patterns during extended periods of intense convection," *Monthly Weather Review*, vol. 110, no. 3, pp. 184–197, 1982.

[24] D. G. Vincent, "Cyclone development in the south pacific convergence zone during fgge, 10-17 January 1979," *Quarterly Journal of the Royal Meteorological Society*, vol. 111, no. 467, pp. 155–172, 1985.

[25] G. Wang, J. Su, Y. Ding, and D. Chen, "Tropical cyclone genesis over the South China Sea," *Journal of Marine Systems*, vol. 68, no. 3, pp. 318–326, 2007.

[26] P. Alpert, B. Neeman, and Y. Shay-El, "Climatological analysis of Mediterranean cyclones using ECMWF data," *Tellus A: Dynamic Meteorology and Oceanography*, vol. 42, no. 1, pp. 65–77, 1990.

[27] A. Bartzokas and D. Metaxas, *Climatic Fluctuation of Temperature and Air Circulation in the Mediterranean*, Commission of the European Communities (CEC), Europe, 1991.

[28] N. Ruiz and W. Vargas, "500 hPa vorticity analyses over Argentina: their climatology and capacity to distinguish synoptic-scale precipitation," *Theoretical and Applied Climatology*, vol. 60, no. 1–4, pp. 77–92, 1998.

[29] E. Xoplaki, J. F. González-Rouco, J. Luterbacher, and H. Wanner, "Wet season Mediterranean precipitation variability: influence of large-scale dynamics and trends," *Climate dynamics*, vol. 23, no. 1, pp. 63–78, 2004.

[30] G. Hadis and P. Mohammad Reza, "Daily vorticity index and it's impacts on precipitation and temperature in Khorassan

region in 1948-2010," *Journal of Applied Researches in Geographical Sciences*, vol. 13, no. 29, pp. 217–236, 2013.

[31] B. Alijani and M. Zahehi, "Statistical and synoptic analysis of Azerbaijan area rainfall," *Iranian Journal of Research in Geography*, pp. 65-66, 2002.

[32] M. Mohammad Hossain and M. Mohammad, "investigating pressure trough of Red Sea from the perspective of the dynamic," *Nivar*, vol. 52, pp. 67–74, 2004.

[33] A. Mofidi, A. Zarrin, and G. R. J. Ghobadi, "Determining the synoptic patterns of autumn-time extreme and severe precipitation over the southern coasts of Caspian Sea," *Journal of the Earth and Space Physics*, vol. 33, no. 3, p. 30, 2008.

[34] E. Kalnay, M. Kanamitsu, R. Kistler et al., "The NCEP/NCAR 40-year reanalysis project," *Bulletin of the American Meteorological Society*, vol. 77, no. 3, pp. 437–471, 1996.

[35] B. E. Doty and J. L. I. Kinter, *Geophysical Data Analysis and Visualization Using the Grid Analysis and Display System*, National Aeronautics and Space Administration, Washington, DC, USA, 1995.

[36] J. E. Martin, *Mid-Latitude Atmospheric Dynamics: a First Course*, John Wiley & Sons, Hoboken, NJ, USA, 2013.

[37] M. R. Kaviani and B. Alijani, *Principles of Climatology*, SAMT Press, Tehran, Iran, 1st edition, 2001.

[38] K. E. Trenberth, "Recent observed interdecadal climate changes in the Northern Hemisphere," *Bulletin of the American Meteorological Society*, vol. 71, no. 7, pp. 988–993, 1990.

[39] B. J. Hoskins and K. I. Hodges, "New perspectives on the Northern Hemisphere winter storm tracks," *Journal of the Atmospheric Sciences*, vol. 59, no. 6, pp. 1041–1061, 2002.

[40] E. K. Chang and D. B. Yu, "Characteristics of wave packets in the upper troposphere. Part I: Northern Hemisphere winter," *Journal of the Atmospheric Sciences*, vol. 56, no. 11, pp. 1708–1728, 1999.

[41] B. Alijani, J. O'Brien, and B. Yarnal, "Spatial analysis of precipitation intensity and concentration in Iran," *Theoretical and Applied Climatology*, vol. 94, no. 1, pp. 107–124, 2008.

[42] H. Lashkari, Z. Mohammadi, and G. Keikhosravi, "Annual fluctuations and displacements of inter tropical convergence zone (ITCZ) within the range of Atlantic Ocean-India," *Open Journal of Ecology*, vol. 7, no. 1, p. 12, 2017.

[43] E. Haghighi, S. Jahanbakhsh, M. R. Banafshe, and I. Rousta, "The study relationship between large-scale circulation patterns of sea level and snow phenomenon in the North West of Iran," *Territory*, vol. 12, no. 48, pp. 19–35, 2016, in Persian.

[44] G. Azizi, H. Mohammadi, M. Karimi Ahmadabad, A. Shamsipour, and I. Rousta, "Identification and analysis of the north atlantic blockings," *International Journal of Current Life Sciences*, vol. 5, no. 4, pp. 577–581, 2015.

[45] G. Azizi, H. Mohammadi, M. Karimi Ahmadabad, A. Shamsipour, and I. Rousta, "The relationship between the Arctic oscillation and North Atlantic blocking frequency," *Open Journal of Atmospheric and Climate Change*, vol. 1, pp. 1–9, 2015.

[46] P. D. Williams, "A proposed modification to the Robert-Asselin time filter," *Monthly Weather Review*, vol. 137, no. 8, pp. 2538–2546, 2009.

[47] Y. G. Rahimi, *Synoptic Analysis with GrADS Software*, vol. 2, SAHA Danesh Press, Tehran-Iran, 1st edition, 2016, in Persian.

[48] G. Keykhosrowi and H. Lashkari, "Analysis of the relationship between the thickness and height of the inversion and the severity of air pollution in Tehran," *Journal of Geography and Planning*, vol. 18, no. 9, pp. 231–257, 2014.

[49] G. R. Janbaz Ghobadi, A. Mofidi, and A. Zarrin, "Identify synoptic patterns of heavy rainfall in the summer on the southern shores of the Caspian Sea," *Geography and Planning*, vol. 22, no. 2, pp. 23–39, 2011.

[50] I. Rousta, F. K. Akhlagh, M. Soltani, and S. S. M. Taheri, "Assessment of blocking effects on rainfall in northwestern Iran," in *Proceedings of COMECAP 2014*, p. 291, Crete University Press, Heraklion, Grecce, 2014.

[51] B. Alijani, A. Mofidi, Z. Jafarpour, and A. Aliakbari-Bidokhti, "Atmospheric circulation patterns of the summertime rainfalls of southeastern Iran during July 1994," *Earth and Space physics*, vol. 37, no. 3, pp. 205–227, 2012.

[52] H. Golmohammadian and M. R. Pishvaei, "Production of daily rotation index and its effect on temperature and precipitation of Khorasan region in the period of 1948-2010," *Journal of Applied researches in Geographical Sciences*, vol. 13, no. 29, pp. 217–236, 2013.

3

Evaluation and Correction of GPM IMERG Precipitation Products over the Capital Circle in Northeast China at Multiple Spatiotemporal Scales

Wei Sun,[1,2,3] **Yonghua Sun,**[1,2,3] **Xiaojuan Li** ⓘ**,**[1,2,3] **Tao Wang,**[1,2,3] **Yanbing Wang,**[1,2,3] **Qi Qiu,**[1,2,3] **and Zhitian Deng**[1,2,3]

[1]*College of Resource Environment and Tourism, Capital Normal University, Beijing 100048, China*
[2]*College of Geospatial Information Science and Technology, Capital Normal University, Beijing 100048, China*
[3]*Beijing Laboratory of Water Resource Security, Capital Normal University, Beijing 100048, China*

Correspondence should be addressed to Xiaojuan Li; lixiaojuan@cnu.edu.cn

Academic Editor: Federico Porcù

Accurate remote-sensed precipitation data are crucial to the effective monitoring and analysis of floods and climate change. The Global Precipitation Measurement (GPM) satellite product offers new options for the global study of precipitation. This paper evaluates the applicability of GPM IMERG products at different time resolutions in comparison to ground-measured data. Based on precipitation data from 107 meteorological stations in the Beijing-Tianjin-Hebei region, GPM products were analysed at three timescales: half-hourly (GPM-HH), daily (GPM-D), and monthly (GPM-M). We use a cumulative distribution function (CDF) model to correct GPM-D and GPM-M products to analyse temporal and spatial distributions of precipitation. We came to the following conclusions: (1) The GPM-M product is strongly correlated with ground station data. Based on five evaluation indexes, NRMSE (Normalized Root Mean Square Error), NSE (Nash-Sutcliffe), FAR (False Alarm Ratio), UR (Underreporting Rate), and CSI (Critical Success Index), the monthly GPM products showed the best performance, better than GPM-HH products and GPM-D products. (2) The performance of GPM products in summer and autumn was better than in winter and spring. However, the GPM satellite's precision in undulating terrain was poor, which could easily lead to serious errors. (3) CDF models were successfully used to modify GPM-D and GPM-M products and improve their accuracy. (4) The range of 0–100 mm precipitation could be corrected best, but the GPM-M products were underestimated. Corrected GPM-M data in the range >100 mm were overestimated. According to this analysis, the GPM IMERG Final Run products at daily and monthly timescales have good detection ability and can provide data support for long-time series analyses in the Beijing-Tianjin-Hebei region.

1. Introduction

Precipitation is an important part of the water cycle of terrestrial ecosystems [1] and has a profound impact on atmospheric and hydrological processes. Precipitation is one of the most important hydrological meteorological variables [2] as it provides the underlying data of most studies on hydrology, climatology, and ecology [1]. At present, the meteorological observations of the national basic meteorological observing stations and the surface precipitation radar are the main observation methods for rainfall. However, while the rainfall data of gauges are highly accurate, only the precipitation data of the corresponding point can be obtained [3]. Therefore, it is hard to fully reflect the spatial distribution and changes in rainfall intensity [2]. Also, there is a lack of data on polar, marine, and remote mountainous areas. Although ground-based rain radars can indirectly obtain precipitation information with high temporal and spatial resolution, their accuracy is affected by the spatial structure of the ground and their observation range is limited [3].

In recent decades, the development of remote sensing and geographic information systems and generation of corresponding remote-sensing product data have provided entirely new methods and means for precipitation observation [4]. The successful launch of the Tropical Rainfall Measuring Mission (TRMM) has created a new era of global satellite rainfall monitoring [2] TRMM carries the first space-borne precipitation radar (PR), which provides 3D precipitation echo information. The Global Precipitation Measurement (GPM), a new generation of satellite precipitation products, is a follow-up to the TRMM. Comprehensive assessments of the TRMM satellite in China have been carried out, including accuracy evaluation of space scale and drought and waterlogging events, based on the different timescales of the TRMM satellites [5–10]. As the new generation of precipitation observing satellites, GPM provides more options for studying precipitation, and currently, scholars from different regions of the world have conducted preliminary assessments of GPM IMERG and TRMM products. Some assessments of the GPM (IMERG) have been conducted in Iran [11, 12], Korea [13], Japan [13], the Blue Nile Basin [14], Southern Canada [15], India [16], Singapore [17], Austria [18], the Main Bolivian Watersheds [19], and Peru [20]. GPM-3IMERGHH is more accurate than TRMM 3B42 V7 in describing the spatial distribution of precipitation [13, 16]. Chiaravalloti [21] assessed GPM and SM2RAIN-ASCAT rainfall products over complex terrain in southern Italy and found that MERG has good performance at the time resolution greater than 6h. However, there are still several uncertainties in different regions, time periods, topography, and precipitation patterns [17]. In China, many scholars have studied the precision of the GPM [22–28, 36]. Comparing GPM products with the TRMM (3B42V7) product, it was found that GPM performed better than TRMM in relatively dry climates. On the monthly scale, the accuracy of GPM in winter precipitation in mainland China is obviously better than that of TRMM because GPM improves the observational ability of weak precipitation and solid precipitation [29, 30]. At the same time, GPM can detect the changes in precipitation day by day [22]. In extreme precipitation events, all the GPM IMERG products are superior to the TRMM series of satellites [17]. In different elevation zones of the Tianshan Mountains, the GPM showed lower error and a higher correlation coefficient with the observation stations [25].

Remote-sensing data include regional and seasonal systematic deviations and random errors. These deviations can be corrected by calibrating the data with rainfall data measured by on-ground weather stations [31, 32]. Many scholars at home and abroad have done a lot of work on the error correction of satellite precipitation products. Common methods are interpolation method, physical model method, and statistical model method. AghaKouchak et al. (2009) corrected the uncertainty of precipitation by establishing a two-parameter stochastic model and used the maximum likelihood to estimate the random error and the multiplication error to correct the TRMM data [33]. Cheema and Bastiaanssen (2012) used the regression analysis (RA) and geographic difference analysis (GDA) to locally correct the TRMM3B43 precipitation data and found that the GDA calibration method performed best in the mountainous area

[34]. However, research on Global Precipitation Measurement (GPM) products mostly focuses on verification of the accuracy of data from different regions; however, research on GPM product data correction is scarce. Guo et al. (2016) conducted accuracy verification on the GPM data before and after the calibration of the Global Precipitation Climate Centre (GPCC) monthly data. It was found that the correlation coefficient of the calibrated GPM data products in the national and regional areas was significantly improved and the relative error was reduced [35]. Jin et al. (2018) used the MERGE error correction method to reduce the precipitation error in high-altitude areas effectively [36]. According to the research into TRMM data correction, cumulative distribution function (CDF) models can be fitted to precipitation distribution data by using the gamma distribution of two parameters [32, 37]. For cumulative precipitation data, the CDF model uses a cumulative distribution function based on multiyear precipitation, which is suitable for spatial and temporal data correction [38–40].

At present, research evaluating the applicability of the GPM products and correction in Beijing, Tianjin, and Hebei Province is relatively scarce. Precipitation is one of the important sources of water resources in Beijing, Tianjin, and Hebei [41]. Rainstorm events can cause serious economic losses in the area, adversely affecting urban operations (roads, transport, etc.) [42]. Therefore, a focus on precipitation events in the Beijing–Tianjin–Hebei region is important. In this paper, our purposes are as follows: (1) to evaluate the performance of the post-real-time GPM IMERG Final Run product at three temporal resolutions (i.e., half-hourly, daily, and monthly) over the Beijing–Tianjin–Hebei region of China; (2) to analyse the applicability of GPM-M products at different rainfall intensities and regions; (3) to correct the GPM-M products and the GPM-D products by using CDF model to improve accuracy; and (4) to analyse the temporal and spatial distribution characteristics of precipitation based on modified GPM-M products. This paper is organized as follows: Section 2 introduces the study area and the different data sets used in this study; Section 3 describes the methodologies used for evaluation and correction; Section 4 presents the results and discussion; and finally, Section 5 provides concluding remarks.

2. Study Area and Data Sources

2.1. Study Area. This article focuses on the areas of the Beijing-Tianjin-Hebei region that called Capital Circle in Northeast China—from the Taihang Mountains in the west to the Zhangbei Plateau in the north, the North China Plain in the south, and the Bohai Sea in the east. Figure 1 shows geographical position, topography, and gauges of the study area. The overall topography is high elevation in the northwest and low in the southeast [43]. The Beijing-Tianjin-Hebei region is situated within a warm-temperate monsoon climate, with dry springs and wet summers.

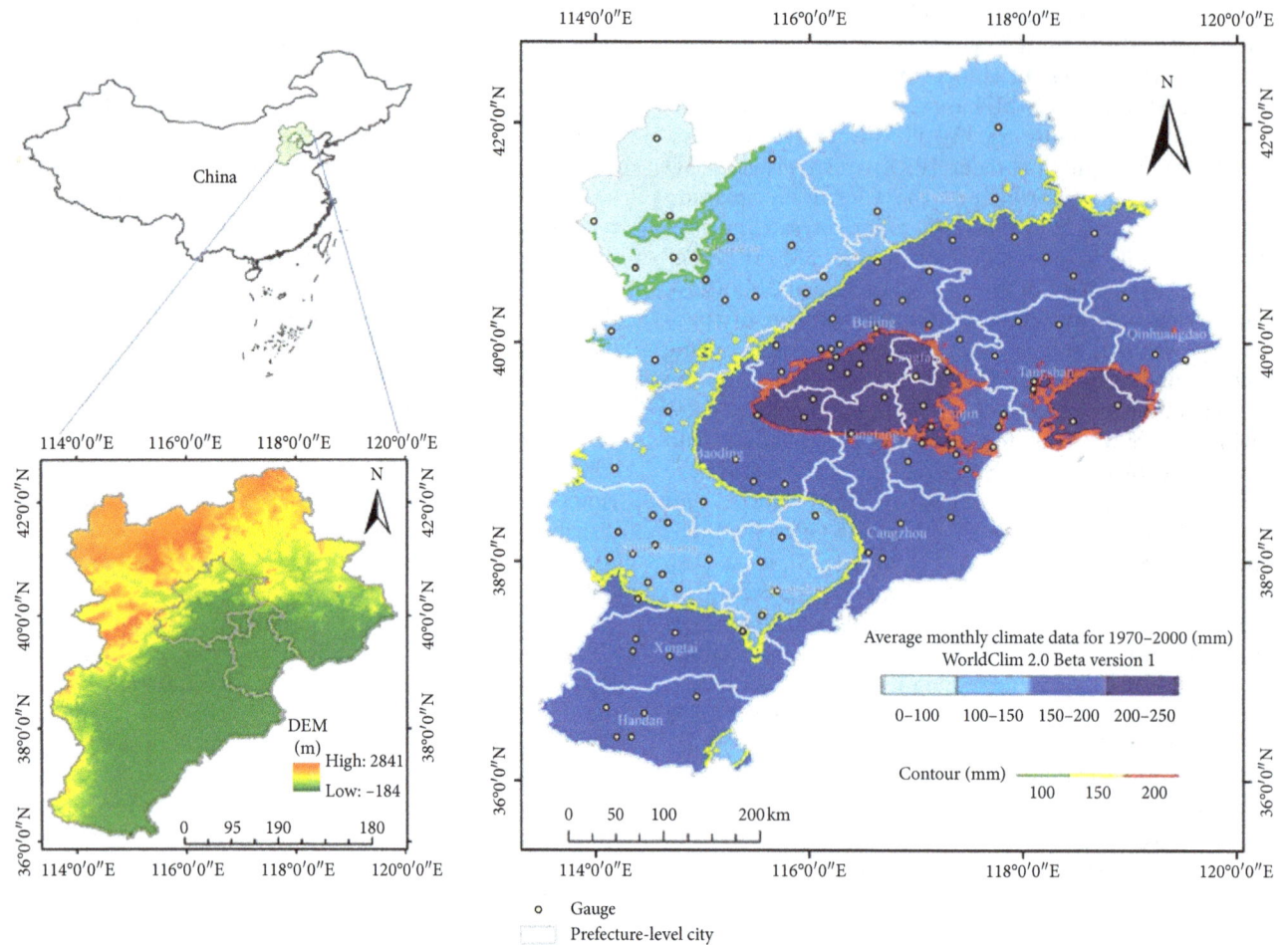

FIGURE 1: Geographical position, topography, and gauges of the study area.

2.2. GPM Products. Global Precipitation Measurement (GPM) is a new generation of satellite precipitation products. Its purpose is to develop the next generation of a space measurement system which can realize frequent and accurate global rainfall measurements [4]. Based on the success of TRMM, GPM focuses on deploying a "core" satellite, under joint development by NASA and the Japan Aerospace Exploration Agency (JAXA) [4]. Launched in February 2014, the Core Observatory combines advanced microwave detection techniques and data correction algorithms to provide more options for studying precipitation. It carries the first space-borne Ku-/Ka-band dual-frequency precipitation radar (DPR) and a multichannel GPM microwave imager (GMI) [4, 44]. The DPR instrument (Ku-band at 13.6 GHz and Ka-band at 35.5 GHz) provides three-dimensional measurements of precipitation structure which are more sensitive to rates of light rain and snowfall than the TRMM precipitation radar [44]. The GMI instrument (frequency from 10 GHz to 183 GHz) is a multichannel, conical-scanning, microwave radiometer, with the aim of measuring precipitation features such as intensity and type [13, 45]. Compared with TRMM which focuses on the observation of precipitation in the tropical and subtropical regions, GPM can capture low precipitation (<0.5 mm·h^{-1}) and solid-state precipitation more accurately, extending measurements to

$\pm 68°$ latitude [4]. The GPM also extends the TRMM sensor load, significantly improving its precipitation observation capability. Its high-resolution precipitation products can reach 0.1° latitude/longitude spatial resolution and half-hourly temporal resolution.

IMERG is a level-3 GPM product and uses the algorithm Day-1 U.S. multi-satellite precipitation estimation which relies on three existing algorithms, namely, TMPA, CMORPH, and PERSIANN [46]. The algorithm aims at intercalibrating and merging "all" satellite microwave precipitation estimators, along with microwave-calibrated infrared (IR) satellite estimates, precipitation gauge analyses, and potentially other precipitation estimators. To obtain rainfall estimates for this study, firstly, all input datasets were processed using the Goddard profiling algorithm 2014 (GPROF2014) and IR rainfall estimates. Then, the rainfall estimates were recalibrated by the Climate Prediction Centre's (CPC) Morphing-Kalman Filter (CMORPHKF) using the Lagrangian time interpolation technique and the PERSIANN-Cloud Classification System (PERSIANN-CCS) to ensure products had 0.1° spatial and 30 min temporal resolutions. Finally, to improve the accuracy of the product, monthly Global Precipitation Climatology Center (GPCC) products were utilized to correct the bias.

These three latency periods, known as "early," "late," and "final," were delayed by about 6 hours, 18 hours, and 4 months, respectively, between the collection of observations and the generation of data products. The GPM (IMERG) final run version 4 product was used, which is the post-real-time research product in the IMERG suite [47].

The "final" IMERG has three kinds of datasets (Table 1): half-hourly scale data (hereafter called GPM-HH), daily scale data (hereafter called GPM-D), and monthly scale data (hereafter called GPM-M). These datasets are useful for long-time series analysis of precipitation events and for disaster risk. Therefore, in this paper, we evaluated the applicability of these three kinds of data in the Beijing-Tianjin-Hebei region. And, we corrected GPM-M and GPM-D products using Cumulative Distribution Function (CDF) model. The IMERG products were downloaded from the PMM website (http://pmm.nasa.gov/data-access/downloads/gpm).

2.3. Gauge Data. We screened the national meteorological stations in the Beijing-Tianjin-Hebei region. Some gauges with missing data were deleted, until 107 national meteorological stations were finally selected. Based on the site precipitation data, the adaptability of GPM products to Beijing, Tianjin, and Hebei at different timescales was evaluated (Figure 1). The dataset time series used was the monthly average precipitation from March 2014 to 2016; and the daily and hourly precipitation from January to December 2016. By obtaining the average monthly precipitation at each gauge, GPM-M products were evaluated. The period of participation in the evaluation was from March 2014 to December 2016. The precipitation was greatest in July–September, making verification more reliable during that time. Therefore, the daily and hourly precipitation data from 1 July to 30 September 2016 were selected for evaluation of GPM-D products and GPM-HH products. Finally, the monthly and daily precipitation data were used to correct GPM-M and GPM-D products using CDF model.

3. Methodology

3.1. Preprocess of GPM Products. The data in this article are referred to in World Time. GPM IMERG products are raster data. Table 1 illustrates the grid unit of the GPM-HH products in mm/h. Therefore, the hourly accumulation of 2 half-hourly precipitations was then multiplied by a factor of 0.5. These three types of data indicated the precipitation rate and therefore needed to be converted into actual rainfall in millimetres. The formulas are as follows:

$$y_{HH} = P \times 0.1 \times 0.5,$$
$$y_D = P \times 0.1 \times 24, \qquad (1)$$
$$y_M = P \times 0.001 \times 24 \times N_D,$$

where y_{HH} is the rainfall of GPM-HH products; y_D is the rainfall of GPM-D products; y_M is the rainfall of GPM-M

products; P is the GPM image grid value; and N_D is the number of days at different months.

3.2. Applicability Evaluation Indices. GPM products at different timescales were compared and analysed according to the grid point values of the corresponding meteorological stations and satellite precipitation data [25]. The quantitative and qualitative correlations between satellite rainfall and ground-based weather station rainfall data were evaluated using two kinds of quantitative indexes (mean) and correlation coefficients (CC). The mean was used to assess the accuracy of satellite precipitation products for rainfall measurements. The average showed the trend of satellite precipitation data in the region; and CC indicated the linear correlation between satellite rainfall data and ground-based weather station rainfall data. The precision evaluation of precipitation products of different magnitudes used Nash-Sutcliffe (NSE with a range of ∞ to 1) and normalized root mean square error (NRMSE) as indices [48]. The higher the NSE value, the closer the GPM value was to the gauged value and the better the simulation effect [49]. The ability to detect rainfall events from satellite rainfall data can be assessed comprehensively in three categories: False Alarm Ratio (FAR), Underreporting Rate (UR), and CSI. The lower the UR, the smaller the false negatives; the lower the FAR, the smaller the empty forecast. In fact, CSI is a more balanced score [22]. RMSE and BIAS are used to evaluate the accuracy of the modified GPM-M products. The root mean square error (RMSE) is used to assess the overall level of error and accurately reflect the accuracy of satellite precipitation products. When the Relative Bias (BIAS) is greater than 0, it shows that the estimated value of satellite is less than the measured value. The closer the BIAS is to 0, the better the agreement between the satellite detection effect and the measured value [49].

3.3. CDF Model. The gamma distribution has strong adaptability and can be fitted to many positive observation datasets. Precipitation distributions are usually skewed and many studies have shown that the gamma distribution is the best model for fitting to precipitation data [38–40]. The gamma distribution, denoted as Γ (α, β), had a probability density function:

$$f(x) = \begin{cases} \dfrac{1}{\beta^\alpha \Gamma(\alpha)} x^{\alpha-1} e^{-x/\beta}, & x > 0, \\ \\ 0, & x = 0, \end{cases} \qquad (2)$$

$$\Gamma(\alpha) = \int_0^\infty x^{\alpha-1} e^{-x} dx,$$

Here, x is the amount of precipitation in mm. Maximum likelihood estimation is used to determine the shape parameter α and the scale parameter β of each gamma distribution.

TABLE 1: GPM final IMERG products summary.

GPM products (latency)	Research contents	Abbreviation	Resolution	Dates	Precipitation units in GIS file
	Adaptability evaluation	GPM-HH	0.1°–30 minutes	2016.07–2016.09	0.1 mm/h
		GPM-D	0.1°–1 day	2016.07–2016.09	0.1 mm/h
Final IMERG (4 months)		GPM-M	0.1°–1 monthly	2014.03–2016.12	0.001 mm/h
	Data correction	GPM-D	0.1°–1 day	2016.07–2016.09	0.1 mm/h
		GPM-M	0.1°–1 monthly	2014.03–2016.12	0.001 mm/h

$$A = \ln\left(\frac{1}{n}\sum_{i=1}^{n} x_i\right) - \frac{1}{n}\sum_{i=1}^{n}\ln x_i,$$

$$\alpha = \frac{1 + \sqrt{1 + 4A/3}}{4A}, \tag{3}$$

$$\beta = \frac{\sum_{i=1}^{n} x_i}{n\alpha},$$

Here, A is the precipitation interval, x_i is the amount of precipitation in the sample, and n is the total number of samples. So, the CDF for precipitation over a given timescale can be expressed as follows:

$$f(x; \alpha, \beta) = \int_0^x f(t)\,dt. \tag{4}$$

The CDFs can be used to model precipitation data measured by both meteorological stations and the GPM satellite. The CDF curve fitted to the GPM data is corrected using the CDF curve for ground-measured precipitation. Since the amount of precipitation in the CDF model cannot be 0, it needs to be treated separately.

Assume that the monthly precipitation recorded by ground gauges is X_{Gauge}, the corresponding GPM precipitation is X_{GPM}, and the corrected precipitation is X'_{GPM}. Then, the specific process is shown in Figure 2.

(a) As shown in Figure 3, data from 107 sites are randomly divided into (1) 75 sites used for construction of the CDF model (training dataset); and (2) 32 sites for model accuracy verification (validation dataset).

(b) When $X_{\text{Gauge}} = 0$, let $X_{\text{GPM}} = 0$ to reduce the GPM satellite's empty reporting rate.

(c) When $X_{\text{Gauge}} > 0$, the CDF curves of the different precipitation ranges are established by using the meteorological station data for the study area, as $\text{CDF}_{\text{Gauge}}$. The corresponding CDF curves of GPM precipitation are recorded as CDF_{GPM}. Using $\text{CDF}_{\text{Gauge}}$ to correct CDF_{GPM}, we obtain the revised GPM precipitation of different precipitation ranges, X'_{GPM}. This can be expressed as follows:

$$X'_{\text{GPM}} = \text{CDF}_{\text{gauge}}^{-1}\left(\text{CDF}_{\text{GPM}}(X_{\text{GPM}})\right). \tag{5}$$

(d) Use the validation dataset to verify the accuracy of the model.

(e) The CDF correction model is applied to the whole research area to achieve revised GPM-D and GPM-M data for the Beijing-Tianjin-Hebei region.

4. Results and Discussion

4.1. Adaptability Evaluation of GPM Products

4.1.1. Results of Adaptability Index. Based on the statistical metrics (Table 2), we evaluated GPM precipitation products based on the observation data of 107 ground meteorological stations. Table 3 shows that the correlation between GPM-HH products and the precipitation data of the gauges was lower, reaching only 0.38. The correlation between GPM-D products and the precipitation data of gauges was higher at 0.75. Compared with GPM-HH and GPM-D products, GPM-M products had strong correlation at 0.90. As the timescale increased, NSE showed a weaker trend while that of NRMSE tended to be stronger. The closer NSE was to 1, the closer were the GPM products and gauged values. The lower the NRMSE value, the smaller the error. Therefore, the GPM-M products show the smallest error and greatest accuracy. The GPM-HH products had the lowest accuracy. There may have been spatial heterogeneity, with great uncertainty [15]. Compared with half-hourly scale products, monthly and daily scale products showed smoother extreme values and lower errors relative to gauge measurements. The ability to detect precipitation events was analysed using FAR, UR, and CSI, with CSI having a more balanced score. As shown in Table 3, the UR, FAR, and CSI at the daily scale were 0.38, 0.20, and 0.53, respectively, better than those of the GPM-HH products. GPM-M products had the highest CSI and best detection ability for precipitation events. According to the trend shown in Table 3, the ability of the tested products to detect precipitation events, in order from high to low, were GPM-M, GPM-D, and GPM-HH.

4.1.2. Spatial Distribution of GPM-M's Adaptability. Based on Tyson polygons, the Beijing-Tianjin-Hebei region was divided into 107 polygons dictated by the locations of the 107 weather stations. Each polygon represented the largest influence range to obtain the spatial distribution of statistical indicators at different timescales and seasons. Figure 4 shows that satellite precipitation products at the monthly scale (GPM-M) were a good reflection of the precipitation events in the Beijing-Tianjin-Hebei region. Most of the region correlation coefficients were greater than 0.9. Meanwhile, GPM-M products had a higher NSE value and a lower NRMSE value with higher accuracy. GPM-M products also had the strongest detection ability for precipitation events. According to the spatial distribution of index, the GPM satellites were poorly represented in precipitation events in the marginal areas of the northwest and southeast of the Beijing-Tianjin-Hebei region. The regions

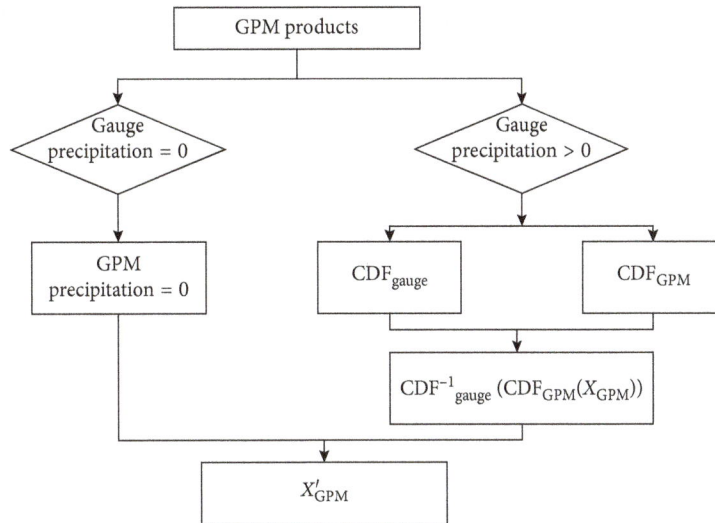

Figure 2: Flowchart of the CDF model for precipitation correction.

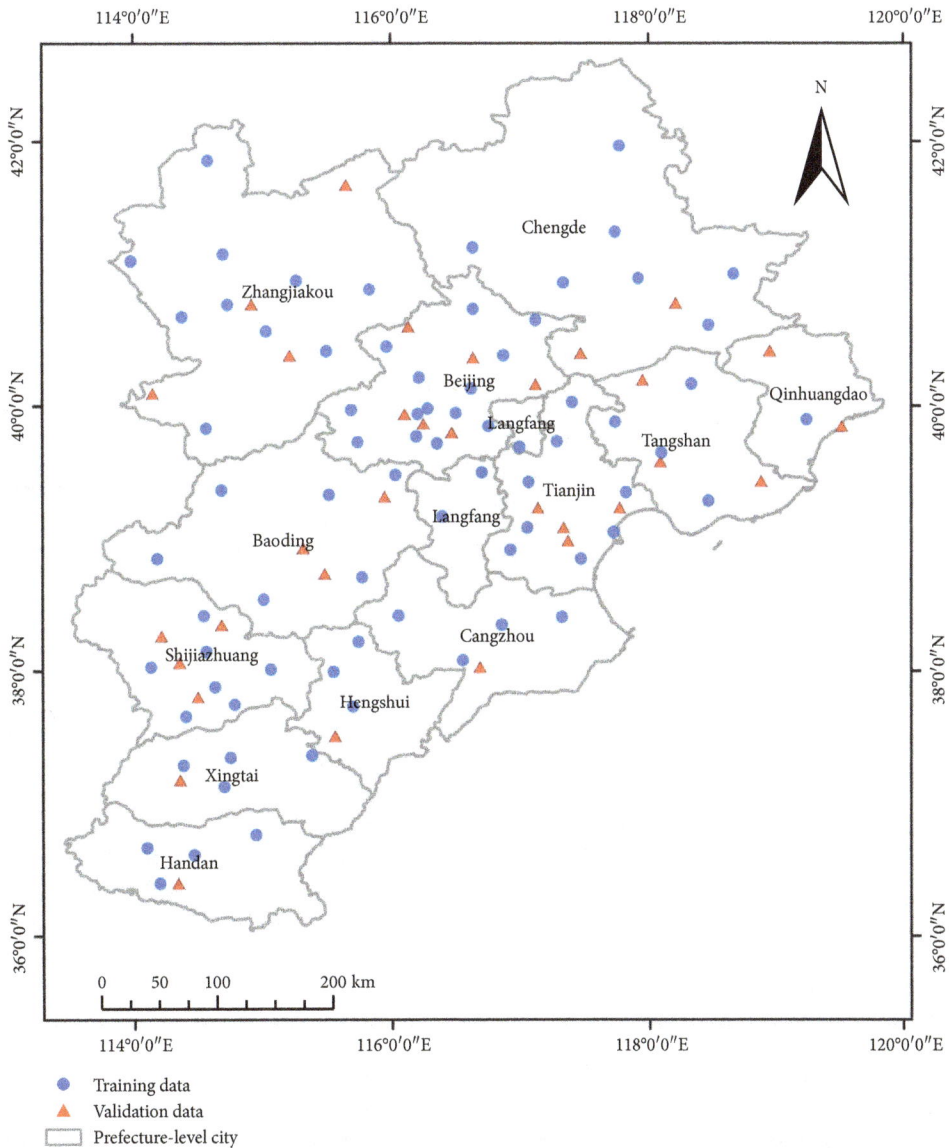

Figure 3: Distribution of training and validation data sources in the Beijing-Tianjin-Hebei region, China.

TABLE 2: List of the statistical metrics used in the evaluation.

Statistic metrics	Equation	Optimum value
Mean (M)	$M_x = 1/N \sum_{n=1}^{n} x_n$; $M_y = 1/N \sum_{n=1}^{N} y_n$	NA
Correlation coefficient (CC)	$CC = \mathrm{Cov}(x_n - y_n)/\delta_x \delta_y$	1
NRMSE	$NRMSE = (\sqrt{1/N \sum_{n=1}^{N} (x_n - y_n)})/(1/N \sum_{n=1}^{N} y_n)$	0
RMSE	$RMSE = \sqrt{\sum_{n=1}^{N} (x_n - y_n)^2 / N}$	0
BIAS	$BIAS = (\sum_{n=1}^{N} (y_n - x_n))/(\sum_{n=1}^{N} y_n) \times 100\%$	0
NSE	$NSE = 1 - (\sum_{n=1}^{N} (x_n - y_n)^2)/(\sum_{n=1}^{N} (y_n - \overline{y_n})^2)$	1
Underreporting rate (UR)	$UR = n_{01}/n_{11} + n_{01}$	0
False alarm ratio (FAR)	$FAR = n_{10}/n_{11} + n_{10}$	0
Critical success index (CSI)	$CSI = n_{11}/n_{11} + n_{01} + n_{10}$	1

Note: n represents number of samples; x_n represents the satellite precipitation estimate; y_n represents the gauge observed precipitation; Cov() represents the covariance; δ_y represents standard deviations of gauge precipitation; δ_x represents standard deviations of satellite precipitation; n_{11} represents the precipitation observed by the gauge and satellite simultaneously; n_{01} represents the precipitation observed by the gauge but not observed by the satellite; n_{10} is contrary to n_{01}; and n_{00} represents the precipitation observed neither by the gauge nor the satellite.

TABLE 3: Adaptability index at different timescales (CC, NRMSE, NSE, UR, FAR, and CSI).

Statistical indicators	Consistency	Precision		Detection ability		
	CC	NRMSE	NSE	UR	FAR	CSI
GPM-HH	0.38	9.78	−0.3	0.46	0.58	0.31
GPM-D	0.75	2.87	0.53	0.38	0.2	0.53
GPM-M	0.90	0.58	0.8	0	0.06	0.94

coinciding with better satellite performance were mostly in the plains. The southeast of the Beijing-Tianjin-Hebei region is located on the coastline of the Bohai Sea. GPM satellites perform poorly in complex calibration systems that distinguish between rain and no rain [13]. With higher precipitation, the error of GPM products increases relatively, resulting in reduced data accuracy. Due to physical errors, it is very difficult for GPM sensors to detect deep convection caused by unstable air mass [13]. The areas in the northwest (mountainous) and southeast regions were unable to accurately detect precipitation events as they were not conducive to the transport of water vapor, so had less precipitation. GPM satellites had difficulties detecting microprecipitation. The contour map of precipitation was drawn according to the average monthly precipitation in July from 1970 to 2000, shown in Figure 1, and the areas of Beijing, Tianjin, and Hebei were divided by different rainfall intensities: 0–100 mm, 100–150 mm, 150–200 mm, and 200–250 mm. According to the correlation between GPM products and meteorological stations in different precipitation intensity areas, it was found that the correlation coefficient was low at 0–100 mm, mostly high between 150 and 200 mm, and higher areas with 200–250 mm were less. It could be seen that with greater precipitation intensity, the trend of the correlation between GPM and gauges increased first and then decreased slightly. This may have been due to the limitations in GPM for detecting microprecipitation, resulting in the lowest correlation in areas with relatively low precipitation. For higher rainfall areas, the GPM was prone to large deviation, with a slight decrease in precipitation of high-value area correlation. The comparison of GPM

product accuracy for different precipitation intensity areas found that GPM-M products had the highest accuracy and performance in the region of 100–200 mm precipitation intensity. For different rainfall intensities, GPM products at different timescales had different performance capabilities (Figure 4). The GPM-M products had the weaker ability to detect precipitation in areas near the 150 mm contour.

4.1.3. Temporal Distribution of GPM-M's Adaptability. The amount of precipitation varies significantly throughout the seasons. Therefore, based on the monthly GPM products (GPM-M), the satellite data and gauge data of the 107 sites in the study area were classified according to the season (spring: March-May; summer: June-August; autumn: September-November; and winter: December-February); therefore, quarterly precipitation data were captured. Based on the correlation coefficients of the different seasons (Figure 5), the correlation coefficient of spring and winter was slightly higher than that of summer and autumn. However, the performance of GPM products at autumn and summer is better than that of spring and winter. In spring and winter, the research area is dry and cold, mainly showing small amounts of precipitation and snowfall, which is a great disturbance to the detection of precipitation by GPM products. The precipitation in summer and autumn is higher than that in spring and winter; that is, the performance of GPM products in wet season is better than that in the dry season. Figure 6 shows the quarterly average rainfall at all stations from the GPM products and gauges from March 2014 to December 2016. Based on the figure, we can see the rainfall distribution of satellite data is in good agreement with the data of ground meteorological stations, which can well reflect the rainfall in each season. However, there are some deviations in the specific quantities and GPM estimates are generally overvalued.

4.2. GPM Products Correction and Accuracy Verification

4.2.1. Data Correction Based on CDF Model. We used precipitation data from 75 national meteorological stations and the GPM products to build CDF curves for correcting

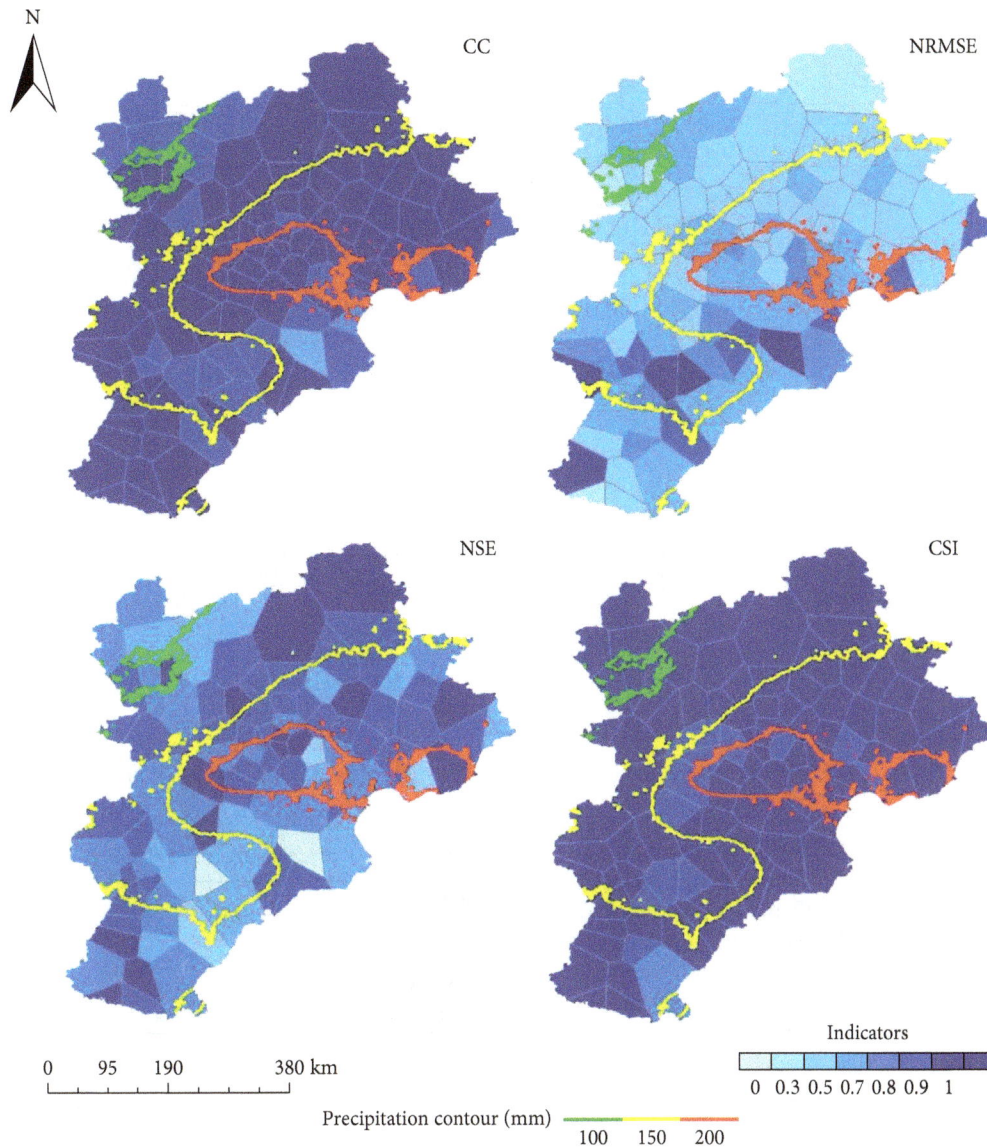

FIGURE 4: Spatial distribution of CC, NRMSE, NSE, and CSI based on GPM-M products.

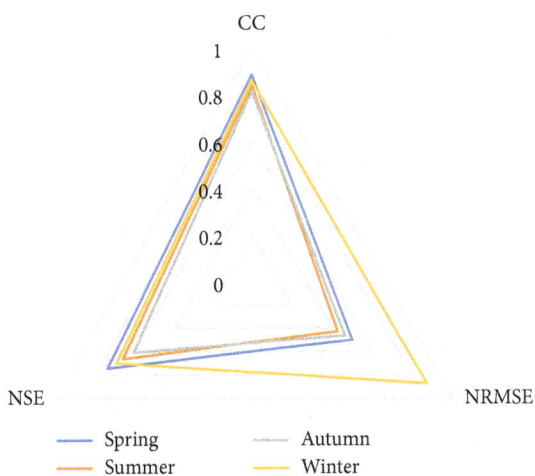

FIGURE 5: Numerical values of CC, NRMSE, and NSE at the seasonal scale.

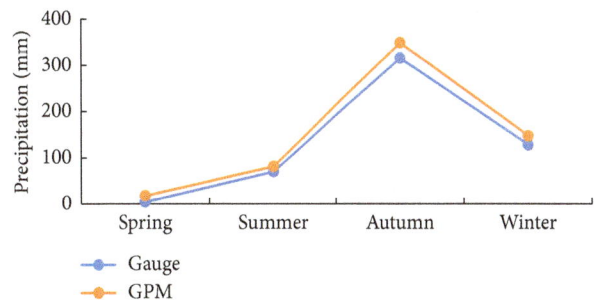

FIGURE 6: The average monthly rainfall of all gauges and GPM monthly data.

the GPM data. Table 4 shows the shape parameter α and scale parameter β for the two CDF models.

Figure 7(a) is a graph based on the CDF models (day), in which the green line is the CDF curve of the gauge data, the

TABLE 4: CDF model shape parameter α and scale parameter β.

	Gauge		GPM	
	α	β	α	β
CDF model (day)	0.6469	75.2344	0.8139	66.8997
CDF model (month)	0.5462	34.4265	0.7976	26.814

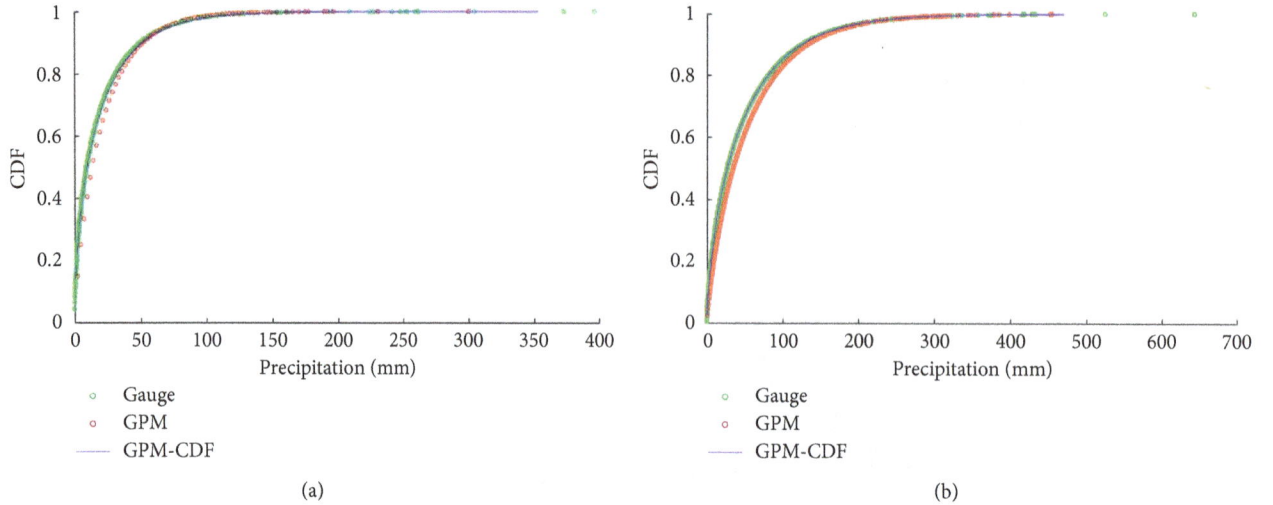

FIGURE 7: CDF curves for gauge, GPM, and GPM-CDF datasets. (a) Day; (b) month.

red line is the CDF curve of GPM-D precipitation training data, and the blue line is the CDF curve of CDF precipitation validation data through the CDF model. Figure 7(b) is a graph based on the CDF models (month), in which the red line is the CDF curve of GPM-M precipitation training data. After calibration, the CDF curve is more and more consistent with the CDF curve of the gauge data, indicating the effectiveness of the modified model. Table 5 shows the precision of the CDF model. The correlation of the CDF model was slightly lower than the correlation of the GPM original precipitation product. However, the RMSE of the CDF model was less than that of the original data, while the precision increased. The BIAS of the CDF model was close to 0, indicating that the agreement between the satellite and ground-measured data was good. Because the BIAS was less than 0, it indicates that the GPM data were underestimated. Therefore, the CDF model is suitable for GPM-M and GPM-D data correction in the study area.

4.2.2. Accuracy Verification of the CDF Model.

To further quantify the accuracy of the CDF correction model, monthly precipitation data from 32 national data stations were used to verify the model (verification dataset). It can be seen from Table 6 that the accuracy of the GPM-D and GPM-M data was improved after correction although it was overestimated. Figure 8 shows the CDF correction results. The abscissa represents the month and the ordinate represents the average monthly precipitation. Figure 8(a) shows that the calibrated result is closer to the real value. When the average monthly precipitation is larger, the correction effect is better. At the same time, when the GPM precipitation is greater than the

TABLE 5: Comparison of the CDF model results.

Statistical indicators	CC	RMSE	BIAS (%)
GPM-D	0.7500	12.3831	−7.58
CDF model (day)	0.7839	10.5738	4.53
GPM-M	0.9031	26.3928	12.31
CDF model (month)	0.9028	25.7803	0.12

TABLE 6: CDF model verification results.

Model validation	CC	RMSE	BIAS (%)
GPM-D	0.7569	12.9672	6.75
GPM_{CDF} (day)	0.7959	11.6452	4.76
GPM-M	0.9159	26.0057	0.56
GPM_{CDF}(month)	0.9167	25.3933	0.92

gauge precipitation, GPM-M products can be better corrected. When the GPM precipitation is less than the site precipitation, the corrected result often deviates more from the real value (Figure 8, red box). The main reason is that when the GPM precipitation is less than the site precipitation, the CDF curve of GPM precipitation is under the gauge CDF curve after trend fitting, which makes the corrected GPM value smaller than the gauge data. Figure 8(b) shows the results of the CDF model (day). After correction, the precipitation of GPM-D product is closer to the real value.

Since the CDF model is better than the original data, we further discuss the correction effect of the correction model in each precipitation interval. Because the daily precipitation range is small, GPM-M data are selected for discussion. As shown in Table 7, correction worked best in the ranges of 0–100 mm and >200 mm. Because of the presence of trace

(a)

(b)

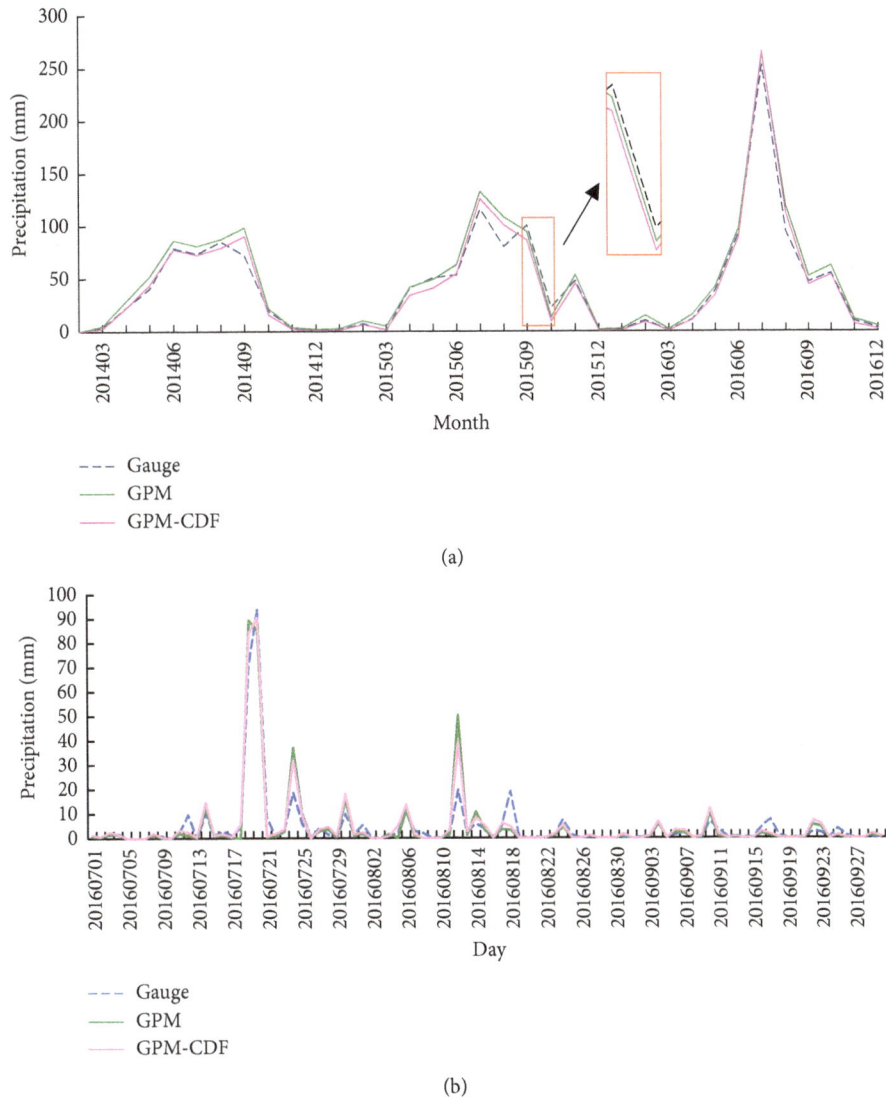

FIGURE 8: The results of the CDF model. (a) Month and (b) day.

precipitation in the 0–100 mm interval, the accuracy of GPM will be affected. High precipitation (>200 mm) results in larger GPM precipitation errors. The CDF-segmented model can effectively correct the errors caused by trace precipitation and extreme precipitation. Among them, the accuracy of corrected GPM-M data was highest in the 0–100 mm interval, but the data were underestimated. The corrected GPM-M products in the >100 mm range were overestimated.

4.3. Analysis of Temporal and Spatial Distribution Characteristics of Precipitation Based on Corrected GPM-M Products

4.3.1. Interannual Variation of Precipitation. We used GPM-M data modified by the CDF model to analyse the spatial and temporal distributions of precipitation in the study area. This paper synthesized corrected GPM-M data and annual data to obtain a mean annual precipitation map

(Figure 9). As can be seen from Figure 10, the highest annual precipitation was concentrated in the northeastern area, where it reached 820 mm. The lowest annual precipitation was concentrated in the northwest. From the northwest to the southeast, the precipitation decreased gradually and showed a clear banded distribution.

4.3.2. Precipitation Changes during the Year. Based on the CDF-corrected GPM-M data, the mean monthly precipitation distribution in the area was calculated (Figure 10). The monthly distribution was approximately the same as the annual one; however, in July, precipitation in the sub-high-value areas of annual precipitation was the highest and had a banded distribution. From November to March of the following year, precipitation in this area was almost <50 mm and precipitation was scarce. From April to June, the monthly precipitation gradually increased, and the high precipitation still occurred in the northeast. Precipitation mostly occurred in July and August, with a maximum of

TABLE 7: CDF model verification results according to precipitation range.

	Indicators	0–100 mm		100–150 mm		150–200 mm		>200 mm	
		GPM	GPM$_{CDF}$	GPM	GPM$_{CDF}$	GPM	GPM$_{CDF}$	GPM	GPM$_{CDF}$
Training dataset	CC	0.8505	0.8359	0.298	0.2976	0.3146	0.3135	0.597	0.5993
	RMSE	21.1972	19.3445	34.9012	37.677	46.3524	50.6796	77.0531	79.635
	BIAS (%)	26.56	7.3	2.79	9.06	8.23	11.68	8.45	8.08
Validation dataset	CC	0.8741	0.8622	0.2196	0.2173	0.4686	0.4686	0.658	0.6582
	RMSE	19.7058	17.5907	36.7289	38.3842	52.6878	58.3426	72.5152	74.9895
	BIAS (%)	26.6	7.4	2.01	4.21	16.38	20.06	10.7	10.24

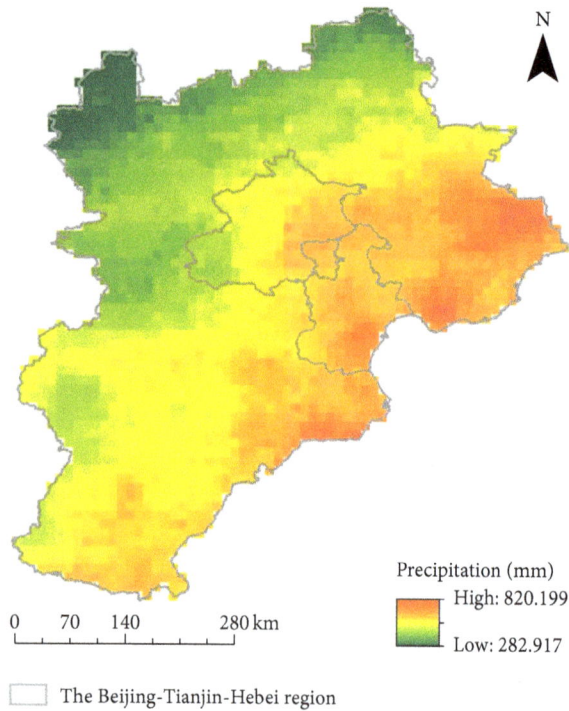

FIGURE 9: Mean annual precipitation based on modified GPM-M data (CDF model).

250 mm. Meanwhile, precipitation in September and October gradually decreased. In summary, the distribution of precipitation in the area shows clear heterogeneity in different months, and the difference in precipitation in different areas is clearer.

5. Conclusion

This study conducted a comparative analysis of the precipitation data of GPM products (Final Run) and the precipitation data of 107 gauges in the Beijing–Tianjin–Hebei region. To begin with, we analysed the applicability and spatial characteristics of GPM products in different time-scales. Next, we used the CDF model to correct GPM-M products. Finally, we analysed the temporal and spatial distribution characteristics of precipitation based on modified GPM-M products and reached the following conclusions:

(1) The GPM-M products have strong applicability and accurately reflected the precipitation events in the Beijing-Tianjin-Hebei region. Based on the comparison of the monthly mean rainfall, the 12-month rainfall distribution of satellite data was in good agreement with the data of ground meteorological stations, which well reflected the rainfall for each month. However, these are still overrated. The performance of GPM products in summer and autumn was better than in winter and spring, and the performance of GPM products in the wet season at a monthly scale was better than in the dry season. Therefore, GPM-M products could be used as the basic data of a hydrological model. However, GPM products have poor detection capability in areas with extremely undulating terrain, such as mountainous areas. Also, the spatial resolution of rainfall data is still relatively coarse, potentially leading to error.

(2) With an increase in precipitation intensity, the correlation between GPM products and sites had a trend of first increasing and then decreasing slightly, highlighting that GPM products may have uncertainty for extreme precipitation and trace precipitation.

(3) The accuracy of the CDF-corrected GPM data is better than that of the original GPM-D and GPM-M product. The GPM product will be more accurately corrected when the GPM-estimated precipitation is greater than the gauge-measured precipitation. When GPM precipitation is less than the gauge-measured precipitation, the corrected results tend to deviate more from the real values. The GPM-M data range of 0–100 mm can be corrected best. Based on the CDF-corrected GPM-M data, the spatial distribution of mean annual precipitation decreases from northwest to southeast across the study area, with a roughly banded distribution. The distribution of monthly precipitation is obviously uneven, but its trend is roughly the same as that of mean annual precipitation.

In general, GPM is a new generation of precipitation observation satellites with high spatial-temporal resolution. GPM-D products and GPM-M products have strong detection capability for rainfall events, and the accuracy of correction GPM products increase, which can provide data support for long-time series analysis in the Beijing-Tianjin-Hebei region. However, due to various factors such as elevation, topography, and longitudinal and latitudinal gradient, it is necessary to further improve the quality and spatial resolution of GPM to provide a more accurate

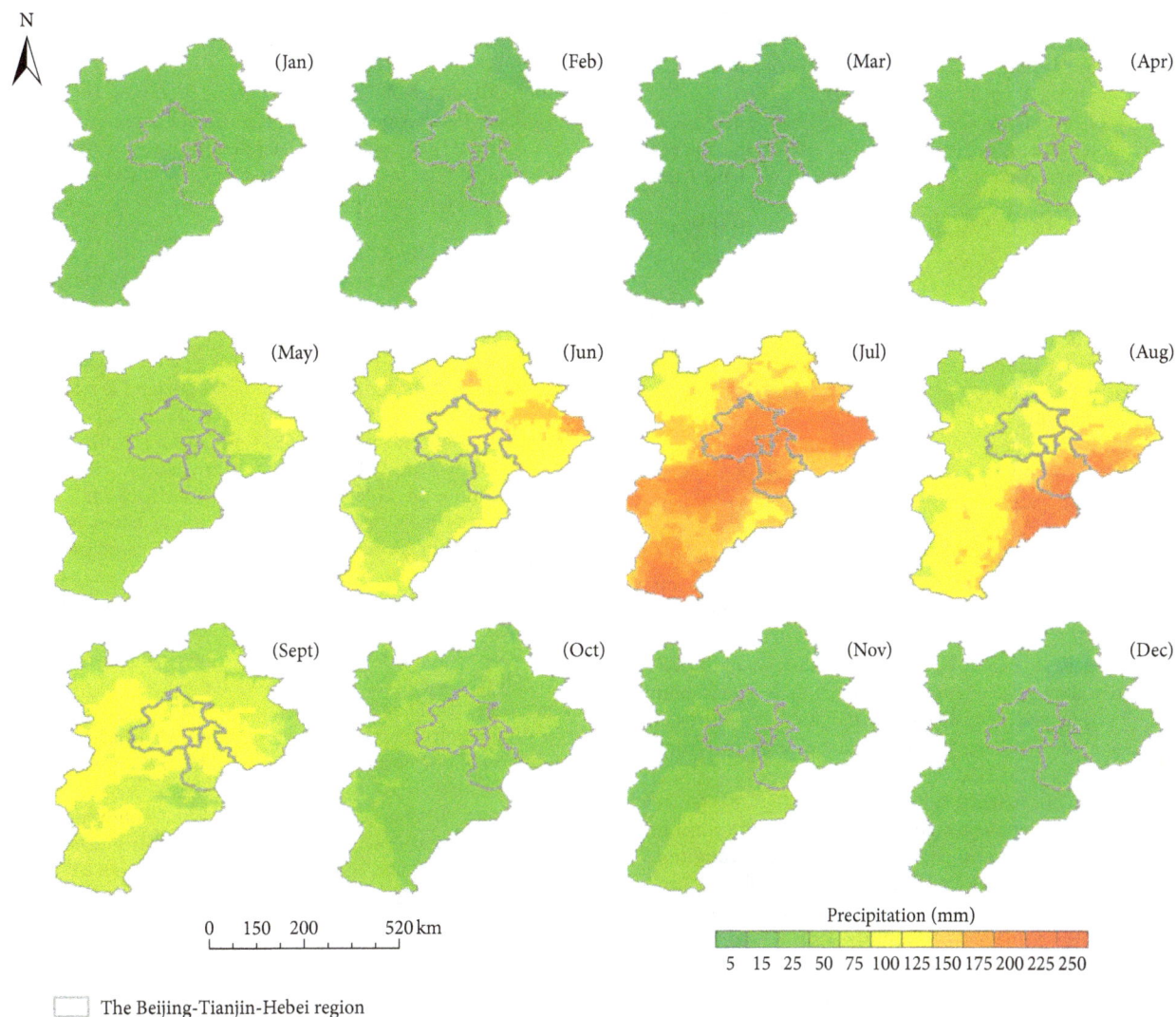

FIGURE 10: Spatial distribution of mean monthly precipitation in the study area.

estimation of future precipitation. It is of great practical significance to reveal the variations and regularity of precipitation and improve the ability of precipitation estimation for areas lacking data. In the study of global climate change, hydrological cycles, the ecological environment, and other scientific research, GPM should be of great value. If it can be applied to agricultural production and disaster prevention, as well as other fields involving disaster assessment and risk prediction, it will have tremendous socioeconomic benefits. Therefore, finding ways to improve the data spatial resolution in the study area will be the main content of our next study.

Conflicts of Interest

The authors declare no conflicts of interest.

Acknowledgments

This paper was jointly supported by the National Key R&D Program of China (2017YFC0406006 and 2017YFC0406004) and Science Foundation of Beijing Municipal Education Commission (SQKM201710028013). The authors would like to thank the China Meteorological Administration for providing ground-based rainfall data. The GPM products were provided by the NASA Goddard Space Flight Centre's Precipitation Processing System (PPS).

References

[1] Y. Yang, J. Du, and L. Cheng, "Evaluation of accuracy and reliability of TRMM satellite precipitation data in Hunan Province," *Journal of Water Resources & Water Engineering*, vol. 27, no. 1, pp. 26–32, 2016.

[2] X. Liu, J. Zhao, H. Zhang, X. Guo, Z. Zhang, and Y. Fu, "Accuracy validation and application of TRMM precipitation data in Northeast China," *Journal of Natural Resources*, vol. 30, no. 6, pp. 1047–1056, 2015.

[3] T. Ji, H. Yang, R. Liu, T. He, and J. Wu, "Applicability analysis of the TRMM precipitation data in the Sichuan-Chongqing region," *Progress in Geography*, vol. 33, no. 10, pp. 1375–1386, 2014.

[4] G. J. Huffman, R. F. Adler, P. Arkin et al., "The global precipitation climatology project (GPCP) combined precipitation dataset," *Bulletin of the American Meteorological Society*, vol. 78, no. 1, pp. 5–20, 1997.

[5] Z. Yao, W. Li, H. Gao, Y. Zhu, B. Zhao, and Q. Zhang, "Remote sensing of flooding using TRMM the microwave imager," *Acta Meteorologica Sinica*, vol. 60, no. 2, pp. 243–249, 2002.

[6] H. Zeng and L. Li, "Accuracy validation of TRMM 3B43 data in Lancang river basin," *Acta Geographica Sinica*, vol. 66, no. 7, pp. 994–1004, 2011.

[7] X.-H. Li, Q. Zhang, and C.-Y. Xu, "Suitability of the TRMM satellite rainfalls in driving a distributed hydrological model for water balance computations in Xinjiang catchment, Poyang lake basin," *Journal of Hydrology*, vol. 426-427, pp. 28–38, 2012.

[8] Y. Yang, G. Cheng, J. Fan, J. Sun, and W. Li, "Accuracy validation of TRMM 3B42 data in Sichuan basin and the surrounding areas," *Journal of the Meteorological Sciences*, vol. 33, no. 5, pp. 526–535, 2013.

[9] G. Zhu, T. Pu, T. Zhang, H. Liu, X. Zhang, and F. Liang, "The accuracy of TRMM precipitation data in Hengduan mountainous region, China," *Scientia Geographica Sinica*, vol. 33, no. 9, 2013.

[10] N. Yang, K. Zhang, Y. Hong et al., "Evaluation of the TRMM multisatellite precipitation analysis and its applicability in supporting reservoir operation and water resources management in Hanjiang basin, China," *Journal of Hydrology*, vol. 549, pp. 313–325, 2017.

[11] E. Sharifi, R. Steinacker, and B. Saghafian, "Assessment of GPM-IMERG and other precipitation products against gauge data under different topographic and climatic conditions in Iran: preliminary results," *Remote Sensing*, vol. 8, 2016.

[12] S. Khodadoust Siuki, B. Saghafian, and S. Moazami, "Comprehensive evaluation of 3-hourly TRMM and half-hourly GPM-IMERG satellite precipitation products," *International Journal of Remote Sensing*, vol. 38, pp. 558–571, 2016.

[13] K. Kim, J. Park, J. Baik, and M. Choi, "Evaluation of topographical and seasonal feature using GPM IMERG and TRMM 3B42 over Far-East Asia," *Atmospheric Research*, vol. 187, pp. 95–105, 2017.

[14] D. Sahlu, E. I. Nikolopoulos, S. A. Moges, E. N. Anagnostou, and D. Hailu, "First evaluation of the day-1 IMERG over the upper blue Nile Basin," *Journal of Hydrometeorology*, vol. 17, pp. 2875–2882, 2016.

[15] Z. E. Asong, S. Razavi, H. S. Wheater, and J. S. Wong, "Evaluation of integrated multisatellite retrievals for GPM (IMERG) over southern Canada against ground precipitation observations: a preliminary assessment," *Journal of Hydrometeorology*, vol. 18, pp. 1033–1050, 2017.

[16] S. Prakash, A. K. Mitra, D. S. Pai, A. AghaKouchak, H. Norouzi, and D. S. Pai, "From TRMM to GPM: how well can heavy rainfall be detected from space?," *Advances in Water Resources*, vol. 88, pp. 1–7, 2016.

[17] M. Tan and Z. Duan, "Assessment of GPM and TRMM precipitation products over Singapore," *Remote Sensing*, vol. 9, 720 pages, 2017.

[18] E. Sharifi, R. Steinacker, and B. Saghafian, "Hourly comparison of GPM-IMERG-final-run and IMERG-real-time (V-03) over a dense surface network in Northeastern Austria," *Geophysical Research Abstracts*, vol. 19, 2017.

[19] F. Satgé, A. Xavier, R. P. Zolá et al., ""Comparative assess-

ments of the latest GPM mission's spatially enhanced satellite rainfall products over the main Bolivian watersheds," *Remote Sensing*, vol. 9, pp. 1–16, 2017.

[20] F. Alexander, A. Véliz, C. Leónidas, T. Ramos, and L. Waldo Sven, "Assessment of tropical rainfall measuring mission (TRMM) and global precipitation measurement (GPM) products in hydrological modeling of the Huancane river basin, Peru," *Scientia Agropecuaria*, vol. 9, no. 1, pp. 53–62, 2018.

[21] F. Chiaravalloti, L. Brocca, A. Procopio, C. Massari, and S. Gabriele, "Assessment of GPM and SM2RAIN-ASCAT rainfall products over complex terrain in southern Italy," *Atmospheric Research*, vol. 206, pp. 64–74, 2018.

[22] G. Tang, Y. Ma, D. Long, L. Zhong, and Y. Hong, "Evaluation of GPM Day-1 IMERG and TMPA Version-7 legacy products over Mainland China at multiple spatiotemporal scales," *Journal of Hydrology*, vol. 533, pp. 152–167, 2016.

[23] R. Xu, F. Tian, L. Yang, H. Hu, H. Lu, and A. Hou, "Ground validation of GPM IMERG and TRMM 3B42V7 rainfall products over southern Tibetan Plateau based on a high-density rain gauge network," *Journal of Geophysical Research: Atmospheres*, vol. 122, pp. 910–924, 2017.

[24] S. Ning, J. Wang, J. Jin, and H. Ishidaira, "Assessment of the latest GPM-era high-resolution satellite precipitation products by comparison with observation gauge data over the Chinese mainland," *Water*, vol. 8, pp. 8–11, 2016.

[25] X. Jin, H. Shao, C. Zhang, and Y. Yan, "The applicability evaluation of three satellite products in Tianshan Mountains," *Journal of Natural Resources*, vol. 31, no. 12, pp. 2074–2085, 2016.

[26] X. Tan, J. Wang, G. Liu, and Z. Zhu, "Study on extreme precipitation monitoring based on multi-satellite remote sensing products," *Geomatics World*, vol. 24, pp. 82–87, 2017.

[27] R. Wang, J. Chen, and X. Wang, "Comparison of IMERG level-3 and TMPA 3B42V7 in estimating typhoon-related heavy rain," *Water*, vol. 9, pp. 1–15, 2017.

[28] Z. Wang, R. Zhong, C. Lai, and J. Chen, "Evaluation of the GPM IMERG satellite-based precipitation products and the hydrological utility," *Atmospheric Research*, vol. 196, pp. 151–163, 2017.

[29] L. Li, W. Zhang, L. Yi, J. Liu, and H. Chen, "Accuracy evaluation and comparison of GPM and TRMM precipitation product over Mainland China," *Advances in Water Science*, vol. 29, no. 3, pp. 303–313, 2018.

[30] Z. Wei, G. Yue, J. Li, and T. Lv, "Comparison study on accuracies of precipitation data using GPM and TRMM product in Haihe river basin," *Bulletion of Soil and Water Conservation*, vol. 37, no. 2, pp. 171–176, 2017.

[31] Y. Tian, C. D. Peters-lidard, and J. B. Eylander, "Real-time bias reduction for satellite-based precipitation estimates," *Journal of Hydrometeorology*, vol. 11, no. 6, pp. 1275–1285, 2010.

[32] X. Zhang and Q. Tang, "Combining satellite precipitation and long-term ground observations for hydrological monitoring in China," *Journal of Geophysical Research: Atmospheres*, vol. 120, no. 12, pp. 6426–6443, 2015.

[33] A. AghaKouchak, N. Nasrollahi, and E. Habib, "Accounting for uncertainties of the TRMM satellite estimates," *Remote Sensing*, vol. 1, no. 3, pp. 606–619, 2009.

[34] M. Jehanzeb, M. Cheema, and W. G. M. Bastiaanssen, "Local calibration of remotely sensed rainfall from the TRMM satellite for different periods and spatial scales in the Indus Basin," *International Journal of Remote Sensing*, vol. 33, no. 8, pp. 2603–2627, 2011.

[35] H. Guo, S. Chen, A. Bao et al., "Early assessment of integrated multi-satellite retrievals for global precipitation measurement over China," *Atmospheric Research*, vol. 176-177, pp. 121–133, 2016.

[36] X. Jin, H. Shao, Y. Qiu, and H. Du, "Correction method of TRMM satellite precipitation data in Tianshan Mountains," *Meteorological Monthly*, vol. 44, no. 7, pp. 882–891, 2018.

[37] W. Xiao and J. Xu, "Applied research of calibration method for SDSM model based on cumulative distribution function," *Acta Agricultures Jiangxi*, vol. 28, no. 1, pp. 74–78, 2016.

[38] Y. Ding, "Research of universality of Γ distribution model of precipitation," *Scinece Atmospherica Sinica*, vol. 18, no. 5, pp. 552–560, 1994.

[39] T. B. Mckee, N. J. Doesken, and J. Kleist, "The relationship of drought frequency and duration to time scales," in *Proceedings of the 8th Conference on Applied Climatology*, pp. 179–184, American Meteorological Society, Anaheim, CA, USA, January 1993.

[40] A. V. M. Ines and J. W. Hansen, "Bias correction of daily GCM rainfall for crop simulation studies," *Agricultural and Forest Meteorology*, vol. 138, pp. 44–53, 2006.

[41] J. Liu, "The temporal-spatial variation characteristics of precipitation in Beijing-Tianjin-Hebei during 1961-2012," *Climate Change Research Letters*, vol. 3, pp. 146–153, 2014.

[42] J. Wang, D. Jiang, and Y. Zhang, "Analysis on spatial and temporal variation of extreme climate events in North China," *Chinese Journal of Agrometeorology*, vol. 33, pp. 166–173, 2012.

[43] A. Zhang, S. Li, and X. Zhao, "Rainstorm risk assessment of Beijing-Tianjin-Hebei region based on TRMM data," *Journal of Natural Disasters*, vol. 26, pp. 160–168, 2017.

[44] A. Y. Hou, R. K. Kakar, S. Neeck et al., "The global precipitation measurement mission," *Bulletin of the American Meteorological Society*, vol. 95, pp. 701–722, 2014.

[45] G. Huffman, D. Bolvin, D. Braithwaite, K. Hsu, and R. Joyce, *Algorithm Theoretical Basis document (ATBD) NASA Global Precipitation Measurement (GPM) Integrated Multi-Satellite Retrievals for GPM (IMERG)*, NASA, Washington, DC, USA, 2013.

[46] G. J. Huffman, N. Gsfc, D. T. Bolvin, D. Braithwaite, K. Hsu, and R. Joyce, *Algorithm Theoretical Basis Document (ATBD) NASA Global Precipitation Measurement (GPM) Integrated Multi-SatellitE Retrievals for GPM (IMERG)*, NASA, Washington, DC, USA, 2014.

[47] G. Huffman, D. Bolvin, D. Braithwaite et al., "First results from the integrated multi-satellite retrievals for GPM (IMERG)," in *Proceedings of EGU General Assembly Conference*, Vienna, Austria, April 2015.

[48] Y. Liu, Y. Wu, Z. Feng, X. Huang, and D. Wang, "Evaluation of a variety of satellite retrieved precipitation products based on extreme rainfall in China," *Tropical Geography*, vol. 37, no. 3, pp. 417–433, 2017.

[49] X. Cai, S. Zou, Z. Lu, B. Xu, and A. Long, "Evaluation of TRMM monthly precipitation data over the inland river basins of Northwest China," *Journal of Lanzhou University*, vol. 49, no. 3, pp. 291–298, 2013.

4

Spatiotemporal Exploration and Hazard Mapping of Tropical Cyclones along the Coastline of China

Shaobo Zhongⓘ**, Chaolin Wang, Zhichen Yu**ⓘ**, Yongsheng Yang**ⓘ**, and Quanyi Huang**

Department of Engineering Physics, Institute of Public Safety Research, Tsinghua University, Beijing 100084, China

Correspondence should be addressed to Shaobo Zhong; zhongshaobo@tsinghua.edu.cn

Academic Editor: Anthony R. Lupo

Spatiotemporal patterns are one of the greatest interests and provide valuable insights into chronological events occurring in space. A tropical cyclone (TC) track is defined as a sequence of successive points, and several different types of analyses are performed to explore the temporal and spatial patterns of the TCs in the Northwest Pacific and along the coastline of China during 1949–2014. Results show that (1) the number of TCs is getting more frequent from April to August and less frequent from August to October with the peak occurring in August almost every year; (2) the mean of the sizes of the annual temporal clusters during 1949–2014 is 52.5 (days), the standard deviation is 17.0 (days), and the average starting point is the 210.5th day; (3) the spatial clusters are located in two areas: the boundary of Guangxi and Guangdong provinces and the boundary of Fujian and Zhejiang provinces; and (4) the within-strata variance is less than the between-strata variance, which implies the locational and seasonal factors are the potential determinants of the heterogeneity of the TCs. Furthermore, several maps representing the hazards of TCs are produced. According to the resultant maps, 12 coastal prefectures (Zhanjiang, Maoming, Fuzhou, Huizhou, Yangjiang, Qinzhou, Ningde, Quanzhou, Jiangmen, Nanning, Zhangzhou, and Hangzhou) have return periods of less than two years, and the two island provinces of Hainan and Taiwan are visited by TCs the most. Guangdong, Guangxi, Fujian, and Zhejiang provinces in particular suffered severely from the destructive TCs.

1. Introduction

A tropical cyclone (TC) is a rapidly rotating storm system characterized by a low-pressure center, strong winds, and a spiral arrangement of thunderstorms that produce heavy rain. Depending on the strength and the location where the storm occurs, tropical cyclones have some synonyms such as tropical storm, hurricane, typhoon, tropical depression, cyclonic storm, and simply a cyclone [1]. An intense tropical cyclone, which brings gales, heavy rain, and storm surge in the course of its landfall, is among the most devastating of all natural hazards and causes huge losses every year all over the world, especially in some coastal areas [2]. China is located in the West Pacific and has a length of continental coastline more than 18,000 km in the east and south parts of the country. There are numerous cities (including Tianjin, Shanghai, Guangzhou, Shenzhen, etc., metropolises) distributed in 14 coastal provinces (autonomous regions, municipalities, and special administrative regions) (12 continental ones and 2 island ones). Due to suitable

natural and climate conditions, more than 600 million people live in these coastal provinces, which amount to 43% of the total population of the country. In terms of the national statistics yearbook of 2014, the accumulated GDP (Gross Domestic Product) of the 10 continental coastal provinces (data are unavailable for Hongkong and Macau) are 36,952 billion RMB (Ren Min Bi) Yuan which is even more than 54% of the total GDP of Mainland China. The nation standard of China "Grade of tropical cyclones" that was put in practice since June 15, 2006 divides tropical cyclones into six scales: TD (tropical depression), TS (tropical storm), STS (severe tropical storm), TY (typhoon), STY (severe typhoon), and SuperTY (super typhoon). Although the fatality of tropical cyclones has been significantly reduced by a highly successful program of warnings and advanced building construction over the last decades, economic losses are escalating rapidly owing to the accelerated construction in TC-prone areas. The main affected areas in China by tropical cyclones are concentrated in coastal cities where population density and economic level are universally higher than other areas. These areas suffer huge losses from typhoons every year. 16 (including 10 coastal and 6 inland) out of 28 provinces in China are affected by TC-induced disasters [3]. In recent decades, the number of casualties caused by hazardous TCs in China has slightly decreased, but the property loss has significantly increased. Since the year of 2000, direct and indirect economic loss from hazardous TCs exceeds tens of billion RMB every year. Densely populated and well-developed coastal areas have suffered high casualty and property loss.

Analysis of spatiotemporal characteristics for TCs is useful to recognize distributional patterns, identify clusters, and predict spatial variables based on historical data, which is able to provide useful support for prevention and response of hazardous TCs, including comprehensive risk assessment, evacuation planning, resource allocation, among others. For example, Wang et al. [4] show the interdecadal variation in TC tracks over the Northwest Pacific taking landfalling locations of TCs in Xiamen as study objects. The results from Gu et al. [5] suggest that since the early 1960s, there has been an overall decreasing trend in the frequency of occurrence, intensity, peak intensity, length of movement, and lifetime of TCs. Chan and Liu [6] explored interannual variations of typhoon activity and concluded these variations appear to be largely constrained by the large-scale atmospheric factors that are closely related to the El Niño—Southern Oscillation (ENSO) phenomenon. Tang et al. [7] analyzed the annual and monthly variations of TC from 1951 to 2006 statistically, including frequencies, intensities, and wind intensity indices. By calculating the TC frequency and wind intensity indices in each $1° \times 1°$ longitude-latitude grid, Tang et al. [7] also analyzed the spatial distribution and the influence extent of TCs. In summary, in spatiotemporal characteristics of TCs, currently existing literature are mainly focused on summary statistics on different scales. Though some temporal and spatial patterns are presented in these papers, it is necessary to get deep insight into spatiotemporal characteristics of TCs with advanced techniques of spatial pattern analysis. Temporal analysis helps ones effectively get insights into chronological phenomena through establishing references of activity and discover periodic patterns in which data are presented by different scales (e.g., hour, day, month, and year). Temporal analyses are often shown in simple linear and circular charts to illustrate chronological order and cycle [8]. Different from the charts, some statistical methods can be used to detect clusters of events along a timeline and assert the confidence level of the results. Whether the occurrences of chronological events have significant anomalies along a timeline is essentially a problem how to identify the clusters of points along a line. There are some existing methods able to solve this problem [9–11]. These kinds of one-dimensional scan statistic methods are also widely applied in cluster detection of temporal data such as disease, drought, and forest fire. There are two kinds of commonly used techniques for exploring spatial patterns: global clustering and local clustering. Global clustering is to judge whether there exists significant spatial clustering trends in spatial phenomena (represented by observed events), which is generally determined by global test statistics [12, 13].

However, global clustering cannot obtain the specific locations of clusters. Local clustering is capable of identifying the centers of clusters and even their sizes and shapes [14–17]. Stratified heterogeneity, which refers to the within-strata variance less than the between-strata variance, is ubiquitous in spatial phenomena. Stratified heterogeneity may imply the determinants of events [18].

To visualize the hazard of disasters, an alternative approach is to evaluate the spatial density of disasters. While point density estimation is widely used to quantify hazard of disasters with concentric impact, a line-based density estimation is imperative to map the hazard of disasters with linear distribution. Density estimation techniques have been applied to examine spatiotemporal dynamics in some domains [19–24]. These types of techniques generally use a density kernel to spread the values of the samples out over a surface. The magnitude at each sample is distributed throughout the study area, and a density value is estimated for each cell in the output surface. Steiniger and Hunter [22] extend the point-based kernel density estimation (KDE) approach to work with sequential GPS-point tracks, the outcome of which is a line-based KDE. Demšar et al. [23] present an alternative geovisualization method for spatiotemporal aggregation of trajectories of tagged animals: stacked space-time densities. Given tracks of tropical cyclones are represented as lines, and in this study, we propose a track density algorithm to map the density of tropical cyclones.

Some GIS tools of mapping and analysis are widely used in manipulation of georeferenced data. Through adding a few geographic elements into the tornado definition and then characterizing tornado density as a density field in GIS, Deng et al. [25] examine how GIS factors function in the process of tornado density mapping. Liu et al. [26] resort to the extension of ArcGIS (a proprietary GIS software package developed by Esri Corporation): Geostatistical Analyst to carry out interpolation analysis of flood risk factors. Also, in those widely used proprietary GIS software packages like ArcGIS, some commonly used methods of spatial analysis and visualization have been devised. Moreover, almost all products also provide script languages to facilitate end-users to glue all kinds of functions (including built-in and customized ones) for domain-oriented problem solving. As will be seen, in our study, we will make full use of these kinds of capability devised by the proprietary GIS.

For purpose of discovering temporal and spatial characteristics of TCs along the coastline of China in the Northwest Pacific, following the exploratory data analysis, this paper mainly explores temporal and spatial clusters with advanced spatial techniques of spatiotemporal cluster analysis and density mapping coupled with GIS based on the TCs record data between 1949 and 2014. In the remainder of this paper, we first explore the temporal clusters of the TCs in the Northwest Pacific during 1949–2014 and the spatial clusters of the landfalling TCs along the coast of China to identify the clustering durations in these years and the clustering areas along the coastal prefectures, respectively. Following these explorations, the degree of temporal and spatial stratified heterogeneity is measured and its significance is tested. Then, the exceedance probabilities (EP) and

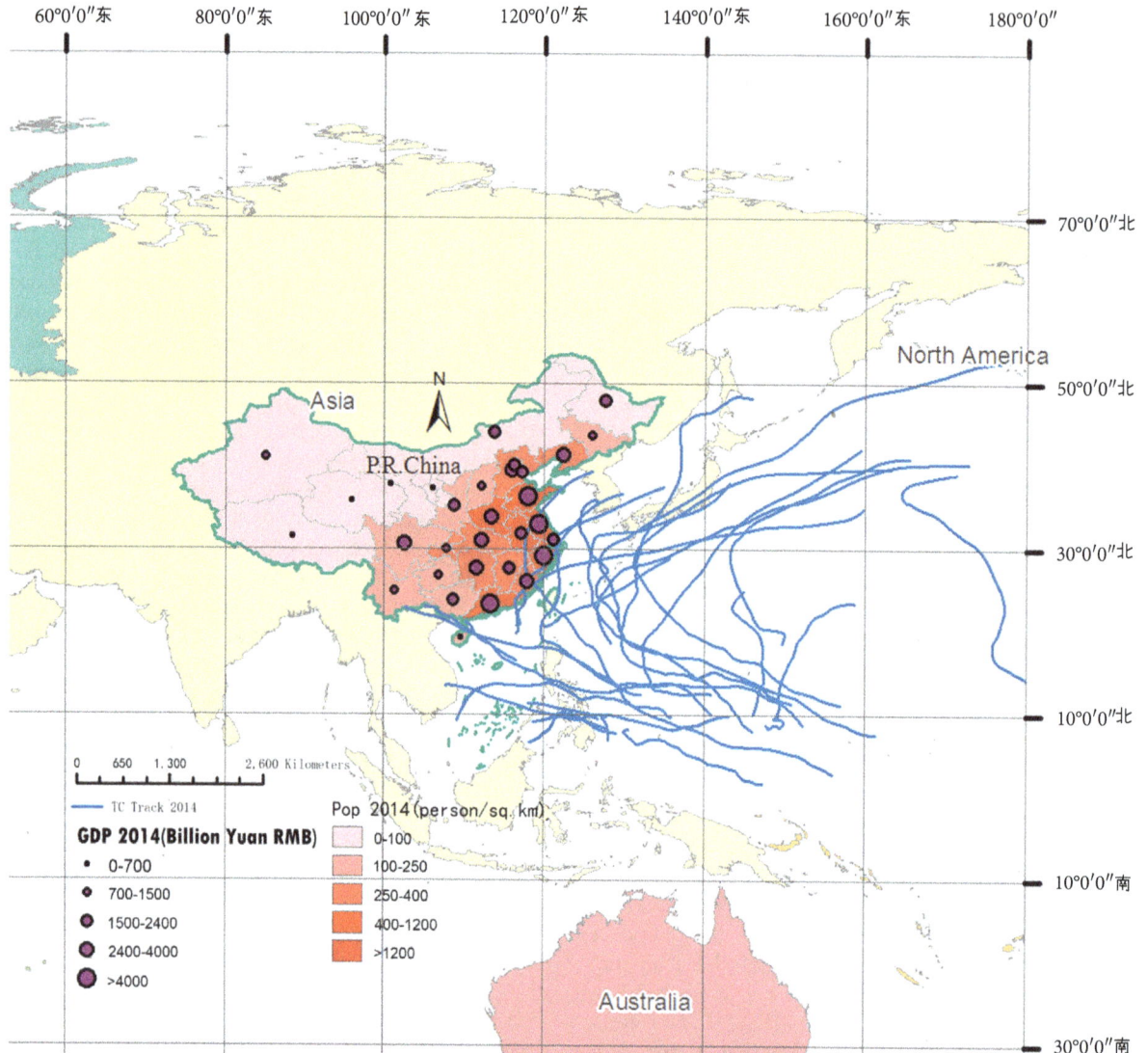

FIGURE 1: The administration map of China and its surrounding areas (the population density and GDP of Mainland China in 2014 are plotted) and tropical cyclones in the Northwest Pacific (here only show the tracks in 2014).

the return periods of the landfalling TCs in the coastal prefectures are rendered. Next, we map the hazards of the TCs in the Northwest Pacific. In the end, the paper summaries the main conclusions, proposes some problems to be solved, and suggests further research aspects.

2. Study Area and Data

2.1. Study Area. The Northwest Pacific is one of the international TC zones with an annually average number of 35 TCs, which amounts to 31% of all the hazardous TCs in the world, two times more than any other zone [27]. Moreover, about 80% of TCs in the Northwest Pacific develop into more severe ones (tropical storm or above). Figure 1 shows the administration map of China at the provincial level and its surrounding areas. For an intuitional recognization of the hazard of TCs, the population and GDP of 2014 are symbolized with graduated colors and graduated symbols, respectively, in Figure 1.

2.2. TC Track Data. The TC track data used in this study are obtained from the China Typhoon Web (http://www.typhoon.gov.cn). These data are published by Shanghai Typhoon Institute (STI) which is entrusted by the China Meteorological Administration (CMA) to compile the Best Track Dataset for Tropical Cyclones over the Northwest Pacific (CMA-STI Best Track Dataset for Tropical Cyclones over the Northwest Pacific) (http://www.typhoon.org.cn). The basin is to the north of the equator and to the west of 180°E, including the South China Sea, and the time span is from 1949 to 2014. In the data set, the data are organized track by track. Each TC track starts with a header and records a sequence of points (the location of the TC head at that time) every 6 hours. Each file in the data set contains the annual data, which is saved as a text format (with a .txt extension). For detailed information, please refer to the data set and its document.

In order to import and analyze the data in ArcGIS (a proprietary GIS software) [28], we wrote a program to parse

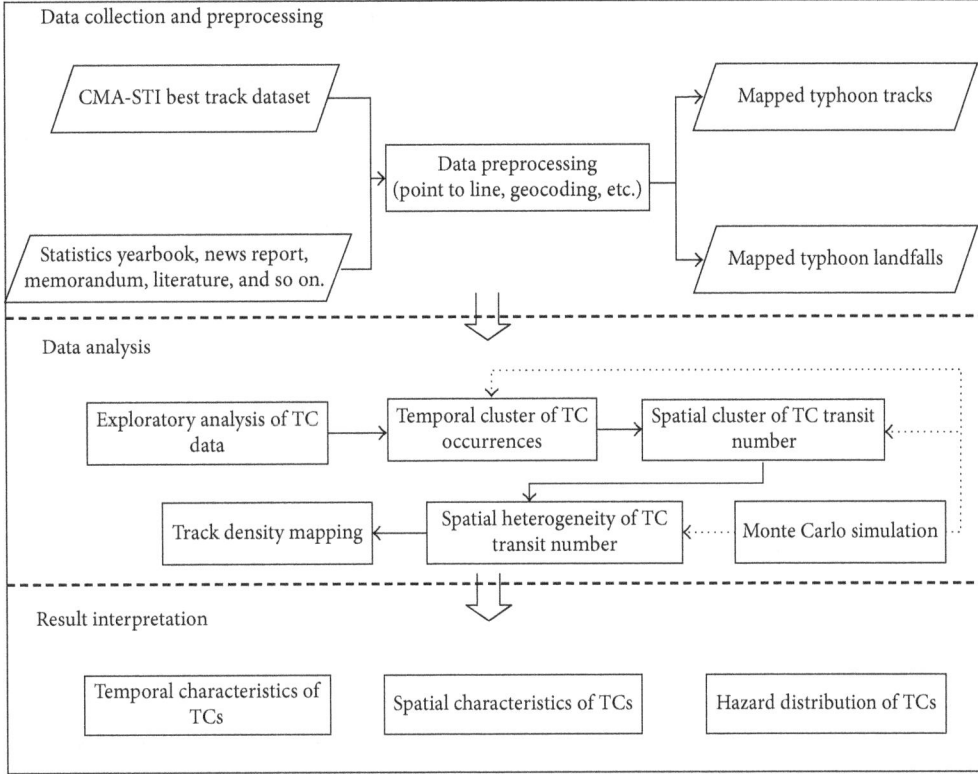

FIGURE 2: Flowchart of the proposed methodology.

the original text files into XY Data files (required by the Add XY Data operation of ArcGIS as input), and then these resultant files are imported into ArcGIS and transferred into point features. Finally, the Points To Line tool is executed to obtain the TC tracks of line features.

2.3. TC Landfall Data. As mentioned above, the time interval of the adjacent track points recorded is 6 hours. Thus, it is difficult to determine the accurate landfall locations of the TCs. To obtain the most accurate landfall locations, some materials are collected from various sources (such as statistics yearbooks, news reports, memorandum, and the literature). In these materials, generally the landfall locations of TCs are recorded as addresses (e.g., Putuo, Zhoushan, Zhejiang Province). The geocoding technique is applied to interpret the addresses into coordinates of longitude and latitude. Given the abundant gazetteer possessed by some online maps like the Baidu map, we created a Baidu map API-based program to automatically geocode the landfall addresses of TCs of interest in batch.

3. Methods

3.1. Workflow. The study of temporal and spatial characteristics of tropical cyclones is performed with the following steps:

(1) *Data collection and preprocessing*: The methods and procedure are explained in the above sections.

(2) *Data analysis*: Exploratory analysis, temporal pattern analysis, and spatial pattern analysis are done with some selected methods. A Monte Carlo simulation is employed to support the hypothesis test of the temporal and spatial cluster methods. To visualize the hazard of TCs, the track density is mapped through a track density estimation algorithm (line-based KDE).

(3) *Result interpretation*: From the results of data analysis, the temporal and spatial characteristics of TCs are summarized.

A brief overview of the methodology is shown in Figure 2.

The main data analysis techniques will be explained in Sections 3.2–3.5.

3.2. Temporal Cluster Detection of TC Occurrences. Let $t_1 \leq t_2 \leq \cdots \leq t_N$ be N TC occurrences on a time period (e.g., one year) ordered according to occurrence dates. A scan statistic with a variable window (interval) is as follows [11]:

$$\Lambda = \sup_{0 < d < n/N, n \geq n_0} \left(\frac{n}{N}\right)^n \left(\frac{N-n}{N}\right)^{N-n} \left(\frac{1}{d}\right)^n \left(\frac{1}{1-d}\right)^{N-n}, \quad (1)$$

where d is the scan window width, n is the number of TC occurrences contained in the scan window, and n_0 is the minimum number of occurrences contained in the clusters to be detected.

This test is the generalized likelihood ratio test for a uniform Null distribution against an alternative of non-random clustering which will detect the most likely cluster (interval) whose observed value of the likelihood ratio is maximum. This method allows for clusters of variable width. For more details, please refer to Appendix A.

3.3. Spatial Clustering Analysis of TC Crossing Number.
Spatial clusters are those regions (prefectures in this study) whose values are significantly higher than their neighbors. Identifying spatial clusters allows one to map "hot" areas and "cold" areas of TCs. The local Moran's I statistic is an index for identifying clusters and outliers and computed according to the following formula:

$$I_i = \frac{x_i - \overline{X}}{S_i^2} \sum_{j=1, j \neq i}^{n} w_{i,j}(x_j - \overline{X}), \qquad (2)$$

where x_i is an attribute of region i, $\overline{X}$ is the mean of the corresponding attribute, $w_{i,j}$ is the spatial weight between region i and j, and

$$S_i^2 = \frac{\sum_{j=1, j \neq i}^{n} \left(x_j - \overline{X}\right)^2}{n-1} - \overline{X}^2, \qquad (3)$$

with n equating to the total number of regions in a study area.

Generally, the distribution of I_i does not meet normality. Though the mean and standard deviation are roughly consistent with those for a standard normal distribution, the kurtosis and the skewness are not [14]. In this study, it is more unlikely that the distribution of the Null hypothesis is normal since we are going to explore the spatial clusters of TCs along the coastal regions that is a zonary area. A Monte Carlo simulation is employed to construct a Null hypothesis (see Section 3.5 for the explanation of the simulation process), and further by a classic hypothesis test procedure, the P value for each region is obtained. The P values represent the statistical significance of the computed index values. A Moran's scatterplot can be used to explore the clusters of attributes of regions in a study area in which the X axis denotes the normalized observed attributes and the Y axis is the spatial lags [14].

3.4. Stratified Heterogeneity of TC Crossing Number.
Wang et al. [29] define a q-statistic to measure stratified heterogeneity. The q-statistic distribution is defined as

$$P(q < x) = P\left(F < \frac{N-L}{L-1} \frac{x}{1-x}\right) = 1 - \alpha, \qquad (4)$$

where N is the number of regions in a study area, L is the number of stratum, F follows a noncentral F-distribution with the 1st df $L-1$, the 2nd df $N-L$, and noncentrality λ, and α is the probability of $q \geq x$. For more information, please refer to Appendix B.

According to the q-statistic, a hypotheses test is carried out by defining the Null hypotheses and alternative hypotheses, respectively. The procedure can be implemented by a software: GeoDetector, which is freely downloadable from http://www.geodetector.org.

3.5. Test Statistical Significance.
To assess the significance of a cluster (including temporal and spatial), a Monte Carlo method of inference is used to implement the hypothesis tests. First, the test statistic value is computed according to the observed data. Secondly, the same statistic is calculated with a given quantity of realizations drawn independently from the Null hypothesis (generally a uniform distribution). Next, a distribution can be fitted by some methods (e.g., histogram and kernel density), which provides an estimate of the distribution of the test statistic under the Null hypothesis. The proportion of test statistic observed for the actual data set provides a Monte Carlo estimate of the upper-tail P value for a one-sided hypothesis test [30]. Specifically, suppose that T_{obs} denotes the test statistic value for the sample data and $T(1) \geq T(2) \geq \cdots \geq T(N_{\text{sim}})$ denote the test statistic values for the simulated data of realizations. If $T(1) \geq \cdots \geq T(l) \geq T_{\text{obs}} \geq T(l+1)$, then the estimated P value is

$$P\left(T \geq T_{\text{obs}}|H_0\right) = \frac{l}{N_{\text{sim}} + 1}, \qquad (5)$$

where one is added to the denominator since the estimate is based on $N_{\text{sim}} + 1$ values $\left(\{T(1), \ldots, T(N_{\text{sim}}), T_{\text{obs}}\}\right)$. The lower-tail P values can be calculated in an analogous manner.

3.6. Track Density Estimation.
To visualize the hazard of TCs, an alternative approach is to evaluate the spatial density of the TC tracks. The algorithm for track density can be considered an extension of the widely used 1D kernel density on point data [31] into 2D kernel density on polyline data (tracks). The contribution of each track to the estimated density is calculated separately, which represents the influence of the particular track on its neighborhood in space. Then, those contributions from all tracks are added up to obtain the total density. The contribution of each track (polyline) to the density is defined within a buffer whose centerline is the projection of the track on the plane space. A value is assigned to each point in the buffer according to the distance of the point to the centerline (the perpendicular distance). This value needs to be normalized using the user-specified kernel size, so that the point on the centerline is assigned a maximum value, and the density value decreases with distance from the centerline and reaches a minimum value at the limit of the kernel. Furthermore, in order to be consistent with the probability density function (pdf), the integral of the kernel function along the perpendicular segment must be equal to 1. For the computation detail, please refer to Appendix C.

4. Results and Discussion

4.1. Temporal Analysis of the TC Occurrences.
There are a total of 2233 recorded TCs in the Northwest Pacific and 816 landfalling TCs along the continental coastline during 1949–2014. The landfalls are more than 12 in an average year. The landfalls amount to 25 in 1961 with 17 as tropical storms or above. The annual number of the TC occurrences and their landfalls are shown in Figure 3.

Several radar plots (also called sample circular plots or spider plots in some literature) are presented in Figures 4 and 5. Figure 4 shows the monthly pattern of the recorded

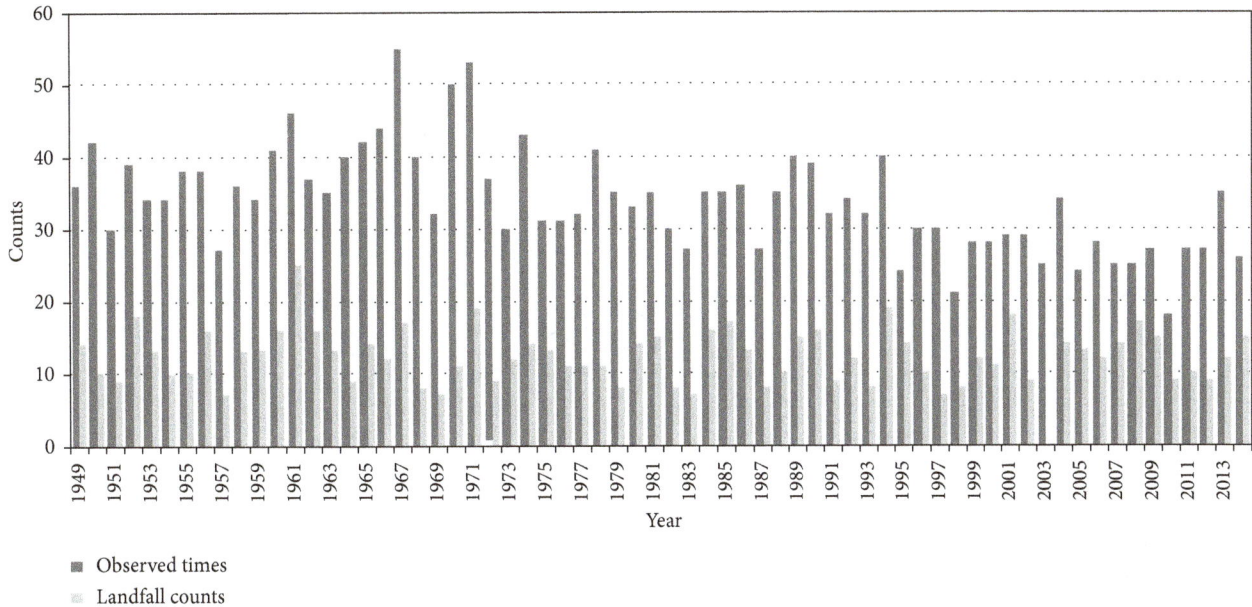

FIGURE 3: The annual number of the observed TC occurrences in the Northwest Pacific and their landfall counts approaching China.

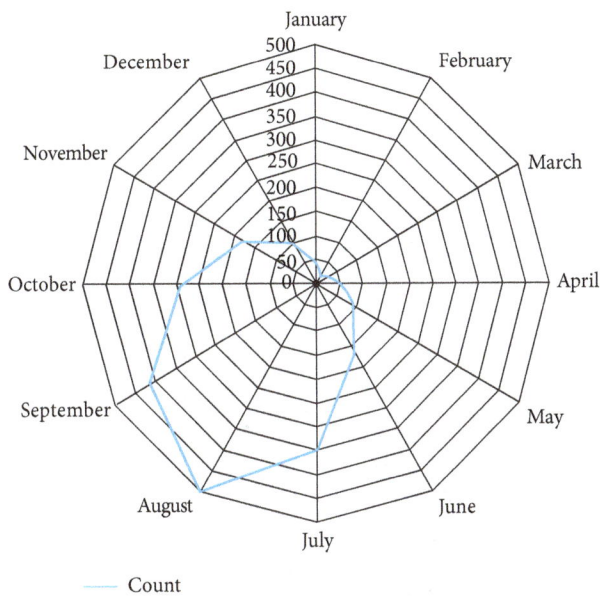

FIGURE 4: Monthly distribution of the observed TC occurrences in the Northwest Pacific during 1949–2014.

TC occurrences in the Northwest Pacific during 1949–2014. Figure 5 shows the monthly pattern of the landfalling TCs along the coastline of China during 1949–2014. These two figures show that the number of TCs increased from April to August and decreased from August to October with the peak occurring in August. According to Figure 5, fortunately, STY and SuperTY, the most destructive TCs, occupy a relatively small proportion in all TCs. Nonetheless, the tremendous consequences caused by STY and SuperTY indicate that the need for prevention and response to them cannot be overemphasized.

Next, we try to detect the most likely clusters of the TC occurrences during 1949–2014 in the Northwest Pacific, which is able to be achieved through a scan over the period with a variable window. For each year, the most likely cluster of points (the times of the TC occurrences) along the whole year (from January 1 to December 31) is detected. The cluster detection of points along line is implemented to look for the most likely clusters. The algorithm presented in (1) and Appendix A is coded with Matlab to find the most likely cluster. Meanwhile, the algorithm of the Monte Carlo simulation is coded to test the significance of the most likely cluster. In our analysis, n_0 is set to 5, the max length of the clusters is limited to 178 (days), and the simulation times are set to 999. After finishing the computation, the starting position, the number of points, and P value are output.

Figure 6 shows the detection result. From the visual examination, these clusters are obviously concentrated in a certain range of time. The mean of the durations of all 66 clusters from 1949 to 2014 is 52.5 (days) and the corresponding standard deviation is 17.0 (days). The average starting point is the 210.5th day. The mean of the centers of these clusters is the 237th day which is consistent with Figure 4.

4.2. Spatial Clusters of Landfalling TCs along the Coastline. The local Moran's I statistic aforementioned is used to analyze clusters of TCs traveling across those prefectures along the continental coastline of China. The landfall times of TCs for every city are summarized. This operation is accomplished through the Spatial Join tool of ArcGIS, which can join the tracks to the prefectures according to the spatial topological relationship of Interaction (meaning that the TC passes through the prefectures). ArcGIS also has been devised with several spatial cluster analysis tools including the Local Moran's I and Getis-Ord G_i^* [15, 32]. The tools produce a map in terms of the calculated results of the statistic itself and P value. Here, we define the spatial weight w_{ij} with Rook contiguity [33]. Figure 7(a) shows the clusters of the times of TCs traveling across the continental coastal prefectures of China, and the corresponding P values are

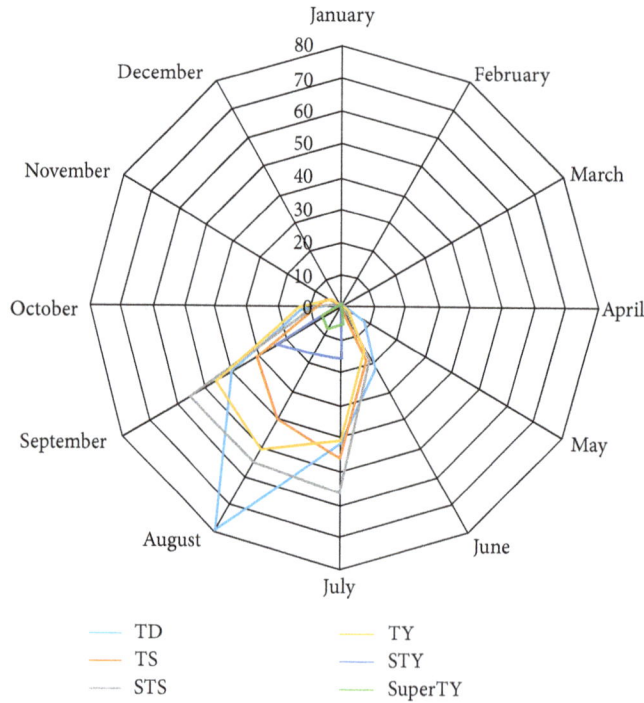

FIGURE 5: Monthly distribution of the landfalling TCs along the coastline of China during 1949–2014.

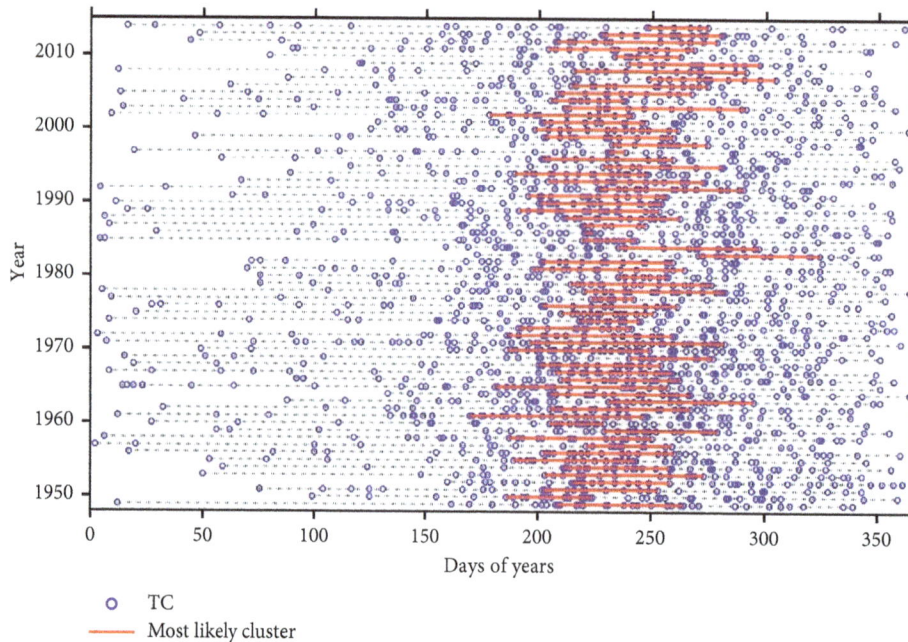

FIGURE 6: The most likely clusters detected for each year during 1949–2014.

shown in Figure 7(b). As seen, the clusters are located in two areas (filled with red in Figure 7(a)): the boundary of Guangxi and Guangdong provinces and the boundary of Fujian and Zhejiang provinces. These two areas are the most frequently stroke ones that can be easily confirmed by comparison with the historical TC records. Other prefectures do not have statistically significant clusters according to the analysis results. A Moran's scatterplot is drawn from the results of local Moran's I (Figure 8). In a Moran's scatterplot, a point is

drawn in one of the four phases according to its X value and Y value. The first phase of the plot contains the HH (High-High) values of the observed data, that is, the attribute of a region is high and the attributes of its neighboring region are also high, which means these regions are a cluster. Similarly, the second phase contains LH (Low-High) values, the third phase contains LL (Low-Low) values, and the fourth phase contains HL (High-Low) values. Here, our interest obviously is on the first phase in order to learn the spatial clusters of the visiting

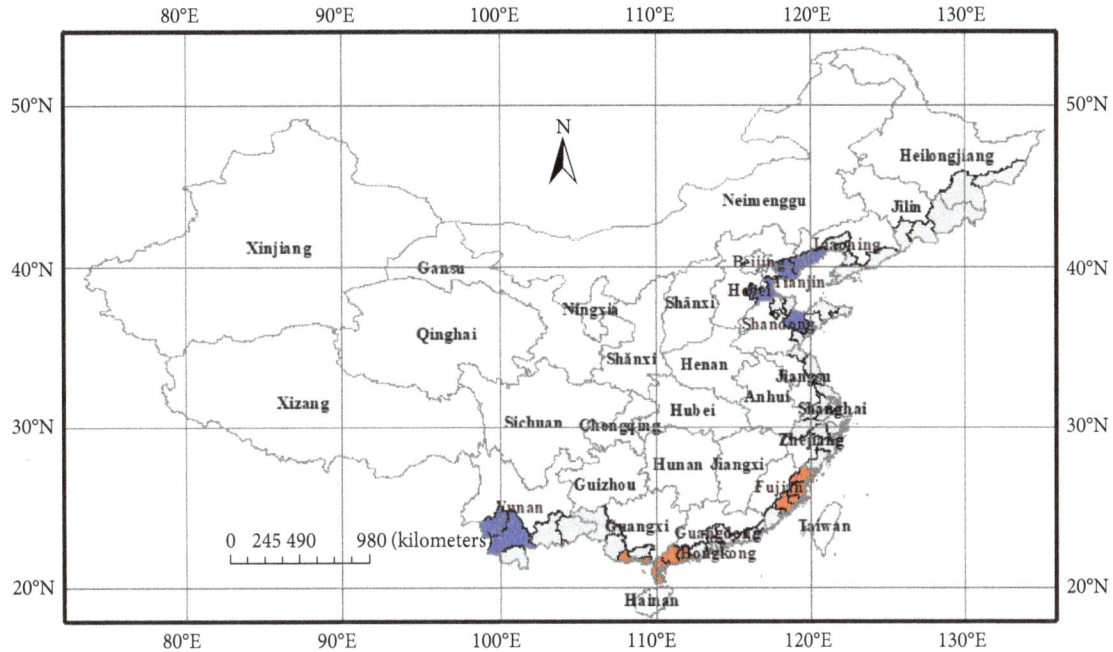

(a)

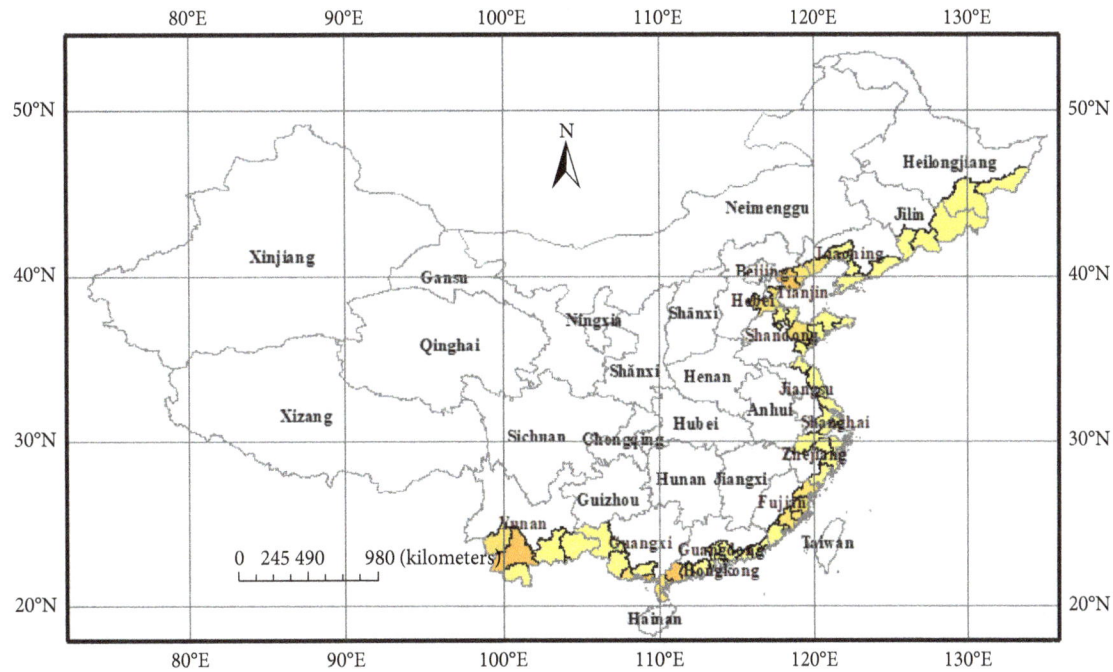

(b)

FIGURE 7: A resultant map of cluster analysis of the landfall times of the TCs along the continental coastal prefectures of China during 1949–2014 according to the observed z-scores. (a) Local Moran's I cluster map; (b) P value map.

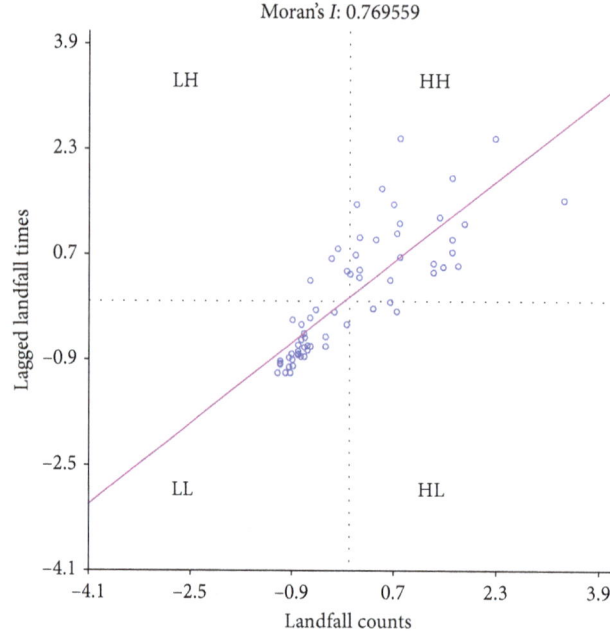

FIGURE 8: The Moran's scatterplot of the landfalling TCs along the continental coastal prefectures of China during 1949–2014 (the first phase means clusters that the landfalling TCs in those prefectures are high and whose neighbors also have high landfall counts).

TABLE 1: Outputs of q-statistic and significance tests for spatial stratified heterogeneity.

N	$\overline{Y}$	L	N_h	$\mu_h = \overline{Y}_h$	σ^2	σ_h^2	λ	F_α	F	q-statistic	P value
66	23.71	3	48	21.02	431.47	259.38	3.91	10.96	24.52	0.44	5.430×10^{-6}
			10	53		389.33					
			8	3.25		10.78					

TABLE 2: Outputs of q-statistic and significance tests for temporal stratified heterogeneity.

N	$\overline{Y}$	L	N_h	$\mu_h = \overline{Y}_h$	σ^2	σ_h^2	λ	F_α	F	q-statistic	P value
36	62.03	4	9	11	2913.8	22	3.25	8.06	54.77	0.87	1.993×10^{-10}
			9	34.67		390.5					
			9	140.33		823.5					
			9	62.11		842.11					

frequencies of TCs. As expected, the dots for detected clusters are drawn in the first phase and have greater observation values and expected mean neighbor values.

4.3. Stratified Heterogeneity of TCs. Two tests of stratified heterogeneity of TCs are carried out. One is for spatial stratified heterogeneity of landfall counts of TCs in the coastal prefectures. In this case, we divide the coastal prefectures into three strata: Low-Low, High-High, and Not Significant according to the analysis results of local Moran's *I*. In the other, we test the temporal stratified heterogeneity of the TC occurrences in different seasons, that is, we divide the TC occurrences into four strata: January–March, April–June, July–September, and October–December. We assessed these assumptions by the q-statistic to the landfall counts and TC occurrences of the spatial and temporal strata. We also tested the significance of the stratified heterogeneity. The results are shown in Tables 1 and 2. In the spatial case, because $F = 24.52 > F_{0.01}\,(2, 63, 3.9082) = 10.96$

and $P = 5.43 \times 10^{-6} \ll 0.01$, we conclude that the spatial stratified heterogeneity is significant. In the temporal case, because $F = 54.77 > F_{0.01}\,(3, 32, 3.2549) = 8.06$ and $P = 1.99 \times 10^{-10} \ll 0.01$, the temporal stratified heterogeneity is significant. We can also identify the most likely boundary of strata by q-statistic to test various stratifications to reach the one with the maximum q value, which should be performed in a future work [29].

4.4. Track Mapping of TCs. The landfall counts of TCs in the coastal prefectures are by year. Assume the landfall counts n_i for a certain prefecture i obey Poisson distribution with intensity λ_i, that is,

$$\Pr(n_i = k) = \frac{(\lambda_i t)^k e^{-\lambda_i t}}{k!}. \tag{6}$$

The exceedance probability (EP) can be calculated as follows:

$$\Pr\left(n_i \geq k\right) = 1 - \sum_{j=0}^{j=k-1} \Pr\left(n_i = j\right). \qquad (7)$$

According to the exceedance probability, the return period of TCs in prefecture i can be obtained by (8)

$$r_i = \frac{1}{\Pr\left(n_i \geq 1\right)}. \qquad (8)$$

The annual intensities for every prefecture are estimated according the landfall counts of TCs from 1949 to 2014.

$$\widetilde{\lambda}_i = \frac{\sum_{j=1949}^{2014} f_{ij}}{2014 - 1949 + 1}, \qquad (9)$$

where f_{ij} is the observed landfall times of TCs for prefecture i in year j.

Figure 9 depicts the exceedance probabilities of the 66 coastal prefectures of China for landfall counts of TCs. According to Figure 9, 12 coastal prefectures (Zhanjiang, Maoming, Fuzhou, Huizhou, Yangjiang, Qinzhou, Ningde, Quanzhou, Jiangmen, Nanning, Zhangzhou, and Hangzhou) have return periods of less than two years. We also observe that: (1) Zhanjiang, Maoming, and Fuzhou rank highest in modeled levels of EP; (2) Huizhou, Yangjiang and Qinzhou are at lesser, but still relative to the remaining prefectures, high EP; and (3) of all 66 selected coastal prefectures, Lincang has no landfalling TCs according to historical records.

Furthermore, a track density map is produced to visualize the continuous impact of TCs around the coastline of China (Figure 10). A Gaussian kernel function is selected here. The formation of the kernel function is defined as $k\left(\|x - x_c\|\right) = e^{-\|x - x_c^2\|/(2\sigma)^2}$, where x_c is the center of the kernel, and σ is window width (a value of $5°$ is specified). Through simply a visual investigation, we can find out that there are two peaks in the track density map. One peak is over the South China Sea, the other is located to the southeast of China in the Western Pacific. Furthermore, there is a dominant northwest gradient of the density. We inferred that large-scale atmospheric circulation causes this result. Further statistical and mechanism analyses are required to ascertain the underlying fact. The analysis results show that the two island provinces of Hainan and Taiwan are located near the peaks. Guangdong, Guangxi, Fujian, and Zhejiang provinces suffered severely from destructive TCs. In comparison, several northern coastal provinces have significantly less TC landfalls. These results show consistency with existing literature [34–36]. The track density map provides a continuous presentation of TC hazards of "places" across China, which will be helpful for preparedness and mitigation of TCs such as hazard hotspot identification and risk assessment.

5. Conclusions

Spatiotemporal patterns of TCs are one of the greatest interests in prevention and control of the destructive disasters. While the accurate behavior of a certain TC is difficult to determine thanks to our limited cognition for the complex dynamic mechanism [37, 38], data-oriented

analysis provides a kind of feasible approach to mining hidden patterns implied in the massive tracks. Time stamps and geographical locations are two types of substantial attributes for TC track data. As stated in the first law of geography, in a spatial context, spatial (spatiotemporal) correlation must be taken into consideration, which is an essential difference between spatial and nonspatial analysis. Since some proprietary GIS software packages are devised with powerful spatiotemporal analysis tools, some traditional methods of identifying TC patterns in time and space are complemented and enhanced when coupled with GIS. Visualization-based approaches provide an intuitive survey for observations and are widely put in practice, while they usually fail to be accepted in some occasions due to their imprecisions and inherent weakness. In order to discover real potential patterns, some robust methods are used in spatial data analysis. When faced with a question like what methods are appropriate for our analysis, the representation of spatial data is also of great concern. As a typical kind of linear features composed of a sequence of points, linear feature specific methods are usually preferred for track analysis of TCs and the like.

The typical periodicity and centralization of TCs encourages us to explore the potential patterns of TCs. Through plotting the circular map on different time scales (month, season), an obvious monthly and seasonal difference is presented. Point cluster analysis along line (temporal interval) is carried out to confirm the exploratory results by a rigorous approach. The cluster analysis demonstrates not only the consistency, but also the detailed analysis of statistical characteristics of these clusters through a Monte Carlo simulation, which is impossible to be perceived by visualization-based analysis methods. The detected temporal clusters are consistent with empirical knowledge. Since the temporal dimension can only describe the marginal features of TCs, several spatial aspects are explored. At this time, our focus is on the landfall locations of TCs. First, a local Moran's I index is used to detect spatial clusters of TCs. The analytical results are consistent with the intuitive realization of TC clustering along the coastline of China. Additionally, statistical significance is calculated according to the spatial Monte Carlo algorithm. This extended analysis helps refine the visual estimation of clusters and provides a rigorous approach to result assessment. Next, to evaluate the stratified heterogeneity of TCs, both spatial and temporal stratified heterogeneity is carried out according to a q-statistic. The tests show that stratified heterogeneity is significant, which means the within-strata variance is less than the between-strata variance. The results imply the locational and seasonal factors are the potential determinants of the heterogeneity of TCs. Sequentially, the exceedance probabilities and the return periods for the coastal prefectures are rendered. Finally, a line-based kernel density estimation is performed on the TC tracks generated in the Northwest Pacific. Coupled with some referable data such as population and GDP, the resultant map is intended to quantitatively evaluate the hazard of TCs and further provide substantial reference for preparedness and mitigation of future TCs along the coastline of China. Furthermore, compared to

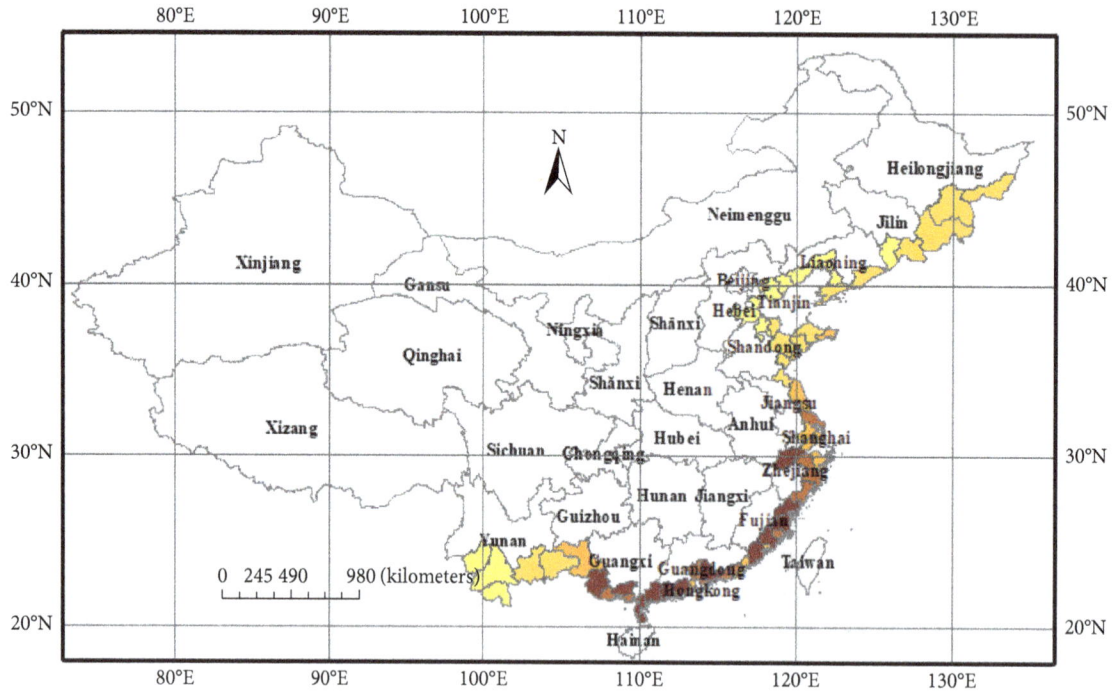

(a)

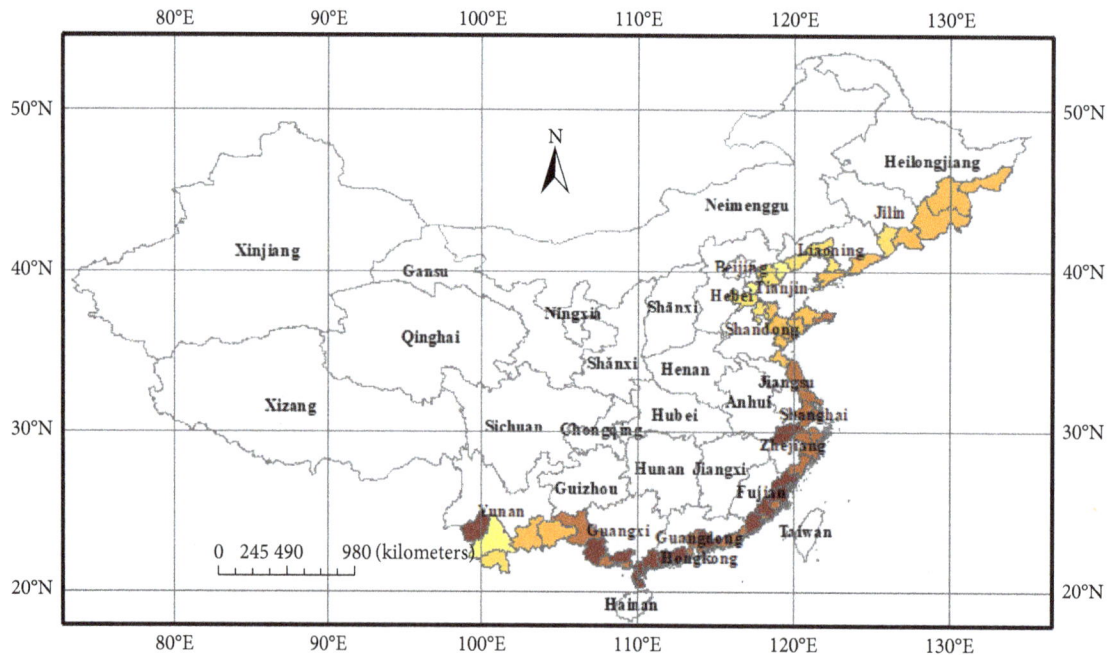

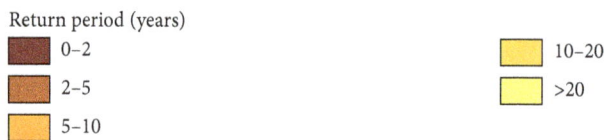

(b)

FIGURE 9: Exceedance probabilities and return periods of the landfall counts of TCs in the coastal prefectures. (a) Exceedance probability; (b) return period.

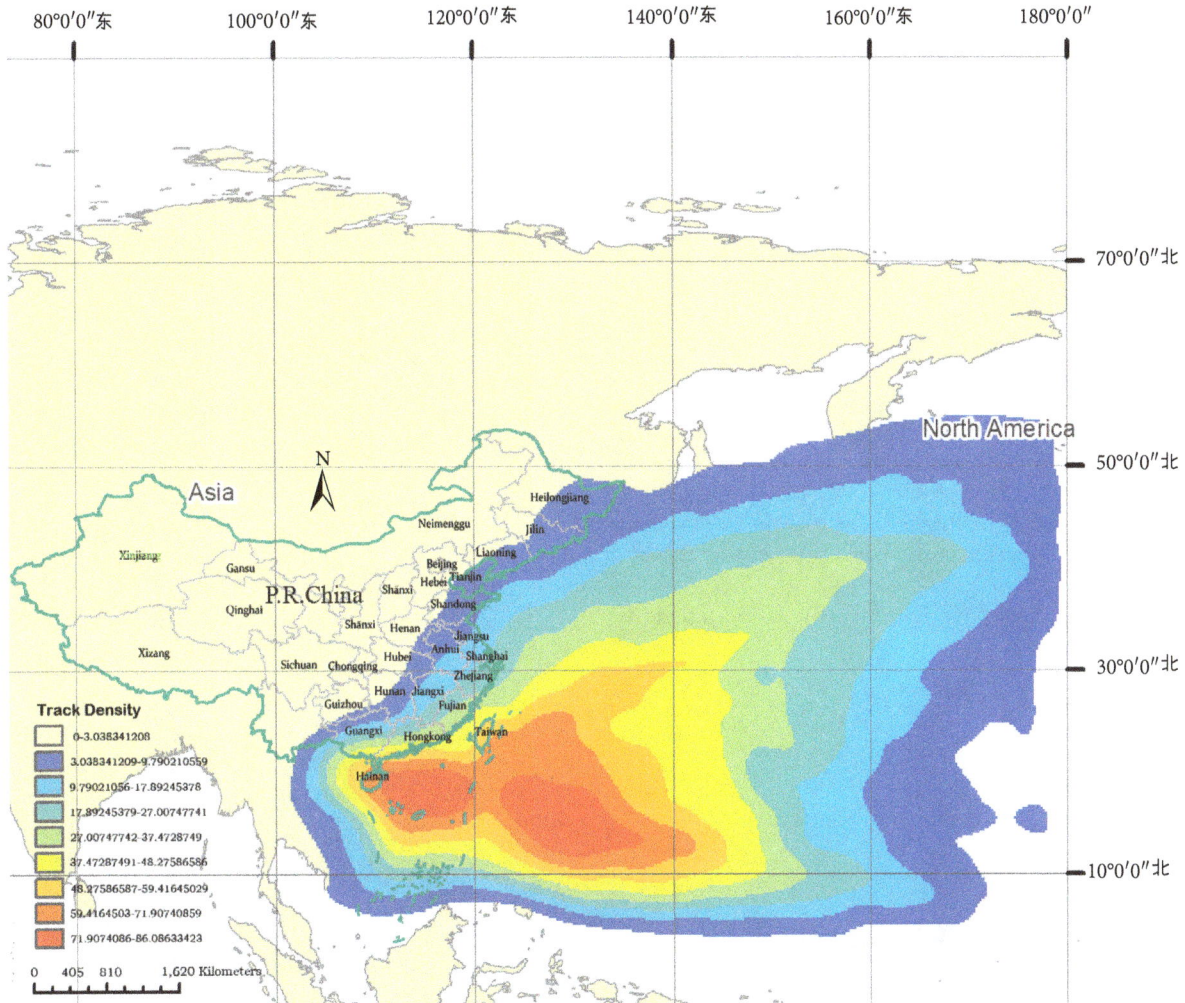

FIGURE 10: Track density map of TCs in the Northwest Pacific.

some findings in existing literature related to the TC mechanisms, these spatial and temporal patterns are consistent with several universally accepted mechanisms that have been deeply studied by some researchers. For example, the TC swarm occurring in space or time is closely related to atmospheric circulation conditions of the tropical Northwest Pacific [39]. The status of the warm pool in the West Pacific largely affects the interannual change of the moving tracks of TCs [40]. Therefore, those comprehensive effects lead to the patterns.

Overall, this paper systematically explored temporal and spatial patterns and mapped the hazards of TCs, through the integrated application of advanced spatial techniques. This work can help obtain insight into temporal and spatial characteristics of TCs compared to previous efforts. These kinds of findings can assist with proactive destructive prevention activities, thereby enhancing the preparedness and mitigation capacity in a predisaster phase. While the present findings offer useful insights into the spatiotemporal patterns of TCs along the coastline of China, further work is clearly required both to reinforce the results and to extend our understanding of the problems. For example, satellite remote sensing techniques enable

retrieval of meteorological parameters from remotely sensed data since 1970s, which makes it necessary to compare between data from meteorological stations and satellites. In prevention and response, a continued effort by researchers, the authorities, and the public is imperative in order to enhance prevention ability and reduce the loss of lives and property damage caused by TCs.

Appendixes

A. Cluster Detection of Points along a Line

Let $x_1 \leq x_2 \leq \cdots \leq x_N$ be N points independently drawn from the uniform distribution on a line segment (or timeline) ordered according to size. The joint p.d.f. at $x_1, x_2, \ldots, x_N$ is $N!$ and $1 - \Pr(n|N; p)$, the probability that no n points are contained in a length p, is the multiple integral of $N! dx_1, \ldots, dx_N$ over $x_1, \ldots, x_N$ in the region: $a \leq x_1 \leq x_2 \leq \cdots \leq x_N \leq b$, and $x_{n+i-1} - x_i > p$, for all $i = 1, 2, \ldots, N - n + 1$. Naus derived $\Pr(n|N; p)$, the c.d.f. of p, and the size of the smallest interval that contains n out of the N points. This calculation gave rise to a test statistic for detecting clusters, namely, the scan statistic.

Assume $x_1, x_2, \ldots, x_N$ be a random sample of N points from the density:

$$f(x) = \begin{cases} a, & \text{for } b \le x \le b + d \\[2mm] \dfrac{1 - ad}{1 - d}, & \text{for } 0 \le x \le b \text{ or } b + d \le x \le 1 \end{cases} \tag{A.1}$$

For the hypothesis test problem,

$$H_0: a = 1,$$

$$H_1: \frac{1 < a \le 1}{d}, \tag{A.2}$$

where d is the scan window width and b is the start location of the scan window. Obviously, if $a = 1$, then the sample is drawn from a uniform distribution in $[0, 1]$. Otherwise, when $a > 1$, the sample has higher density in $[b, b + d]$ than in $[0, b] \cup [b + d, 1]$. This is the basic principle of the hypothesis test.

According to the above notation, the scan statistic is formulated as follows:

$$\Lambda = \sup_{0 < d < n/N, n \ge n_0} \left(\frac{n}{N}\right)^n \left(\frac{N - n}{N}\right)^{N-n} \left(\frac{1}{d}\right)^n \left(\frac{1}{1 - d}\right)^{N-n}, \tag{A.3}$$

where n_0 is the minimum number of points of the clusters to be detected.

B. Test for Stratified Heterogeneity

Assume that a study area is composed of N regions and is stratified into $h = 1, 2, \ldots, L$ stratum; stratum h is composed of N_h regions; and Y_i and Y_{hi} denote the value of region i in the population and in stratum h, respectively.

The q-statistic is defined as follows:

$$q = 1 - \frac{\sum_{h=1}^{L} \sum_{i=1}^{N_h} \left(Y_{hi} - \overline{Y}_h\right)^2}{\sum_{i=1}^{N} \left(Y_i - \overline{Y}\right)^2} = 1 - \frac{\sum_{h=1}^{L} N_h \sigma_h^2}{N \sigma^2} = 1 - \frac{\text{SSW}}{\text{SST}}, \tag{B.1}$$

where the total sum of squares

$$\text{SST} = \sum_{i=1}^{N} \left(Y_i - \overline{Y}\right)^2 = N\sigma^2 \tag{B.2}$$

and the within sum of squares

$$\text{SSW} = \sum_{h=1}^{L} \sum_{i=1}^{N_h} \left(Y_{hi} - \overline{Y}_h\right)^2 = \sum_{h=1}^{L} N_h \sigma_h^2. \tag{B.3}$$

The value of the statistic is within $[0, 1]$ (0 if there is no stratified heterogeneity and 1 if the population is fully stratified) and increase monotonously with the increase of stratified heterogeneity.

Let

$$F \stackrel{\text{def}}{=} \frac{\text{SSB}/(L - 1)}{\text{SSW}/(N - L)}, \tag{B.4}$$

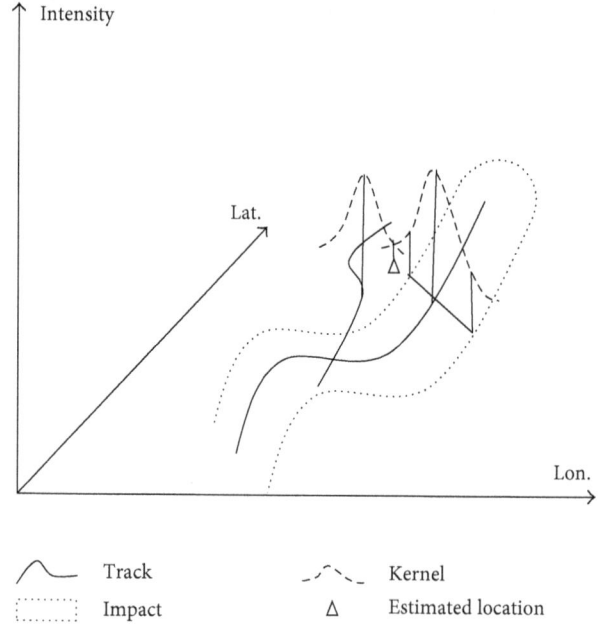

FIGURE 11: A schematic of the calculation of track density.

where $\text{SSB} = \sum_{h=1}^{L} N_h (\overline{Y}_h - \overline{Y})^2$ is the between-strata sum of squares, then F follows a noncentral F-distribution with the 1st df $L - 1$, the 2nd df $N - L$, and noncentrality λ as follows:

$$F \sim F(L - 1, N - L; \lambda). \tag{B.5}$$

C. Calculation of Track Density

Let $l_1, l_2, \ldots, l_N$ are the N tracks and for a certain location p, the distances (perpendicular distances) from p to these tracks are $d_{p1}, d_{p2}, \ldots, d_{pN}$, then the track density of p can be calculated as follows:

$$\tilde{f}(p) = \sum_{i=1}^{N} f(d_{pi}), \tag{C.1}$$

where $f(\cdot)$ is a distance decay function (kernel function). Figure 11 illustrates how to calculate the track density at a certain location (two tracks are shown in Figure 11).

Kernel estimation methods originally are used to estimate a probability density function. With respect to the track data set, a density function defines the probability of observing a track at a location p. In this sense, only a function conforming to some certain conditions can be adopted as a kernel [30].

The pseudocode in Table 3 presents a more detailed description of the calculation procedure of the track density. For each cell in a tessellation of the area to be kernelled, first calculate the normalized distances (TrackDensity) to its neighboring tracks, second, add the densities for each separate track up to a summarized density, and finally, the summarized density (SumDensity) is normalized with the number of tracks. Once all cells are processed according to the steps, a raster map of track density will be produced.

TABLE 3: Pseudocode for calculating the track density.

```
SumDensity = 0
For each Track
{
TrackDensity = 0
    calculate NeighboringArea around the track
    for each Cell in the NeighboringArea
    {
        calculate DistanceToTrack
        TrackDensity = normalized DistanceToTrack
    }
    SumDensity = SumDensity + TrackDensity
}
normalize SumDensity with number of tracks
```

In the algorithm, a track is represented as a list of successive points in 2D space, that is, a 2D polyline consisting of points $p_1, p_2, p_n, p_{n+1}, \ldots, p_m$, where $p_n = (x_n, y_n)$, x_n, and y_n are geographic coordinates. Though a track in the real world is always a continuous line, this discretized representation of the track is a kind of widely used data structure in GIS, which enables storage and analysis of real-world objects in computers. We adopt this standard assumption in our study: a series of straight-line segments between sample points are able to represent an original continuous track approximately.

Conflicts of Interest

No potential conflicts of interest were reported by the authors.

Acknowledgments

The authors acknowledge the support from the National Natural Science Foundation of China (Grant nos. 70901047 and 91224004) and Project in the National Science and Technology Pillar Program during the Twelfth Five-Year Plan Period (Grant nos. 2015BAK10B01 and 2015BAK12B03). The authors also appreciate support for this paper from the Collaborative Innovation Center of Public Safety.

References

[1] NOAA, What is the Difference Between a Hurricane, a Cyclone, and a Typhoon?, National Oceanic and Atmospheric Administration, Silver Spring, MD, USA, July 2016, http://oceanservice.noaa.gov/facts/cyclone.html.

[2] P. J. Webster, G. J. Holland, J. A. Curry, and H. R. Chang, "Changes in tropical cyclone number, duration, and intensity in a warming environment," Science, vol. 309, no. 5742, pp. 1844–1846, 2005.

[3] D. F. Liu, L. Pang, and B. T Xie, "Typhoon disaster in China: prediction, prevention, and mitigation," Natural Hazards, vol. 49, no. 3, pp. 421–436, 2009.

[4] L. Wang, G. H. Chen, and R. H. Huang, "Spatiotemporal distributive characteristics of tropical cyclone activities over the Northwest Pacific in 1979-2006," Journal of Nanjing Institute of Meteorology, vol. 32, no. 2, pp. 182–188, 2009.

[5] C. L. Gu, J. C. Kang, G. D. Yan et al., "Spatial and temporal variability of northwest pacific tropical cyclone activity in a global warming scenario," Journal of Tropical Meteorology, vol. 22, no. S1, pp. 15–23, 2016.

[6] J. C. L. Chan and K. S. Liu, "Global warming Western North Pacific typhoon activity from an observational perspective," Journal of Climate, vol. 17, no. 23, pp. 4590–4602, 2010.

[7] L. Tang, H. U. Deyong, and L. I. Xiaojuan, "Spatiotemporal characteristics of tropical cyclone activities in Northwestern Pacific from 1951 to 2006," Journal of Natural Disasters, vol. 21, no. 1, pp. 31–38, 2012.

[8] R. B. Santos, Crime Analysis with Crime Mapping, Sage Publications, Thousand Oaks, CA, USA, 2016.

[9] J. I. Naus, "The distribution of the size of the maximum cluster of points on a line," Journal of the American Statistical Association, vol. 60, no. 310, pp. 532–538, 1965.

[10] R. J. Huntington and J. I. Naus, "A simpler expression for kth nearest neighbor coincidence probabilities," Annals of Probability, vol. 3, no. 5, pp. 894–896, 1975.

[11] N. Nagarwalla, "A scan statistic with a variable window," Statistics in Medicine, vol. 15, no. 7–9, pp. 845–850, 1996.

[12] P. A. P. Moran, "Notes on continuous stochastic phenomena," Biometrika, vol. 37, no. 1-2, pp. 17–23, 1950.

[13] A. D. Cliff and J. K. Ord, Spatial Processes: Models and Applications, Pion, London, UK, 1981.

[14] L. Anselin, "Local indicators of spatial association—LISA," Geographical Analysis, vol. 27, no. 2, pp. 93–115, 1995.

[15] J. K. Ord and A. Getis, "Local spatial autocorrelation statistics: distributional issues and an application," Geographical Analysis, vol. 27, no. 4, pp. 286–306, 1995.

[16] M. Kulldorff, "A spatial scan statistic," Communications in Statistics-Theory and Methods, vol. 26, no. 6, pp. 1481–1496, 1997.

[17] M. A. Costa and M. Kulldorff, "Maximum linkage space-time permutation scan statistics for disease outbreak detection," International Journal of Health Geographics, vol. 13, no. 1, p. 20, 2014.

[18] J. F. Wang, X. H. Li, G. Christakos et al., "Geographical detectors-based health risk assessment and its application in the neural tube defects study of the Heshun region, China," International Journal of Geographical Information Science, vol. 24, no. 1, pp. 107–127, 2010.

[19] A. Asgary, A. Ghaffari, and J. Levy, "Spatial and temporal analyses of structural fire incidents and their causes: a case of Toronto, Canada," Fire Safety Journal, vol. 45, no. 1, pp. 44–57, 2010.

[20] U. Demšar and K. Virrantaus, "Space–time density of trajectories: exploring spatio-temporal patterns in movement data," International Journal of Geographical Information Science, vol. 24, no. 10, pp. 1527–1542, 2010.

[21] P. Timothée, L. B. Nicolas, S. Emanuele et al., "A network based kernel density estimator applied to Barcelona economic activities," in Proceedings of the Computational Science and Its Applications (ICCSA 2010), pp. 32–45, Fukuoka, Japan, March 2010.

[22] S. Steiniger and A. J. S. Hunter, "A scaled line-based kernel density estimator for the retrieval of utilization distributions and home ranges from GPS movement tracks," Ecological informatics, vol. 13, pp. 1–8, 2013.

[23] U. Demšar, K. Buchin, E. E. van Loon, and J. Shamoun-Baranes, "Stacked space-time densities: a geovisualisation approach to explore dynamics of space use over time," Geo-Informatica, vol. 19, no. 1, pp. 85–115, 2015.

[24] L. Tang, Z. Kan, X. Zhang, F. Sun, X. Yang, and Q. Li, "A network kernel density estimation for linear features in space–time analysis of big trace data," International Journal of Geographical Information Science, vol. 30, no. 9, pp. 1717–1737, 2015.

[25] Y. Deng, B. Wallace, D. Maassen, and J. Werner, "A few GIS clarifications on tornado density mapping," *Journal of Applied Meteorology and Climatology*, vol. 55, no. 2, pp. 283–296, 2016.

[26] Y. L. Liu, G. R. Feng, Y. Xue, H. M. Zhang, and R. G. Wang, "Small-scale natural disaster risk scenario analysis: a case study from the town of Shuitou, Pingyang County, Wenzhou, China," *Natural Hazards*, vol. 75, no. 3, pp. 2167–2183, 2015.

[27] A. J. Colbert, B. J. Soden, and B. P. Kirtman, "The impact of natural and anthropogenic climate change on western North Pacific tropical cyclone tracks," *Journal of Climate*, vol. 28, no. 5, pp. 1806–1823, 2015.

[28] Esri, *What is ArcGIS?*, Esri, Redlands, CA, USA, June 2016, https://www.arcgis.com/features/index.html.

[29] J. F. Wang, T. L. Zhang, and B. J. Fu, "A measure of spatial stratified heterogeneity," *Ecological Indicators*, vol. 67, pp. 250–256, 2016.

[30] L. A. Waller and C. A. Gotway, *Applied Spatial Statistics for Public Health Data*, Wiley, Hoboken, NJ, USA, 2004.

[31] B. W. Silverman, *Density Estimation for Statistics and Data Analysis*, Chapman & Hall, New York, NY, USA, 1986.

[32] A. Getis and J. K. Ord, "The analysis of spatial association by use of distance statistics," *Geographical Analysis*, vol. 24, no. 3, pp. 189–206, 1992.

[33] B. J. L. Berry and D. F. Marble, *Spatial Analysis: A Reader in Statistical Geography*, Prentice-Hall, Englewood Cliffs, NJ, USA, 1968.

[34] F. C. Guan and Q. H. Xie, "The statistical characteristics of typhoon in the south China sea," *Marine Science Bulletin*, vol. 3, no. 4, pp. 21–29, 1984.

[35] N. Shi and J. D. Zhou, "A statistical analysis of typhoon activities over south China sea and ENSO," *Meteorological Monthly*, vol. 15, no. 4, pp. 9–14, 1989, in Chinese.

[36] D. S. Wu, X. Zhao, W. Z. Feng et al., "The statistical analysis to the local harmful typhoon of South China Sea," *Journal of Tropical Meteorology*, vol. 21, no. 3, pp. 309–314, 2005, in Chinese.

[37] P. J. Vickery and L. A. Twisdale, "Prediction of hurricane wind speeds in the United States," *Journal of Structural Engineering*, vol. 121, no. 11, pp. 1691–1699, 1995.

[38] K. Emanuel, S. Ravela, E. Vivant, and C. Risi, "A statistical deterministic approach to hurricane risk assessment," *Bulletin of the American Meteorological Society*, vol. 87, no. 3, pp. 299–314, 2006.

[39] H. Wang, Y. H. Ding, and J. H. He, "Influence of western north Pacific summer monsoon changes on typhoon genesis," *Acta Meteorologica Sinica*, vol. 64, no. 3, pp. 345–356, 2006.

[40] R. H. Huang and G. H. Chen, "Research on interannual variations of tracks of tropical cyclones over Northwest Pacific and their physical mechanism," *Acta Meteorologica Sinica*, vol. 65, no. 5, pp. 683–694, 2007.

5

Solar Radiation Models and Gridded Databases to Fill Gaps in Weather Series and to Project Climate Change in Brazil

Fabiani Denise Bender ⓘ and Paulo Cesar Sentelhas ⓘ

Department of Biosystems Engineering, ESALQ, University of São Paulo, 13418-900 Piracicaba, SP, Brazil

Correspondence should be addressed to Paulo Cesar Sentelhas; pcsentel.esalq@usp.br

Academic Editor: Anthony R. Lupo

The quantification of climate change impacts on several human activities depends on reliable weather data series, without gaps and long enough to build up future climate. Based on that, this study aimed to evaluate the performance of temperature-based models for estimating global solar radiation and gridded databases (AgCFSR, AgMERRA, NASA/POWER, and XAVIER) as alternative ways for filling gaps in historical weather series (1980–2009) in Brazil and to project climate change scenarios based on measured and gridded weather data. Projections for mid- and end-of-century periods (2040–2069 and 2070–2099), using seven global climate models from CMIP5 under intermediate (RCP4.5) and high (RCP8.5) emission scenarios, were performed. The Bristow–Campbell model was the one that best estimated solar radiation, whereas the XAVIER gridded database was the closest to observed weather data. Future climate projections, under RCP4.5 and RCP8.5 scenarios, as expected, showed warmer conditions for all scenarios over Brazil. On the contrary, rainfall projections are more uncertain. Despite that, the rainfall amounts will be reduced in the North-Northeast region and increased in Southern Brazil. No significant differences between projections using the observed and XAVIER gridded database were observed; therefore, such a database showed to be reliable for both to fill gaps and to generate climate change scenarios.

1. Introduction

Given the projections of global climate changes, simulation models can be used to estimate the impact of historical and future climates on human activities, mainly in crop growth and yield and food availability [1]. For proper simulations, these models require high-quality and long-term historical daily weather data [2]. However, the major difficulty regarding historical weather data in Brazil is the low density of weather stations, associated with the reduced number of measured variables and the large amount of missing data [3–5].

To overcome the lack of reliable weather data series, missing data can be filled in with estimated or interpolated data. Among the different approaches used to fill weather data gaps in, the main methods are climatic generators, which generate stochastic sequences of daily data, such as WGEN [6] and SIMMETEO [7] generators; empirical correlations using commonly measured meteorological variables present in the observed data [8–10]; and the use of the gridded weather database, based on satellite and/or surface data [2, 4, 11].

Once the historical data series have been filled, these can be used for generating future climate scenarios, derived from projections of climate models, which can be global (GCMs) or regional (RCMs). Despite the finer resolution of RCMs, considering the continental dimension of Brazil, GCMs (which would provide the RCM boundary conditions) offer insight into the general characteristics of future climate [12, 13].

Due to the uncertainties associated with the GCM projections, different models can indicate different climate responses, and one way to reduce such an uncertainty is by considering an ensemble modeling approach [14], with the projections being obtained from multiple models, resulting in more reliable scenarios than if the models are considered individually [15].

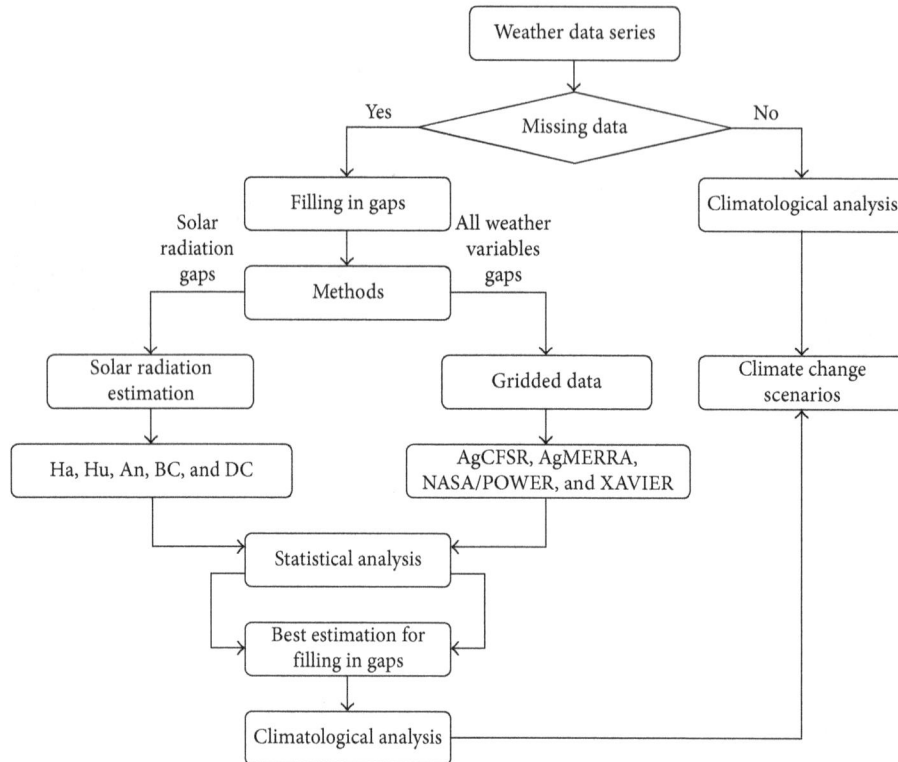

FIGURE 1: Flow chart of the steps used in the present study for filling in gaps in historical weather series and to generate future climate scenarios.

These future changes can be projected based on GCMs generated by the Coupled Model Intercomparison Project Phase 5 (CMIP5 [16]), under different greenhouse gases emissions that follow distinct representative concentration pathways (RCPs) [17–19], assessed in the Fifth Assessment Report (AR5) of the Intergovernmental Panel on Climate Change (IPCC) [20]. For South America and specifically for Brazil, the first projections have indicated an increase in temperatures and an uncertain pattern in the rainfall distribution [12, 13]. Such patterns have been confirmed in the more recent studies of Chou et al., Sánchez et al., and Salviano et al. [21–23].

Given the great importance of historical weather data for assessing the impacts of climate change on human activities, mainly agriculture, in addition to the fact that Brazil has a low weather station density, with a large amount of missing data [3–5], the general objective of this study was to evaluate the performance of different alternatives to fill in weather data gaps and, based on that, to create climate change scenarios for Brazil. More specifically, this study aimed (i) to evaluate the performance of temperature-based models for estimating solar radiation and gridded databases, such as AgCFSR, AgMERRA, NASA/POWER, and XAVIER, as procedures to fill in gaps of weather data (maximum and minimum air temperature, solar radiation, rainfall, wind speed, and relative humidity) for the period of 1980–2009; (ii) to generate, from the complete historical weather data, climate change scenarios, over the medium-term (2040–2069) and long-term (2070–2099) periods, based on seven GCMs of CMIP5, under intermediate (RCP4.5) and high

(RCP8.5) emission scenarios; and (iii) to identify patterns of climate change in air temperature and rainfall in different Brazilian regions to define the expected trends in relation to the historical climate.

2. Materials and Methods

The present study was developed according to different steps and in a logical sequence presented in the flow chart of Figure 1 and in the following sections.

2.1. Sites and Weather Data. Historical daily measured weather data of maximum and minimum air temperature, sunshine hours, rainfall, wind speed, and relative humidity, from 1980 to 2009, were obtained from the Brazilian National Institute of Meteorology (INMET). Thirty-one sites well distributed in the country were considered, as presented in Figure 2. More detailed description about the percentage of missing values for each weather variable is in Table S1 of Supplementary Materials.

2.2. Filling Gaps in the Meteorological Database. Due to the large percentage of missing data in the historical weather databases, ranging from 1 to 46% (Figure 2 and Table S1 of Supplementary Materials), weather variables were generated by temperature-based models (solar radiation) and gridded databases (all variables), as alternatives to fill these gaps in.

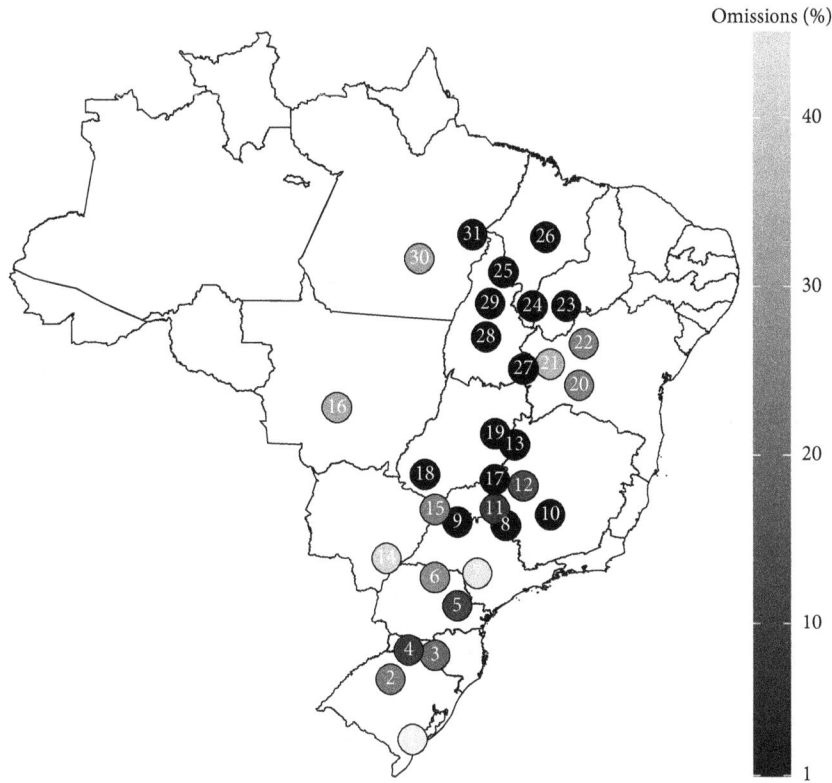

FIGURE 2: Weather stations from the Brazilian National Institute of Meteorology used in the present study, with total percentage of missing data (maximum and minimum air temperature, sunshine hours, rainfall, relative humidity, and wind speed), in the period from 1980 to 2009.

TABLE 1: Solar radiation-estimating models based on maximum and minimum air temperature.

Model	Equation*	Coefficients	Reference
Ha	$Q_g = a \times \Delta T^{0.5} \times Q_0$	a	[9]
Hu	$Q_g = b \times \Delta T^{0.5} \times Q_0 + c$	b and c	[26]
An	$Q_g = d \times (1 + 2.7 \times 10^{-5} \times \mathrm{Alt}) \times \Delta T^{0.5} \times Q_0$	d	[27]
BC	$Q_g = e \times (1 - \exp(-f \times \Delta T^g)) \times Q_0$	e, f, and g	[10]
DC	$Q_g = h \times (1 - \exp(-i \times \Delta T^j / \Delta T_m)) \times Q_0$	h, i, and j	[28]

*Q_g: solar radiation ($\mathrm{MJ \cdot m^{-2} \cdot d^{-1}}$); Q_0: extraterrestrial solar radiation ($\mathrm{MJ \cdot m^{-2} \cdot d^{-1}}$); ΔT: thermal amplitude ($T_{max} - T_{min}$) (°C); ΔT_m: 7-day moving average of ΔT (°C); a, b, c, d, e, f, g, h, and j: coefficients to be adjusted for each model. For BC and DC models, e and h coefficients were determined by a relationship with altitude, $a = 0.75 + 2 \times 10^{-5} \times \mathrm{Alt}$, as proposed by Allen et al. [29]. Q_0 and N are astronomical values calculated according to Allen et al. [30]. Ha: Hargreaves; Hu: Hunt; An: Annandale; BC: Bristow–Campbell; DC: Donatelli–Campbell.

2.2.1. Temperature-Based Solar Radiation Models. As solar radiation is not commonly recorded by conventional weather stations, its values were calculated from sunshine hours (n) data, following the model proposed by Ängström [8] and Prescott [24], with coefficients as suggested by Glover and McCulloch [25], and then admitted as the reference values (Table 1). The temperature-based models for estimating solar radiation use maximum and minimum air temperatures as inputs to estimate atmospheric transmissivity [10], which is affected by cloudiness. Five solar radiation models (Hargreaves (Ha), Hunt (Hu), Annandale (An), Bristow–Campbell (BC), and Donatelli–Campbell (DC)) were assessed as presented in Table 1.

2.2.2. Daily Gridded Database. Gaps in measured weather data (maximum and minimum air temperature, solar

radiation, rainfall, wind speed, and relative humidity) were also filled in with data from the following four gridded databases:

(a) AgCFSR and AgMERRA datasets [11], developed as a part of the Agricultural Model Intercomparison and Improvement Project (AgMIP) [31], to provide consistent, daily time series with global coverage of climate variables. They are result of a combination of NCEP's reanalysis of the Climate Forecast System Reanalysis (CFSR) [32] and NASA's Modern-Era Retrospective Analysis for Research and Applications (MERRA) [33] with observed datasets from weather stations' networks and satellites, available on a daily temporal scale, for the period between 1980 and 2010, at $0.25° \times 0.25°$ horizontal resolution.

(b) National Aeronautics and Space Administration database developed by the Prediction of Worldwide Energy Resource (NASA/POWER) [34], composed by satellite data, radiosondes, surface observations, and numerical modeling from data assimilation. The meteorological variables are available on a daily world scale, but in a grid of lower resolution, that is, of greater horizontal spacing, with $1° \times 1°$ horizontal resolution, for the period from 1983 to the near present. Just for rainfall, the historical series started in 1997.

(c) Gridded dataset developed by Xavier et al. [4], referred to as XAVIER that includes only daily observed data from rain gauges and conventional and automatic weather stations for the period of 1980–2013, available on a spatial resolution of $0.25° \times 0.25°$ only for Brazil.

2.2.3. Evaluation of Solar Radiation Models and Daily Gridded Database for Filling in Weather Data Gaps. Concerning the solar radiation models, two independent datasets were considered with two years each, for the calibration and evaluation of the adjusted coefficients. To avoid inconsistencies in the analysis, two consecutive years with less than 2% of missing data (temperature and sunshine hours) were chosen. For the evaluation of the gridded weather data, the entire database was employed for the period between 1980 and 2009.

The performance of temperature-based solar radiation models and gridded databases for filling in daily data gaps was assessed by comparing estimated and measured data on a daily basis, using the common model performance evaluation indices, such as the coefficient of determination (r^2) as a measure of precision; agreement index (d) [35] as a measure of accuracy; confidence index (c) [36] (being classified as great for values higher than 0.85, very good for values between 0.76 and 0.85, good between 0.66 and 0.75, median between 0.61 and 0.65, suffering between 0.51 and 0.60, bad between 0.41 and 0.50, and terrible for values lower than 0.41); mean error or bias (Bias) that indicates the tendency of error; and mean absolute error (MAE), which gives the magnitude of the errors [37].

2.3. Climate Change Projections. Climate change scenarios, based on measured weather data fulfilled with the best alternative, were projected by models that are publicly available through the CMIP5 [16], based on two RCPs [18]: intermediate emission scenario (RCP4.5) and high emission scenario (RCP8.5). As suggested by Ward et al. [38], the intermediate scenario appears as the most likely future for planning purposes, in which observed fossil fuel trajectories show up to be consistent, whereas the high emission scenario represents the extreme conditions.

The future scenarios were generated based on the delta method [39], in which simulated mean monthly changes are imposed for the baseline for all sites by adding temperature changes and multiplying precipitation changes, without

changing the variability within a month (e.g., the number of rainy days), following the procedure as described by Hudson and Ruane [40]. All other variables were kept unchanged.

Projections were performed for mid-of-century (2040–2069) and end-of-century (2070–2099) periods, for the following CMIP5 GCMs: CNRM-CM5 [41], CSIRO-Mk3-6-0 [42], GISS-E2-R [43], HadGEM2-ES [44, 45], INMCM4 [46], MIROC-ESM [47], and MPI-ESM_LR [48]. The use of seven different GCMs was adopted since the uncertainties are inherent to the climate system, as a result of nonlinear interactions and the intrinsic complexity of the natural atmospheric phenomena [49]. Therefore, for the same emission scenario, different models produce diverse projections of climate change, and one way to minimize these uncertainties is through a set of global and/or regional models, known as an ensemble approach [15]. In this sense, the climate projection presented here for each variable is an average of the outputs of seven GCMs.

As an alternative to the use of gridded historical climate data for future climate projections, we analyzed climate projections based on measured weather data compared to the climatology provided by the best alternative method, considering only the nine sites which had a percentage of missing data on air temperature and rainfall lower than 10%, as presented in Table S1 of Supplementary Materials.

3. Results

3.1. Filling Gaps in Measured Weather Data

3.1.1. Solar Radiation Models. Table 2 presents the average daily annual coefficients of the temperature-based solar radiation models for all Brazilian locations assessed. The Ha model displayed adjusted coefficients varying from $0.10°C^{-0.5}$ to $0.18°C^{-0.5}$, differing from the original values of $0.16°C^{-0.5}$ and $0.19°C^{-0.5}$ obtained by Hargreaves and Samani [50] for continental and coastal regions, respectively. The adjusted b coefficient for the Hu model ranged from 0.04 to 0.22. However, the c coefficient of this model showed quite distinct values, ranging from −7.70 and 9.98. The coefficients e of the BC model and h of the DC model were similar, ranging from 0.75 to 0.77 in both models; however, the coefficients f and g of the BC model were smaller than the coefficients i and j of the DC model, whereas f and g of the BC model were, in average, 0.03 and 1.63 and i and j of the DC model were 0.07 and 2.24.

Statistical indices for each temperature-based model assessed are presented in Figure 3. For more detailed results, see Tables S2 and S3 of Supplementary Materials. As presented in Figure 3, r^2 for the BC model ranges between 0.32 and 0.79, with a mean value of 0.62. For the DC model, r^2 values range from 0.26 to 0.76, with an average value of 0.59.

The estimated solar radiation values presented d between 0.44 and 0.93 for the Ha and Hu models and from 0.55 to 0.92 for the An model, with a mean value of 0.79, for all of them. For the BC and DC models, this index ranged from 0.62 to 0.93 and from 0.60 to 0.93, respectively, with average values of 0.86 and 0.85 (Figure 3; Tables S2 and S3).

TABLE 2: Average daily annual coefficients of Hargreaves (Ha), Hunt (Hu), Annandale (An), Bristow–Campbell (BC), and Donatelli–Campbell (DC) temperature-based solar radiation models for each of the Brazilian locations considered in this study.

Model	Ha	Hu		An	BC				DC	
Coefficients	a	b	c	d	e	f	g	h	i	j
RSPE	0.18	0.19	−1.12	0.18	0.75	0.07	1.29	0.75	0.14	1.98
RSCA	0.17	0.19	−2.08	0.17	0.76	0.02	1.73	0.76	0.06	2.29
SCCN	0.17	0.18	−1.49	0.17	0.77	0.04	1.52	0.77	0.14	1.94
SCCH	0.17	0.18	−1.06	0.17	0.76	0.03	1.64	0.76	0.11	2.08
PRCA	0.10	0.11	−0.63	0.10	0.77	0.05	1.02	0.77	0.15	1.57
PRLO	0.12	0.13	−1.46	0.12	0.76	0.03	1.30	0.76	0.09	1.88
SPAV	0.17	0.16	0.66	0.16	0.77	0.04	1.44	0.77	0.16	1.89
SPFR	0.18	0.12	1.79	0.13	0.77	0.04	1.32	0.77	0.11	1.86
SPVP	0.17	0.16	1.70	0.17	0.76	0.03	1.65	0.76	0.08	2.20
MGBA	0.14	0.16	−1.86	0.14	0.76	0.05	1.26	0.76	0.09	2.00
MGUB	0.17	0.16	0.99	0.17	0.76	0.01	1.98	0.76	0.03	2.57
MGPM	0.17	0.16	0.94	0.16	0.77	0.02	1.79	0.77	0.04	2.42
MGUN	0.16	0.16	−0.51	0.16	0.76	0.02	1.75	0.76	0.04	2.46
MSIV	0.17	0.19	−2.00	0.17	0.76	0.02	1.82	0.76	0.07	2.27
MSPA	0.18	0.20	−3.19	0.17	0.76	0.02	1.79	0.76	0.06	2.39
MTDI	0.15	0.14	1.56	0.15	0.76	0.04	1.39	0.76	0.10	2.03
GOCA	0.17	0.17	0.23	0.17	0.77	0.02	1.81	0.77	0.04	2.46
GOJA	0.15	0.16	−0.76	0.15	0.76	0.05	1.32	0.76	0.10	2.02
DFBR	0.18	0.15	3.08	0.17	0.77	0.03	1.71	0.77	0.06	2.36
BABJ	0.18	0.18	0.12	0.18	0.76	0.03	1.72	0.76	0.06	2.39
BABA	0.17	0.18	−1.26	0.17	0.76	0.03	1.60	0.76	0.05	2.38
BARR	0.18	0.21	−3.08	0.18	0.76	0.01	2.23	0.76	0.02	2.94
PIBJ	0.11	0.04	9.98	0.11	0.76	0.03	1.25	0.76	0.08	1.88
MAAP	0.17	0.19	−2.88	0.17	0.76	0.02	1.79	0.76	0.04	2.52
MACA	0.18	0.20	−2.27	0.18	0.75	0.03	1.71	0.75	0.05	2.42
MABC	0.16	0.22	−7.70	0.16	0.75	0.01	2.04	0.75	0.02	2.75
TOTA	0.17	0.16	1.05	0.17	0.76	0.02	1.83	0.76	0.04	2.52
TOPN	0.17	0.20	−3.77	0.16	0.75	0.03	1.58	0.75	0.06	2.31
TOPA	0.17	0.20	−2.89	0.17	0.75	0.03	1.68	0.75	0.07	2.30
PASF	0.13	0.11	2.84	0.13	0.75	0.03	1.35	0.75	0.07	2.02
PAMA	0.17	0.22	−5.29	0.17	0.75	0.02	1.92	0.75	0.04	2.51
Mean	0.16	0.17	−0.66	0.16	0.76	0.03	1.62	0.76	0.07	2.24

The confidence index (c) ranged from 0.31 to 0.81, with an average of 0.61 for the Ha and An models, and from 0.25 to 0.82 for the Hu model, with an average of 0.62 (Figure 3). For the BC model, c ranged from 0.35 to 0.82, while for the DC model, c ranged from 0.32 to 0.80, with an average of 0.68 and 0.66, respectively. Considering the average values for all sites, the models of Ha, Hu, and An presented performances classified as "median," whereas the performances of BC and DC models were classified as "good," according to the Camargo and Sentelhas [36] classification.

3.1.2. Gridded Database. Table 3 presents the performance of the different daily gridded databases used to fill the gaps in the historical weather series. All databases showed high accuracy ($d \geq 0.89$) for maximum air temperature (T_{max}), with XAVIER also showing very high precision ($r^2 = 1$). Except for AgCFSR, all models underestimated T_{max}. Among all databases, XAVIER was the best one for estimating T_{max}, with MAE = 0.17°C, whereas NASA/POWER presented the highest MAE of 2.46°C.

All databases showed high accuracy ($d \geq 0.93$) and good precision ($r^2 \geq 0.77$) for minimum air temperature (T_{min}). As to T_{max}, XAVIER showed the best performance, with the

lowest Bias (0.06°C) and MAE (0.30°C). On the contrary, NASA/POWER presented the worst performance, with Bias = 0.76°C and MAE = 1.74°C. Both AgCFSR and AgMERRA presented similar Bias and MAE, as well as similar c index, respectively, of 0.84 and 0.86 (Table 3).

For global solar radiation (Q_g), NASA/POWER and XAVIER presented the best performance, with the latter presenting the highest accuracy ($d = 0.97$) and precision ($r^2 = 0.94$), resulting in a c index of 0.94, classified as great [36]. NASA/POWER showed $r^2 = 0.76$ and $d = 0.93$. All databases underestimated Q_g, with Bias ranging from −0.58 to −1.32 MJ·m^{-2}·d^{-1}. In terms of MAE, XAVIER was the database with the best performance, with MAE = 1.57 MJ·m^{-2}·d^{-1}.

For the rainfall (Rain), AgCFSR, AgMERRA, and NASA/POWER showed poor performance with $r^2 \leq 0.25$, $d \leq 0.67$, $c \leq 0.33$, and MAE ≥ 4.48 mm·d^{-1}. On the contrary, XAVIER presented good precision ($r^2 = 0.88$) and high accuracy ($d = 0.96$), resulting in an optimum performance ($c = 0.90$), with a slight underestimation tendency (Bias = −0.10 mm·d^{-1}) and the lowest error magnitude (MAE = 1.51 mm·d^{-1}).

XAVIER also presented the best performance for estimating relative humidity (RH), with high precision

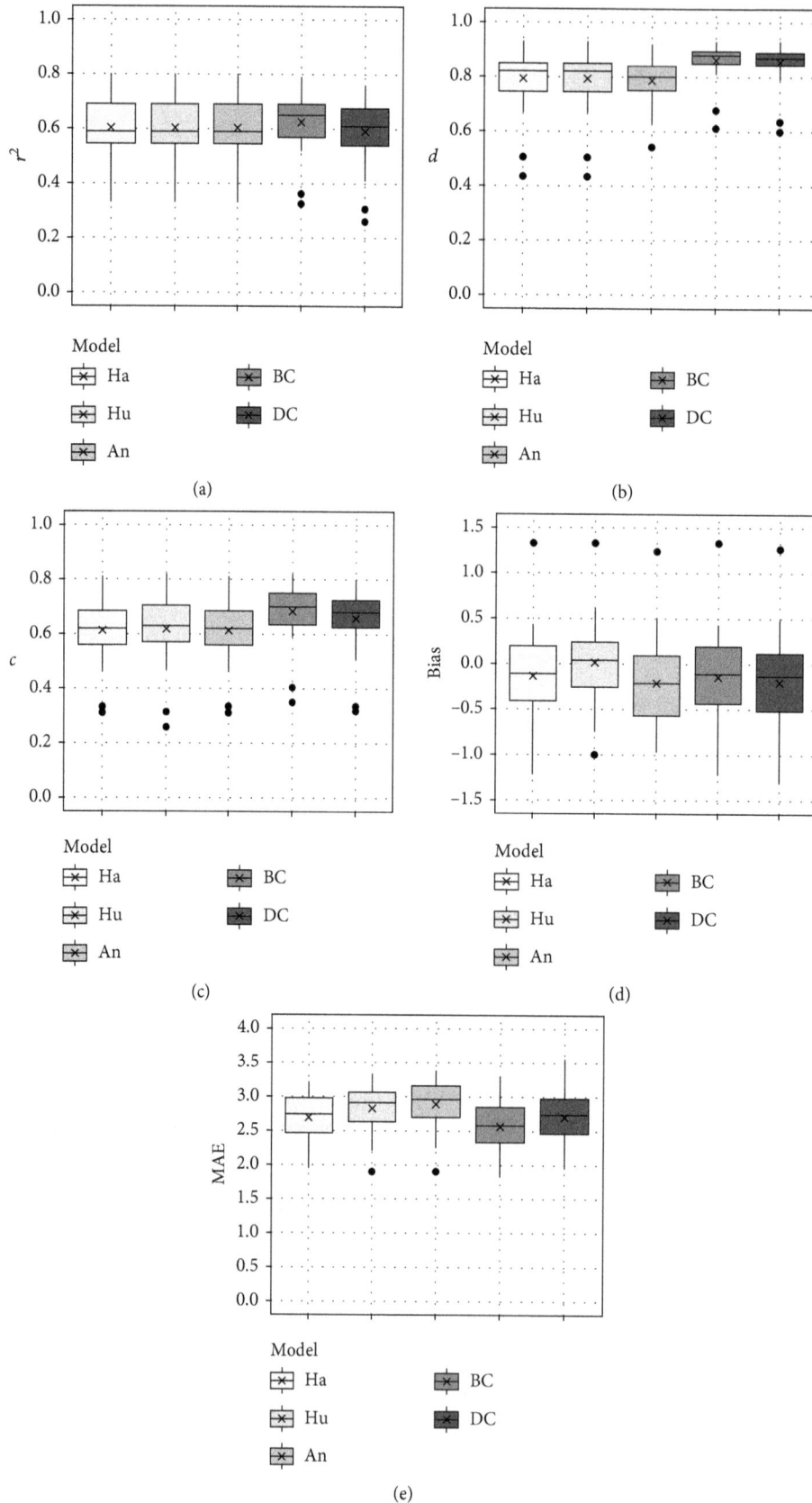

FIGURE 3: Boxplot of the statistical indices and errors of Hargreaves (Ha), Hunt (Hu), Annandale (An), Bristow–Campbell (BC), and Donatelli–Campbell (DC) temperature-based solar radiation models, when compared to measured solar radiation data of 31 Brazilian sites. Boxes denote the lower (25%) to upper quartile (75%) values, with a horizontal line at the median and crosses at mean values.

TABLE 3: Statistical evaluation of daily gridded databases for maximum air temperature (T_{max}), minimum air temperature (T_{min}), solar radiation (Q_g), rainfall (Rain), relative humidity (RH), and wind speed (WS_{2m}), considering 31 locations in Brazil.

Variable	Database	Indexes					
		r^2	d	c	Classification	Bias (°C)	MAE (°C)
T_{max}	AgCFSR	0.77	0.93	0.82	Very good	0.47	1.88
	AgMERRA	0.76	0.93	0.81	Very good	−0.37	1.86
	NASA/POWER	0.68	0.89	0.73	Good	−1.18	2.46
	XAVIER	**1.00**	**1.00**	**1.00**	Great	**−0.03**	**0.17**
		r^2	d	c	Classification	Bias (°C)	MAE (°C)
T_{min}	AgCFSR	0.80	0.94	0.84	Very good	0.71	1.63
	AgMERRA	0.82	0.95	0.86	Great	0.08	1.45
	NASA/POWER	0.77	0.93	0.82	Very good	0.76	1.74
	XAVIER	**0.98**	**0.99**	**0.98**	Great	**0.06**	**0.30**
		r^2	d	c	Classification	Bias (MJ·m^{-2}·d^{-1})	MAE (MJ·m^{-2}·d^{-1})
Q_g	AgCFSR	0.64	0.89	0.71	Good	−0.66	2.65
	AgMERRA	0.64	0.88	0.71	Good	−1.32	2.79
	NASA/POWER	0.76	0.93	0.81	Very good	**−0.58**	2.19
	XAVIER	**0.94**	**0.97**	**0.94**	Great	−1.19	**1.57**
		r^2	d	c	Classification	Bias (mm·d^{-1})	MAE (mm·d^{-1})
Rain	AgCFSR	0.16	0.60	0.24	Terrible	−0.03	4.88
	AgMERRA	0.25	0.67	0.33	Terrible	−0.03	4.48
	NASA/POWER	0.20	0.65	0.29	Terrible	0.12	4.86
	XAVIER	**0.88**	**0.96**	**0.90**	Great	**−0.10**	**1.51**
		r^2	d	c	Classification	Bias (%)	MAE (%)
RH	AgCFSR	0.52	0.79	0.57	Suffering	−2.43	14.43
	AgMERRA	0.38	0.72	0.45	Bad	−6.67	17.06
	NASA/POWER	0.55	0.82	0.61	Median	−6.75	11.14
	XAVIER	**0.90**	**0.97**	**0.92**	Great	**0.18**	**3.76**
		r^2	d	c	Classification	Bias (m·s^{-1})	MAE (m·s^{-1})
WS_{2m}	AgCFSR	0.10	0.56	0.18	Terrible	0.17	0.83
	AgMERRA	0.08	0.53	0.15	Terrible	0.54	0.98
	NASA/POWER	0.14	0.69	0.21	Terrible	0.64	0.86
	XAVIER	**0.47**	**0.79**	**0.54**	Suffering	**0.22**	**0.49**

($r^2 = 0.90$) and accuracy ($d = 0.97$) and small errors (Bias = 0.18% and MAE = 3.76%), whereas the other systems underestimated RH, with MAE higher than 11%.

Despite the poor performance of all databases for estimating wind speed (WS_{2m}), XAVIER displayed the best statistical indices, with $r^2 = 0.47$, $d = 0.79$, and $c = 0.54$, and the smallest error, with MAE = 0.49 m·s^{-1}, which, however, is still classified as suffering according to Camargo and Sentelhas [36].

3.2. Climate Change Projections. Based on the historical measured weather data fulfilled with the XAVIER gridded database, the ensemble of climate change projections was performed for RCP4.5 and RCP8.5 emission scenarios on 31 sites from 1980 to 2009, from mid- to end-of-century periods. Annual maximum and minimum temperatures showed an increase in tendency, while for rainfall, the South region will mostly experience increases (annually), and the North and Northeast regions will experience decreases, as presented in Figures 4–6. More details can be found in Tables S4 and S5 of Supplementary Materials.

Annual average changes, for all 31 sites, of maximum temperature showed increases in medium- and long-term projections of 2.01 and 2.52°C for RCP4.5 and 2.70 and

4.61°C for RCP8.5, while for minimum temperature, the increases will be of 1.79 and 2.25°C for RCP4.5 and 2.56 and 4.45°C for RCP8.5 (Table 4). Under the same emission scenarios and future projected periods, higher increases will occur for maximum than for minimum temperatures. As expected, increases under the RCP8.5 scenario will be higher than those under RCP4.5. However, such increases are much more pronounced in the long-term projections, with the mean increase achieved between 2.39 and 4.48°C, under intermediate and high emission scenarios.

Rainfall projections for the 31 sites showed a decrease of −6.18 and −6.68% for RCP4.5 and −4.34 and −8.62% for RCP8.5 for the medium- and long-term projections (Table 4); however, these changes must be analyzed carefully, since rainfall is a variable of high spatial variability and with distinct distribution patterns over the country.

The monthly climate changes projected for all 31 sites for the RCP8.5 scenario in a long term (2070–2099) are presented in Figure 7. Temperature changes will vary between 2 and 7°C for T_{max} (Figure 7(a)) and between 2 and 5.5°C for T_{min} (Figure 7(b)). The highest temperature increases will occur in the second semester of the year, mainly in October, for both. Therefore, as shown before, higher temperatures are expected on future climate projections, with increases

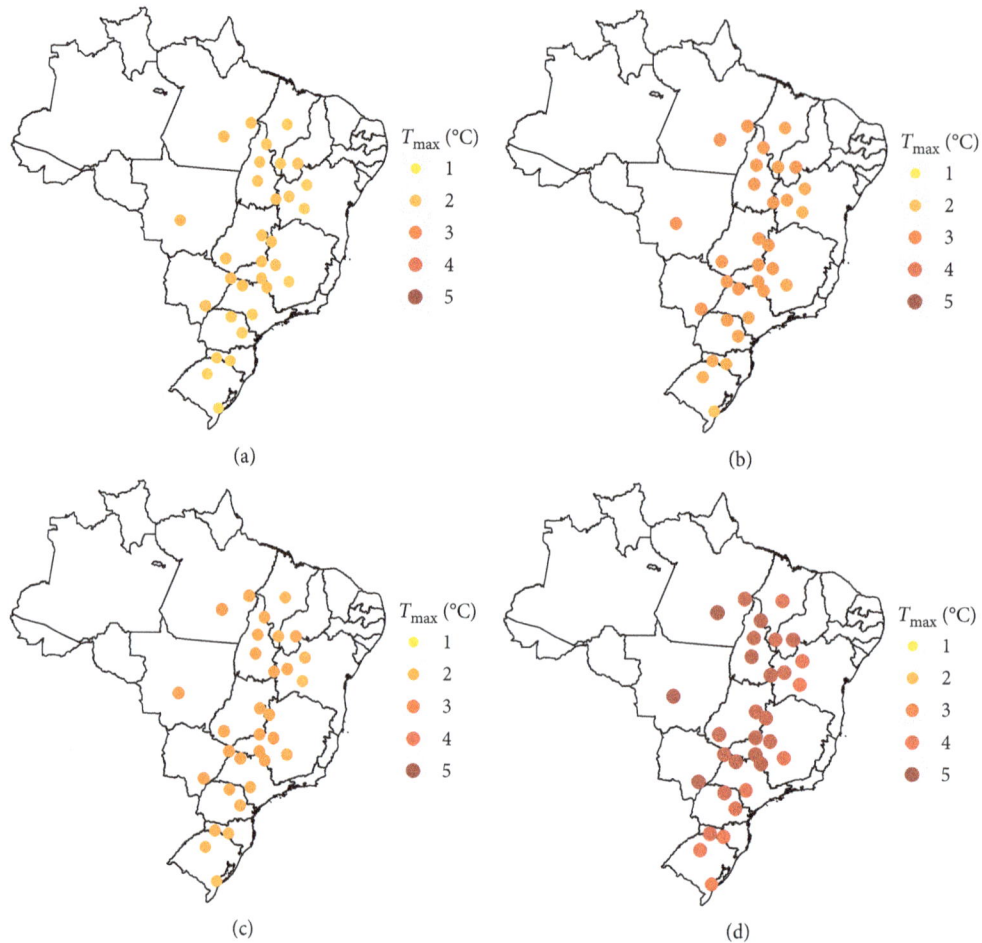

FIGURE 4: Maximum air temperature average change from seven global climate models (GCMs), for 31 sites in Brazil, for mid-of-century (2040–2069) and end-of-century (2070–2099) periods, under intermediate (RCP4.5) and high (RCP8.5) emission scenarios, having as reference the historical (1980–2009) period: (a) RCP4.5 2040–2069; (b) RCP8.5 2040–2069; (c) RCP4.5 2070–2099; (d) RCP8.5 2070–2099.

that will persist every month [13, 22]. Rainfall reduction especially in North and Northeast regions will occur mainly from August to October, which coincides with the dry season and the period of higher temperatures.

Analyzing the future climate projections, by comparing the observed and XAVIER gridded database as a reference for climatology, the projected annual average of maximum and minimum temperature and rainfall was similar, with about the same variability for both databases (Figure 8). For air temperature projections, based on the observed and gridded climatology, the differences were not greater than 0.06 and 0.08°C, respectively, for maximum and minimum temperatures, in both emission scenarios and future periods considered. Similarly, for rainfall, the differences between the two databases did not exceed 1%, considering all scenarios and periods.

4. Discussion

4.1. Filling Gaps in Measured Weather Data

4.1.1. Solar Radiation Models. In general, the temperature-based models for estimating Q_g presented very similar performance after their calibration for 31 sites in Brazil

(Figure 3). However, the models which were based on three coefficients, BC and DC, had a subtle better performance, improving the general confidence index c above 0.6 for most simulations. As this is the first attempt to calibrate these models considering several locations around the country, the calibrated coefficients (a for Ha; b and c for Hu; d for An; e, f, and g for BC; and h, i, and j for DC) were quite different from those obtained by other authors for specific locations or locations within the same state, such as those presented by Barbosa et al. [51] for the state of Minas Gerais (MG), by Conceição and Marin [52] in the northwest of the state of São Paulo, and by Massignam [53] in the state of Santa Catarina. Also, the performances of these models when considering several locations spread in the country were a bit worse than those reported by specific locations [51–53], which is mainly caused by the greater Q_g variability observed around the country with the different atmospheric transmissivity caused by diverse cloud types.

Despite the differences in performance reported above, the present study confirmed that BC and DC are the best temperature-based methods for estimating Q_g. The performance of these methods, however, can vary according to the region and the season of the year, as reported by Rivington

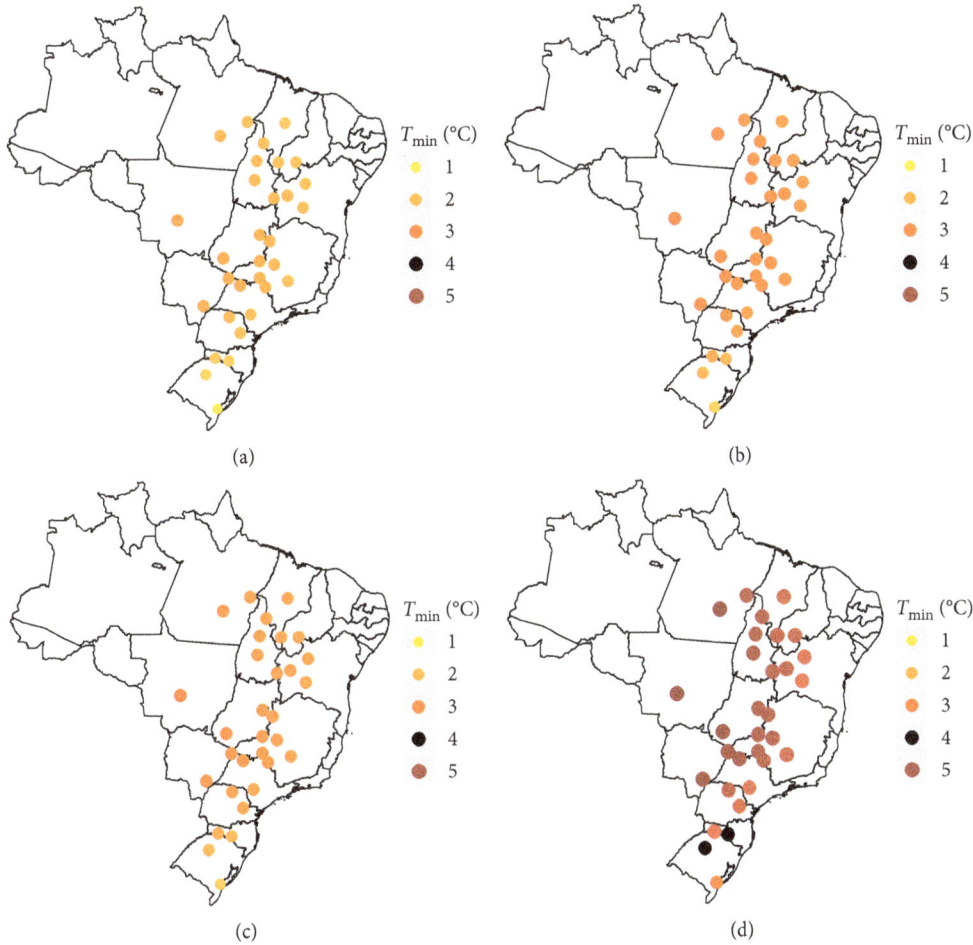

FIGURE 5: Minimum air temperature average change from seven global climate models (GCMs), for 31 sites in Brazil, for mid-of-century (2040–2069) and end-of-century (2070–2099) periods, under intermediate (RCP4.5) and high (RCP8.5) emission scenarios, having as reference the historical (1980–2009) period: (a) RCP4.5 2040–2069; (b) RCP8.5 2040–2069; (c) RCP4.5 2070–2099; (d) RCP8.5 2070–2099.

et al. [54]. In this study, it was found that the best Q_g estimates were found in Southern and Southeastern Brazil, where it seems to be a better correlation between nebulosity and daily thermal amplitude. In these regions, the confidence index was classified between good and very good, as can be seen in Tables S2 and S3 of Supplementary Materials.

4.1.2. Gridded Database. The gridded data provided by difference sources presented distinct performances for simulating weather conditions and variability in different parts of Brazil. For T_{max} and T_{min}, as well as for Q_g, the four systems assessed presented good to great performance, according to the classification of Camargo and Sentelhas [36], with $r^2 \geq 0.64$, $d \geq 0.88$, and c index always above 0.71. In general, XAVIER was the system that presented the best performance for these three variables, with c always above 0.90. On the contrary, for Rain, RH, and WS_{2m}, the performances were quite variable, with AgCFSR, AgMERRA, and NASA/POWER presenting the worst estimates, with c equal to or below 0.33, 0.61, and 0.21, respectively, whereas XAVIER presented great performance for Rain ($c = 0.90$) and RH ($c = 0.92$). For WS_{2m}, XAVIER also had a better

performance than the other sources, however, with lower indices when compared to the other weather variables ($r^2 = 0.47$, $d = 0.79$, and $c = 0.54$).

Similar results were found by Monteiro et al. [55] and by Battisti et al. [5] when using NASA/POWER, XAVIER, and AgMERRA gridded databases in several Brazilian locations. Despite the similar performances observed by these authors regarding the gridded data they used, both of them concluded that the differences between observed and gridded data were not enough to lead to significant differences for estimating the potential yield of sugarcane [55] and soybean [5]. However, when simulating the attainable yield, which depends on the rainfall, Monteiro et al. [55] realized that the use of observed data improved the estimates substantially, once NASA/POWER did not represent rainfall spatial and temporal variability very well, as also observed in the present study (Table 3). Following the same strategy, Battisti et al. [5] also observed that the use of rainfall data from AgMERRA did not provide reliable results of the soybean attainable yield, whereas XAVIER data did.

Regarding rainfall data, the major limitation for their spatial interpolation based on satellite data, as done by AgCFSR, AgMERRA, and NASA/POWER, is the low or

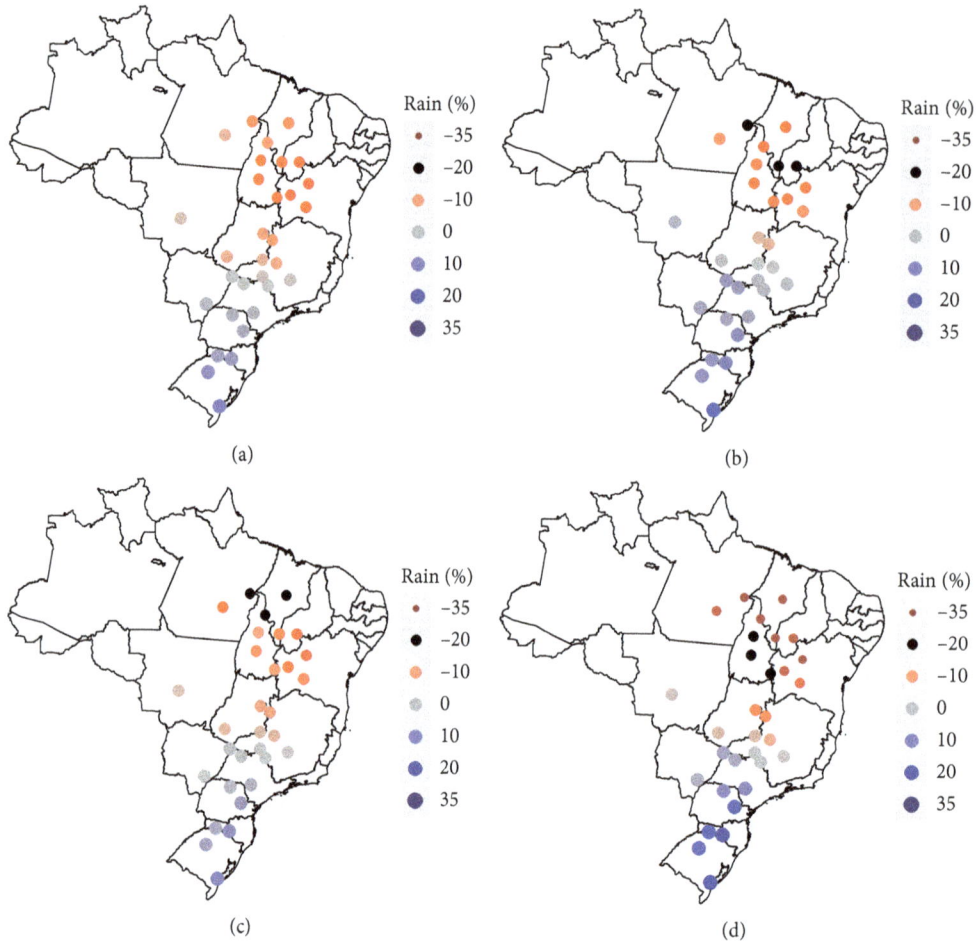

FIGURE 6: Rainfall average change from seven global climate models (GCMs), for 31 sites in Brazil, for mid-of-century (2040–2069) and end-of-century (2070–2099) periods, under intermediate (RCP4.5) and high (RCP8.5) emission scenarios, having as reference the historical (1980–2009) period: (a) RCP4.5 2040–2069; (b) RCP8.5 2040–2069; (c) RCP4.5 2070–2099; (d) RCP8.5 2070–2099.

TABLE 4: Overall changes of maximum, minimum, and mean air temperature and rainfall, averaged from seven global climate models (GCMs) for 31 Brazilian sites for mid-of-century (2040–2069) and end-of-century (2070–2099) periods, under intermediate (RCP4.5) and high (RCP8.5) emission scenarios, when compared to the historical climate conditions (1980–2009).

| Period | RCP4.5 | | | | RCP8.5 | | | |
| | Temperature (°C) | | | Rain (%) | Temperature (°C) | | | Rain (%) |
	Maximum	Minimum	Mean		Maximum	Minimum	Mean	
2040–2069	2.01	1.79	1.90	−6.18	2.70	2.56	2.63	−4.34
2070–2099	2.52	2.25	2.39	−6.68	4.61	4.35	4.48	−8.62

inadequate resolution of the images which is not good enough to capture extreme events [56, 57] and local spatial variability associated with the topography [58, 59]. Similarly, the poor performance of all databases to estimate WS_{2m} is related to two main aspects: the small magnitude of this variable, which leads to large errors even with small deviations, and its high spatial variability associated with the topography and land cover [60]. Finally, the median to bad AgCFSR, AgMERRA, and NASA/POWER performance to estimate RH is related to the fact that the former two provide RH at the time of maximum daily temperature, which is not the daily average, which resulted in MAE between 14 and

17% in the assessed regions. NASA/POWER estimates RH based on similar procedures employed by AgCFSR and AgMERRA, which resulted in errors of similar magnitude, about 11%, very close to those reported by Stackhouse et al. [34] for several locations in the United States for a historical weather series of 31 years.

From the results presented in Table 3, the XAVIER gridded database was the best one to represent spatial and temporal weather data variability in Brazil, once it is based on data from ground stations from several sources. In addition, its high spatial resolution (0.25°) allows a reasonable characterization of the topography and land cover effects on

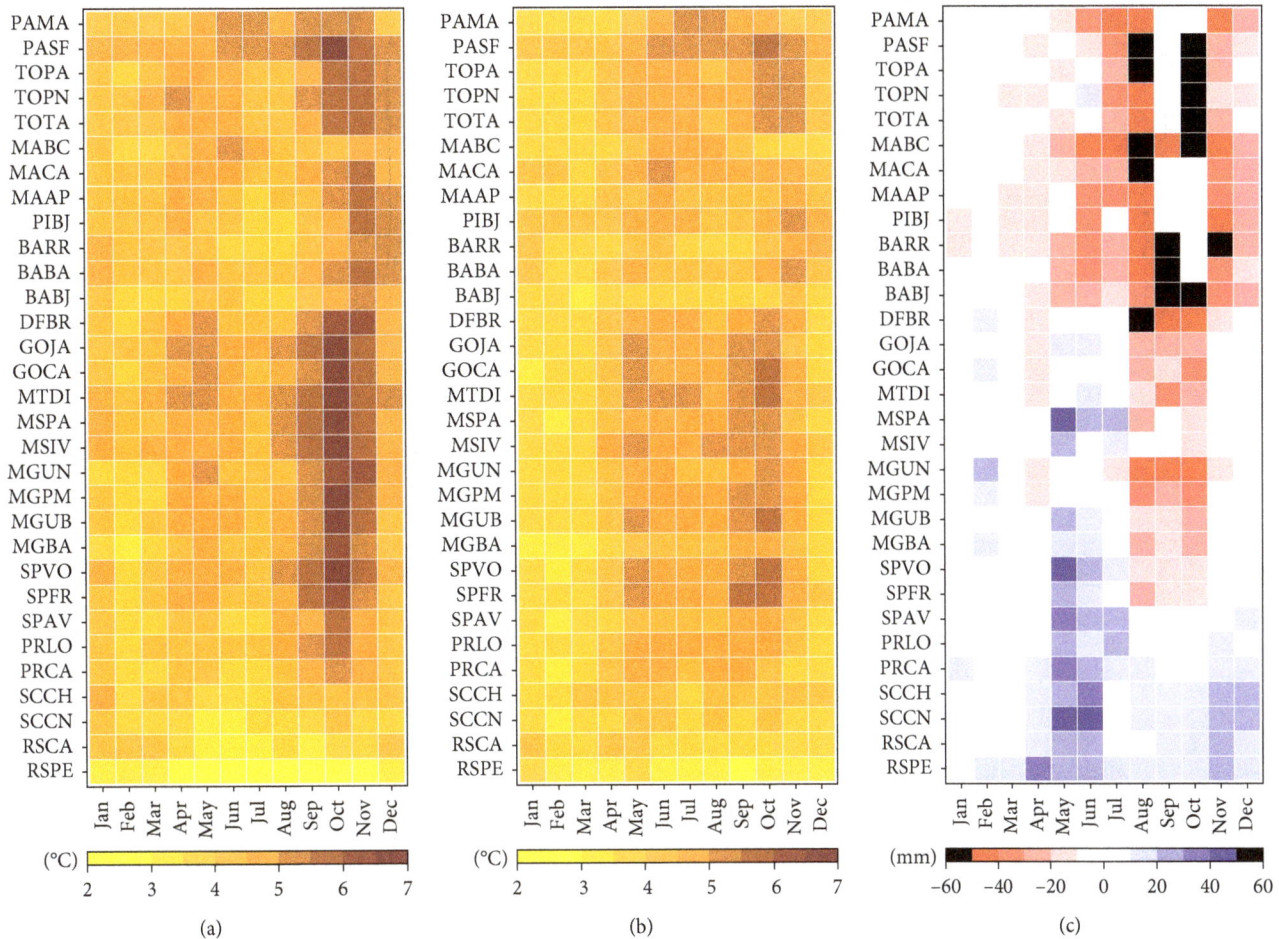

FIGURE 7: Monthly projected changes of maximum air temperature (a), minimum air temperature (b), and rainfall (c), averaged from seven global climate models (GCMs), in 31 Brazilian locations, at the end-of-century (2070–2099) period and under a high emission scenario (RCP8.5), when compared to the historical climate (1980–2009).

surface weather variables, which are difficult to be captured by satellite estimates, as done by AgCFSR, AgMERRA, and NASA/POWER.

4.2. *Climate Change Projections.* The temperature increases presented in this study are in line with the projections performed by Chou et al., Sánchez et al., Torres and Marengo, and Reboita et al. [21, 22, 61, 62]. For air temperature, Torres and Marengo [61] projected increases exceeding 2°C by the end of the present century in South America with more than 90% of probability, which was confirmed by our results (Figures 4 and 5; Table 4). For rainfall, decreases will be expected in the northern part of the country, whereas in the center-southern part, rainfall increase will prevail; these results are comparable to those obtained by Sánchez et al. and PMBC [22, 49]. The rainfall reduction in Northern Brazil will occur mainly from August to October, which coincides with the dry season, and when high temperatures predominate, it leads to higher water deficits, increasing the risks for rainfed perennial crops as well as for annual and perennial irrigated crops by increasing the crop water demand and irrigation requirements [63, 64].

Comparing the future climate projections generated from observed and XAVIER gridded databases, considered as the historical basis for future climate projections, the results did not show any substantial difference in the projected scenarios of temperature and rainfall, which makes possible to use the XAVIER database for studying the impacts of climate change on agriculture or any other human activity.

5. Conclusions

This study assessed the potential use of temperature-based solar radiation models and gridded databases as options to fill gaps in weather series and to project climate change scenarios in Brazil. Among the temperature-based solar radiation models, the one with the best performance was the BC model, which presented the lowest errors and highest precision and accuracy. In relation to the gridded data, the XAVIER database was the best one to represent observed weather series in Brazil, showing up to be reliable for both to fill gaps in and to be used as a reference to agricultural planning and agroclimatic risk studies for the present and future climates. Due to its outstanding performance, the

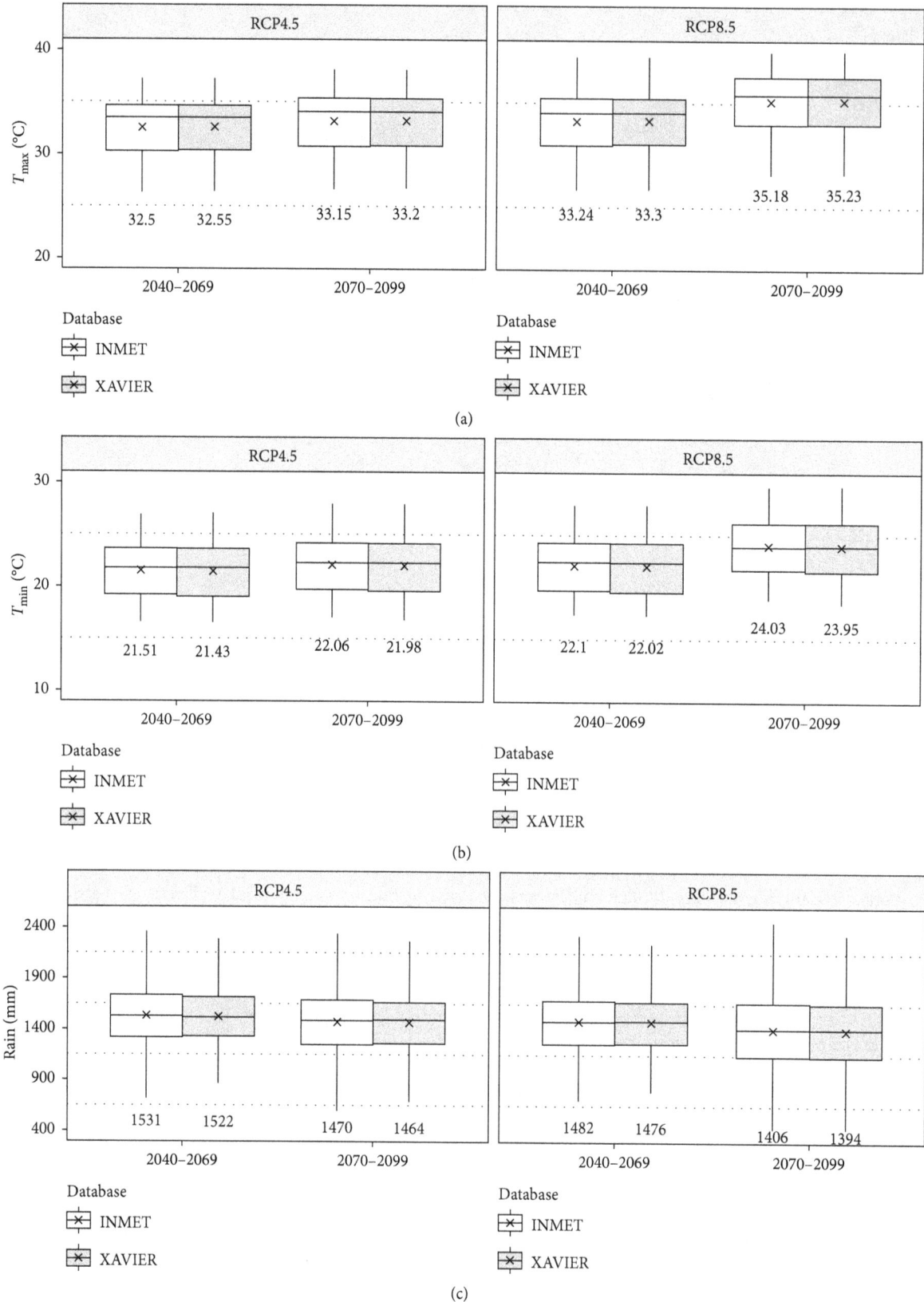

FIGURE 8: Boxplot of the projected annual average of maximum air temperature (a), minimum air temperature (b), and rainfall (c), based on seven global climate models (GCMs), for mid-of-century (2040–2069) and end-of-century (2070–2099) periods, under intermediate (RCP4.5) and high (RCP8.5) emission scenarios, based on the INMET and XAVIER historical database (1980–2009). Boxes denote the lower (25%) to upper (75%) values, with a horizontal line at the median and crosses at the mean values.

XAVIER database can also be used for studies related to the impact of climate variability and climate change on other human activities in Brazil.

Conflicts of Interest

The authors declare that there are no conflicts of interest regarding the publication of this article.

Acknowledgments

The authors would like to thank the Brazilian National Council for Scientific and Technological Development (CNPq) for the funds to support this project (Ph.D. scholarship for Fabiani Denise Bender and research fellowship for Paulo Cesar Sentelhas).

Supplementary Materials

The supplementary material gives additional information for this paper, with a detailed description of weather stations considered in this study, and the percentage of missing data for each weather variable (Table S1); a detailed statistical performance for Hargreaves (Ha), Hunt (Hu), Annandale (An), Bristow–Campbell (BC), and Donatelli–Campbell (DC) temperature-based models employed for estimating solar radiation (Tables S2 and S3); and the annual changes of maximum, minimum, and mean air temperature and rainfall projected by seven global climate models (GCMs), for mid-of-century (2040–2069) and end-of-century (2079–2099) periods, under intermediate (RCP4.5) and high (RCP8.5) emission scenarios, when compared to the historical climate (1980–2009), for 31 Brazilian sites (Tables S4 and S5). (*Supplementary Materials*)

References

[1] S. Asseng, Y. Zhu, E. Wang, and W. Zhang, "Crop modeling for climate change impact and adaptation," in *Crop Physiology: Applications for Genetic Improvement and Agronomy*, V. O. Sadras and D. F. Calderini, Eds., pp. 505–546, Academic Press, San Diego, CA, USA, 2015.

[2] J. van Wart, P. Grassini, and K. G. Cassman, "Impact of derived global weather data on simulated crop yields," *Global Change Biology*, vol. 19, no. 12, pp. 3822–3834, 2013.

[3] A. Heinemann, M. Dingkuhn, D. Luquet, J. C. Combres, and S. Chapman, "Characterization of drought stress environments for upland rice and maize in central Brazil," *Euphytica*, vol. 162, no. 3, pp. 395–410, 2008.

[4] A. C. Xavier, C. W. King, and B. R. Scanlon, "Daily gridded meteorological variables in Brazil (1980-2013)," *International Journal of Climatology*, vol. 36, no. 6, pp. 2644–2659, 2015.

[5] R. Battisti, F. D. Bender, and P. C. Sentelhas, "Assessment of different gridded weather data for soybean yield simulations in Brazil," *Theoretical and Applied Climatology*, vol. 83, no. 1, pp. 1–11, 2018.

[6] C. W. Richardson and D. A. Wright, *WGEN: A Model for Generating Daily Weather Variables*, Vol. ARS-8, U. S. Department of Agriculture, Agricultural Research Service, Washington, DC, USA, 1984.

[7] S. Geng, F. W. T. Penning de Vries, and I. Supit, "A simple method for generating daily rainfall data," *Agricultural and Forest Meteorology*, vol. 36, no. 4, pp. 363–376, 1986.

[8] A. Ängström, "Solar and terrestrial radiation. Report to the international commission for solar research on actinometric investigations of solar and atmospheric radiation," *Quarterly Journal of the Royal Meteorological Society*, vol. 50, no. 210, pp. 121–126, 1964.

[9] G. H. Hargreaves, "Responding to tropical climates," in *The 1980-81 Food and Climate Review, the Food and Climate Forum*, L. E. Slater, Ed., pp. 29–32, Aspen Institute for Humanistic Studies, Boulder, CO, USA, 1981.

[10] K. L. Bristow and G. S. Campbell, "On the relationship between incoming solar radiation and daily maximum and minimum temperature," *Agricultural and Forest Meteorology*, vol. 31, no. 2, pp. 159–166, 1984.

[11] A. C. Ruane, R. Goldberg, and J. Chryssanthacopoulos, "Climate forcing datasets for agricultural modeling: merged products for gap-filling and historical climate series estimation," *Agricultural and Forest Meteorology*, vol. 200, pp. 233–248, 2015.

[12] J. A. Marengo, C. A. Nobre, E. Salati, and T. Ambrizzi, *Caracterização do Clima de Referência e Definição das Alterações Climáticas Para o Território Brasileiro ao Longo do Século XXI*, Ministério do Meio Ambiente. Secretaria de Biodiversidade e Florestas, Ministério do Meio Ambiente (MMA), Brasília, Brazil, 2007.

[13] J. A. Marengo, T. Ambrizzi, R. P. da Rocha et al., "Future change of climate in South America in the late twenty-first century: intercomparison of scenarios from three regional climate models," *Climate Dynamics*, vol. 35, no. 6, pp. 1073–1097, 2010.

[14] C. Tebaldi and R. Knutti, "The use of the multi-model ensemble in probabilistic climate projections," *Philosophical Transactions of the Royal Society A: Mathematical, Physical and Engineering Sciences*, vol. 365, no. 1857, pp. 2053–2075, 2007.

[15] G. Sampaio and P. L. da Silva Dias, "Evolução dos modelos climáticos e de previsão de tempo e clima," *Revista USP*, vol. 103, pp. 41–54, 2014.

[16] K. E. Taylor, R. J. Stouffer, and G. A. Meehl, "An overview of CMIP5 and the experiment design," *Bulletin of the American Meteorological Society*, vol. 93, no. 4, pp. 485–498, 2012.

[17] M. Meinshausen, S. J. Smith, K. Calvin et al., "The RCP greenhouse gas concentrations and their extensions from 1765 to 2300," *Climatic Change*, vol. 109, no. 1-2, pp. 213–241, 2011.

[18] R. H. Moss, J. A. Edmonds, K. A. Hibbard et al., "The next generation of scenarios for climate change research and assessment," *Nature*, vol. 463, no. 7282, pp. 747–756, 2010.

[19] D. P. van Vuuren, J. Edmonds, M. Kainuma et al., "The representative concentration pathways: an overview," *Climatic Change*, vol. 109, no. 1-2, pp. 5–31, 2011.

[20] Intergovernmental Panel on Climate Change (IPCC), *Climate Change 2013: The Physical Science Basis, Contribution of Working Group I to the Fifth Assessment Report of the Intergovernmental Panel on Climate Change*, IPCC, Geneva, Switzerland, 2013.

[21] S. C. Chou, A. Lyra, C. Mourão et al., "Evaluation of the eta simulations nested in three global climate models," *American Journal of Climate Change*, vol. 3, no. 5, pp. 438–454, 2014.

[22] E. Sánchez, S. Solman, A. R. C. Remedio et al., "Regional climate modelling in CLARIS-LPB: a concerted approach towards twentyfirst century projections of regional temperature and precipitation over South America," *Climate Dynamics*, vol. 45, no. 7-8, pp. 2193–2212, 2015.

[23] M. F. Salviano, J. D. Groppo, and G. Q. Pellegrino, "Análise de tendências em dados de precipitação e temperatura no Brasil," *Revista Brasileira de Meteorologia*, vol. 31, no. 1, pp. 64–73, 2016.

[24] J. A. Prescott, "Evaporation from water surface in relation to solar radiation," *Transactions of the Royal Society of South Australia*, vol. 64, pp. 114–118, 1940.

[25] J. Glover and J. S. G. McCulloch, "The empirical relation between solar radiation and hours of bright sunshine in the high-altitude tropics," *Quarterly Journal of the Royal Meteorological Society*, vol. 84, no. 359, pp. 56–60, 1958.

[26] L. A. Hunt, L. Kuchar, and C. J. Swanton, "Estimation of solar radiation for use in crop modelling," *Agricultural and Forest Meteorology*, vol. 91, no. 3-4, pp. 293–300, 1998.

[27] J. G. Annandale, N. Z. Jovanic, N. Benade, and R. G. Allen, "Software for missing data error analysis of Penman-Monteith reference evapotranspiration," *Irrigation Science*, vol. 21, no. 2, pp. 57–67, 2002.

[28] M. Donatelli and G. S. Campbell, "A simple model to estimate global solar radiation," in *Proceedings of the 5th Congress of the European Society for Agronomy*, Nitra, Slovakia, June-July 1998.

[29] R. G. Allen, R. W. Hill, and V. Srikanth, "Evapotranspiration parameters for variably-sized wetlands," *American Society of Agricultural Engineers*, vol. 22, no. 6, pp. 725–735, 1994.

[30] R. G. Allen, L. S. Pereira, D. Raes, and M. Smith, "Crop evapotranspiration: guidelines for computing crop water requirements," in *FAO Irrigation and Drainage Paper 56*, FAO, Rome, Italy, 1998.

[31] C. Rosenzweig, J. W. Jones, J. L. Hatfield et al., "The Agricultural Model Intercomparison and Improvement Project (AgMIP): protocols and pilot studies," *Agricultural and Forest Meteorology*, vol. 170, pp. 166–182, 2013.

[32] S. Saha, S. Moorthi, H. Pan et al., "The NCEP climate forecast system reanalysis," *Bulletin of the American Meteorological Society*, vol. 91, no. 8, pp. 1015–1057, 2010.

[33] M. M. Rienecker, M. J. Suarez, R. Gelaro et al., "MERRA: NASA's Modern-Era Retrospective Analysis for Research and Applications," *Journal of Climate*, vol. 24, no. 14, pp. 3624–3648, 2011.

[34] P. W. Stackhouse, D. Westberg, J. M. Hoell, W. S. Chandler, and T. Zhang, *Prediction of Worldwide Energy Resource (POWER)—Agroclimatology Methodology—(1.0° Latitude by 1.0° Longitude Spatial Resolution), v 1.0.2*, NASA, Washington, DC, USA, 2015.

[35] C. J. Willmott, S. G. Ackleson, R. E. Davis et al., "Statistics for the evaluation and comparison of models," *Journal of Geophysical Research*, vol. 90, no. C5, pp. 8995–9005, 1985.

[36] A. P. Camargo and P. C. Sentelhas, "Avaliação do desempenho de diferentes métodos de estimativa da evapotranspiração potencial no estado de São Paulo, Brasil," *Revista Brasileira de Agrometeorologia*, vol. 5, no. 1, pp. 89–97, 1997.

[37] C. J. Willmott and K. Matsuura, "Advantages of the mean absolute error (MAE) over the root mean square error (RMSE) in assessing average model performance," *Climate Research*, vol. 30, no. 1, pp. 79–82, 2005.

[38] J. D. Ward, A. D. Werner, W. P. Nel, and S. Beecham, "The influence of constrained fossil fuel emissions scenarios on climate and water resource projections," *Hydrology and Earth System Sciences*, vol. 15, no. 6, pp. 1879–1893, 2011.

[39] R. L. Wilby, S. P. Charles, E. Zorita et al., "Guidelines for use of climate scenarios developed from statistical downscaling methods," in *Proceedings of the Supporting Material of the Intergovernmental Panel on Climate Change (IPCC)Task Group on Data and Scenario Support for Impacts and Climate Analysis*, Leipzig, Germany, October 2004.

[40] N. I. Hudson and A. C. Ruane, "Appendix 2. Guide for running AgMIP climate scenario generation tools with R in Windows, Version 2.3," in *Handbook of Climate Change and Agroecosystems: The Agricultural Model Intercomparison and Improvement Project (AgMIP) Integrated Crop and Economic Assessments, Part 1*, C. Rosenzweig and D. Hillel, Eds., ICP Series on Climate Change Impacts, Adaptation, and Mitigation, vol. 3, pp. 387–440, Imperial College Press, London, UK, 2015.

[41] A. Voldoire, E. Sanchez-Gomes, D. Salas y Mélia et al., "The CNRM-CM5.1 global climate model: description and basic evaluation," *Climate Dynamics*, vol. 40, no. 9-10, pp. 2091–2121, 2013.

[42] L. D. Rotstayn, S. J. Jeffrey, M. A. Collier et al., "Aerosol- and greenhouse gas-induced changes in summer rainfall and circulation in the Australasian region: a study using single-forcing climate simulations," *Atmospheric Chemistry and Physics*, vol. 12, no. 14, pp. 6377–6404, 2012.

[43] G. A. Schmidt, M. Kelley, L. Nazarenko et al., "Configuration and assessment of the GISS ModelE2 contributions to the CMIP5 archive," *Journal of Advances in Modeling Earth Systems*, vol. 6, no. 1, pp. 141–184, 2014.

[44] W. J. Collins, N. Bellouin, M. Doutriaux-Boucher et al., "Development and evaluation of an Earth-system model–HadGEM2," *Geoscientific Model Development*, vol. 4, no. 4, pp. 1051–1075, 2011.

[45] G. M. Martin, N. Bellouin, W. J. Collins et al., "The HadGEM2 family of Met Office Unified Model climate configurations," *Geoscientific Model Development*, vol. 4, no. 3, pp. 723–757, 2011.

[46] E. M. Volodin, N. A. Dianskii, and A. V. Gusev, "Simulating present-day climate with the INMCM4.0 coupled model of the atmospheric and oceanic general circulations," *Izvestiya, Atmospheric and Oceanic Physics*, vol. 46, no. 4, pp. 414–431, 2010.

[47] S. Watanabe, T. Hajima, K. Sudo et al., "MIROC-ESM 2010: model description and basic results of CMIP5-20c3m experiments," *Geoscientific Model Development*, vol. 4, no. 4, pp. 845–872, 2011.

[48] M. A. Giorgetta, J. Jungclaus, C. H. Reick et al., "Climate and carbon cycle changes from 1850 to 2100 in MPI-ESM simulations for the coupled model intercomparison project phase 5," *Journal of Advances in Modeling Earth Systems*, vol. 5, no. 3, pp. 572–597, 2013.

[49] Painel Brasileiro de Mudanças Climáticas (PBMC), *Sumário Executivo, Base Científica das Mudanças Climáticas, Contribuição do Grupo de Trabalho I ao Primeiro Relatório de Avaliação Nacional de Mudanças Climáticas*, PBMC, Brazil, 2013.

[50] G. H. Hargreaves and Z. A. Samani, "Estimating potential evapotranspiration," *Journal of Irrigation and Drainage Engineering*, vol. 108, no. 3, pp. 225–230, 1982.

[51] L. A. Barbosa, C. R. Silva, M. T. G. Paula et al., "Estimativa da radiação solar com base na temperatura do ar na região Sul, Sudeste, Oeste de Minas e Campo das Vertentes," in

Proceedings of the XVII Congresso Brasileiro de Agrometeorologia, SBAGRO 2011, pp. 4–8, Guarapari, ES, Brazil, 2011.

[52] M. A. F. Conceição and F. R. Marin, "Avaliação de modelos para a estimativa de valores diários da radiação solar global com base na temperatura do ar," *Revista Brasileira de Agrometeorologia*, vol. 15, no. 1, pp. 103–107, 2007.

[53] A. M. Massignam, "Estimativa da radiação solar em função da amplitude térmica," in *Proceedings of the XV Congresso Brasileiro de Agrometeorologia, CBAGRO 2011*, pp. 1–5, Aracajú, SE, Brazil, 2007.

[54] M. Rivington, G. Bellochi, K. B. Matthews, and K. Buchan, "Evaluation of three model estimations of solar radiation at 24 UK stations," *Agricultural and Forest Meteorology*, vol. 132, no. 3-4, pp. 228–243, 2005.

[55] L. A. Monteiro, P. C. Sentelhas, and G. U. Pedra, "Assessment of NASA/POWER satellite-based weather system for Brazilian conditions and its impact on sugarcane yield simulation," *International Journal of Climatology*, vol. 38, no. 3, pp. 1571–1581, 2017.

[56] R. S. V. Teegavarapu, "Statistical corrections of spatially interpolated missing precipitation data estimates," *Hydrological Processes*, vol. 28, no. 11, pp. 3789–3808, 2014.

[57] B. Liebmann and D. Allured, "Daily precipitation grids for South America," *Bulletin of the American Meteorology Society*, vol. 86, no. 11, pp. 1567–1570, 2005.

[58] H. Ezzine, A. Bouziane, D. Ouazar, and M. D. Hasnaoui, "Downscaling of open coarse precipitation data through spatial and statistical analysis, integrating NDVI, NDWI, ELEVATION, and distance from sea," *Advances in Meteorology*, vol. 2017, Article ID 8124962, 20 pages, 2017.

[59] X. Zheng and J. Zhu, "A methodological approach for spatial downscaling of TRMM precipitation data in North China," *International Journal of Remote Sensing*, vol. 36, no. 1, pp. 144–169, 2015.

[60] L. Yu, S. Zhong, X. Bian, and W. E. Heilman, "Temporal and spatial variability of wind resources in the United States as derived from the climate forecast system reanalysis," *Journal of Climate*, vol. 28, no. 3, pp. 1166–1183, 2015.

[61] R. R. Torres and J. A. Marengo, "Uncertainty assessments of climate change projections over South America," *Theoretical and Applied Climatology*, vol. 112, no. 1-2, pp. 253–272, 2013.

[62] M. S. Reboita, R. P. da Rocha, C. G. Dias, and R. Y. Ynoue, "Climate projections for South America: RegCM3 driven by HadCM3 and ECHAM5," *Advances in Meteorology*, vol. 2014, Article ID 376738, 17 pages, 2014.

[63] N. Mancosu, R. Snyder, G. Kyriakakis, and D. Spano, "Water scarcity and future challenges for food production," *Water*, vol. 7, no. 12, pp. 975–992, 2015.

[64] J. Hatfield, K. J. Boote, B. A. Kimball et al., "Climate impacts on agriculture: implications for crop production," *Agronomy Journal*, vol. 103, no. 2, pp. 351–370, 2011.

6

Predicting Microbursts in the Northeastern U.S. using Lightning Flash Rates and Simple Radar Parameters

Stephen M. Jessup ⓘ **and Amanda L. Burke** ⓘ

Department of the Earth Sciences, The College at Brockport, State University of New York, Brockport, NY, USA

Correspondence should be addressed to Stephen M. Jessup; sjessup@brockport.edu

Academic Editor: Tomeu Rigo

Convective storms that produce microburst winds are difficult to predict because the strong surface winds arise in a short time period. Previous research suggests that timing and patterns in cloud height, echo top height, vertical integrated liquid (VIL), intracloud (IC) lightning, and cloud-to-ground (CG) lightning may identify and predict microbursts. Eleven quasi-cellular microburst cases and eight non-microburst severe wind cases were identified from New York, Pennsylvania, and New Jersey between 2012 and 2016. Total lightning data (IC + CG) were obtained from Vaisala's National Lightning Detection Network (NLDN), and radar parameters were obtained from the Thunderstorm Identification Tracking Analysis and Nowcasting (TITAN) software. Values of VIL, echo top height, and cloud height were tracked through time along with total lightning strikes within a 15 km radius of the storm center. These parameters were plotted with respect to their mean and standard deviation for the 45 minutes leading up to event occurrence. Six of eleven cases featured peaks in total and IC lightning within 25 minutes prior to the microburst. These were the only variables among those examined to peak more than half the time for either the microburst cases or the null cases. The results suggest that microbursts behave somewhat differently than severe wind events, particularly in terms of lightning and VIL timing. The results dispute previous research that suggests that microbursts are highly predictable by the behavior of lightning and radar parameters.

1. Introduction

Microbursts are strong winds exhibiting a divergent damage pattern across an area 4 km wide or less [1]. Although dry microbursts (reflectivity <35 dBZ) are possible in other climates, all observed microbursts in the Northeast U.S. are wet microbursts, produced by rapid changes in thunderstorm environments. They present a serious hazard to life and property. Prediction of microburst winds is difficult owing to the rapid lifecycle associated with many convective storms. Such storms can produce intense downbursts, usually >5 minutes in duration, with limited warning. Previous studies [2–4] have suggested that microbursts might be predicted by changes in radar parameters and/or lightning frequency. However, these studies have focused on only a handful of cases, mostly in the southern U.S. Microbursts have not been extensively studied in the Northeast U.S., such that the validity of these earlier findings remains a question for the Northeast. Furthermore, other studies have not compared microburst-producing storms with non-microburst storms that produced wind damage to determine the likelihood of microburst false alarms.

Thunderstorm downdrafts occur in response of the concentration and/or phase changes of water. This generates negative buoyancy in convective environments. In the presence of steep ambient lapse rates or continued diabatic cooling, the air maintains its negative buoyancy while descending. With large amounts of negative buoyancy, microbursts spreading out at the ground can be produced [5].

Microburst formation is often attributed to hail core collapse, and the cooling of midlevel air caused by the rapid melting of hail and subsequent evaporative cooling is associated with many microbursts because it can rapidly generate large negative buoyancy [5–7]. Thunderstorm collapse occurs when a storm's downdraft overtakes the updraft, causing mixed-phase precipitation particles to fall out of the cloud. The rapidly descending downdraft is capable of creating a microburst, or strong diverging winds at the surface [8]. Not all downdrafts associated with collapsing thunderstorms produce microbursts, and not all thunderstorms that produce microbursts decay rapidly following the microburst. High reflectivities (>50 dBZ) must be above the melting level (−10°C) for the downdraft to be strong enough to create a microburst. Microbursts are unlikely with storms that exhibit high reflectivities below the melting level [3]. Lightning production often increases as the updraft strengthens, particularly between 0°C and −20°C, suggesting that lightning peaks may be important predictors of microburst events [9].

Goodman et al. [2] discovered lightning indicators of a microburst in Alabama that resulted in >15 m·s⁻¹ winds. A peak, followed by a sharp decrease, in the intracloud (IC) lightning flash rate occurred six minutes before thunderstorm collapse and the subsequent microburst event. An abrupt increase in cloud-to-ground (CG) lightning activity occurred five minutes before the microburst, directly after the peak in IC lightning. Williams et al. [3], examining several microbursts also in Alabama, similarly found that a peak in IC lightning preceded a peak in CG lightning, both of which occurred before the microburst. Kane [4] studied one downburst in Massachusetts and found that five-minute CG lightning peaked just a few minutes before the downburst. Metzger and Nuss [10] examined lightning activity associated with wind, hail, and mixed severe reports. They found that wind-type lightning jumps were characterized by increasing CG strike rates and either increasing (12 of 18 cases) or steady or decreasing (6 of 18 cases) IC flash rates.

Other signatures of a vigorous updraft, such as peaks in vertical integrated liquid (VIL), cloud height, and echo top height, may presage microburst development. Strong updrafts allocate low-level moisture into a storm, where the moisture then condenses and freezes when lifted above the melting level, creating graupel and hail. Hail is associated with high values of VIL [11]. VIL is sensitive to reflectivity, such that higher reflectivity values (>40 dBZ, and especially >50 dBZ) are associated with higher VIL [12]. Reflectivity values above 55 dBZ are often contaminated by hail, and VIL is typically capped near this value, assuming that hail is present above this value and not all reflectivity is produced by liquid water.

Echo top height is defined as the height of the radar beam at which a certain reflectivity threshold is exceeded [13]. A reflectivity threshold >18 dBZ is typically considered the minimum threshold for echo top height of a given storm cell [14]. High echo top heights indicate that hail is likely present throughout a storm, especially in the "charging zone" between 0°C and −20°C [9]. Convective storms with strong updrafts cause greater hail production and higher echo top heights. Vigorous updraft signatures are precursors to thunderstorm collapse, possible microburst development, and damaging outflow winds.

Tall cloud heights are also suggestive of a vigorous updraft. In order for a convective cloud to grow, the updraft must ingest and lift moisture. The stronger the updraft, the higher the moisture that can be lifted, resulting in higher cloud heights. The 0 dBZ reflectivity echo can be used as a proxy for cloud height [2].

Peak values of IC lightning flashes, VIL, echo top heights, and cloud heights about six minutes before the outflow imply a strong updraft in a convective storm [2]. The six-minute prediction interval for a strong updraft is reflective of a convective storm's lifecycle. In a convective storm, peaks in updraft strength are rapidly followed by the downdraft becoming dominant [3]. CG lightning forms as particles rapidly descend, collide, and build up charge as a result of a storm's downdraft [15]. A sharp decrease in IC lightning activity and increases in cloud-to-ground (CG) lightning strikes indicate the downdraft is overcoming the updraft.

In this study, we examine quasi-cellular microbursts and a collection of non-microburst quasi-cellular severe wind events in the Northeast U.S. from 2012 to 2016 to determine how well peaks in various radar and lightning parameters perform in predicting microburst occurrence.

2. Data and Methodology

Quasi-cellular microburst cases during the years 2012–2016 were identified through the National Centers for Environmental Information (NCEI) Storm Events Database [16]. Reports of thunderstorm winds >36 m·s⁻¹ (70 kt), with a summary indicating a National Weather Service- (NWS-) confirmed microburst, were recorded. Level II Next Generation Weather Radar (NEXRAD) data were acquired to determine whether the microburst was quasi-cellular. Quasi-cellular storms included isolated cells, small clusters of cells, and cells that later merged with larger convective features. In addition, the quasi-cellular storm center had to be easily tracked to be included in this study.

A second dataset consisting of non-microburst-producing wind damage reports from the NCEI SED was included for comparison with the microburst events. High wind events were chosen rather than ordinary thunderstorms because most days on which quasi-cellular microbursts occur feature widespread wind damage. Thus, a primary forecast challenge on these days is determining whether a given thunderstorm cell will produce "ordinary" wind damage or microburst wind damage. Quasi-cellular events that produced reported wind speeds ≥26 m/s (50 kt) were selected from the microburst days. These events were subject to the same storm tracking and lightning analyses as the microburst events.

The radar data were processed by the Thunderstorm Identification Tracking Analysis and Nowcasting (TITAN) system [17], to track the storm centroid. TITAN also produced estimates of vertical integrated liquid (VIL), cloud top height (0 dBZ), and echo top height (18 dBZ). TITAN was run for reflectivity thresholds every 5 dBZ from 30 dBZ to 50 dBZ and for hail caps of 53 dBZ and 56 dBZ. The results in this paper are presented for the 45 dBZ/56 dBZ runs, which produced the most reliable tracks in some cases where cells were clustered or merged with linear features. Others have preferred to use a slightly lower threshold of 40 dBZ [14, 18, 19], but these studies were primarily interested in heavy precipitation and flash flooding, while the present study is more interested in tracking high reflectivity cores that produced severe weather.

Reflectivity centroids from the TITAN output were ingested into the GR2Analyst software (Gibson Ridge) and compared with the Level II NEXRAD data to identify the complex and simple track numbers associated with each microburst-producing storm. The VIL, echo top height, and cloud height values for each verified storm cell center along the microburst-producing storm track were plotted with time, and trends were distinguished by plotting time-series graphs of each parameter.

Lightning data (IC and CG) were obtained from Vaisala's National Lightning Detection Network [20, 21] for the period from 2012 to 2016 across most of Pennsylvania, New Jersey, and New York (see Figure 1 for a general map of the study area; all but the eastern half of Long Island was included). Detection efficiency (DE) for NLDN cloud-to-ground (CG) lightning was found to be approximately 90–95% from 2002 to 2012 [22], and following an upgrade to the network, it improved to over 95% since 2013 [23]. Prior to 2013, the NLDN had a cloud-to-cloud (CC) lightning, or intracloud (IC) lightning, and DE of 15%–25%. Following the upgrade in 2013, this DE increased to about 50% [23, 24].

Lightning data were paired with the reflectivity centroids to determine lightning flash rates for each microburst-producing storm. Total lightning strikes within 15 km of each reflectivity centroid from TITAN were aggregated and assumed to be related to the storm cell that produced the microburst. One-minute and five-minute flash rates were computed for IC, CG, and total (IC + CG) lightning.

All of the lightning and radar parameters, including the null datasets, were examined using standard Z-scores. The Z-score normalizes the deviations in a dataset from the mean with the standard deviation of the said data. This manipulates a dataset towards zero mean and unit variance, allowing for equal comparisons across data with multiple scales, and indicates how far each point is from the mean. Deviations of lightning data greater than 2σ, or two times the standard deviation, were found to detect severe weather with a high probability of detection (POD), and a low false alarm rate (FAR) [25]. These large increases in lightning activity have been termed "lightning jumps" [25, 26]. For consistency, and to analyze local maxima of values in a dataset relative to the time of a microburst, the radar parameter deviations were also standardized to identify large deviations from the mean.

The mean and standard deviations for all datasets were calculated over data for 45 minutes before a microburst occurred and then employed over the entire 60-minute time period. Deviations from the total, cloud-to-ground, and intracloud datasets were plotted together, as well as plots of VIL, echo top, and cloud top. In addition, 2σ, 1σ, -1σ, and -2σ, as well as the time of the microburst, were highlighted to examine any changes in the datasets. A "peak" in a parameter is considered to be the maximum data value exceeding 2σ at any time during the 60-minute time period examined for each event.

A separate analysis was performed using the "2σ" algorithm from the study of Schultz et al. [25]. Detected large increases in lightning activity have been termed "lightning jumps" [25, 26]. The "2σ" algorithm introduces a minimum lightning flash threshold of 10 flashes per minute, to decrease the number of lightning jump false alarms. The average minute flash rate is then calculated between two time steps (two minutes in our case) where the threshold is exceeded. For each averaged flash rate, the standard deviation is calculated from the previous five averaged lightning observations (or 10 minutes prior), not including the time being investigated. The calculated standard deviation is doubled and becomes the "jump threshold," where the averaged flash rates that exceed the jump threshold is a "lightning jump." This algorithm creates a moving lightning jump threshold, rather than a static threshold over a total time period [25]. We follow the original paper in comparing total lightning using this algorithm, as it performed the best in the original paper.

3. Results

The procedures above yielded 11 quasi-cellular microburst cases to be examined. The 11 microburst cases were spread across Pennsylvania and New York, with local maxima in the Mohawk River Valley and Hudson River Valley of New York and the Valley and Ridge region of Pennsylvania (Table 1 and circle points in Figure 2). It appears that orography may play a role in producing microbursts in the Northeastern U.S. A twelfth case (not shown) was located along the northern shore of central Long Island, but its track ran east of the available lightning data. This case was excluded from the analyses. No microburst cases meeting the case selection criteria were found in New Jersey during the period for which lightning data were available.

The eight null cases do not show a strong preference for complex topography (Table 2 and triangle points in Figure 2), and they are scattered across New York and Pennsylvania. They generally occurred earlier in the day than the microbursts (Tables 1 and 2) and produced weaker winds.

The total lightning data from storm cells associated with microbursts indicated that total lightning peaked within 20 minutes before the microburst in five of eleven cases and within 25 minutes before the microburst in six of eleven cases (Figures 3–5 and Table 3). CG lightning peaked within 25 minutes before the microburst in five of eleven cases, and IC lightning peaked within 25 minutes before the microburst in six of the eleven cases. About one-third of the time, lightning peaked after the microburst. The mean lead time for total and CG lightning peaks was about 8 to 10 minutes, and for IC lightning peaks, it was about 14 minutes.

Using instead the Schultz 2σ algorithm (Figures 6 and 7), four of the eleven cases (20140708NY, 20150623, 20150630, and 20160616) met the lightning jump threshold for total lightning, and three others (20120724, 20140703, and 20150612) equaled or closely approached the threshold. For the null cases (Figure 8), only the 20150612 case saw the time rate of change in flash rate exceeding the lightning jump threshold.

The radar parameters were found to be less significant than the lightning parameters, often just barely exceeding the 2σ threshold used to determine whether a particular value had peaked (Figures 9–11 and Table 3). The radar parameter peaks had shorter mean lead times than the

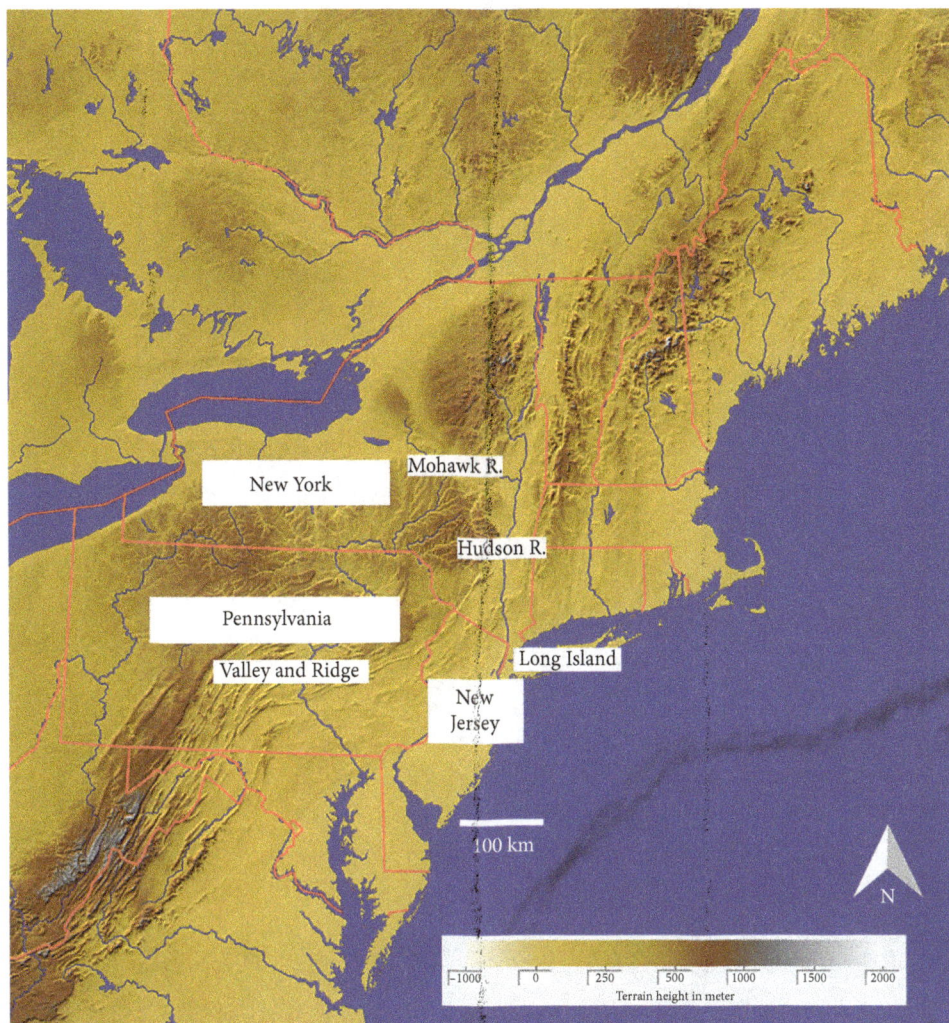

FIGURE 1: General map of the study area, highlighting key locations.

TABLE 1: List of microbursts examined in this study.

Case	Date	Time (UTC)	Latitude (°)	Longitude (°)	Wind speed (m/s)
1	7/24/2012	04:12	43.13	−75.32	49
2	6/24/2013	22:08	42.658	−73.726	45
3	7/3/2014	19:55	42.8842	−75.2009	45
4	7/8/2014NY	23:10	43.31	−75.29	43
5	7/8/2014PA	22:52	40.46	−76.98	43
6	6/12/2015	21:05	41.66	−74.19	45
7	6/23/2015	16:35	40.9068	−77.4522	36
8	6/30/2015	18:35	40.6631	−75.5153	36
9	8/3/2015	23:33	42.9344	−74.6262	43
10	6/16/2016	18:55	39.938	−78.6537	36
11	7/1/2016	20:17	42.802	−73.9144	40

lightning parameters, particularly VIL, with a mean lead time of about four minutes. VIL peaked within 25 minutes of the microburst in four of the eleven cases. Three of those peaks were within six minutes of the microburst. Cloud tops peaked after the microburst in three of the eleven cases and did not peak before the microburst. Echo tops peaked within 25 minutes of the microburst in two of the eleven cases, one with 2-minute lead time and the other with 17-minute lead

time. VIL thus appears to be the most useful predictor of the radar parameters, yet it peaks before the microburst less than half the time.

The mean lightning lead times for the null events were similar to those for the microburst events, but the individual values are either far before the severe wind event or after it occurred (Figures 12 and 13 and Table 4). Total lightning peaked within 25 minutes before the severe wind event only

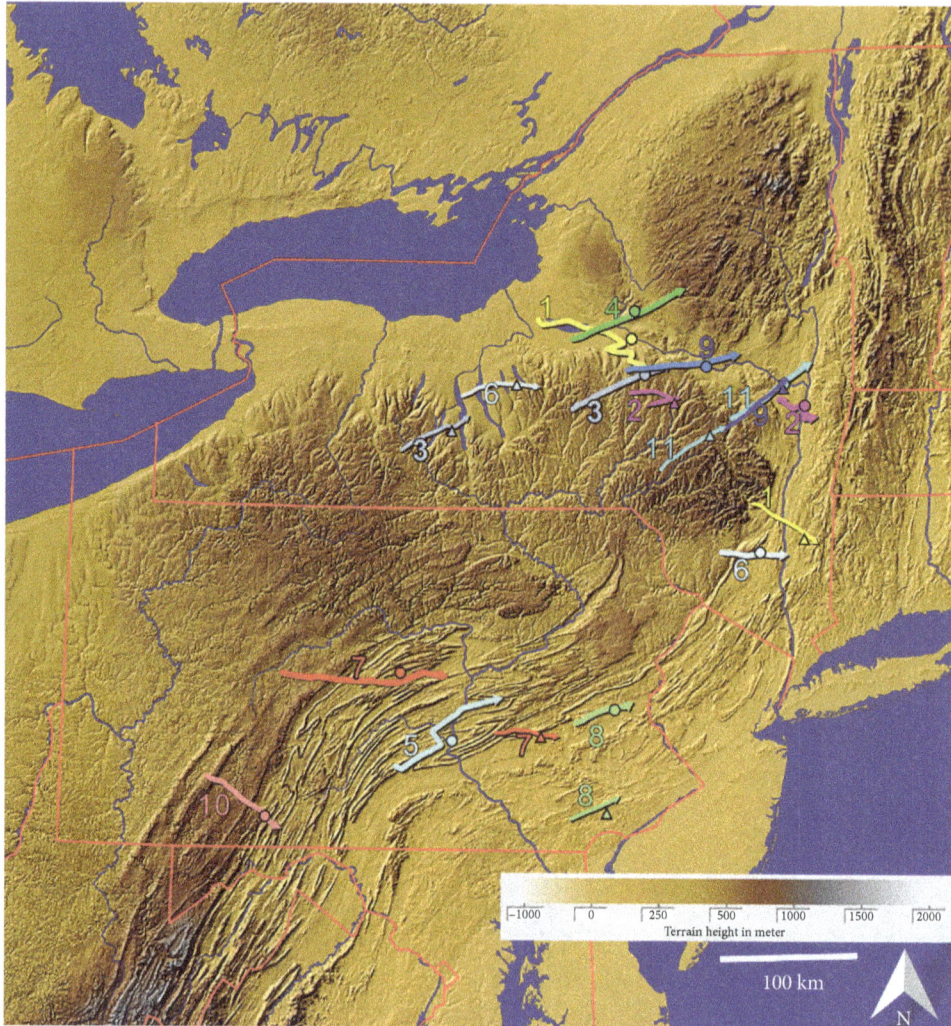

FIGURE 2: Terrain map of the eleven quasi-cellular microburst storm tracks (lines) and locations (circles) and the eight quasi-cellular null event storm tracks (lines) and locations (triangles).

TABLE 2: List of null severe wind events examined in this study.

Case	Date	Time (UTC)	Latitude (°)	Longitude (°)	Wind speed (m/s)
1	7/24/2012	20:29	41.7747	−73.768	31
2	6/24/2013	19:28	42.7	−74.93	26
3	7/3/2014	17:50	42.52	−77.00	26
6	6/12/2015	19:15	42.83	−76.40	33
7	6/23/2015	20:12	40.4917	−76.185	27
8	6/30/2015	19:17	39.979	−75.5917	27
9	8/3/2015	19:00	42.7801	−73.9222	26
11	7/1/2016	18:42	42.4813	−74.6101	26

once, and IC lightning peaked within 25 minutes before the severe wind event in three of eight cases. Neither total nor IC lightning had a single peak within 10 minutes of the null event. CG lightning, on the contrary, showed some predictive capability, with four cases peaking within 25 minutes of the null events; three of these cases peaked within 10 minutes.

For the radar parameters of the null events, the mean lead times were close to the time of the microburst Figures 14 and 15. Except for echo tops, the majority of the 2σ maxima in the radar parameters were after the severe wind event:

three of eight events for cloud tops and six of eight events for VIL "peaked" after the event. For the three cases where the positive lead time was small, this may be indicative of small errors in the reported timing of the events. It could also indicate some microphysical processes amplifying liquid water in the cloud after evacuating a large gust of entrained dry air. For VIL and cloud tops, none of the eight events saw a 2σ maximum within 25 minutes before the event. Echo tops showed the best potential as a predictor, with two of the eight events maximized at greater than 2σ within ten

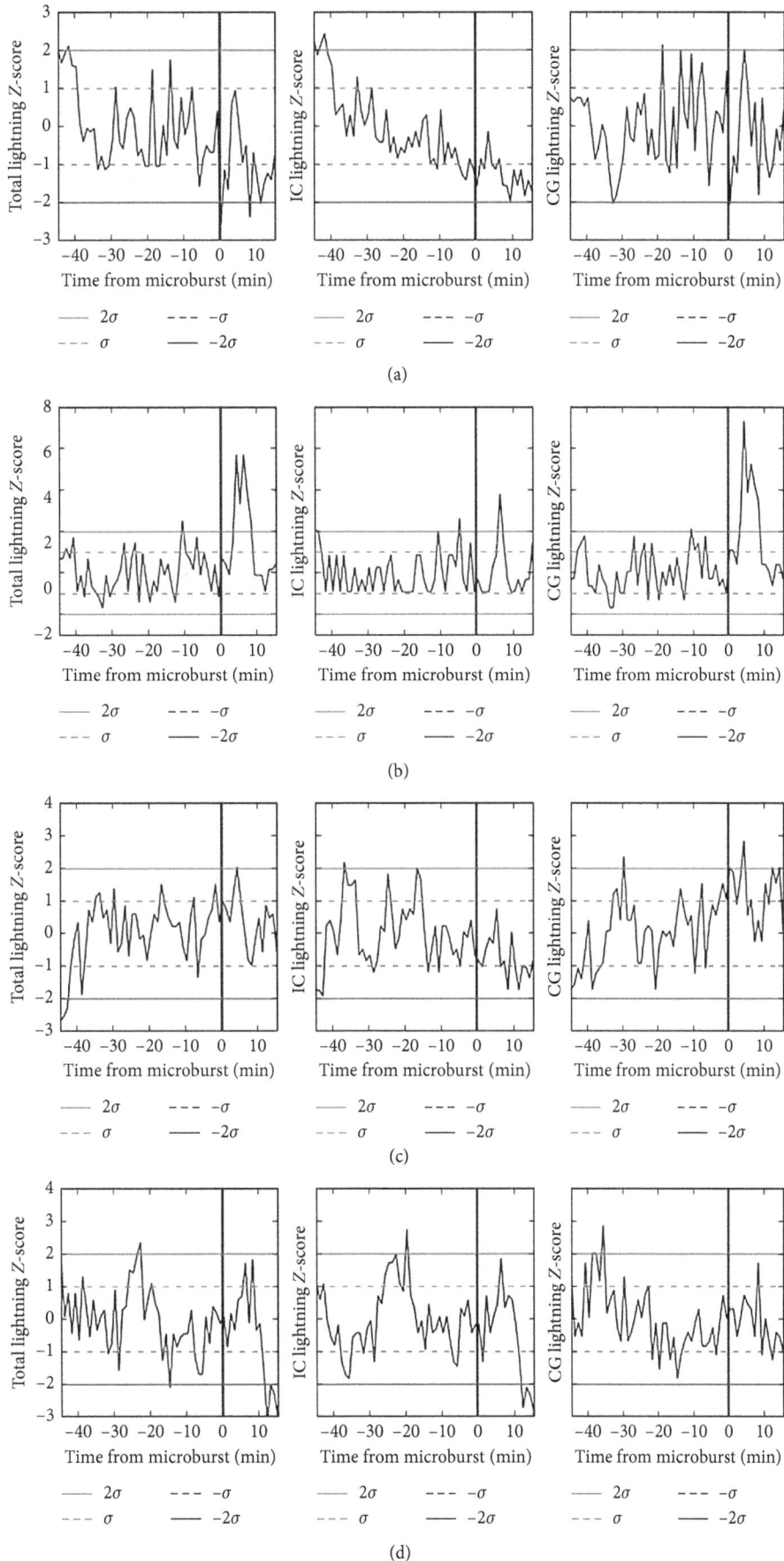

FIGURE 3: Plots of total lightning, IC lightning, and CG lightning for the first four microburst cases. (a) 20120724 microburst. (b) 20130624 microburst. (c) 20140703 microburst. (d) 20140708NY microburst.

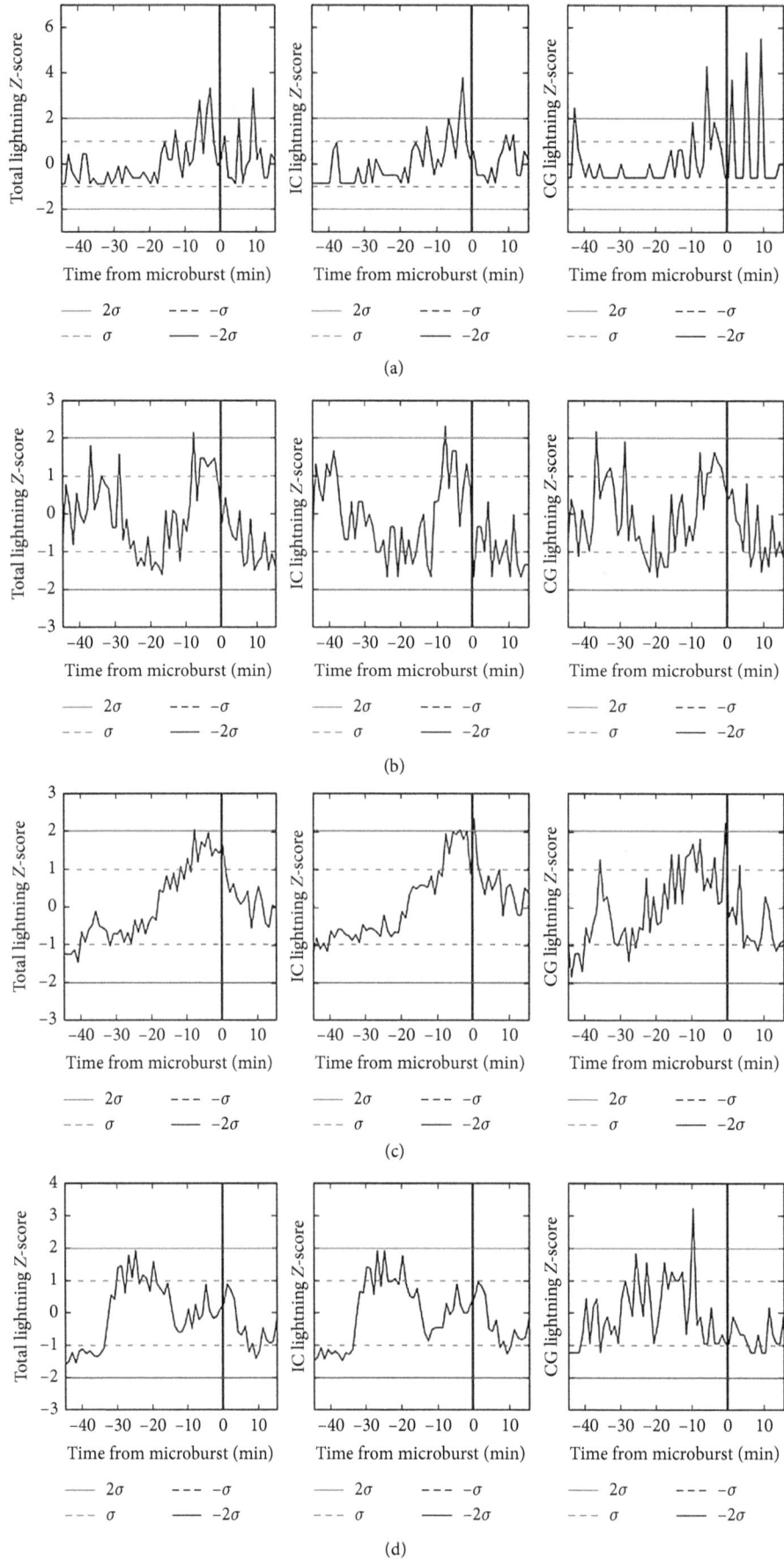

FIGURE 4: As in Figure 3, except for the next four cases. (a) 20140708PA microburst. (b) 20150612 microburst. (c) 20150623 microburst. (d) 20150630 microburst.

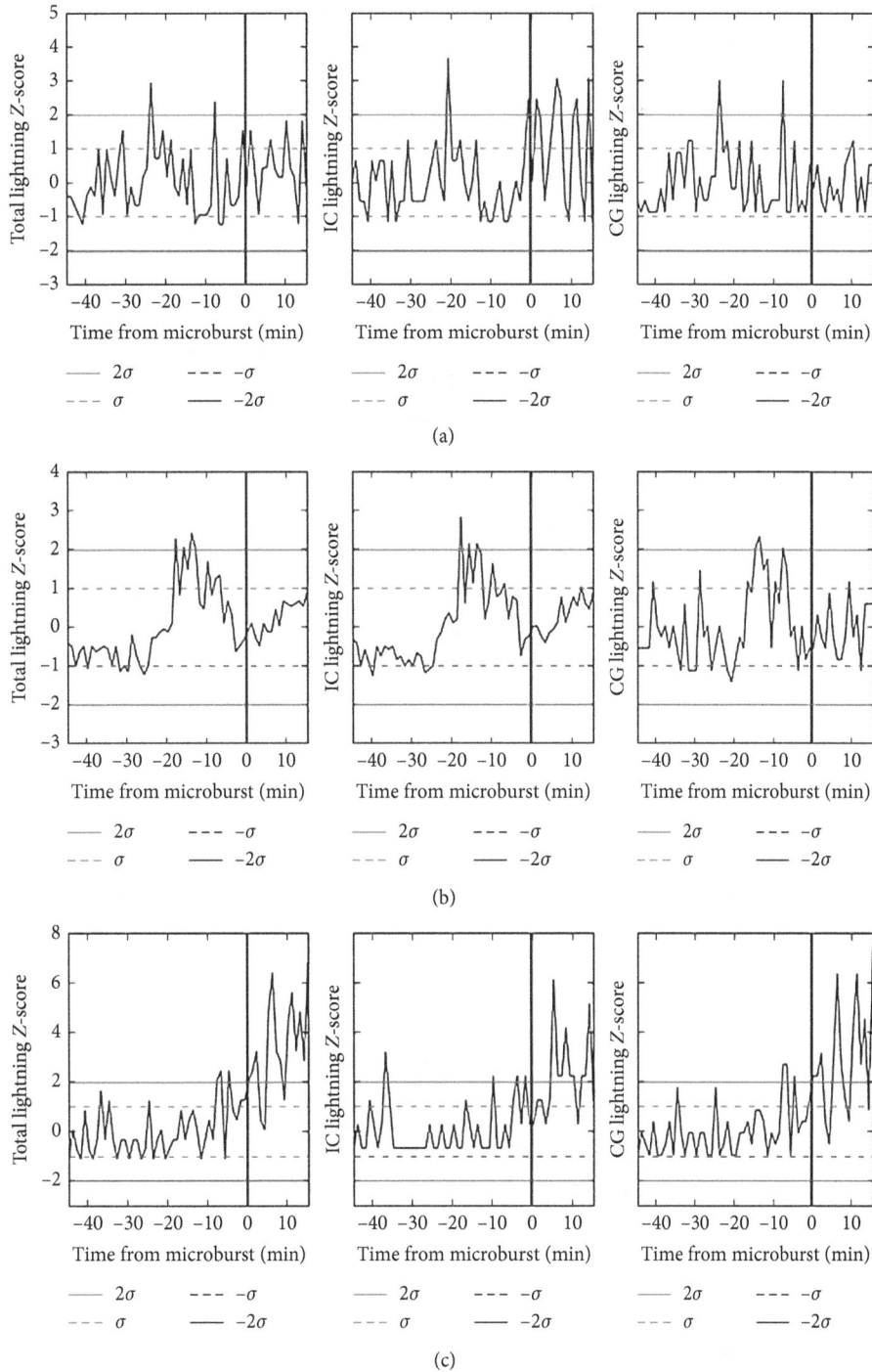

FIGURE 5: As in Figure 3, except for the last three cases. (a) 20150803 microburst. (b) 20160616 microburst. (c) 20160701 microburst.

minutes of the event. In short, trends in radar parameters for the null events show little predictive capability, as many of the maxima fall short of the 2σ maximum, and many of the 2σ maxima, especially for VIL, happen after the event.

4. Discussion

Previous studies [2, 3] have linked microburst formation to the development of downdraft dominance. Downdraft dominance is characterized by an increase in intracloud (IC) lightning, followed by a peak and sharp decrease in IC flashes. Peaks in IC lightning flashes were found to occur six minutes prior to outflow winds for a case in Alabama [3]. Thus, we hypothesize that IC lightning activity is related to a storm's microphysical and convective states before thunderstorm collapse. IC lightning forms in the temperature region between $0°C$ and $-20°C$, where ice, graupel particles, and supercooled water are produced.

TABLE 3: Time of 2σ maxima of each parameter with respect to the microburst.

Date	Total ltng. (minutes)	CC (minutes)	CG (minutes)	VIL (minutes)	Cloud top (minutes)	Echo top (minutes)
20120724	−42	−42	−20	N/A	N/A	N/A
20130624	+4	+6	+4	+2	+2	N/A
20140703	+4	−41	+5	−14	N/A	N/A
20140708NY	−23	−20	−36	−1	N/A	−17
20140708PA	−3	−3	+9	−5	N/A	N/A
20150612	−8	−8	−37	−42	N/A	N/A
20150623	−6	0	−1	N/A	+15	N/A
20150630	N/A	N/A	−10	0	N/A	N/A
20150803	−8	−21	−8	+12	N/A	N/A
20160616	−14	−18	−14	N/A	N/A	N/A
20160701	+15	+5	+15	+15	N/A	−2
Mean/SD	−8.1/15.9	−15.2/16.8	−14.2/17.4	−4.1/17.8	10.7/7.5	−9.5/10.6

Negative numbers indicate before the microburst, and positive numbers indicate after the microburst. N/A indicates maximum value was less than 2σ.

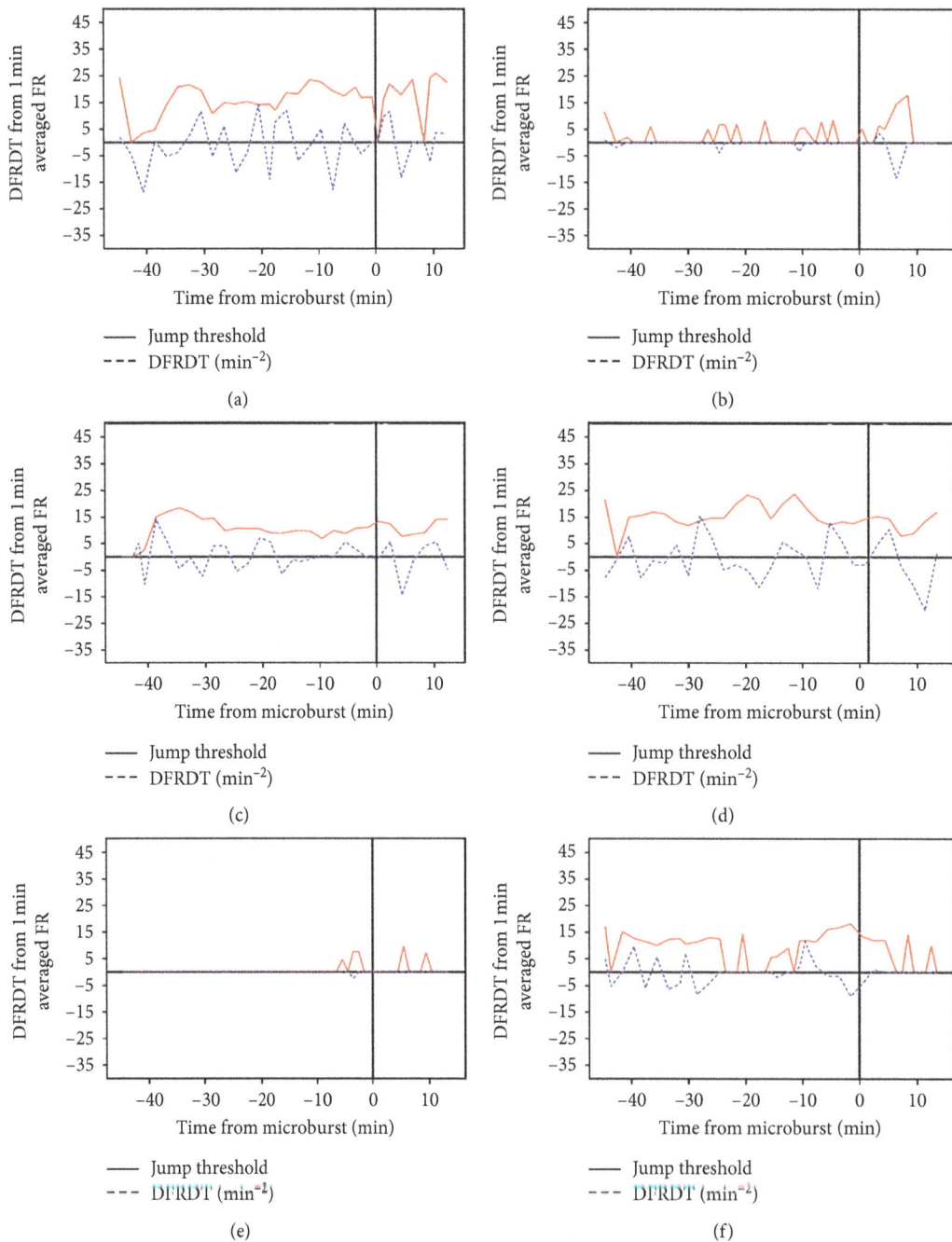

(a)

(b)

(c)

(d)

(e)

(f)

FIGURE 6: Continued.

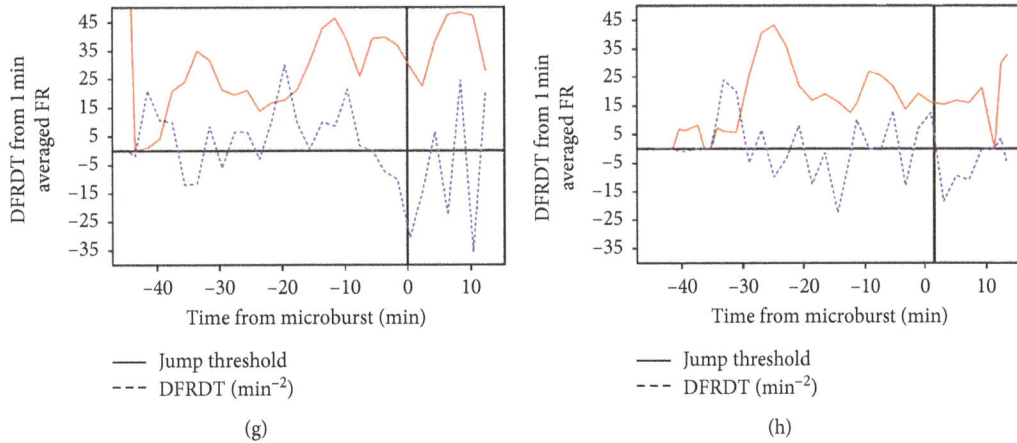

FIGURE 6: Plots of total lightning jumps using the 2σ analysis of Schultz et al. [25] for the first eight microburst cases. (a) 20120724 total lightning. (b) 20130624 total lightning. (c) 20140703 total lightning. (d) 20140708NY total lightning. (e) 20140708PA total lightning. (f) 20150612 total lightning. (g) 20150623 total lightning. (h) 20150630 total lightning.

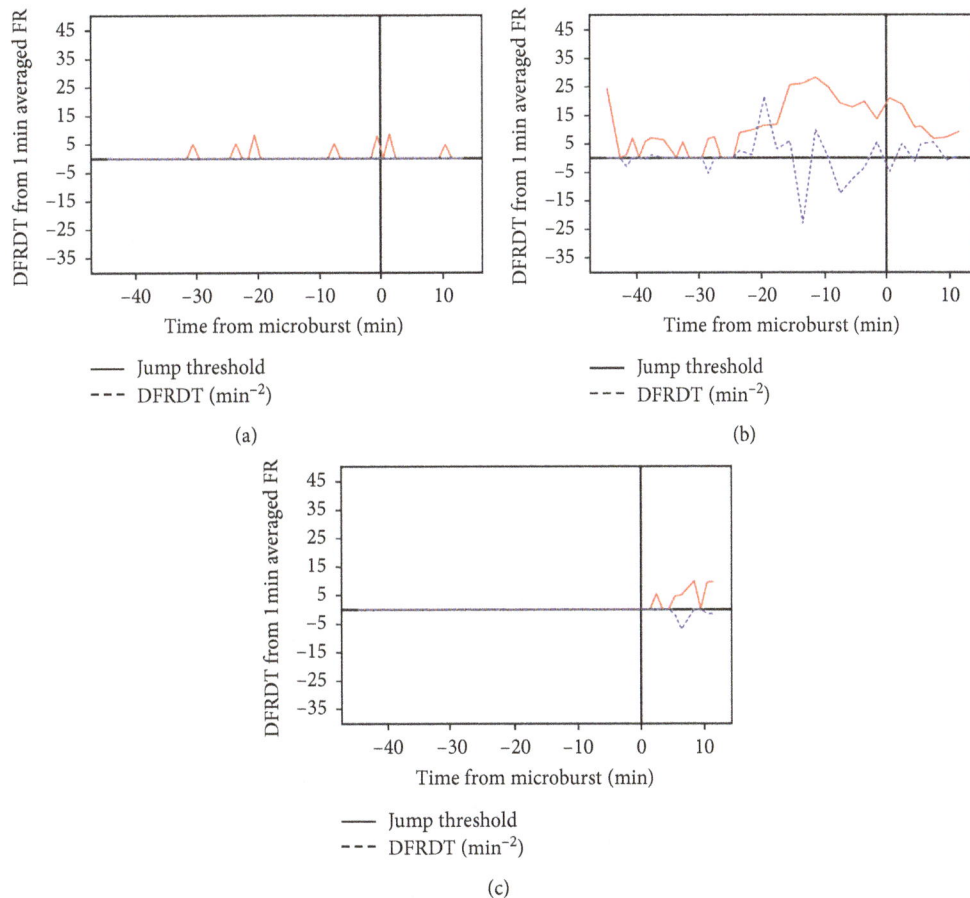

FIGURE 7: As in Figure 6, but for the last three microburst cases. (a) 20150803 total lightning. (b) 20160616 total lightning. (c) 20160701 total lightning.

Collisions between particles cause noninductive charging and the release of electricity in the form of IC flashes [9]. In order for a saturated $0°C$ to $-20°C$ region to exist, a vigorous, deep updraft must be present to ingest water vapor into the storm. A peak in IC lightning activity indicates a vigorous updraft that extends into this $0°C$ to $-20°C$ layer, an indicator of imminent thunderstorm collapse, and a possible microburst. In the present study, IC lightning was found to peak and rapidly decline as much as 41 minutes before microburst occurrence, suggesting that perhaps

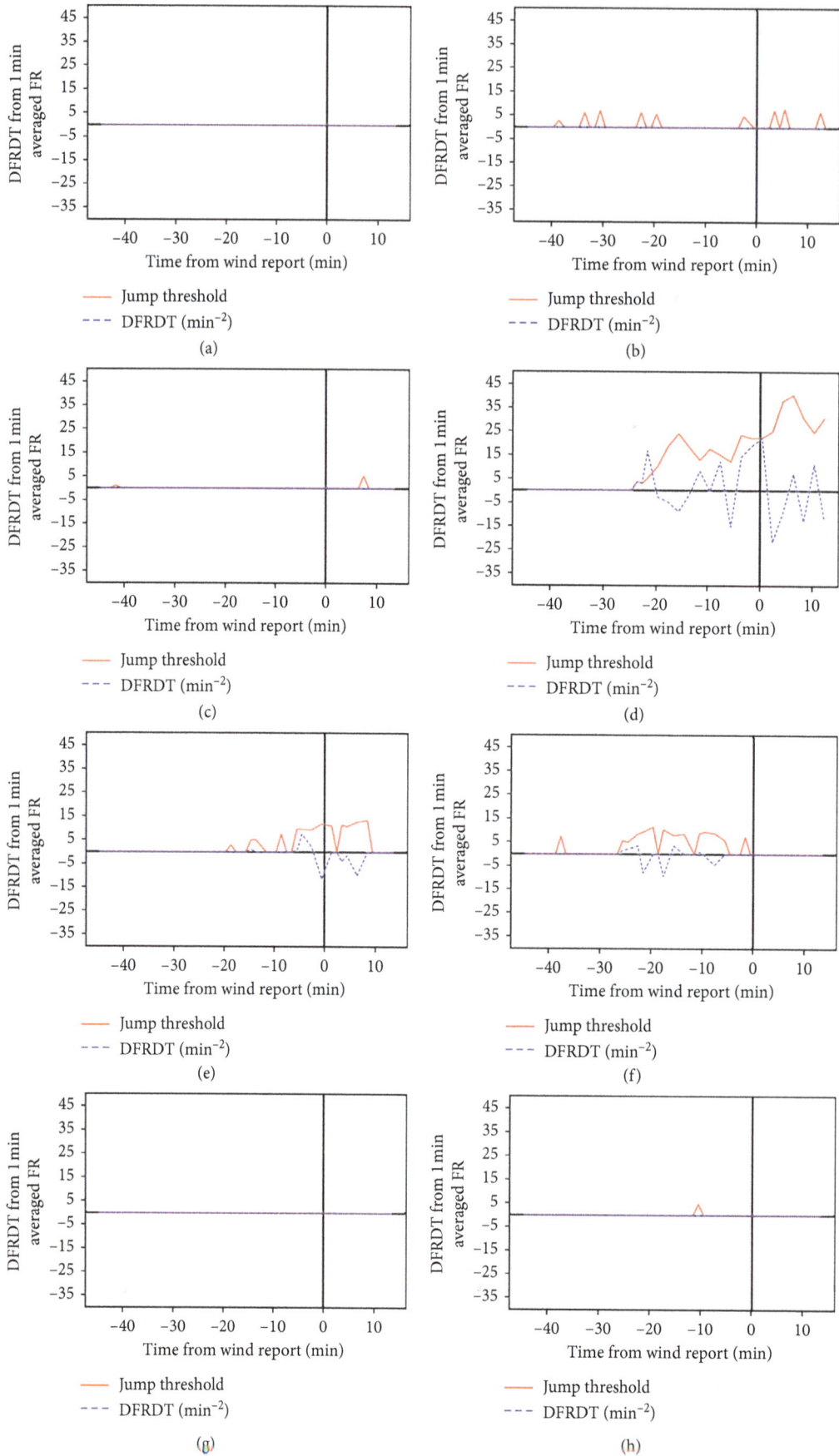

FIGURE 8: Plots of total lightning using the Schultz et al. [25] 2σ analysis for the eight null cases. (a) 20120724 total lightning. (b) 20130624 total lightning. (c) 20140703 total lightning. (d) 20150612 total lightning. (e) 20150623 total lightning. (f) 20150630 total lightning. (g) 20150803 total lightning. (h) 20160701 total lightning.

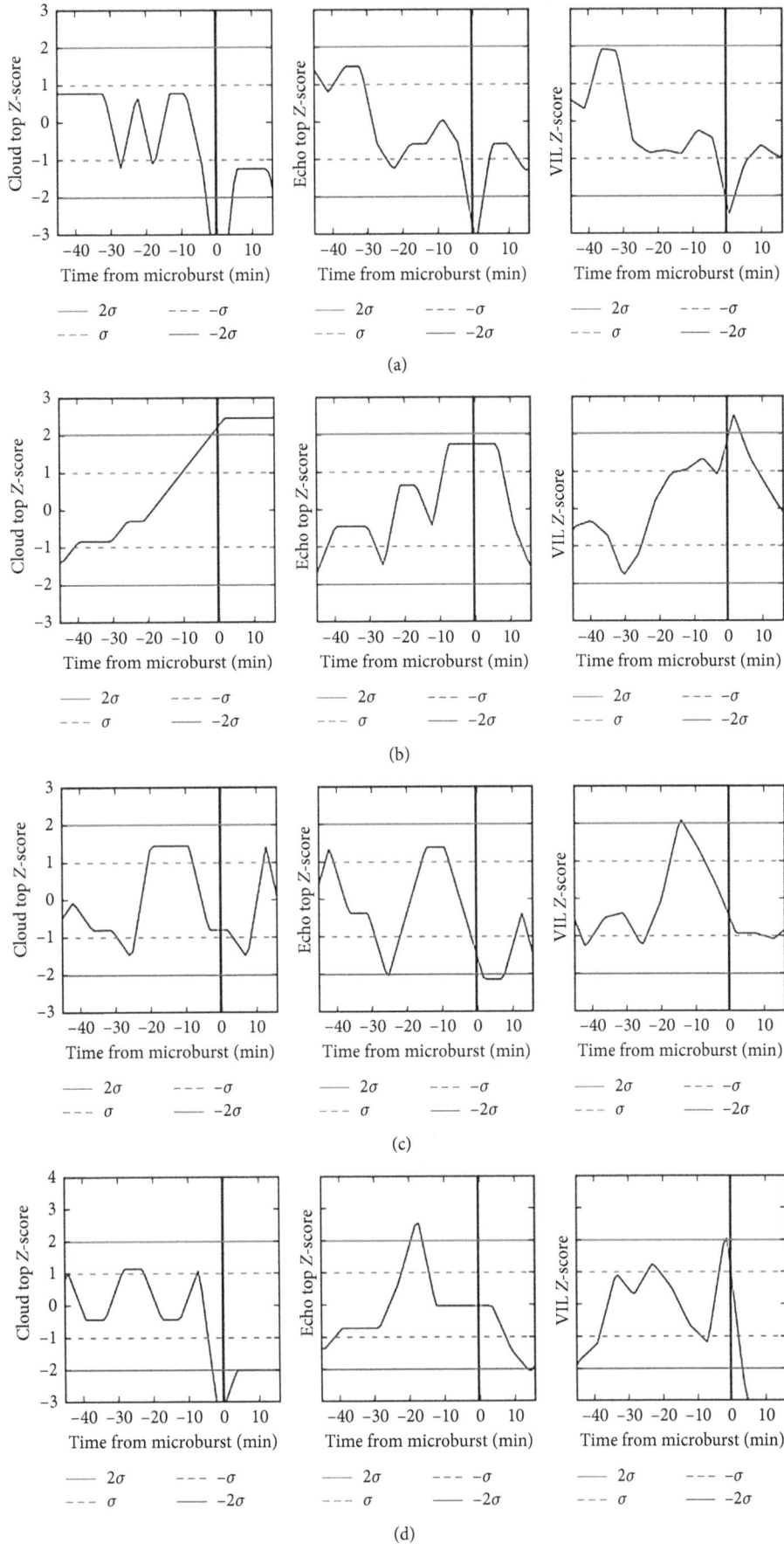

FIGURE 9: Radar parameters—cloud top, echo top, and VIL—for the first four microburst cases. (a) 20120724 microburst. (b) 20130624 microburst. (c) 20140703 microburst. (d) 20140708NY microburst.

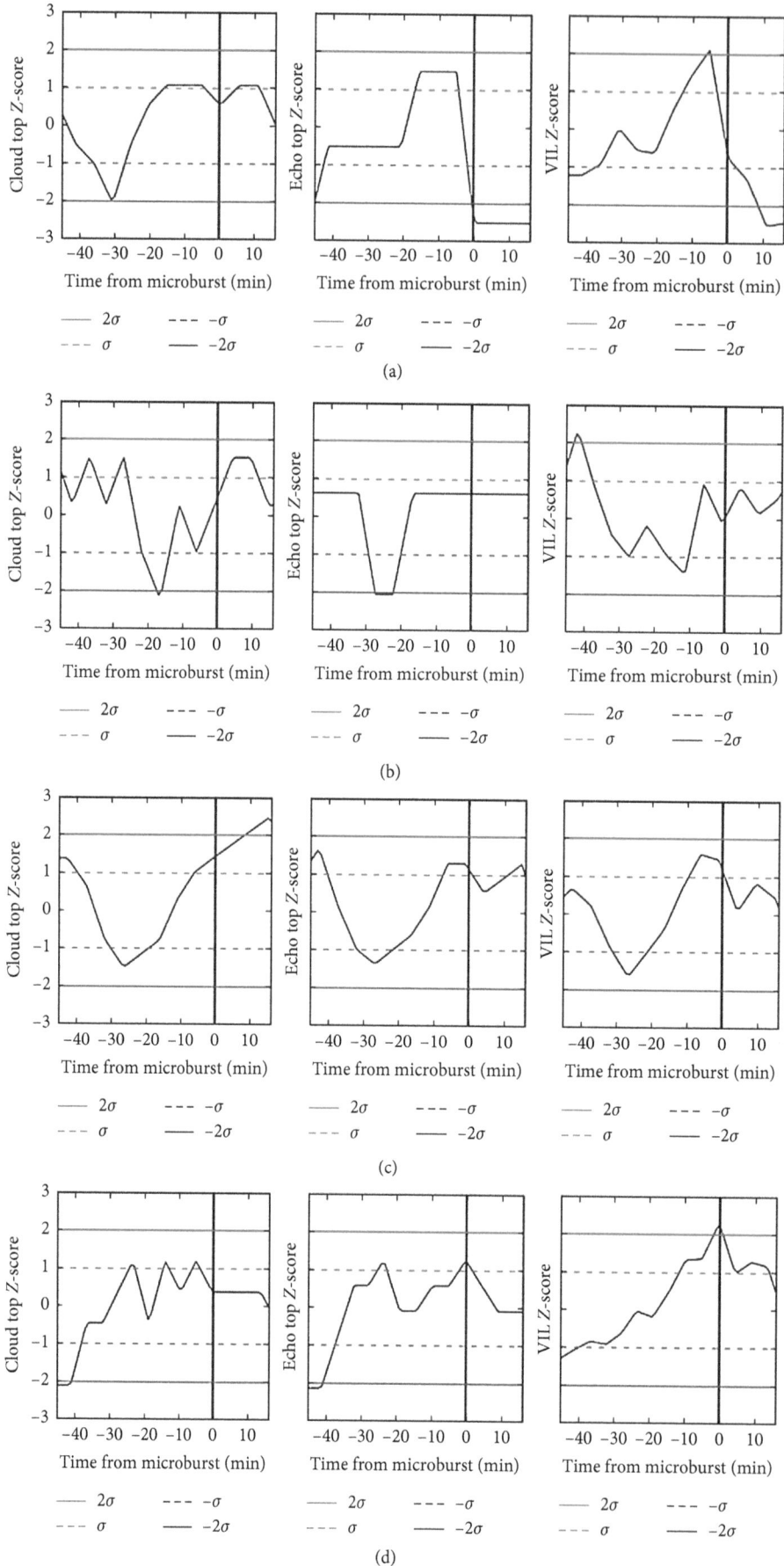

FIGURE 10: As in Figure 9, except for the next four cases. (a) 20140708PA microburst. (b) 20150612 microburst. (c) 20150623 microburst. (d) 20150630 microburst.

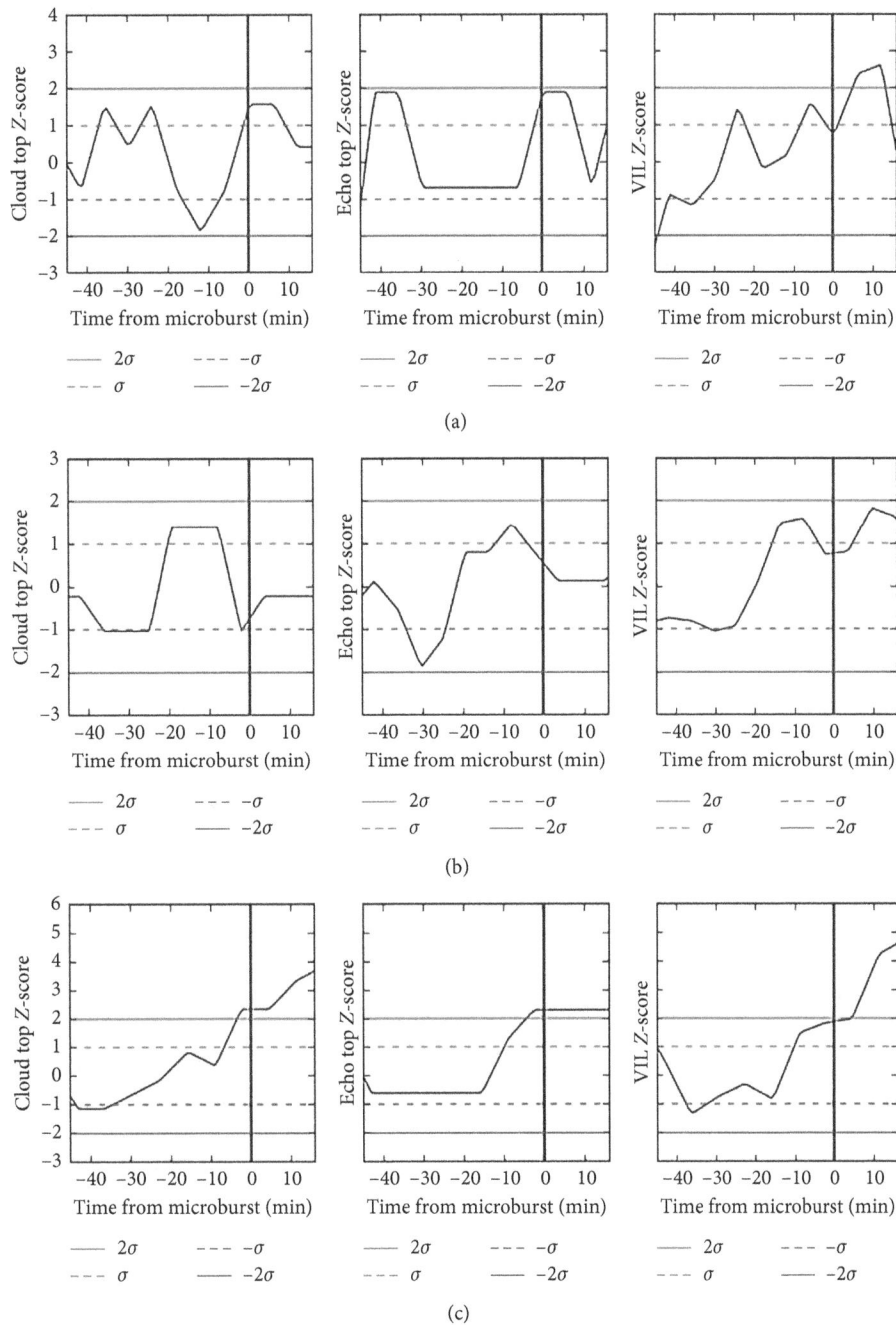

FIGURE 11: As in Figure 9, except for the last three cases. (a) 20150803 microburst. (b) 20160616 microburst. (c) 20160701 microburst.

another mechanism is generating IC lightning peaks in some of these cases. IC lightning was found to peak within 25 minutes before the microburst in six of eleven microburst cases but only in one of eight null cases. This suggests that the process of downdraft dominance is more common in microbursts than in severe wind events, and detection of a peak in IC lightning in real time may distinguish potential microburst scenarios from less dangerous, ordinary severe wind events.

The transition from strong updraft to strong downdraft is accompanied by a transition in lightning from IC to CG lightning [2]. CG lightning forms as particles rapidly descend, collide, and build up charge as a result of a storm's downdraft [15]. A sharp decrease in IC lightning activity and increases in cloud-to-ground (CG) lightning strikes indicate the downdraft is overcoming the updraft. Kane [4] found that increasing CG lightning occurred about 10 minutes before a downburst in the northeast. In the present study, peaks in IC and CG lightning did not behave as consistently as the literature might suggest. Rather than a transition from an IC lightning peak to a CG lightning peak, in three microburst cases, CG lightning presaged the IC lightning, and in three more microburst cases, their peaks occurred during the same five-minute window.

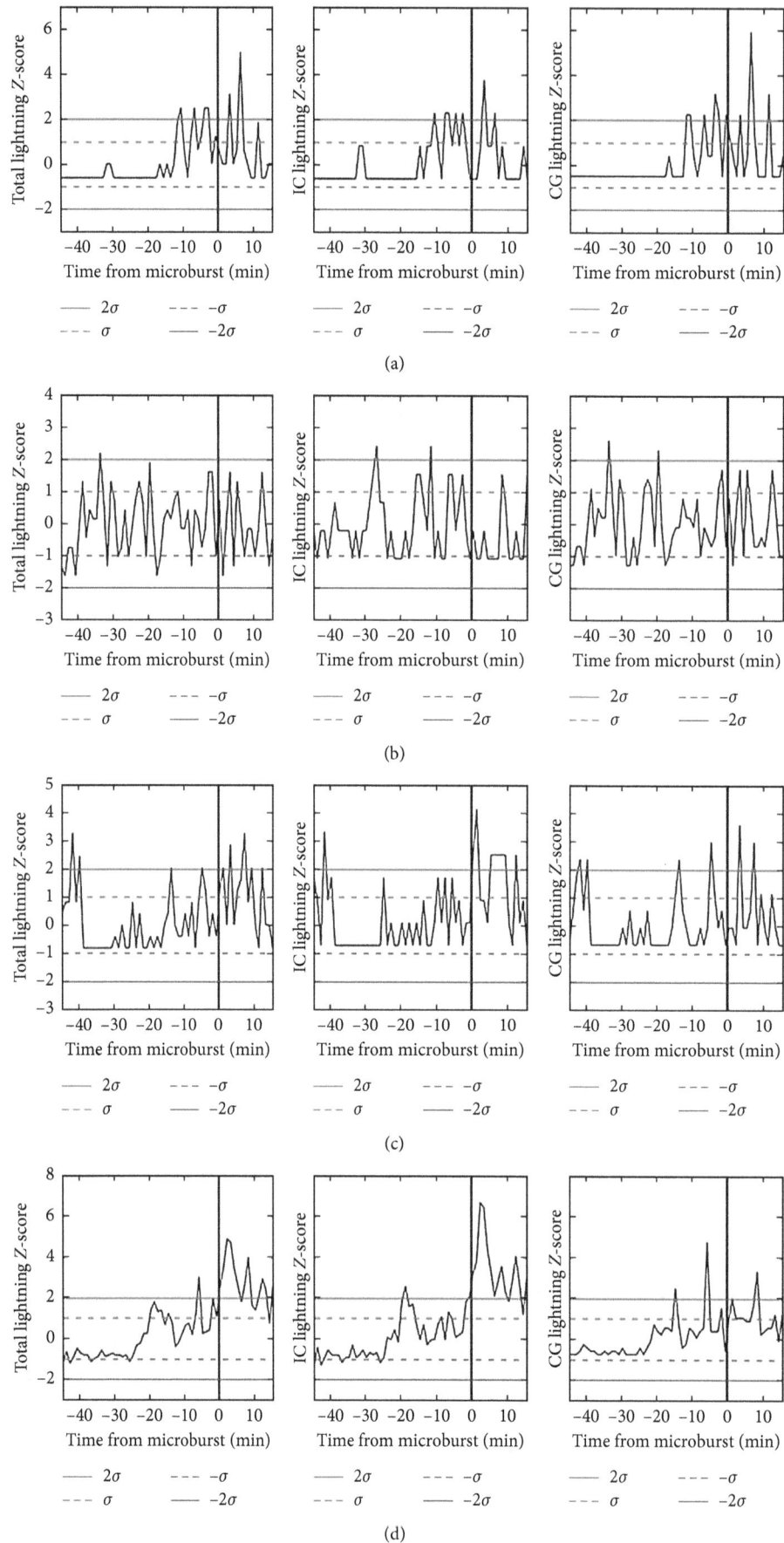

FIGURE 12: Plots of total lightning, IC lightning, and CG lightning for the first four severe wind cases. (a) 20120724 null. (b) 20130624 null. (c) 20140703 null. (d) 20150612 null.

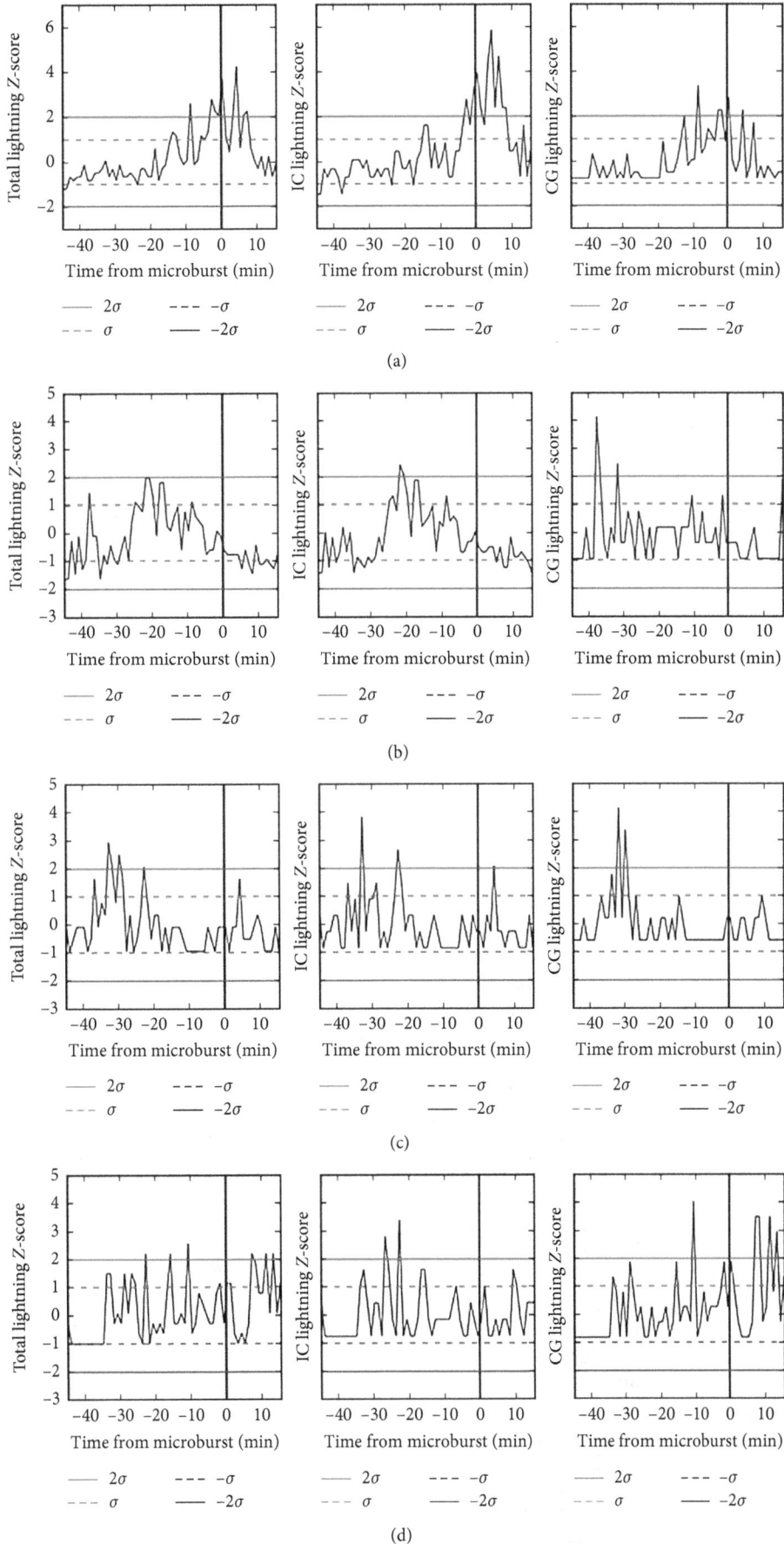

FIGURE 13: As in Figure 12, but for the last four severe wind cases. (a) 20150623 null. (b) 20150630 null. (c) 20150803 null. (d) 20160701 null.

TABLE 4: Time of 2σ maxima of each parameter with respect to the severe wind report.

Date	Total ltng. (minutes)	CC (minutes)	CG (minutes)	VIL (minutes)	Cloud top (minutes)	Echo top (minutes)
20120724	+6	+3	+6	+1	N/A	N/A
20130624	−34	−12	−34	+13	+13	N/A
20140703	+7	+1	+3	+1	N/A	0
20150612	+2	+2	−6	+6	N/A	N/A
20150623	+4	+4	−9	N/A	N/A	N/A
20150630	−21	−22	−2	+5	+6	N/A
20150803	−33	−33	−32	−31	+15	−5
20160701	−11	−23	−23	+15	−31	−3
Mean/SD	−8.4/18.1	−10.0/14.5	−10.6/14.9	1.4/15.3	−3.3/24.4	−2.7/2.5

Negative numbers indicate before the report, and positive numbers indicate after the report. N/A indicates maximum value was less than 2σ.

Metzger and Nuss [10] found that severe wind events tend to be characterized by increasing CG strike rates. However, in eight of eleven microburst cases in the present study, CG lightning was decreasing or steady at the time of the microburst. They also found that IC lightning was increasing in two-thirds of wind cases and decreasing or steady in one-third of wind cases. This ratio was lower in the present study, with six cases increasing and five cases decreasing or steady. The severe wind events from the present study contained five of eight events with increasing CG lightning and four of eight events with increasing IC lightning. It appears that decreasing CG lightning behavior may distinguish microbursts from many more ordinary severe wind events.

Metzget and Nuss [10] also examined radar parameters (VIL, VIL density, and 55 dBZ height) and determined that two of these parameters must decrease by a certain threshold for severe wind events. For microbursts in the present study, VIL and echo tops were decreasing for six of twelve cases, while cloud top height was increasing for six of twelve cases. The behavior of echo tops was quite different in the severe wind cases, where they were found to be decreasing at the time of the event for six of eight cases. VIL was decreasing for four of the eight cases, and cloud tops were increasing for four of eight cases. In short, our study suggests that the particular radar parameters' behavior is too erratic to make robust conclusions.

Five out of eleven microburst cases and five out of eight null cases had more CG lightning strikes than IC lightning flashes (Tables 5 and 6). The null events did not all fall on the same day as the microbursts, suggesting that there can be appreciable differences in the IC : CG ratio from storm to storm. The analysis of the IC : CG ratio is complicated by the very different DEs of IC and CG lightning by the Vaisala network. In reality, IC lightning is likely much more common than these data suggest, owing to its 50% DE. Correcting for this deficiency would likely lower the IC : CG ratios appreciably. Severe storms have been found to typically have higher IC flash rates than CG lightning strikes [27]. However, Boccippio et al. [28] found anomalously low ratios of IC : CG flashes around the Appalachian Mountains. This anomaly has been attributed to differences in storm morphology over the Appalachian Mountains and elevation differences [28]. Murray and Colle [29] further discuss how convection is favored on the leeward side of the

TABLE 5: Summary of IC : CG and CG+ : CG− ratios for microburst events.

Date	IC : CG	CG+ : CG−
20120724	0.67	0.29
20130624	0.25	0.19
20140703	0.87	0.66
20140708NY	2.01	0.88
20140708PA	2.07	0.67
20150612	0.32	1.39
20150623	1.69	2.36
20150630	8.16	1.34
20150803	0.94	1.20
20160616	3.63	0.60
20160701	0.34	1.04

TABLE 6: Summary of IC : CG and CG+ : CG− ratios for null severe wind events.

Date	IC : CG	CG+ : CG−
20120724	0.64	0.22
20130624	0.25	0.19
20140703	1.03	1.19
20150612	2.79	0.81
20150623	1.80	1.39
20150630	4.66	2.38
20150803	2.10	0.39
20160701	0.65	1.16

Appalachians in the evening, when the majority of microburst and null cases occurred.

Five microburst cases and four null cases had more CG+ lightning than CG− lightning. Three dates—20150623, 20150630, and 20160701—had more CG+ lightning for both the null case and the microburst case. In general, CG+ : CG− lightning ratios were closer to one than IC : CG ratios. This disputes the results of previous studies. For example, Carey et al. [30] found that the eastern United States sees predominantly negative lightning nearby severe weather (large hail and tornadoes) regions more often than in other regions of the country.

5. Conclusion

This study examined quasi-cellular microburst activity and concurrent wind damage events in the Northeast U.S.,

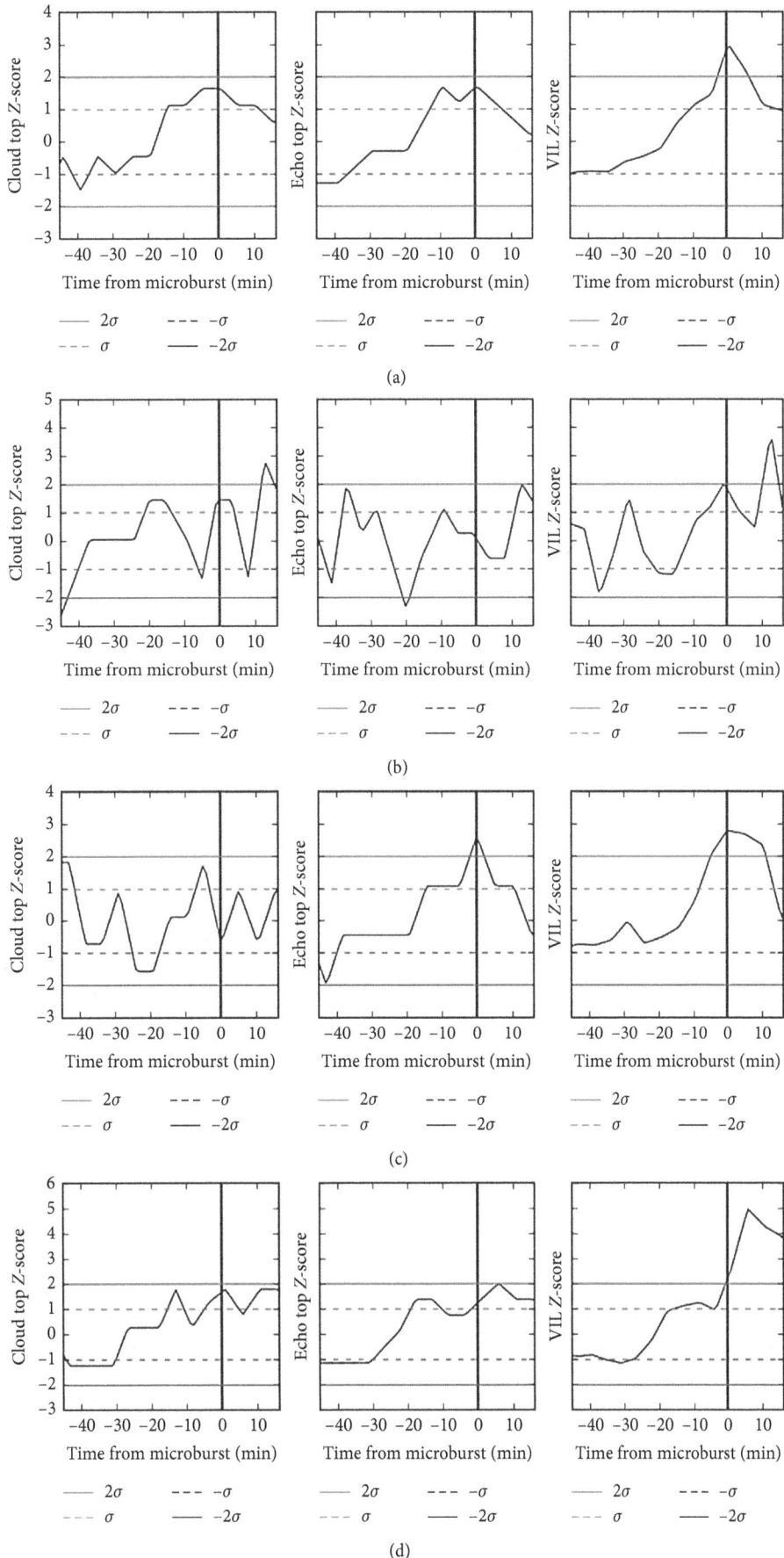

FIGURE 14: Radar parameters—cloud top, echo top, and VIL—for the first four severe wind cases. (a) 20120724 null. (b) 20130624 null. (c) 20140703 null. (d) 20150612 null.

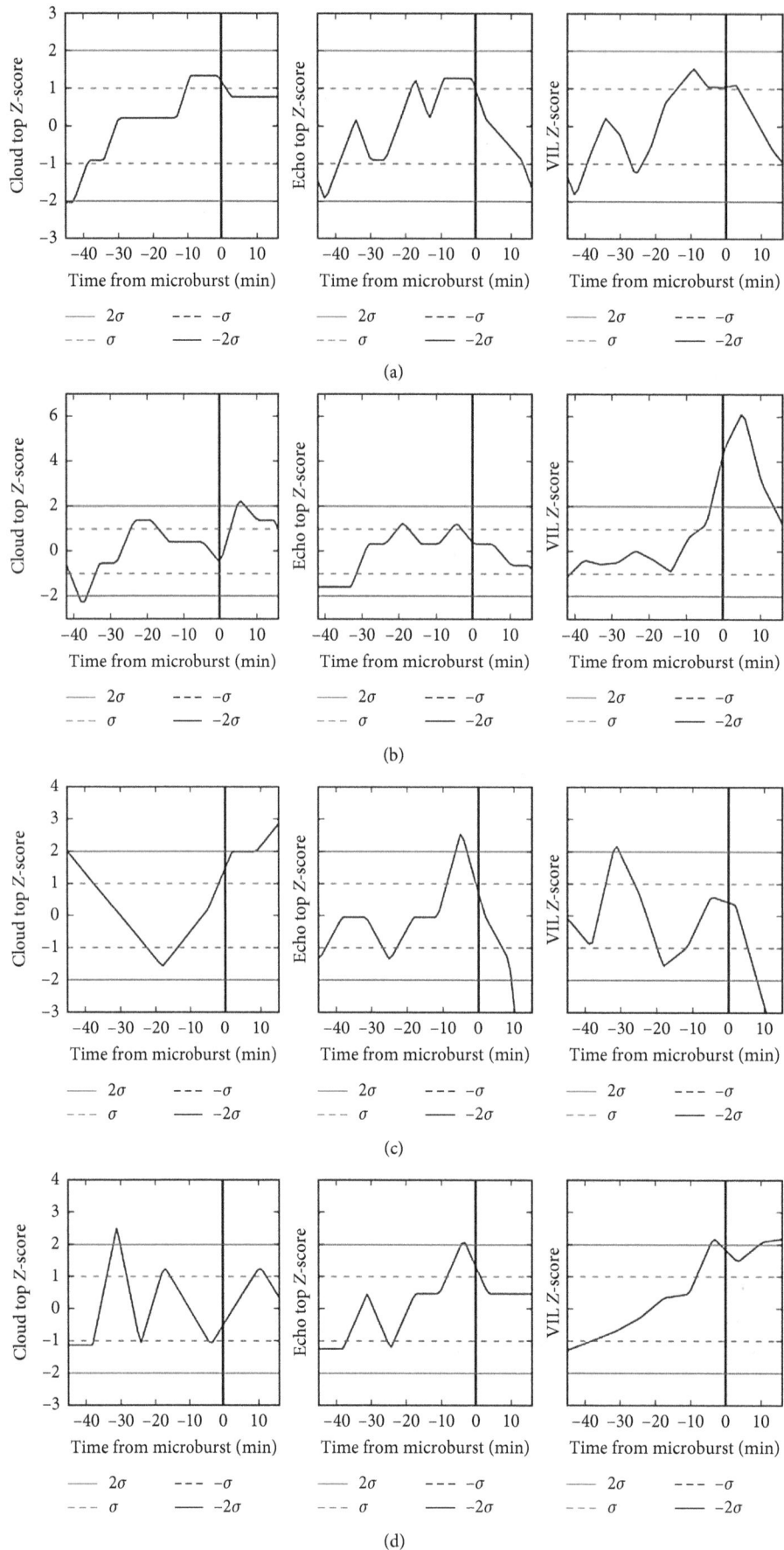

FIGURE 15: As in Figure 14, but for the last four severe wind cases. (a) 20150623 null. (b) 20150630 null. (c) 20150803 null. (d) 20160701 null.

seeking precursor signals for microbursts in the radar and lightning data and looking to determine the likelihood of false alarms. Standard deviations relative to the mean of the 45 minutes prior to microburst or severe wind report were used to determine whether lightning and radar parameters peaked prior to microburst or wind damage occurrence. Total lightning was the best-performing variable, with six of eleven cases showing a peak within 25 minutes before the microburst. IC lightning peaked within 25 minutes before the microburst six times, and CG lightning peaked within this same time window five times. The 2σ requirement for a "peak" eliminated echo top and cloud top from consideration most of the time. VIL peaked at the time of the microburst or within fifteen minutes before the microburst in four of eleven cases. The null events showed erratic lightning behavior, with no peaks within 10 minutes of the microburst for either total or IC lightning. CG lightning provided some predictive values for the null cases, peaking within eleven minutes of the severe wind report in four of eight cases. VIL often peaked after the wind damage (six of eight cases). As in the microburst cases, cloud top and echo top were eliminated from more than half the cases as they failed to produce a 2σ peak. The "2σ" algorithm from Schultz et al. [25] was adopted, and lightning jumps were less common than using a static 2σ, with only three microburst cases and one wind case experiencing a lightning jump before the event.

While the overall mean of these variables suggests that the variables peak with several minutes of lead time to predict microbursts, there is a large amount of variability about the mean. The results call some findings of previous studies into question, at least for the region of study. First, there was wide variability in the timing of lightning and radar parameter microburst peaks, spanning nearly the entire 60-minute window examined in this study. Relying on peaks in the parameters alone may generate early, false warnings in some cases and may cause late warnings or no warning at all in other cases. Second, previous studies have suggested that IC lightning peaks preceded CG lightning peaks. This was true for less than half of the cases in this study. Instead, CG peaks preceded IC peaks in three cases, and the two peaked nearly simultaneously in three other cases. And finally, peaks had been observed in the Southeast U.S. in cloud height and echo tops prior to declines in these parameters. In the present study, echo tops and cloud height were typically constant before declining, leading to a lack of 2σ peaks.

The results of this study offer important implications for weather forecasters. Radar parameters are less consistent than lightning as short-term predictors of microburst activity, although the timing of peaks in VIL seems to discriminate fairly well between microburst and null events. In some portions of the Northeast U.S., beam blockage becomes an issue, limiting the availability of radar data. Lightning as a forecast and verification tool may provide insight into where radar data are absent, but the inconsistency of lightning peak timing relative to microburst time complicates the use of lightning data in the absence of radar data.

Disclosure

Amanda L. Burke is currently at the Department of Atmospheric and Geographic Sciences, University of Oklahoma, 120 David L Boren Blvd., Norman, OK, USA. This work was presented at the 29th Conference on Severe Local Storms in Stowe, Vermont.

Conflicts of Interest

The authors declare that they have no conflicts of interest.

Acknowledgments

Thanks are due to Randy Chase for writing a script to interpolate the radar and lightning data, thanks are due to Mary Lynn Baeck for assistance in running TITAN, and thanks are due to Scott Rochette for reviewing the manuscript. Appreciation is due to Vaisala, Inc. for providing data from its National Lightning Detection Network. This work was supported by the first author's employment at the College at Brockport, State University of New York.

References

[1] R. M. Wakimoto, "Convectively driven high winds," in *Severe Convective Storms. Meteoroological Monographs*, vol. 50, pp. 255–298, American Meteorological Society, Boston, MA, USA, 2001.

[2] S. J. Goodman, D. E. Buechler, P. D. Wright, and W. D. Rust, "Lightning and precipitation history of a microburst-producing storm," *Geophysical Research Letters*, vol. 15, no. 11, pp. 1185–1188, 2012.

[3] E. R. Williams, M. E. Weber, and R. E. Orville, "The relationship between lightning type and convective state of thunderclouds," *Journal of Geophysical Research*, vol. 94, no. D11, pp. 13213–13220, 1989.

[4] R. J. Kane, "Correlating lightning to severe local storms in the northeastern United States," *Weather and Forecasting*, vol. 6, no. 1, pp. 3–12, 1991.

[5] E. L. Kuchera and M. D. Parker, "Severe convective wind environments," *Weather and Forecasting*, vol. 21, no. 4, pp. 595–612, 2006.

[6] D. Atlas, C. W. Ulbrich, and C. R. Williams, "Physical origin of a wet microburst: observations and theory," *Journal of the Atmospheric Sciences*, vol. 61, no. 10, pp. 1186–1195, 2004.

[7] R. C. Srivastava, "A simple model of evaporatively driven dowadraft: application to microburst downdraft," *Journal of the Atmospheric Sciences*, vol. 42, no. 10, pp. 1004–1023, 1985.

[8] M. R. Hjelmfelt, "Structure and life cycle of microburst outflows observed in Colorado," *Journal of Applied Meteorology*, vol. 27, no. 8, pp. 900–927, 1988.

[9] Y. Yair, B. Lynn, C. Price et al., "Predicting the potential for lightning activity in Mediterranean storms based on the weather research and forecasting (WRF) model dynamic and microphysical fields," *Journal of Geophysical Research*, vol. 115, article D04205, 2010.

[10] E. Metzger and W. A. Nuss, "The relationship between total cloud lightning behavior and radar-derived thunderstorm structure," *Weather and Forecasting*, vol. 28, no. 1, pp. 237–253, 2013.

[11] S. A. Amburn and P. L. Wolf, "VIL density as a hail indicator," *Weather and Forecasting*, vol. 12, no. 3, pp. 473–478, 1997.

[12] J. C. Brimelow, G. W. Reuter, A. Bellon, and D. Hudak, "A radar-based methodology for preparing a severe thunderstorm climatology in central Alberta," *Atmosphere-Ocean*, vol. 42, no. 1, pp. 13–22, 2004.

[13] V. Lakshmanan, K. Hondl, C. K. Potvin, and D. Preignitz, "An improved method for estimating radar echo-top height," *Weather and Forecasting*, vol. 28, no. 2, pp. 481–488, 2013.

[14] L. Yang, J. Smith, M. L. Baeck, B. Smith, F. Tian, and D. Niyogi, "Structure and evolution of flash flood producing storms in a small urban watershed," *Journal of Geophysical Research: Atmospheres*, vol. 121, no. 7, pp. 3139–3152, 2016.

[15] K. L. Cummins, M. J. Murphy, and J. V. Tuel, "Lightning detection methods and meteorological applications," in *Proceedings of 4th International Symposium on Military Meteorology, Hydro-Meteorological Support of Allied Forces and PfP Members Tasks Realization*, Malbork, Poland, September 2000.

[16] NOAA, *Storm Events Database*, NOAA, Silver Spring, MD, USA, June 2018, https://www.ncdc.noaa.gov/stormevents/.

[17] M. Dixon and G. Wiener, "TITAN: thunderstorm identification, tracking, analysis, and nowcasting-a radar-based methodology," *Journal of Atmospheric and Oceanic Technology*, vol. 10, no. 6, pp. 785–797, 1993.

[18] R. S. Schumacher and R. H. Johnson, "Organization and environmental properties of extreme-rain-producing mesoscale convective systems," *Monthly Weather Review*, vol. 133, no. 4, pp. 961–976, 2004.

[19] M. Steiner, R. A. Houze, and S. E. Yuter, "Climatological characterization of three-dimensional storm structure from operational radar and rain gauge data," *Journal of Applied Meteorology*, vol. 34, no. 9, pp. 1978–2007, 1995.

[20] R. L. Holle, K. L. Cummins, and W. A. Brooks, "Seasonal, monthly, and weekly distributions of NLDN and GLD360 cloud-to-ground lightning," *Monthly Weather Review*, vol. 144, no. 8, pp. 2855–2870, 2016.

[21] G. Medici, K. L. Cummins, D. J. Cecil, W. J. Koshak, and S. D. Rudlosky, "The intracloud lightning fraction in the contiguous United States," *Monthly Weather Review*, vol. 145, no. 11, pp. 4481–4499, 2017.

[22] K. L. Cummins and M. J. Murphy, "An overview of lightning locating systems: history, techniques, and data uses, with an in-depth look at the U.S. NLDN," *IEEE Transactions on Electromagnetic Compatibility*, vol. 51, no. 3, pp. 499–518, 2009.

[23] A. Nag, M. J. Murphy, K. L. Cummins, A. E. Pifer, and J. A. Cramer, "Recent evolution of the U.S. National Lightning Detection Network," in *Proceedings of 23rd International Lightning Detection Conference*, Vaisala, Tucson, AZ, USA, 2014, http://www.vaisala.com/en/events/ildcilmc/Pages/ILDC-2014-archive.aspx.

[24] M. J. Murphy and A. Nag, "Cloud lightning performance and climatology of the U.S. based on the upgraded U.S. National Lightning Detection Network," in *Proceedings of Seventh Conference on the Meteorological Applications of Lightning Data*, American Meteor Society, Phoenix, AZ, USA, 2015, https://ams.confex.com/ams/95Annual/webprogram/Paper262391.html.

[25] C. J. Schultz, W. A. Petersen, and L. D. Carey, "Preliminary development and evaluation of lightning jump algorithms for the real-time detection of severe weather," *Journal of Applied Meteorology and Climatology*, vol. 48, no. 12, pp. 2543–2563, 2009.

[26] C. Farnell, T. Rigo, and N. Pineda, "Lightning jump as a nowcast predictor: application to severe weather events in Catalonia," *Atmospheric Research*, vol. 183, pp. 130–141, 2017.

[27] S. J. Goodman, R. Blakeslee, H. Christian et al., "The North Alabama lightning mapping array: recent severe storm observations and future prospects," *Atmospheric Research*, vol. 76, no. 1–4, pp. 423–437, 2005.

[28] D. J. Boccippio, K. L. Cummins, H. J. Christian, and S. J. Goodman, "Combined satellite- and surface-based estimation of the intracloud-cloud-to-ground lightning ratio over the continental United States," *Monthly Weather Review*, vol. 129, no. 1, pp. 108–122, 2001.

[29] J. C. Murray and B. A. Colle, "The spatial and temporal variability of convective storms over the northeast United States during the warm season," *Monthly Weather Review*, vol. 139, no. 3, pp. 992–1012, 2011.

[30] L. D. Carey, S. A. Rutledge, and W. A. Petersen, "The relationship between severe storm reports and cloud-to-ground lightning polarity in the contiguous United States from 1989 to 1998," *Monthly Weather Review*, vol. 131, no. 7, pp. 1211–1228, 2003.

7

Improving TIGGE Precipitation Forecasts using an SVR Ensemble Approach in the Huaihe River Basin

Chenkai Cai ⓘ, Jianqun Wang ⓘ, and Zhijia Li

College of Hydrology and Water Resources, Hohai University, Nanjing 210098, China

Correspondence should be addressed to Jianqun Wang; wangjq@hhu.edu.cn

Academic Editor: Mario M. Miglietta

Recently, the use of the numerical rainfall forecast has become a common approach to improve the lead time of streamflow forecasts for flood control and reservoir regulation. The control forecasts of five operational global prediction systems from different centers were evaluated against the observed data by a series of area-weighted verification and classification metrics during May to September 2015–2017 in six subcatchments of the Xixian Catchment in the Huaihe River Basin. According to the demand of flood control safety, four different ensemble methods were adopted to reduce the forecast errors of the datasets, especially the errors of missing alarm (MA), which may be detrimental to reservoir regulation and flood control. The results indicate that the raw forecast datasets have large missing alarm errors (MEs) and cannot be directly applied to the extension of flood forecasting lead time. Although the ensemble methods can improve the performance of rainfall forecasts, the missing alarm error is still large, leading to a huge hazard in flood control. To improve the lead time of the flood forecast, as well as avert the risk from rainfall prediction, a new ensemble method was proposed on the basis of support vector regression (SVR). Compared to the other methods, the new method has a better ability in reducing the ME of the forecasts. More specifically, with the use of the new method, the lead time of flood forecasts can be prolonged to at least 3 d without great risk in flood control, which corresponds to the aim of flood prevention and disaster reduction.

1. Introduction

With rapid development of the economy and population growth, as well as the impact of climate change and increasing demand for freshwater, many regions of China are facing a shortage of water resources. At the same time, due to the influence of monsoon climate, flood disasters frequently occur in China during the rainy season, causing enormous economic losses to society [1]. Streamflow forecasting and flood regulation play a key role in China's flood control and water resource management system [2–4]. Traditionally, the flood forecast approach takes precipitation from ground sites as its input; its forecast lead time is limited and cannot satisfy the demand for flood prevention and water resource utilization [5, 6]. Therefore, it is a most urgent and important task to make efforts to lengthen the lead time and improve the accuracy of flood forecasting. As many previous studies have mentioned, numerical rainfall forecasting is an effective way to solve this problem [7, 8].

Recently, the revolution of computer technology and the progress of meteorological and climate models have brought about the continuous development of numerical weather forecasts [9–11]. As a major component of The Observing System Research and Predictability Experiment (THORPEX), TIGGE (the THORPEX Interactive Grand Global Ensemble) dataset consists of ensemble forecast datasets from eleven main forecasting centers worldwide, starting in 2006, aimed at improving the accuracy of high-impact weather forecasts within two weeks [12, 13]. A number of studies on TIGGE precipitation forecasts have been extensively carried out for extreme rain events and hydrological forecasting. Clark and Hey [14] applied the medium range numerical weather prediction model output from NCEP (National Centers for Environmental Prediction) to streamflow forecasting in the United States, but the results showed that the model apparently has low skill in predicting precipitation and temperature. Pappenberger et al. [15] improved the lead time of flood forecasts to 8 days in advance by using TIGGE data as a meteorological input to the European Flood Alert System. Additionally, another case in the Upper Huaihe Basin also showed that a reliable flood warning is available as early as 10 days in advance [16]. He et al. [17] found that the uncertainties in precipitation would dominate and propagate through a coupled atmospheric-hydrologic-hydraulic cascade system. Su et al. [18] evaluated the errors of quantitative precipitation forecasts and probabilistic quantitative precipitation forecasts from six operational global ensemble prediction systems in TIGGE during June to August 2008–2012 in the Northern Hemisphere. Sagar et al. [19] assessed the skill of a numerical weather prediction model for rainstorms over India.

Due to the low skill of precipitation forecasts, many researchers have suggested that it is possible to improve the accuracy of the forecasts through downscaling and ensemble forecasting [20–24]. Although several ensemble methods have been used to reduce the error of rainfall forecasts, most of these methods established the relationship between different forecasts and the observed value to reduce the overall errors such as mean absolute errors (MAEs) or root mean square errors (RMSEs) which neglected the different effects of false alarm (FA, event forecasted to occur but did not occur) and missing alarm (MA, event forecasted not to occur but did occur) on flood control security [23, 25]. A FA only reduces the benefits of flood resources, while an MA is disadvantageous to flood control safety. As safety is the most important target in flood control, it is a main task to find a method which can provide reliable precipitation forecast to improve the lead time of flood forecast without bringing huge flood risk by large MEs (missing alarm errors).

In this paper, we evaluated the predictions of five forecasting centers in the Huaihe River Basin of China during May to September 2015–2017. Both linear and nonlinear system analysis methods were applied to ensemble forecasting with the purpose of reducing the prediction errors. Furthermore, since flood control safety is the first priority in water resource management, specific attention will be paid to the errors of MAs to avoid increasing the risk in flood control; a new objective function will also be proposed. In the next section, we detail the evaluation indicators along with the correction methods, which include both linear and nonlinear methods. Section 3 introduces an overview of the datasets and the study area. The results and discussion are described in Section 4, and finally, the conclusions are provided in Section 5.

2. Methodology

The main assumption in post-processing the forecasts is that the observation and forecast are correlated, and the future behavior of the system will remain the same. The purpose of this study is to lessen the ME of rainfall forecasts by means of system analysis so that rainfall prediction can be better utilized in flood forecasting without bringing about huge risk in flood control.

2.1. Verification Methods. The accuracy of precipitation prediction is usually used as an evaluation criterion to determine whether rainfall prediction information is available. To estimate the skill of the forecasts with different lead times, multiple classic statistical characteristics were assessed to compare the forecasts and observations, including MAE and RMSE. Meanwhile, a qualitative analysis was also adopted to complement the statistical characteristics based on the classification standard for daily rainfall formulated by the meteorological department of China. According to the total amount of rainfall in 24 h, this standard divides the daily precipitation into seven types: no rain, light rain, medium rain, heavy rain, rainstorm, heavy rainstorm, and extreme rainstorm. Generally, for meteorological researchers, no rain means that the daily precipitation is 0; however, due to the limited influence of rainfall of less than 1 mm on the formation of floods, the no rain standard was changed to less than 1 mm in this study, and each classification standard is detailed in Table 1.

On the basis of the magnitudes of forecasts and observations, three combinations of the results are as follows:

> Hit: the magnitude of the forecast is the same as that of the observation
>
> Miss alarm: the magnitude of the forecast is smaller than that of the observation
>
> False alarm: the magnitude of the forecast is larger than that of the observation

The skill of the forecasts can be assessed based on a contingency table that contains the frequency of the combinations. From the contingency table, the rates of hit, MA and FA can be computed for each magnitude:

Hit rate (HR):

$$\alpha_i = \left(\frac{n_{i,j}}{N_i}\right) \times 100\%, \quad i = j. \tag{1}$$

MA rate (MAR):

$$\beta_i = \left(\frac{n_{i,j}}{N_i}\right) \times 100\%, \quad i < j. \tag{2}$$

FA rate (FAR):

$$\gamma_i = \left(\frac{n_{i,j}}{N_i}\right) \times 100\%, \quad i > j, \tag{3}$$

where $n_{i,j}$ is the number of cases with the predicted magnitude of i and the measured magnitude of j; N_i is the number of the forecast for the magnitude of i.

2.2. Ensemble Methods. Due to the determinacy and randomness of the atmospheric motion, it is impossible to obtain the initial field of numerical prediction objectively and accurately, and a perfect weather prediction model does not exist [26]. Ensemble forecasting is an efficient method to eliminate the uncertainty caused by observation error and analysis error by using data from different sources [27, 28]. Several cases have proven that it is possible to improve the accuracy of precipitation forecasts by using ensemble methods. In former studies, the researchers usually used

TABLE 1: Classification standard of daily precipitation.

Magnitude	Classification standard of precipitation	Amount of daily precipitation (mm)
1	No rain	0–0.9
2	Light rain	1.0–9.9
3	Medium rain	10.0–24.9
4	Heavy rain	25.0–49.9
5	Rainstorm	50.0–99.9
6	Heavy rainstorm	100.0–249.9
7	Extreme rainstorm	>250.0

perturbed forecasts from a single dataset or control forecasts from several datasets as the input of ensemble forecast [29–31].

In this paper, we used the control forecast of five selected datasets and compared three kinds of linear ensemble forecasting methods, namely, the ensemble mean (EM), bias-removed ensemble mean (BREM), and linear regression (LR) methods [32, 33]. Meanwhile, since the previous studies showed that the prediction errors of different magnitudes exhibit a nonlinear variation, support vector machine, as an effective nonlinear regression method, was also employed for ensemble forecasting.

2.2.1. Bias-Removed Ensemble Mean. The bias-removed ensemble mean is developed from the ensemble mean, and the calculation formula is as follows:

$$F = \overline{O} + \frac{1}{m}\sum_{i=1}^{m}\left(F_i - \overline{F_i}\right), \qquad (4)$$

where $\overline{O}$ represents the average value of observations during the training period, m represents the number of members used in the ensemble forecast, F_i represents the forecast value of member i, $\overline{F_i}$ represents the average value of member i during the training period, and F is the ensemble forecast value.

2.2.2. Support Vector Regression. In formula (4), the relationship between the ensemble forecast value F and the forecast value of member $F_1, F_2, \ldots, F_m$ is linear. The relationship can be considered to be nonlinear as follows:

$$F = f\left(F_1, F_2, \ldots, F_m\right). \qquad (5)$$

In formula (5), f represents the nonlinear relationship. A BP artificial neural network model or a support vector regression model can be used to approximate f.

Support vector machine (SVM) is a mechanical learning method for classification and regression proposed by Vanpik et al. in 1995 based on the structural risk minimization (SRM) principle [34]. SVR (support vector regression), with good generalization and nonlinear processing ability, has been widely employed in hydrology [35–38]. The methodology of SVR is briefly described below.

The basic idea of SVR for nonlinear case is to map the original problem to a linear problem in a high-dimensional feature space by nonlinear transformation to approximate f with input vector $x \in R^n$, output vector $y \in R$ and sample set $(x_1, y_1), \ldots, (x_l, y_l)$:

$$f(x) = \langle w, \phi(x) \rangle + b. \qquad (6)$$

In formula (6), $f(x)$ is the regression function, $\phi(x)$ is the nonlinear transformation function, w is the weight vector, and b is the threshold. Then, the ε-SVR model is built by solving the optimization problem:

$$\min_{w, \xi_i, \xi_i^*, b} \quad \frac{1}{2} w^T w + C \frac{1}{l}\sum_{i=1}^{l}\left(\xi_i + \xi_i^*\right),$$

$$\text{s.t.} \quad (\langle w, \phi(x_i) \rangle + b) - y_i \le \varepsilon + \xi_i, \qquad (7)$$

$$y_i - (\langle w, \phi(x_i) \rangle + b) \le \varepsilon + \xi_i^*,$$

$$\xi_i, \xi_i^* \ge 0., \quad i = 1, 2, \ldots, l.$$

The dual form of the problem can be expressed as

$$\max_{\alpha, \alpha^*} \quad \sum_{i=1}^{l} \alpha_i(y_i - \varepsilon) - \alpha_i^*(y_i + \varepsilon)$$

$$- \frac{1}{2}\sum_{i=1}^{l}\sum_{j=1}^{l}\left(\alpha_i - \alpha_i^*\right)\left(\alpha_j - \alpha_j^*\right)K(x_i \cdot x_j), \qquad (8)$$

$$\text{s.t.} \quad \sum_{i=1}^{l}\left(\alpha_i - \alpha_i^*\right) = 0,$$

$$0 \le \alpha_i, \alpha_i^* \le C, \quad i = 1, \ldots, l,$$

where, $K(x_i, x_j) = \langle \phi(x_i), \phi(x_j) \rangle$ represents the kernel function. After the Lagrange multipliers, α_i and α_i^*, in formula (8) have been determined, the ε-SVR regression function $f(x)$ can be established as

$$f(x) = \sum_{SVs}\left(\overline{\alpha}_i - \overline{\alpha}_i^*\right)K(x_i, x) + b. \qquad (9)$$

In this study, the radial basis function kernel with a parameter σ was selected for nonlinear transformation:

$$K(x_i, x) = \exp\left(\frac{-|x - x_i|^2}{(2\sigma^2)}\right). \qquad (10)$$

The parameter ε in formula (8) should be determined beforehand, but in many practical cases, it is hard to determine the value of ε before training. In order to solve this problem, a new parameter ν was introduced by Schölkopf (ν-SVR) [39]. Subsequently, the optimization problem can be changed as

$$\min_{w, b, \xi, \xi^*, \varepsilon} \quad \frac{1}{2} w^T w + C\left(\nu\varepsilon + \frac{1}{l}\sum_{i=1}^{l}\left(\xi_i + \xi_i^*\right)\right),$$

$$\text{s.t.} \quad (\langle w, \phi(x_i) \rangle + b) - y_i \le \varepsilon + \xi_i, \qquad (11)$$

$$y_i - (\langle w, \phi(x_i) \rangle + b) \le \varepsilon + \xi_i^*,$$

$$\xi_i, \xi_i^* > 0, \quad i = 1, \ldots, l, \varepsilon > 0.$$

Also, the dual form of formula (11) is

$$\max_{\alpha,\alpha^*} \sum_{i=1}^{l} \alpha_i \left(y_i - \varepsilon\right) - \alpha_i^* \left(y_i + \varepsilon\right)$$

$$- \frac{1}{2} \sum_{i=1}^{l} \sum_{j=1}^{l} \left(\alpha_i - \alpha_i^*\right) \left(\alpha_j - \alpha_j^*\right) K\left(x_i \cdot x_j\right),$$

$$\text{s.t.} \quad \sum_{i=1}^{l} \left(\alpha_i - \alpha_i^*\right) = 0, \quad\quad\quad (12)$$

$$\sum_{i=1}^{l} \left(\alpha_i + \alpha_i^*\right) \leq C \cdot \nu,$$

$$0 \leq \alpha_i, \alpha_i^* \leq \frac{C}{l}, \quad i = 1, \ldots, l.$$

The capability of ν-SVR depends on the parameter C, ν, and σ. The cost constant C is a compromise between the complexity and generalization of the model. It is used to adjust the ratio of confidence range and empirical risk in the sample space, which determines the penalty degree for the sample whose loss is greater than ε. The parameter ν represents the lower bound of support vector and the upper bound of gap error. Additionally, since ν is introduced to optimize ε which controls the width of insensitive band and affects the number of support vectors, ν also indirectly affects the number of support vectors and prediction accuracy, and the reduction of ν leads to the reduction of support vectors. Accordingly, the quadratic programming problem (formula (12)) can be solved by determining the three parameters and the MAE of the ensemble forecast can be calculated by the following formula:

$$\text{MAE} = \frac{1}{l} \sum_{i=1}^{l} \left| f\left(x_i\right) - y_i \right|. \quad\quad\quad (13)$$

In formula (13), $f(x_i)$ and y_i indicate the ith ensemble forecast and observed values, and l is the sample size.

Usually, in order to get the best approximation ability, nonlinear optimization algorithm is used to optimize the parameters C, ν, and σ by minimizing the MAE. However, the number of support vectors can be very large if the ν-SVR model is solved by optimization algorithm. Therefore, the range of parameter ν should be analyzed with determined C and σ by perturbation method.

In this paper, the ν-SVR model was used to approximate f and the parameters of SVR were optimized by the PSO (particle swarm optimization) algorithm with the objective function of the minimization of the MAE [40, 41]. Furthermore, the number of support vectors was controlled to be less than $l/2$ by controlling the range of ν.

All the SVR models used in this research are based on the open source software LIBSVM developed by Lin Chih-Jen. More information about the model is available at https://www.csie.ntu.edu.tw/~cjlin/.

3. Data and Study Area

3.1. Study Area. The Huaihe River Basin is located in the middle east of China between the Yangtze River Basin and the Yellow River Basin, with an area of 270 thousand km^2. Similar to many Chinese basins affected by the monsoon climate, the annual rainfall in the basin is uneven, with drought in spring and winter and rain in summer and autumn. The annual average rainfall of the basin is 875 mm, and the rainy season occurs from May to September, during which 50%–75% of the annual precipitation is concentrated. Therefore, floods are frequent and flood resources are abundant in the rainy season of the basin. However, only a small amount of flood resources is being used to solve the problem of water supply in the drought season. The continuous increase of freshwater demand in the basin has imposed new challenges on the flood control and hazard alleviation. The Xixian Catchment is one of the most important subcatchments in the upper Huaihe River Basin, which is located in the headstream area of the main stream of the Huaihe River. In this paper, the Xixian Catchment was selected as a typical subwatershed of the Huaihe River Basin to evaluate and correct precipitation forecasts of TIGGE, making it possible to improve the lead time of flood forecasts. Additionally, to meet the demand of flood forecasting, we divided the Xixian Catchment into six subcatchments, as shown in Figure 1.

3.2. Forecast Data and Observation Data. Eleven main operational forecast centers from different countries and regions participate in the TIGGE program, including the Bureau of Meteorology of Australia (BoM), China Meteorological Centre (CMA), Centre for Weather Forecasting and Climate Studies (CPTEC), Environment and Climate Change Canada (ECCC), European Centre for Medium-Range Weather Forecast (ECMWF), Japan Meteorological Agency (JMA), Korea Meteorological Agency (KMA), National Meteorological Service of France (Météo-France), United Kingdom Meteorological Office (UKMO), National Centers for Environmental Prediction (NCEP), and National Centre for Medium Range Weather Forecasting (NCMRWF). Five centers were selected in this study: CMA, JMA, KMA, ECMWF, and UKMO. The other six centers were not included in this study for various reasons. The data from BoM ended in 2010. As the location of CPTEC is in the Southern Hemisphere, its initial perturbations are not included in the Northern Hemisphere midlatitude climate. The dataset of the ECCC is missing several months in 2017. Météo-France only provides ensemble forecasts at 6:00 and 18:00 UTC. For NCEP, the dataset is missing many days in 2017. NCMRWF started providing data to TIGGE on August 2017. More details of the datasets used in this paper are briefly given in Table 2.

As shown in Table 2, on account of the differences in the ensemble forecasts of the 5 centers in terms of the horizontal resolution and forecast length, a series of methods were applied to make the datasets consistent. (1) We only selected the forecast with a base time at 00:00UTC; (2) the forecast

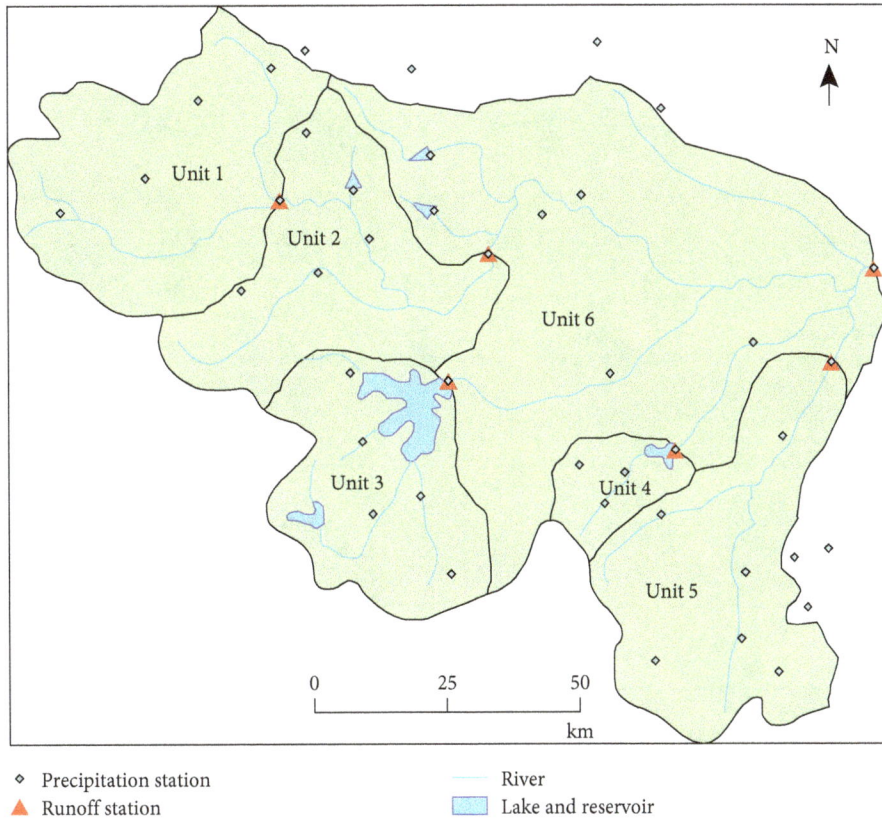

FIGURE 1: The Xixian Catchment in the upper Huaihe River Basin and the location of the six subcatchments.

TABLE 2: Configurations of the five TIGGE ensemble forecasts investigated in this study.

Centre	Number of ensemble members	Horizontal resolution	Forecast length (h)	Base time (UTC)	Steps (h)
CMA	20 + 1	TL639	240	00:00; 12:00	6
JMA	50 + 1	T479	240	00:00; 12:00	6
KMA	23 + 1	N320	288	00:00; 12:00	6
ECMWF	50 + 1	TL639 (0–240 h) TL319 (240–360 h)	360	00:00; 12:00	6
UKMO	23 + 1	N640	360	00:00; 12:00	6

length used in this paper was 168 h (7 days); and (3) the spatial resolutions of the centers were converted into $0.5^0 \times 0.5^0$. Additionally, several stations located in different subcatchments of the Xixian Catchment in the Huaihe River Basin were chosen as the validation dataset, as shown in Figure 1.

4. Results and Discussion

The results obtained are discussed in the four subsections below: (1) the evaluation of raw precipitation forecasting for different lead times and subcatchments; (2) the rain forecast skills of different ensemble methods; (3) a new ensemble method based on SVR and ME minimization; and (4) comparison of the performances obtained using different ensemble methods.

4.1. Evaluation of the Raw Forecasts. As a first step of this study, we estimated the accuracy of raw TIGGE precipitation forecasts from the selected five centers through their

relationships with the measured data. The performance of the precipitation forecasts from different centers is illustrated in Figure 2. Significantly, JMA has the best performances in all the subcatchments, especially for a longer lead time. For most of the subcatchments, the forecast skills of ECMWF are close to that of JMA when the lead time is less than 5 d, but the RMSE of ECMWF rises dramatically as the lead time becomes longer. In general, the prediction skills of ECMWF, UKMO, and JMA exhibit more obvious downward trends with the increase of the lead time, while the other two forecasts fluctuate at different lead times. Moreover, the performance of KMA in the study area is very unusual, as the RMSE of the +1 d forecast is almost two times that of other centers with the same lead time. However, the RMSE of KMA reduces to a normal level when the lead time increases, and it has the most volatile performance of the five datasets.

The daily predicted rainfall versus observed rainfall during 2015–2017 in the whole Xixian Catchment for +1 d, +2 d, +5 d and +7 d is illustrated in Figure 3. The low skill of

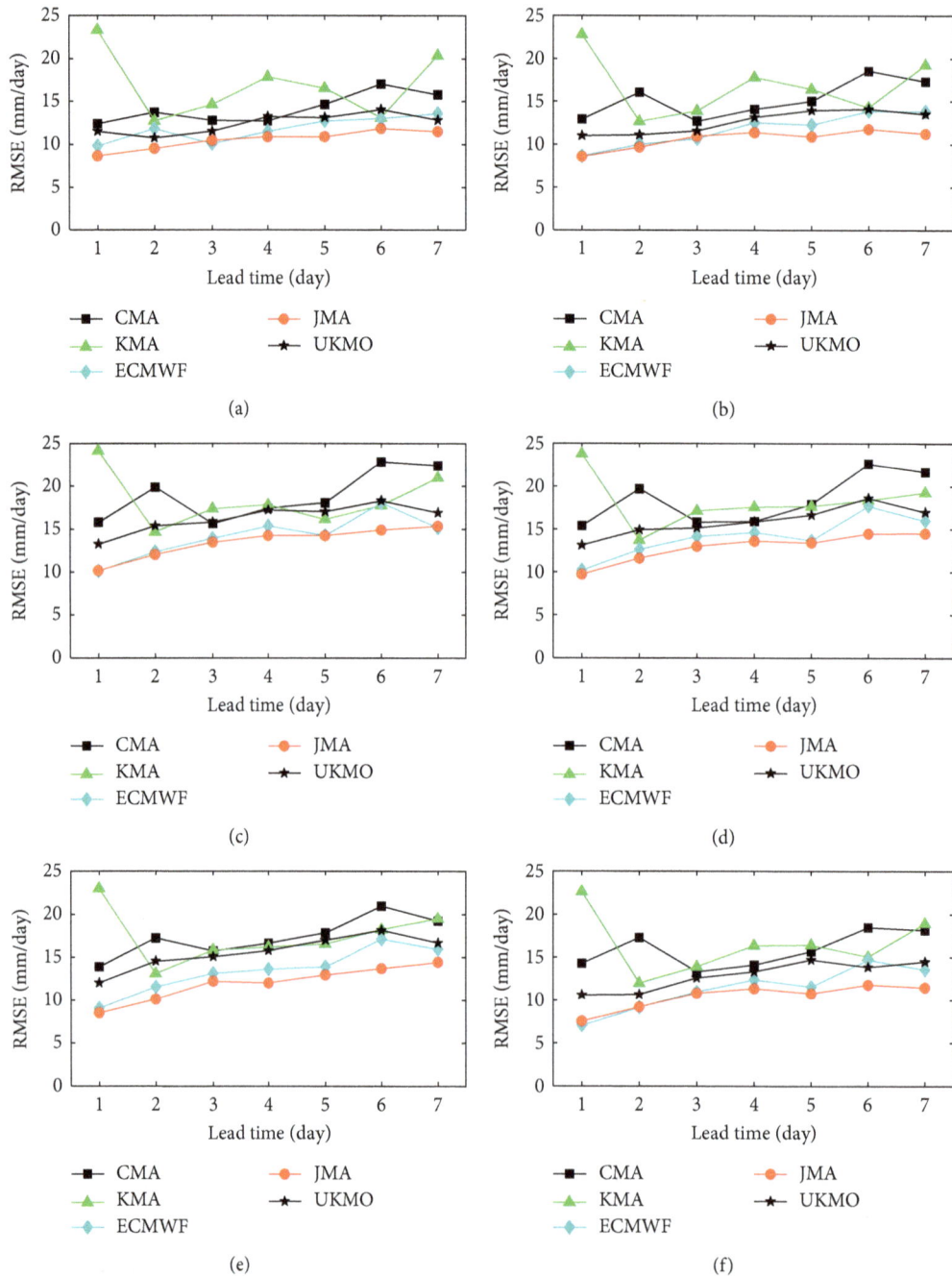

FIGURE 2: The RMSEs of the forecasts in the six different subcatchments during May to September 2015–2017, where (a–f) represent subcatchments 1–6.

KMA in the +1 d forecast is clearly identified in Figure 3(a), which is caused by the overestimates of rainfall. Table 3 lists the three rates of the Xixian Catchment for different lead times. From the table, the HRs vary inversely to the lead times, consistent with the results of Figure 2. In addition, it is obvious that all FARs with different lead times are much larger than the MARs, indicating that these forecasts may overestimate the actual rainfall overall. However, Figure 4 implies a converse conclusion for JMA and ECMWF in that the forecasts may underestimate the rainfall, as the outlier values of MA are more numerous and greater in value than those of FA. The results of forecast error distributions show that JMA and ECMWF underestimate the rainfall, while CMA, KMA, and UKMO overestimate the rainfall. Although the three rates and RMSEs indicate that JMA and ECMWF are the most skillful datasets in the Xixian Catchment, as mentioned above, they are still unfavorable for extending the lead time of flood forecasting. The analysis results prove that the raw rainfall forecast data cannot meet the demand for flood control, so it is necessary to improve the forecast ability by ensemble methods.

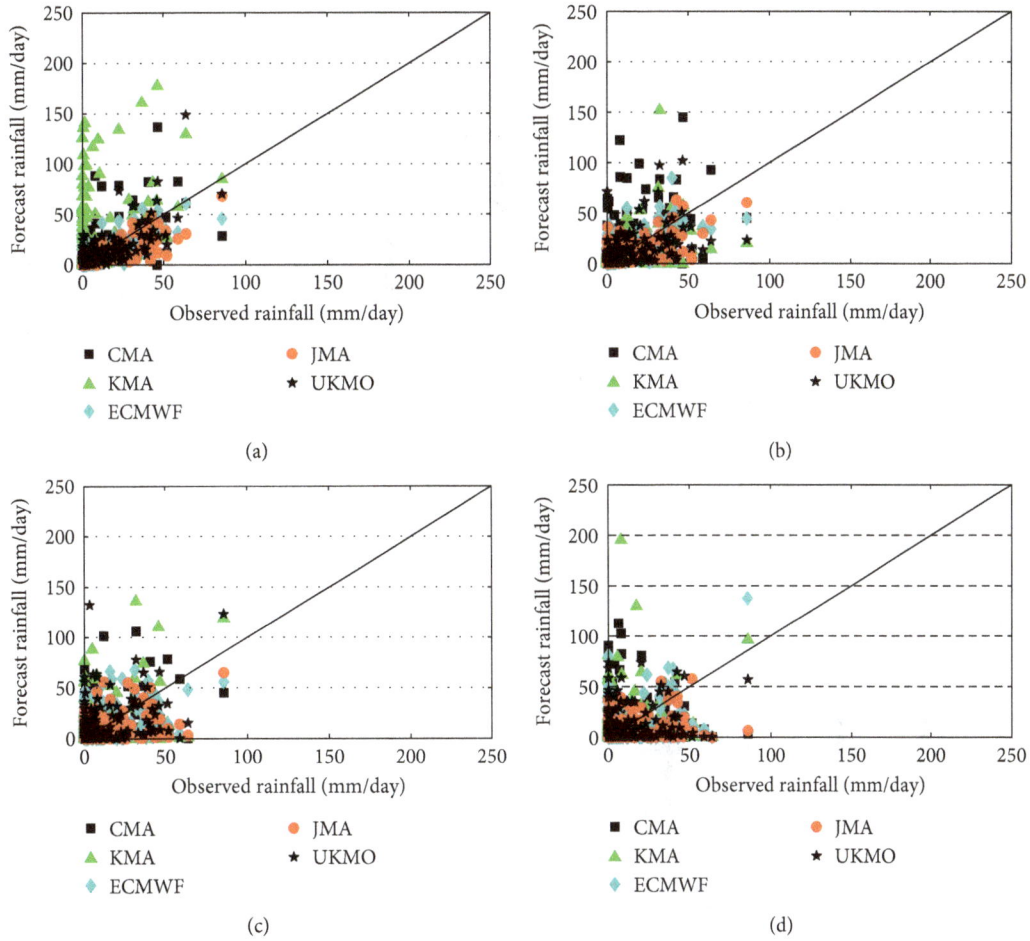

FIGURE 3: Xixian catchment observations versus raw (a) +1 d (b) +2 d (c) +5 d and (d) +7 d forecasts for five centers during the rainy seasons in 2015 to 2017.

TABLE 3: The three rates of the forecasts in Xixian catchment.

Lead time (days)	CMA			KMA			ECMWF			JMA			UKMO		
	HR	FAR	MAR	HR	FAR	MAR	HR	FAR	MAR	HR	FAR	MAR	HR	FAR	MAR
1	61.44	29.19	9.37	57.73	33.33	8.93	67.97	22.88	9.15	70.37	20.04	9.59	65.58	25.71	8.71
2	65.36	24.84	9.80	64.49	21.79	13.73	70.81	19.39	9.80	66.45	22.66	10.89	66.88	20.92	12.20
3	62.53	24.84	12.64	61.44	24.18	14.38	66.23	23.31	10.46	61.44	28.76	9.80	64.27	23.09	12.64
4	61.00	25.71	13.29	61.87	22.22	15.90	64.71	22.88	12.42	56.86	31.15	11.98	63.18	23.31	13.51
5	58.61	25.71	15.69	52.29	30.50	17.21	60.57	26.80	12.64	57.73	28.10	14.16	59.26	24.62	16.12
6	54.47	27.67	17.86	57.30	24.40	18.30	60.35	25.27	14.38	55.99	28.54	15.47	57.30	25.05	17.65
7	54.03	27.23	18.74	57.52	22.22	20.26	57.73	27.45	14.81	56.86	25.93	17.21	59.91	22.00	18.08

4.2. Evaluation of Ensemble Forecasts. As shown formerly, ensemble forecasting is a feasible way to reduce the errors of the initial field and system by assembling different datasets. In this paper, we employed the control forecast datasets from five centers for ensemble forecasting. Additionally, the performances of the ensemble methods were evaluated by the RMSE, three rates, and ME distributions from the aspects of the training and verification periods. Four different ensemble methods were used in this study, including both linear and nonlinear methods: EM, BREM, LR and SVR. The data during May to September 2015–2016 were selected as

the training period, while the data in the flood season of 2017 were used for validation.

The datasets of JMA were chosen as the best raw material, while those of CMA were selected as the worst for comparison with the ensemble forecasts during the training period. The four ensemble methods have better performances than CMA, and only few cases are inferior to JMA (Figure 5), which proves that the ensemble prediction results are better than the original datasets. The RMSEs of EM and BREM are very close, and the differences in RMSEs between LR and the two are not large. In summary, the improvements

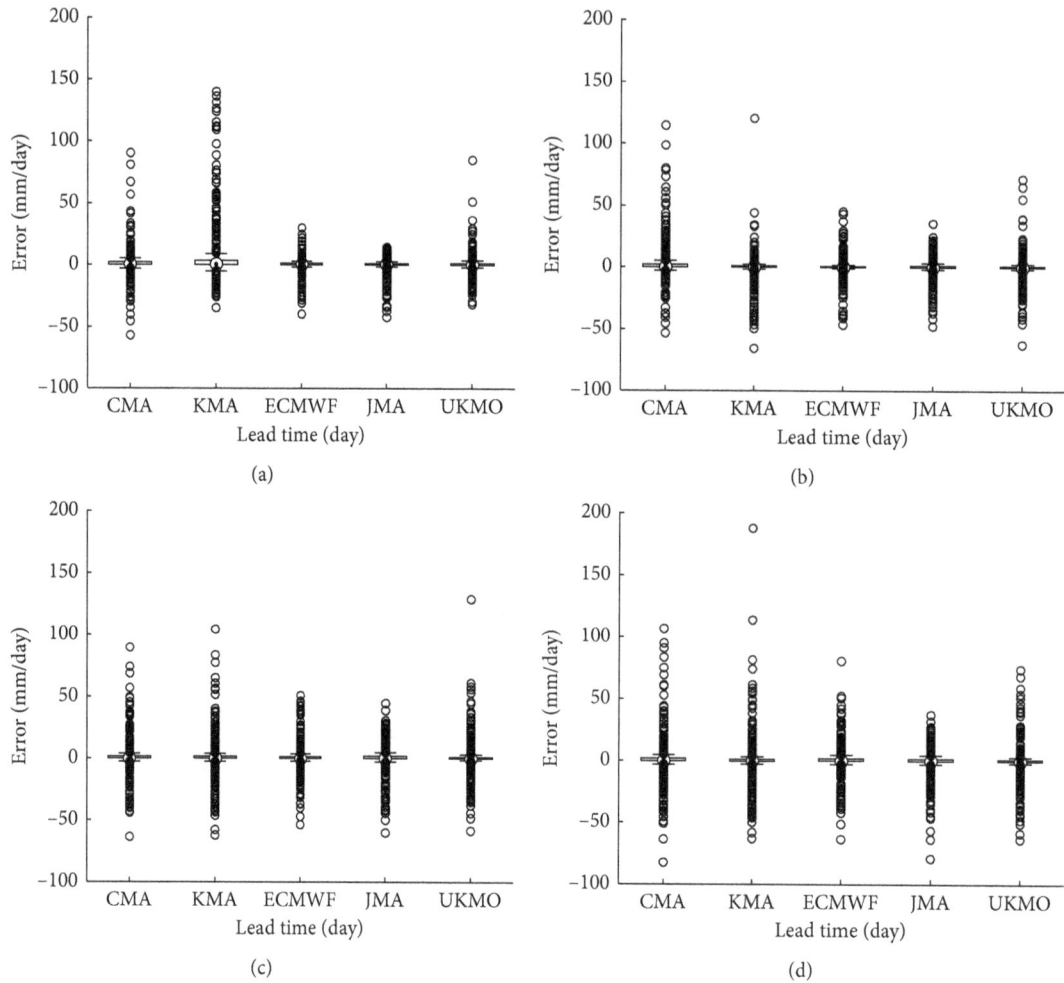

FIGURE 4: Box and whisker plots of forecast errors in the Xixian watershed during 2015–2017. The cross represents the mean value, the horizontal line in the box represents the median of the distribution (50% of the data are greater than this value), and the upper and lower box limits represent the upper and lower quartiles (25% of data greater/lower than the value), respectively. Maximum and minimum values are indicated by the top and bottom horizontal lines. The outlier points show values more than two-thirds of the quantile. (a) +1 d. (b) +2 d. (c) +5 d. (d) +7 d

of the three linear methods are extremely similar, that is, LR is slightly better than EM and BREM. SVR, as a nonlinear method, has a significantly better performance than linear methods in the training period. Additionally, since the accuracy of ensemble forecasts depends on the accuracy of the raw datasets, there is still an inverse relationship between the prediction accuracy and lead time.

Table 4 reveals a quite different conclusion that although the three linear methods perform similarly in RMSE, the HR of BREM is significantly higher than those of the other two methods, which have even lower HRs than the raw JMA dataset. Meanwhile, the HR of LR has a sharp decrease as the lead time reaches +6 d. This phenomenon is caused by the sudden drop of the hit number for no rain, from 127 at +5 d to 8 at +6 d. The main reason for the drop is that the linear equation enlarges all the values to the same extent due to the overall underestimation of the forecasts. Additionally, the SVR has the best result, with an increase of almost 15% in HR; however, compared with FAR, the decline of MAR is evidently smaller.

Since this paper focuses on the application of numerical precipitation forecasts to improve the lead time of flood forecasts and flood control safety, the error of MAs (ME) in the forecasts should be taken into consideration. The ME distribution of the four methods in the Xixian Catchment for different lead times in the training period is presented in Figure 6. The distribution of the four methods indicates the performance of the ensemble results in terms of correcting the ME. Both BREM and SVR have a range of MEs within 5 mm at different lead times during the training period, but there are still some large MEs, with a maximum of more than 70 mm. The performance of SVR at +3 d is acceptable for use in flood forecasting, with few MEs over 20 mm.

SVR has a remarkable advantage in promoting the accuracy of forecasts in the training period; however, many former cases suggested that some correction approaches perform well in the training period but have a poor performance in the validation period. Therefore, we evaluated the accuracy in the verification period in the Xixian

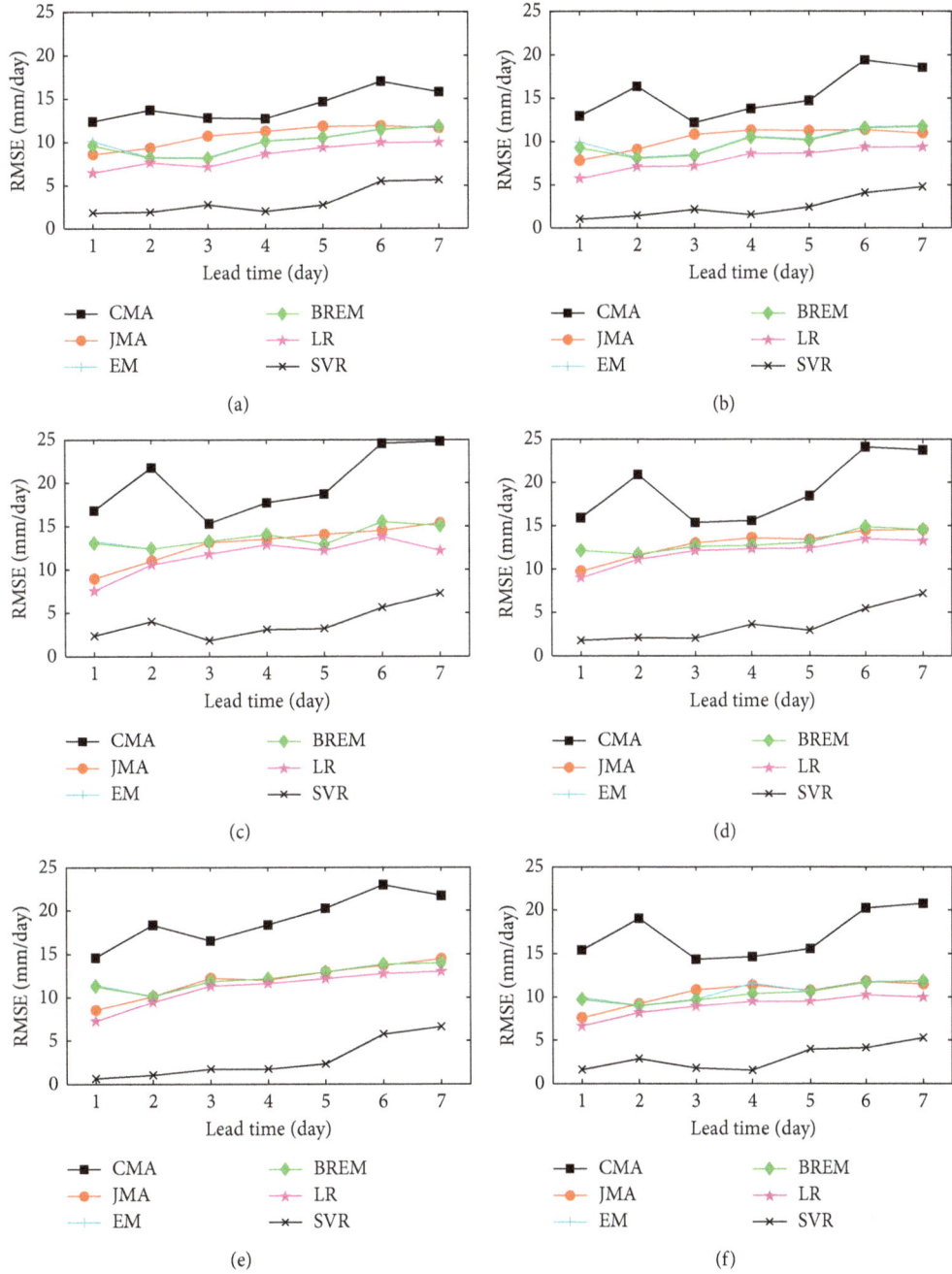

FIGURE 5: The RMSEs of different ensemble methods in the six different subcatchments during the training period, where (a–f) represent subcatchments 1–6.

TABLE 4: The three rates of the forecasts obtained using different ensemble methods in the Xixian catchment during the training period.

Lead time (days)	EM			BREM			LR			SVR		
	HR	FAR	MAR	HR	FAR	MAR	HR	FAR	MAR	HR	FAR	MAR
1	60.78	34.97	4.25	70.59	17.97	11.44	67.97	24.51	7.52	91.18	4.58	4.25
2	66.01	27.78	6.21	72.22	18.95	8.82	68.95	22.88	8.17	87.58	7.84	4.58
3	62.42	31.70	5.88	69.93	21.24	8.82	66.67	24.18	9.15	79.41	13.40	7.19
4	54.58	35.95	9.48	66.01	23.20	10.78	60.13	28.76	11.11	90.52	3.92	5.56
5	54.58	34.97	10.46	62.75	25.16	12.09	51.96	37.91	10.13	88.89	5.23	5.88
6	51.96	37.58	10.46	62.42	26.14	11.44	16.67	74.84	8.50	84.64	6.86	8.50
7	50.00	38.24	11.76	62.09	25.82	12.09	14.38	76.80	8.82	84.64	7.19	8.17

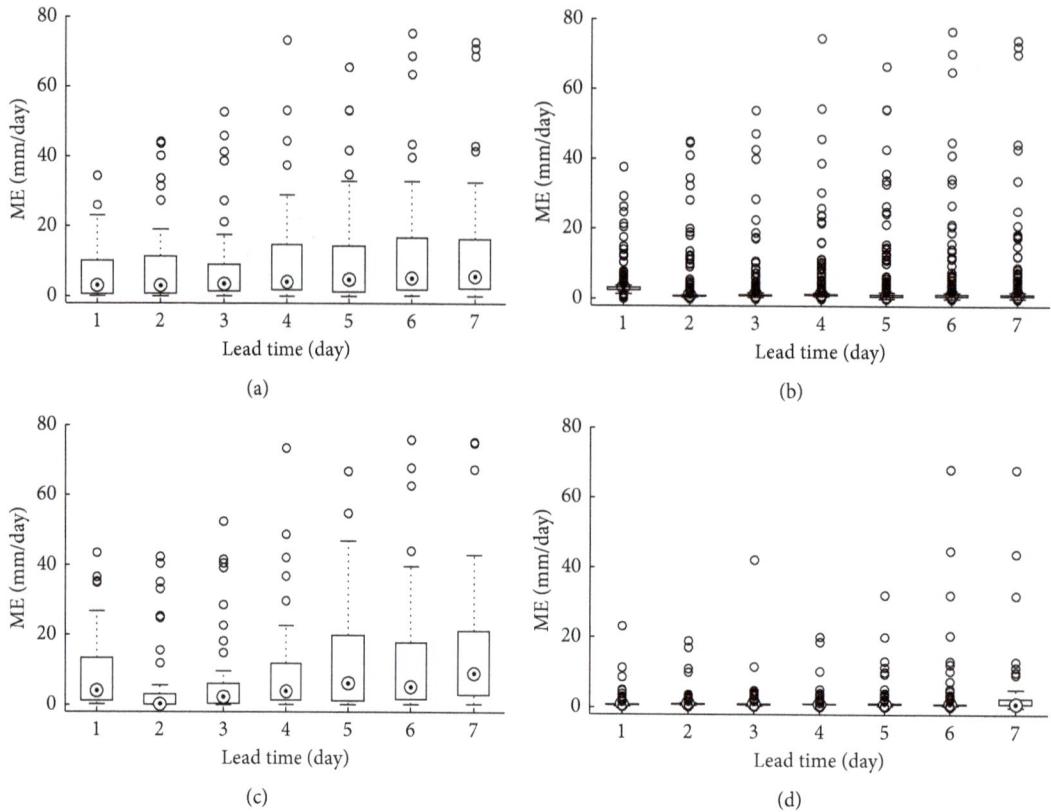

FIGURE 6: Box and whisker plots showing the performance in terms of the ME for 1–7 d rainfall obtained using different methods in the Xixian Catchment during the training period. (a) FM. (b) PREM. (c) LR. (d) SVR.

Catchment from three aspects: the RMSE, MAR, and ME distributions.

Additionally, the RMSE and three rates show different results in assessing the forecast ability during the verification period (Figure 7 and Table 5). Figure 7 indicates that the performance of LR is worse than those of EM and BREM in the verification period, which has advantages in the training period. Moreover, the RMSE of SVR presents a sharp increase at +6 d, indicating the worst forecasting ability of the four methods, while the three rates show that this method has advantages at each lead time. Both views imply that SVR has the best performance within +5 d. In addition, we compared and analyzed the ME of each method to carry out a more comprehensive assessment.

The ME distribution of BREM is the most uniform during the verification period (Figure 8), with most of the MEs between 0 and 5 mm except for the outlier values. However, the outlier values of BREM are greater than those of SVR, which is disadvantageous to flood control safety. Although the ME distribution ranges of SVR are larger, its outlier values are fewer in number and smaller in value within +3 d. To sum up, for the three linear methods, BREM has an obvious advantage over the other two, but the nonlinear method SVR is better than BREM. The result of SVR presents high precision in terms of the RMSE and three rates, but the distribution of the ME indicates that it still cannot avoid the huge disadvantages conferred to flood control security caused by MAs. Consequently, it is

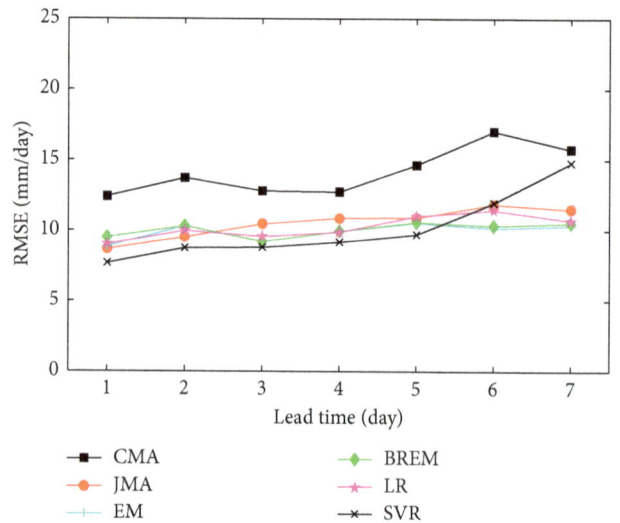

FIGURE 7: The RMSEs of different ensemble methods in the Xixian catchment during the verification period.

necessary to find a new ensemble method with smaller ME for flood forecast.

4.3. A New Ensemble Method for MA Rate Minimization. Although the results of SVR ensemble forecasting have suggested that it can effectively improve the accuracy of precipitation forecasts from the aspect of RMSE and the

TABLE 5: The three rates of the forecasts obtained using different ensemble methods in the Xixian catchment during the verification period.

Lead time (days)	EM			BREM			LR			SVR		
	HR	FAR	MAR	HR	FAR	MAR	HR	FAR	MAR	HR	FAR	MAR
1	67.97	22.88	9.15	78.43	7.19	14.38	68.63	18.95	12.42	81.70	10.46	7.84
2	67.97	24.18	7.84	74.51	16.99	8.50	68.63	20.26	11.11	73.20	14.38	12.42
3	52.94	37.91	9.15	66.01	24.18	9.80	58.82	27.45	13.73	71.90	18.30	9.80
4	56.86	33.33	9.80	66.67	21.57	11.76	59.48	26.80	13.73	77.12	11.11	11.76
5	54.25	34.64	11.11	60.78	26.14	13.07	52.94	32.03	15.03	73.86	15.69	10.46
6	50.98	34.64	14.38	63.40	21.57	15.03	7.84	74.51	17.65	75.82	12.42	11.76
7	47.06	36.60	16.34	57.52	24.84	17.65	15.69	69.28	15.03	71.90	12.42	15.69

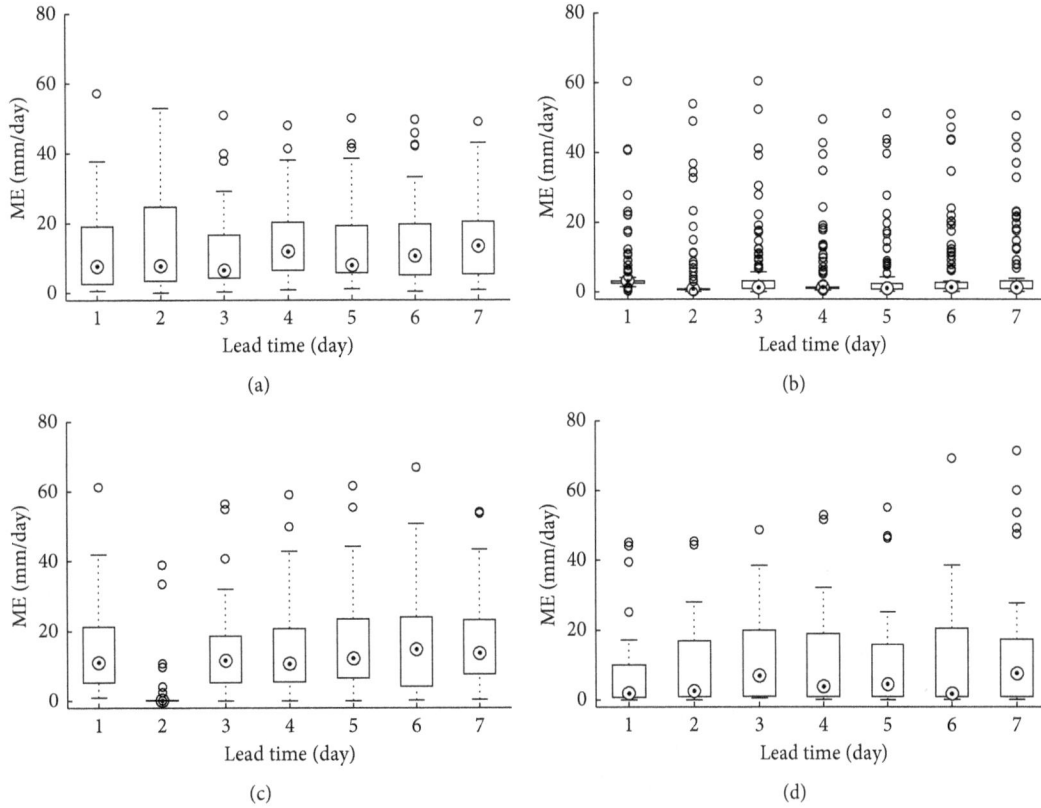

FIGURE 8: Box and whisker plots showing the performance in terms of the ME for 1–7 d rainfall forecasts obtained using different methods in the Xixian catchment during the verification period. (a) FM. (b) PREM. (c) LR. (d) SVR.

three rates, there are still large MEs (around 50 mm), which may generate great risk in flood control. However, since the FE may only lead to loss of economic benefits, while the ME may cause MAs of floods along with gigantic losses of life and economic damage, the results still cannot meet the requirement of prolonging the lead time of flood forecast without bringing huge flood risk. The traditional ensemble methods with minimizing MAE considers the whole deviation, which implies equal weights of ME and FE, but it is not appropriate for flood control where safety is the most important. Therefore, the ME should be reduced as much as possible in the ensemble forecast, and a modest loss of FE is acceptable. For this purpose, a new objection function was proposed for the ν-SVR model by minimizing the RMSE of MAs (SVR-MA).

$$\text{RMSEma} = \frac{1}{m} \sum_{i=1}^{l} (D_i)^2, \tag{14}$$

$$D_i = \begin{cases} y_i - f(x_i), & f(x_i) < y_i, \\ 0, & f(x_i) \geq y_i, \end{cases} \tag{15}$$

$$m = \sum_{i=1}^{l} I(D_i),$$

$$I(D_i) = \begin{cases} 1, & D_i > 0, \\ 0, & D_i \leq 0, \end{cases} \tag{16}$$

where $f(x_i)$ represents the function established by ν-SVR. Here, formula (14) was employed to optimize the parameters

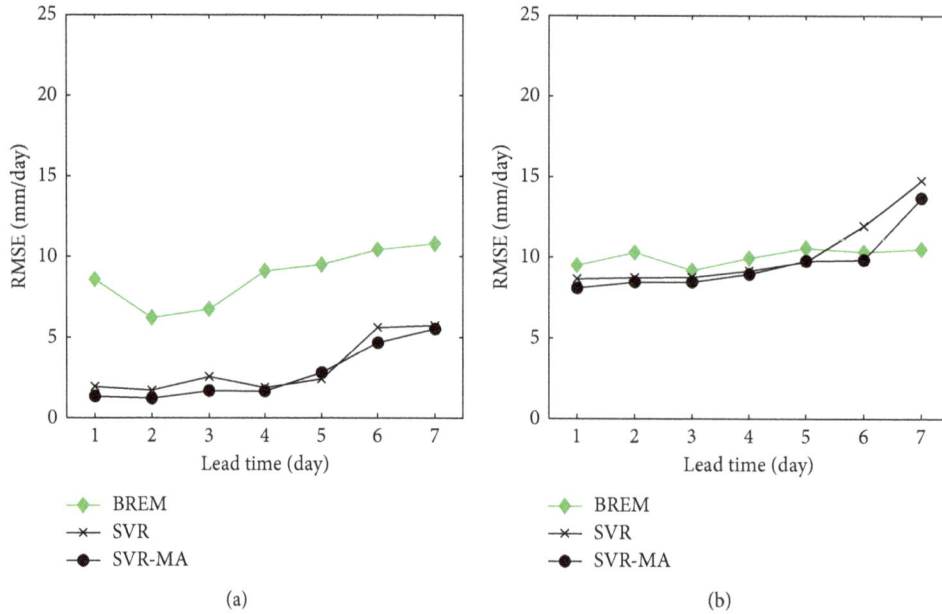

FIGURE 9: The RMSEs of different ensemble methods in the Xixian catchment. (a) Training period. (b) Verification period.

C, σ, and ν by minimizing the ME instead of the MAE in formula (13). Also, the number of support vector was controlled around $2/l$ by perturbing ν after the optimization.

Additionally, the improvement of the new method was analyzed compared with SVR and BREM for both the training period and the verification period. The RMSEs of SVR and SVR-MA are obviously smaller than that of BREM, and the RMSE of SVR-MA implies the best performance of the three methods within +3 d (Figure 9). From the aspect of the three rates (Table 6), all the HRs from different lead times are less than 50%, with a sharp increase in FAR, while the decrease of MAR is not clear. Table 7 illustrates the numbers of hits, MAs, and FAs from +1 d in the Xixian Watershed. Compared with the original SVR, the number of hits at the no rain magnitude using SVR-MA is far less than that with SVR, while the FAR number has a dramatic increase at light rain. Additionally, the MA number presents a slight decrease at each magnitude. For the purpose of flood control safety, it is a viable solution to exchange the accuracy of some FAs for the decrement in the MA number, especially for the FAs at no rain, which have a limited influence on the benefits of water resource management.

As we have discussed in the preceding part of this paper, it is biased to evaluate the rainfall forecast using only the three rates and RMSE. The distribution of MEs obtained using SVR-MA with different lead times is shown in Figure 10. The performance of SVR-MA is better than that of SVR, with a smaller distribution range and fewer outlier values for both the training period and verification period, especially when the lead time is 2 d. As most of the distribution ranges of MEs within +3 d in the verification period are less than 15 mm, several outlier values are expected. Moreover, most of the outlier values within +3 d are less than 30 mm.

TABLE 6: The three rates of SVR-MA in the Xixian catchment.

Lead time (days)	Training period			Verification period		
	HR	FAR	MAR	HR	FAR	MAR
1	47.71	49.02	3.27	39.22	52.29	8.50
2	78.10	16.99	4.90	71.90	20.92	7.19
3	32.68	64.05	3.27	26.14	64.05	9.80
4	49.67	44.12	6.21	43.14	47.71	9.15
5	36.60	59.48	3.92	26.14	61.44	12.42
6	33.01	61.76	5.23	26.80	62.09	11.11
7	33.99	60.78	5.23	27.98	63.62	8.50

TABLE 7: The numbers of hit, FAs, and MAs in the Xixian catchment obtained by different methods at +1 d during the whole period.

Magnitude	SVR-RM			SVR		
	Hit	FA	MA	Hit	FA	MA
No rain	121	0	11	320	0	16
Light rain	43	220	5	43	23	5
Medium rain	23	6	4	20	4	3
Heavy rain	15	4	3	16	3	1
Rainstorm	4	0	0	5	0	0
Heavy rainstorm	0	0	0	0	0	0
Extreme rainstorm	0	0	0	0	0	0

In conclusion, compared with the original ν-SVR, the new ensemble method (SVR-MA) has a similar performance in RMSE and three rates, which proves that the new model still has good approximation and generation ability. Meanwhile, SVR-MA is an efficient way to reduce the number and value of ME, especially for the number of ME over 20 mm. Most of the MEs are less than 10 mm in 3 d, and the maximum is around 30 mm. The new model not only reduces the overall error, but also makes the ME under

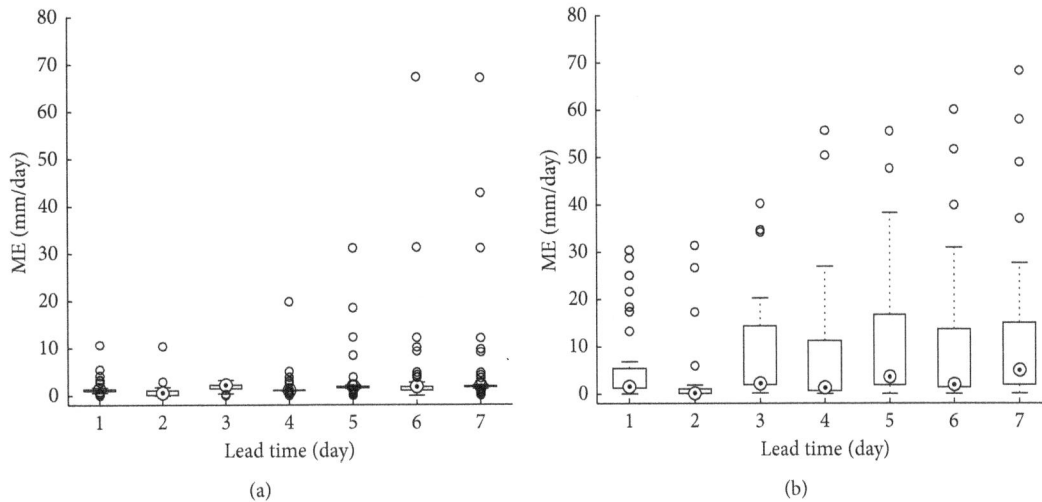

FIGURE 10: Box and whisker plots showing the distribution of the ME for 1–7 d rainfall forecasts obtained by SVR-MA in the Xixian Catchment. (a) Training period. (b) Verification period.

special control, which is beneficial to flood control safety and can be applied to prolong the forecast period of flood forecasting with small flood control risk.

5. Conclusions

How to extend the lead time of flood forecasts is essential for flood control safety and the development of society and the economy. Although numerical forecasting provides a possible way to solve this issue, it is still an urgent problem to fairly evaluate the accuracy of a rainfall forecast. Simple summary scores and classifications have limitations and cannot reflect the risk to flood control that precipitation forecasts may produce. In addition, more attention should be paid to the different impacts of MAs and FAs for flood control safety. In this study, we evaluated the ability of meteorological models of TIGGE to forecast the rainfall in six subcatchments with small areas during the flood seasons. Five numerical datasets were retained over the rainy seasons of 2015–2017 and compared with the observed dataset. We focused on the possibility for using rainfall forecast datasets to improve the lead time of flood forecasting. Additionally, we attempted to improve the performance of precipitation forecasts by employing different multimodal ensemble methods.

The forecast skill of most datasets from TIGGE has an inverse relationship with the lead time, which is in accordance with previous studies. Comparison between different datasets is undertaken from different aspects to evaluate the accuracy of the rainfall forecasts. The results show that the forecasts of JMA have the best performances in each subcatchment, and KMA and CMA exhibit irregular fluctuations at different lead times. Although the HRs of rainfall forecasts from the five centers are over 50% at +7 d, with high overestimations in CMA, KMA, and UKMO, as well as obvious underestimations by JMA and ECMWF, the raw datasets cannot meet the demand for flood forecasting.

Four ensemble methods were used in this study, including both linear and nonlinear methods. The results of SVR within 5 d are more reliable and accurate in comparison to those of the other three linear methods. Since, in this study, we placed more emphasis on the decrement of the ME, which leads to the risk of flood control, a new ensemble approach was proposed based on SVR. Compared with the original SVR, SVR-MA has a better performance in terms of ME at the expense of the FE, which is linked to the loss of benefits. However, most of the newly increased FAs are augmented from the no rain to light rain levels, which have a relatively low influence on the efficiency of water resource management. Meanwhile, we found that the performance of SVR-MA depends more on the accuracy of the datasets; thus, the results of SVR-MA after +3 d exhibit low skill in terms of the ME. It is worth mentioning that SVR-MA helps correcting the rainfall forecast and the ME of SVR-MA within +3 d may be an acceptable result for flood forecasting without causing much extra risk. These results provide us a new way to improve the lead time of flood forecasts under the prerequisite of flood control safety.

Conflicts of Interest

The authors declare that they have no conflicts of interest.

Acknowledgments

This study is supported by The National Key Research and Development Program of China (grant number: 2016YFC0400909). This study benefited from the TIGGE dataset provided by European Centre for Medium-Range Weather Forecasts (ECMWF), Reading, UK. We are grateful to the reviewers of the manuscript for their constructive comments and useful suggestions. We also wish to acknowledge Prof. Lin and the development team of LIBSVM.

References

[1] Y. Zhang, G. Wang, Y. Peng, and H. Zhou, "Risk analysis of dynamic control of reservoir limited water level by considering flood forecast error," *Science China-Technological Sciences*, vol. 54, no. 7, pp. 1888–1893, 2011.

[2] P. Liu, L. Li, S. Guo et al., "Optimal design of seasonal flood limited water levels and its application for the Three Gorges Reservoir," *Journal of Hydrology*, vol. 527, pp. 1045–1053, 2015.

[3] D. Liu, X. Li, S. Guo, D. Rosbjerg, and H. Chen, "Using a Bayesian probabilistic forecasting model to analyze the uncertainty in real-time dynamic control of the flood limiting water level for reservoir operation," *Journal of Hydrologic Engineering*, vol. 20, no. 2, article 04014036, 2015.

[4] Y. Zhou, S. Guo, F.-J. Chang, P. Liu, and A. B. Chen, "Methodology that improves water utilization and hydro-power generation without increasing flood risk in mega cascade reservoirs," *Energy*, vol. 143, pp. 785–796, 2018.

[5] E. Roulin and S. Vannitsem, "Skill of medium-range hydrological ensemble predictions," *Journal of Hydrometeorology*, vol. 6, no. 5, pp. 729–744, 2005.

[6] J. Thielen, J. Bartholmes, M.-H. Ramos, and A. de Roo, "The European flood Alert system-part 1: concept and development," *Hydrology and Earth System Sciences*, vol. 13, no. 2, pp. 125–140, 2009.

[7] H. J. Bao, L.-N. Zhao, Y. He et al., "Coupling ensemble weather predictions based on TIGGE database with Grid-Xinanjiang model for flood forecast," *Advances in Geosciences*, vol. 29, pp. 61–67, 2011.

[8] H. L. Cloke and F. Pappenberger, "Ensemble flood forecasting: a review," *Journal of Hydrology*, vol. 375, no. 3-4, pp. 613–626, 2009.

[9] P. Bauer, A. Thorpe, and G. Brunet, "The quiet revolution of numerical weather prediction," *Nature*, vol. 525, no. 7567, pp. 47–55, 2015.

[10] E. E. Ebert and J. L. McBride, "Verification of precipitation in weather systems: determination of systematic errors," *Journal of Hydrology*, vol. 239, no. 1, pp. 179–202, 2000.

[11] A. J. Simmons and A. Hollingsworth, "Some aspects of the improvement in skill of numerical weather prediction," *Quarterly Journal of the Royal Meteorological Society*, vol. 128, no. 580B, pp. 647–677, 2002.

[12] P. Bougeault, Z. Toth, C. Bishop et al., "The thorpex interactive grand global ensemble," *Bulletin of the American Meteorological Society*, vol. 91, no. 8, pp. 1059–1072, 2010.

[13] R. Swinbank, M. Kyouda, P. Buchanan et al., "The TIGGE project and its achievements," *Bulletin of the American Meteorological Society*, vol. 97, no. 1, pp. 49–67, 2016.

[14] M. P. Clark and L. E. Hay, "Use of medium-range numerical weather prediction model output to produce forecasts of streamflow," *Journal of Hydrometeorology*, vol. 5, no. 1, pp. 15–32, 2004.

[15] F. Pappenberger, J. Bartholmes, J. Thielen, H. L. Cloke, R. Buizza, and A. de Roo, "New dimensions in early flood warning across the globe using grand-ensemble weather predictions," *Geophysical Research Letters*, vol. 35, no. 10, 2008.

[16] Y. He, F. Wetterhall, H. Bao et al., "Ensemble forecasting using TIGGE for the July-September 2008 floods in the upper Huai catchment: a case study," *Atmospheric Science Letters*, vol. 11, no. 2, pp. 132–138, 2010.

[17] Y. He, F. Wetterhall, H. L. Cloke et al., "Tracking the uncertainty in flood alerts driven by grand ensemble weather predictions," *Meteorological Applications*, vol. 16, no. 1, pp. 91–101, 2009.

[18] X. Su, H. Yuan, Y. Zhu, Y. Luo, and Y. Wang, "Evaluation of TIGGE ensemble predictions of northern Hemisphere summer precipitation during 2008-2012," *Journal of Geophysical Research: Atmospheres*, vol. 119, no. 12, pp. 7292–7310, 2014.

[19] S. K. Sagar, M. Rajeevan, S. V. B. Rao, and A. K. Mitra, "Prediction skill of rainstorm events over India in the TIGGE weather prediction models," *Atmospheric Research*, vol. 198, pp. 194–204, 2017.

[20] R. L. Wilby and T. Wigley, "Downscaling general circulation model output: a review of methods and limitations," *Progress in Physical Geography*, vol. 21, no. 4, pp. 530–548, 1997.

[21] T. N. Krishnamurti, A. K. Mishra, A. Chakraborty, and M. Rajeevan, "Improving global model precipitation forecasts over India using downscaling and the FSU superensemble. Part I: 1-5-day forecasts," *Monthly Weather Review*, vol. 137, no. 9, pp. 2713–2735, 2009.

[22] N. Voisin, F. Pappenberger, D. P. Lettenmaier, R. Buizza, and J. C. Schaake, "Application of a medium-range global hydrologic probabilistic forecast scheme to the Ohio River Basin," *Weather and Forecasting*, vol. 26, no. 4, pp. 425–446, 2011.

[23] S. Louvet, B. Sultan, S. Janicot, P. H. Kamsu-Tamo, and O. Ndiaye, "Evaluation of TIGGE precipitation forecasts over West Africa at intraseasonal timescale," *Climate Dynamics*, vol. 47, no. 1-2, pp. 31–47, 2016.

[24] J. Vuillaume and S. Herath, "Improving global rainfall forecasting with a weather type approach in Japan," *Hydrological Sciences Journal-Journal Des Sciences Hydrologiques*, vol. 62, no. 2, pp. 167–181, 2017.

[25] D. Demeritt, S. Nobert, H. L. Cloke, and F. Pappenberger, "The European flood alert system and the communication, perception, and use of ensemble predictions for operational flood risk management," *Hydrological Processes*, vol. 27, no. 1, pp. 147–157, 2013.

[26] T. N. Krishnamurti, C. M. Kishtawal, T. E. LaRow et al., "Improved weather and seasonal climate forecasts from multimodel superensemble," *Science*, vol. 285, no. 5433, pp. 1548–1550, 1999.

[27] R. S. Ross and T. N. Krishnamurti, "Reduction of forecast error for global numerical weather prediction by the Florida State University (FSU) Superensemble," *Meteorology and Atmospheric Physics*, vol. 88, no. 3-4, pp. 215–235, 2005.

[28] T. J. Cartwright and T. N. Krishnamurti, "Warm season mesoscale superensemble precipitation forecasts in the southeastern United States," *Weather and Forecasting*, vol. 22, no. 4, pp. 873–886, 2007.

[29] L. Wu, D.-J. Seo, J. Demargne, J. D. Brown, S. Cong, and J. Schaake, "Generation of ensemble precipitation forecast from single-valued quantitative precipitation forecast for hydrologic ensemble prediction," *Journal of Hydrology*, vol. 399, no. 3-4, pp. 281–298, 2011.

[30] A. J. Clark, "Generation of ensemble mean precipitation forecasts from convection-allowing ensembles," *Weather and Forecasting*, vol. 32, no. 4, pp. 1569–1583, 2017.

[31] V. V. Kharin and F. W. Zwiers, "Climate predictions with multimodel ensembles," *Journal of Climate*, vol. 15, no. 7, pp. 793–799, 2002.

[32] T. N. Palmer, A. Alessandri, U. Andersen et al., "Development of a European multimodel ensemble system for seasonal-to-interannual prediction (DEMETER)," *Bulletin of the American Meteorological Society*, vol. 85, no. 6, pp. 853–872, 2004.

[33] X. Zhi, H. Qi, Y. Bai, and C. Lin, "A comparison of three kinds of multimodel ensemble forecast techniques based on the TIGGE data," *Acta Meteorologica Sinica*, vol. 26, no. 1, pp. 41–51, 2012.

[34] Y. B. Dibike, S. Velickov, D. Solomatine, and M. B. Abbott, "Model induction with support vector machines: introduction and applications," *Journal of Computing in Civil Engineering*, vol. 15, no. 3, pp. 208–216, 2001.

[35] P. Yu, S. Chen, and I. Chang, "Support vector regression for real-time flood stage forecasting," *Journal of Hydrology*, vol. 328, no. 3-4, pp. 704–716, 2006.

[36] C. Lai and M. Tseng, "Comparison of regression models, grey models, and supervised learning models for forecasting flood stage caused by typhoon events," *Journal of the Chinese Institute of Engineers*, vol. 33, no. 4, pp. 629–634, 2010.

[37] G. Lin, Y. Chou, and M. Wu, "Typhoon flood forecasting using integrated two-stage support vector machine approach," *Journal of Hydrology*, vol. 486, pp. 334–342, 2013.

[38] N. S. Raghavendra and P. C. Deka, "Support vector machine applications in the field of hydrology: a review," *Applied Soft Computing*, vol. 19, pp. 372–386, 2014.

[39] B. Scholkopf, S. Mika, C. J. C. Burges et al., "Input space versus feature space in kernel-based methods," *IEEE Transactions on Neural Networks*, vol. 10, no. 5, pp. 1000–1017, 1999.

[40] B. Yadav, S. Ch, S. Mathur, and J. Adamowski, "Estimation of in-situ bioremediation system cost using a hybrid Extreme Learning Machine (ELM)-particle swarm optimization approach," *Journal of Hydrology*, vol. 543, pp. 373–385, 2016.

[41] G. Xu and G. Yu, "On convergence analysis of particle swarm optimization algorithm," *Journal of Computational and Applied Mathematics*, vol. 333, pp. 65–73, 2018.

8

High-Resolution Climate Simulations in the Tropics with Complex Terrain Employing the CESM/WRF Model

Tomi Afrizal (ID)[1] **and Chinnawat Surussavadee** (ID)[2,3]

[1]*Interdisciplinary Graduate School of Earth System Science and Andaman Natural Disaster Management,*
Prince of Songkla University, Phuket Campus, Phuket 83120, Thailand
[2]*Telecommunications Engineering Department, Faculty of Engineering, King Mongkut's Institute of Technology Ladkrabang,*
Bangkok 10520, Thailand
[3]*Research Laboratory of Electronics, Massachusetts Institute of Technology, Cambridge, MA 02139, USA*

Correspondence should be addressed to Chinnawat Surussavadee; pop@alum.mit.edu

Academic Editor: Mario M. Miglietta

This study evaluates the high-resolution climate simulation system CESM/WRF composed of the global climate model, Community Earth System Model (CESM) version 1, and the mesoscale model, Weather Research and Forecasting Model (WRF), for simulating high-resolution climatological temperature and precipitation in the tropics with complex terrain where temperature and precipitation are strongly inhomogeneous. The CESM/WRF climatological annual and seasonal precipitation and temperature simulations for years 1980–1999 at 10 km resolution for Sumatra and nearby regions are evaluated using observations and the global climate reanalysis ERA-Interim (ERA). CESM/WRF simulations at 10 km resolution are also compared with the downscaled reanalysis ERA/WRF at 10 km resolution. Results show that while temperature and precipitation patterns of the original CESM are very different from observations, those for CESM/WRF agree well with observations. Resolution and accuracies of simulations are significantly improved by dynamically downscaling CESM using WRF. CESM/WRF can simulate locations of very cold temperature at mountain peaks well. The high-resolution climate simulation system CESM/WRF can provide useful climate simulations at high resolution for Sumatra and nearby regions. CESM/WRF-simulated climatological temperature and precipitation at 10 km resolution agree well with ERA/WRF. This suggests the use of CESM/WRF for climate projections at high resolution for Sumatra and nearby regions.

1. Introduction

Climate change is an important threat to humanity. The global average temperature has been rising and has been projected to increase up to 2–5°C by the end of the twenty-first century [1]. Global warming causes glacier retreat, sea level rise, extreme weather intensification, and changes in precipitation amount and pattern [2]. Accurate numerical systems for climate simulation and projection at sufficiently high resolution are required for effective climate change mitigation and adaptation, which will improve resilience of the society.

To obtain climate simulations and projections, global climate models (GCMs), which model physical processes of atmosphere, ocean, land surface, and cryosphere by taking into the account of different scenarios of increasing greenhouse gases, can be employed [1]. However, the coarse resolutions of GCMs are generally not sufficient to provide useful climate change information and impacts for a specific area, particularly where climate and weather are inhomogeneous, e.g., the areas with complex terrain and in the tropics where precipitation is strongly driven by convection at finer scales than those can be resolved by GCMs.

Indonesia's Sumatra Island is in the tropics with the equator running through it and is one of the rainiest areas on Earth. Weather-related disasters, i.e., floods, landslides, and severe storms often affect Sumatra. Since Sumatra has complex terrain with high mountains and volcanoes and temperature and precipitation are very inhomogeneous, numerical systems for simulating and projecting climate at high resolution are required for appropriately adapting to climate change and reducing climate change impacts. Although there are some previous climate simulation studies for Southeast Asia [3, 4] and for Indonesia [5], their simulations at 60 km resolution are too coarse to resolve climatological temperature and precipitation of Sumatra, as the results from this study will clearly show.

High-resolution climate simulations and projections can be obtained by dynamically downscaling GCM outputs using a mesoscale model [6, 7]. However, such previous studies for Sumatra do not exist. Since climate is different for different areas and GCMs and mesoscale models perform differently for different areas, this study is the work towards the main goal of developing a numerical system that is capable of providing useful high-resolution climate simulations for Sumatra. If the system performs well, this study could be extended to the development of a numerical system for providing climate projections at high resolution for Sumatra.

The Community Earth System Model (CESM) version 1 [8] is a GCM participating in the fifth Coupled Model Intercomparison Project (CMIP5) [9]. CESM climate simulations and projections have been widely used for climate studies [10–13]. Since CESM's three-dimensional outputs are publicly available and are ready to be used to drive a mesoscale model, CESM is employed in this study. Several mesoscale models are publicly available, e.g., the fifth-generation NCAR/Penn State Mesoscale Model (MM5) [14, 15] and the next-generation Weather Research and Forecasting Model (WRF) [16, 17]. WRF is employed in this study.

The numerical climate simulation system employed in this study is composed of CESM and WRF and is called CESM/WRF. Although the preliminary study [18] has evaluated the performance of CESM/WRF for simulating climatological temperature and precipitation for Sumatra by comparing CESM/WRF simulations for years 1980–1999 with observation and reanalysis datasets, it has evaluated only simulated annual temperature and precipitation and the highest resolution of observation and reanalysis datasets used for evaluation is only 0.5°. The performance of CESM/WRF is evaluated in further details in this study and is compared with that of the original CESM.

Several climate simulation studies [19–21] have employed WRF to downscale the ERA-Interim reanalysis dataset (ERA) [22]. Soares et al.[19] have evaluated the performances for simulating climatological temperature and precipitation of ERA downscaled using WRF (ERA/WRF) and the original ERA using observations from dense ground stations in Portugal and have shown that ERA/WRF performs better than ERA, particularly for high-resolution precipitation simulations in the rainiest regions with crucial orographic enhancement. Huang and Gao [20] have evaluated the performances for simulating climatological temperature and precipitation of WRF dynamical downscaling forced by ERA and the National Centers for Environmental Prediction (NCEP) Global Final

Analysis (FNL) and have shown that they perform very differently.

Observations for temperature and precipitation at 10 km resolution or better for the 20-year period of 1980–1999 and dense ground stations in Sumatra do not exist [23]. Sparse ground stations cannot provide accurate information about spatial distributions of temperature and precipitation, particularly in the tropics with complex terrain, where temperature and precipitation are very inhomogeneous. From results of [19], ERA dynamically downscaled using WRF should provide reasonable high-resolution temperature and precipitation for evaluating CESM/WRF-simulated climatological temperature and precipitation at 10 km resolution, particularly for the regions with the strong orographic effect. In this study, ERA is dynamically downscaled using WRF to obtain downscaled reanalysis at 10 km resolution (ERA/WRF). CESM/WRF-simulated climatological temperature and precipitation at 10 km resolution are compared with those of ERA/WRF. The comparison of CESM/WRF and ERA/WRF also benefits climate projection studies. Since ERA does not provide climate projection while CESM does, if CESM/WRF performs comparably or better than ERA/WRF, it will gain confidence of using CESM/WRF for climate projections.

Section 2 describes the research methodology employed in this study, which includes the study area, observation and reanalysis datasets, the high-resolution numerical climate simulation system CESM/WRF, and the downscaled reanalysis ERA/WRF. The evaluation results are presented in Section 3. The study is summarized and concluded in Section 4.

2. Research Methodology

2.1. Study Area. Figure 1 shows the study area, which covers Indonesia's Sumatra Island and part of Java Island, Malaysia, and Singapore. Sumatra is one of the major islands in Indonesia with the equator crossing near its center. It stretches along a diagonal northwest-southeast axis in western Indonesia with a total land area of about 473,606 km^2. Sumatra and Java islands have complex terrain with several high mountains and volcanoes stretching along the west coast of the island due to the Sunda Megathrust where the Indo-Australian plate subducts beneath the Eurasian plate. Malaysia also has high mountains near its center.

The climate of Sumatra is tropical with hot and humid weather. There are 2 major seasons, including the rainy season approximately from October to April of each year and the dry season approximately from May to September of each year. Sumatra has significant precipitation amount throughout the year. Precipitation is mostly convective and varies greatly from area to area. Intense convective precipitation could lead to landslides and floods. Sumatra has often been affected by floods, landslides, and severe storms. High-resolution numerical climate simulations and projections are hence important for better understanding of climate change and its potential impacts for Sumatra.

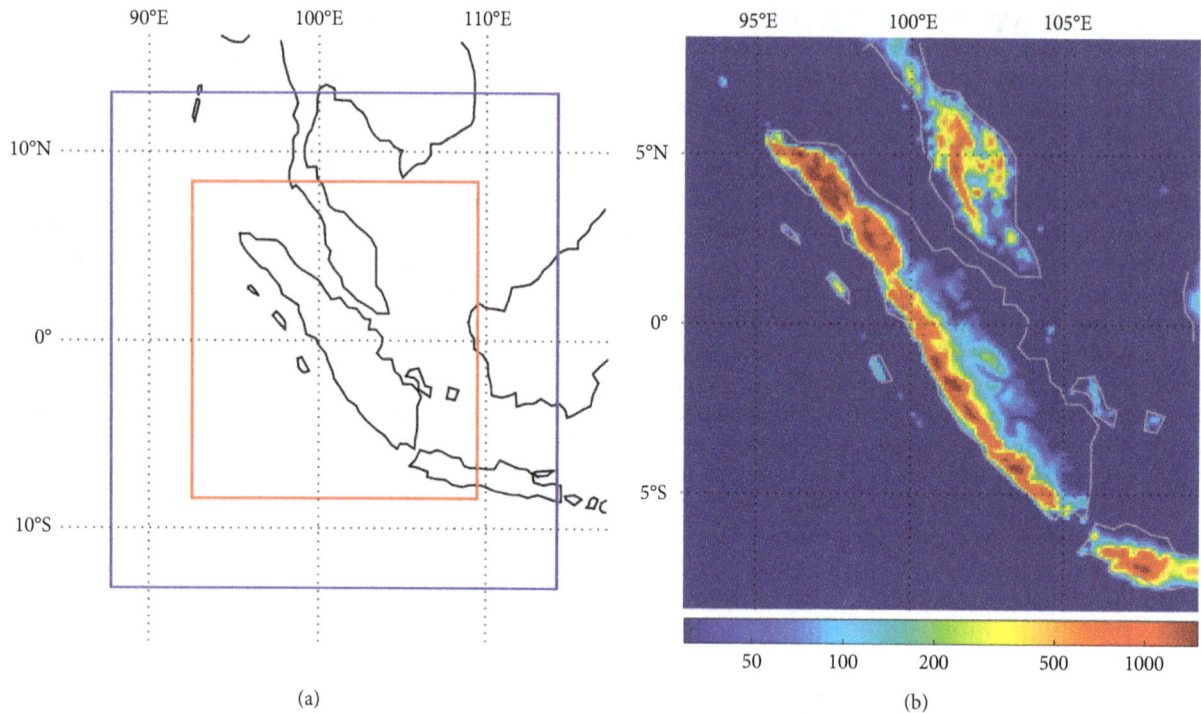

(a)

(b)

FIGURE 1: (a) Coverages of WRF's 2 co-centered domains employed in this study. (b) Topography (m) above mean sea level of the WRF's inner domain

2.2. Observation and Reanalysis Datasets. Four observation datasets including the University of Delaware Air Temperature and Precipitation version 3.01 (UD) [24], the University of East Anglia Climatic Research Unit TS3.10 (CRU) [25], the Global Precipitation Climate Center (GPCC) [26], and the NOAA Climate Prediction Center Unified Gauge-Based Analysis of Global Daily Precipitation (CPC) [27] and the ERA-Interim global atmospheric reanalysis dataset (ERA) [22] covering years 1980–1999 are employed in this study for evaluating CESM and CESM/ WRF climatological simulations.

UD monthly global gridded data for air temperature and precipitation are available from 1900 to 2010 and are produced using data both from the Global Historical Climate Network and the archive of Legates and Willmott. CRU monthly global gridded data for air temperature and precipitation are available from 1901 to 2009 and are produced using daily or subdaily data by National Meteorological Services and other external agents. GPCC monthly global gridded precipitation data are available from 1901 to present and are produced using quality-controlled data from 67,200 global stations. CPC daily global gridded precipitation data are available from 1979 to present and are produced using quality-controlled gauge reports from over 30,000 global stations with consideration of orographic effects. UD, CRU, GPCC, and CPC are available only over land and are on regular 0.5° grids.

ERA reanalysis is produced using the four-dimensional variational analysis (4D-Var) system that relies on both observations and model-based forecasts and is available from 1979 to present. ERA is available for both land and sea

and is on a regular 0.75° grid with 60 vertical levels from the surface up to 0.1 mb. ERA outputs are available every 6 h, i.e., 00, 06, 12, and 18Z and can be used as initial and boundary conditions for WRF. ERA precipitation is computed using accumulated precipitation available every 12 h. This study treats ERA reanalysis data as observations.

2.3. CESM/WRF Climate Simulations. The numerical climate simulation system used in this study is composed of the global climate model Community Earth System Model (CESM) version 1 [8] and the next-generation mesoscale numerical weather prediction model Weather Research and Forecasting (WRF) Model [16, 17]. CESM has been developed by the National Center for Atmospheric Research (NCAR) and is a fully-coupled global climate model (GCM). It is a GCM that participates in the fifth Coupled Model Intercomparison Project (CMIP5). CESM outputs are at ~1-degree resolution with 26 pressure levels and are available every 6 h, including 00, 06, 12, and 18Z.

WRF has been widely employed for research and operations. There are several WRF versions. This study employs WRF with the Advanced Research WRF core version 3.7.1. WRF's initial and boundary conditions are CESM outputs. Figure 1 shows coverages of WRF's 2 co-centered domains employed in this study. The one-way nesting strategy is used. WRF's vertical levels are terrain following. Each domain has 35 vertical levels extending from the ground up to 50 mb. The outer domain has the size of 100 × 100 grid points with 30 km resolution and covers latitudes of 13°S–13°N and longitudes of 87°E–114°E. The inner domain has the size of 190 × 190 grid points with 10 km resolution and covers latitudes of 8.5°S–8.5°N and longitudes of 92.5°E–110°E. The spectral nudging is employed for the outer domain for altitudes above planetary boundary layer so that WRF simulations at large scales are consistent with CESM outputs. The wavenumber for spectral nudging is set to 3, which corresponds to the adjustment only for waves greater than ~1,000 km [28]. The frequency of the adjustment is 24 h.

Several WRF physics options are available. Surussavadee and Aonchart [29] have found that the combination of the WRF Double-Moment 6-class microphysics scheme [30] and the Betts–Miller–Janjic cumulus parameterization scheme [31] provides the best agreement between high-resolution weather forecasts and satellite observations for Thailand and nearby regions. Surussavadee [17] has found that the combination of the Bretherton and Park (UW) planetary boundary layer scheme [32], the Revised MM5 Monin-Obukhov surface layer scheme [33], and the Unified Noah land surface model [34] provides the best agreement between simulated near-surface winds and ground measurements for Thailand. The WRF physics options employed in this study follow the best physic options found in References [17, 29].

Sumatra's climate for 20 y covering 1980–1999 is simulated by CESM/WRF. Two separate WRF integrations for each decade are used in order to optimize computational resources and time. The spin up time of 1 y is employed. CESM/WRF outputs from the inner domain at 10 km resolution are employed in this study. Since CESM/WRF outputs are at 10 km resolution, whereas all observation datasets, i.e., CPC, CRU, GPCC, and UD are at 0.5° resolution and the ERA reanalysis is at 0.75° resolution, to generate CESM/WRF simulations at a resolution comparable to those of observations and reanalysis; CESM/WRF simulations at 10 km resolution are convolved with a Gaussian function having full width at half maximum (FWHM) of 50 km before they are evaluated. All observations and reanalysis are bilinearly interpolated on the grid of the CESM/WRF inner domain. The simulation performances of CESM and CESM/WRF are evaluated using the performance metrics, including root-mean-squared errors (RMSEs), mean errors (MEs), which is E[observations − simulations], and correlation coefficients (CCs) of simulations and observations.

2.4. ERA/WRF Climate Simulations. The ERA-Interim global atmospheric reanalysis dataset (ERA) [22] is dynamically downscaled using WRF to obtain downscaled reanalysis at 10 km resolution and is called ERA/WRF. The WRF domain configurations and physics options and the strategy for downscaling employed for ERA/WRF are the same as those employed for CESM/WRF. ERA/WRF outputs from the inner domain at 10 km resolution are employed.

3. Results

3.1. Evaluation of Climatological Temperature Simulated by CESM and CESM/WRF Using Observations and Reanalysis.

Figure 2 compares 20-year average temperature (°C) simulated by the original CESM and the WRF-downscaled CESM (CESM/WRF) with CRU, UD, and ERA and the average of CRU, UD, and ERA. All are on the same CESM/WRF inner grid. CESM/WRF in this section is the original 10 km resolution CESM/WRF convolved with a Gaussian function having full width at half maximum (FWHM) of 50 km. Figure 2 shows that all observations, including CRU, UD, and ERA, are different from one another. UD appears to have the highest spatial resolution among the three observations, whereas the spatial resolution of ERA is the lowest. CRU's temperature over eastern Sumatra is obviously higher than that of UD and ERA. Although all observations show the cold temperature patterns over high mountains along the west coasts of Sumatra and Java islands and in the center areas of Malaysia, only UD can resolve cold spots at mountain peaks.

Simulated temperature of the original CESM and CESM/WRF is very different for both pattern and intensity. CESM's temperature over sea is obviously higher than all observations and CESM/WRF. CESM's temperature over land is also very different from all observations and CESM/WRF. CESM/WRF agrees with observations much better than CESM does, while CESM obviously cannot resolve temperature for high mountains along the west coasts of Sumatra and Java islands; CESM/WRF does it well. CESM/WRF can also simulate cold spots over mountain peaks well. The locations of CESM/WRF's cold spots agree well with those of UD. CESM/WRF's temperature over sea is obviously improved over that of CESM. Downscaling CESM using WRF significantly improves the simulated climatological temperature for both land and sea.

Figure 3(a) shows the scatter plots comparing annual average temperature (°C) for individual years of 1980–1999 simulated by CESM and CESM/WRF with the average of CRU, UD, and ERA. Points in the scatter plots are samples of all grid cells of the CESM/WRF inner domain. Simulated annual average temperature for the 20 y of CESM/WRF is more accurate than that of CESM and agrees well with observations. Downscaling CESM using WRF improves correlation coefficient (CC) between simulations and observations from 0.62 to 0.83. CESM's simulated temperature at the low end appears to be clipped at ~24°C. CESM also overestimates temperatures higher than ~28°C. CESM/WRF does not have these two issues. Figure 3(b) shows scatter plots comparing 20-y average temperature (°C) simulated by CESM and CESM/WRF with the average of CRU, UD, and ERA. Results are similar to those shown in Figure 3(a). CESM/WRF 20 y average temperature simulations are obviously more accurate than CESM simulations. Downscaling CESM using WRF improves correlation coefficient between simulations and observations from 0.66 to 0.87.

Table 1 shows RMSEs, MEs (E[observations − simulations]), and CCs for annual average temperature (°C) for 20 y of 1980–1999 simulated by CESM and CESM/WRF evaluated using the average of CRU, UD, and ERA for land only, sea only, and both land and sea. Boldface highlights the model that performs best for each performance metric. Table 2 shows the same as that shown in Table 1, but

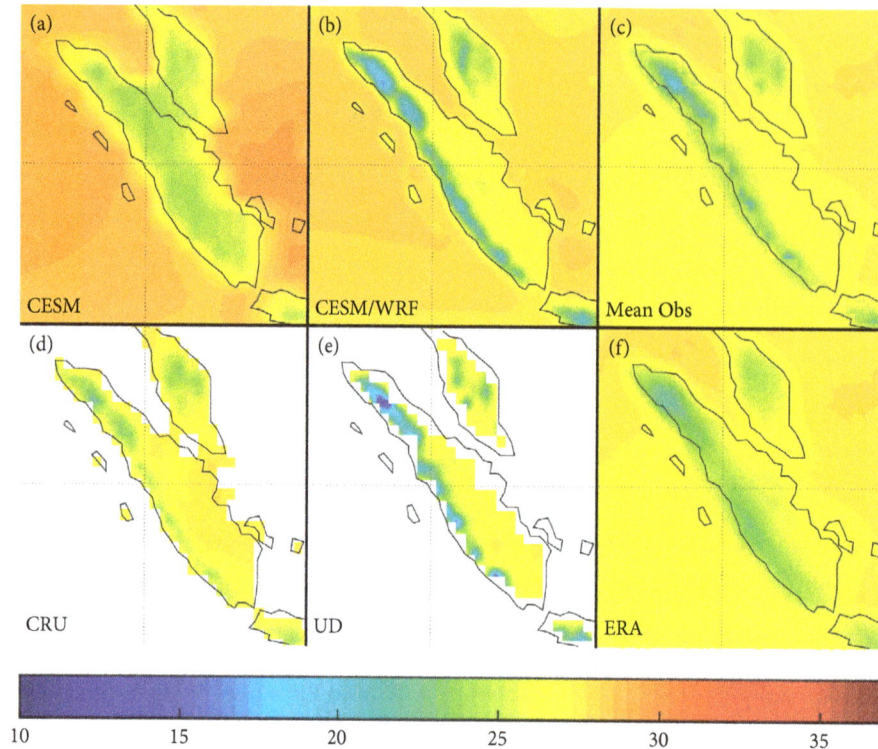

FIGURE 2: Comparisons of 20-y average temperature (°C) simulated by CESM and CESM/WRF with CRU, UD, and ERA and the average of CRU, UD, and ERA. CESM and ERA are at 1° and 0.75° resolutions, respectively. CRU and UD are at 0.5° resolution. CESM/WRF is at 50 km resolution.

for 20-y average temperature. Results are consistent with those shown in Figures 2 and 3. Accuracies of CESM/WRF simulations are significantly better than those of CESM simulations for both land and sea. Dynamic downscaling improves overall RMSE from 1.69 to 1.03°C for annual average temperature for 20 y and from 1.62 to 0.94°C for 20-y average temperature. ME reduces from −1.18 to −0.57°C. CC improves from 0.62 to 0.83 for annual average temperature for 20 y and improves from 0.66 to 0.87 for 20 y average temperature.

3.2. Comparison of High-Resolution Climatological Temperature Simulated by CESM/WRF and ERA/WRF.
Climatological temperatures simulated by CESM/WRF and ERA/WRF at 10 km resolution are compared in this section. Figure 4 shows CESM/WRF- and ERA/WRF-simulated 20-y average temperature simulations (°C). The pattern and intensity of CESM/WRF-simulated 20-y average temperature agree well with those of ERA/WRF. Both CESM/WRF and ERA/WRF show low temperature pattern along the west coasts of Sumatra and Java and in the middle of Malaysia. Both also show fine-scale cold spots over mountain peaks, which are consistent with UD shown in Figure 2. Their areas of high temperature also agree well. Comparison of ERA/WRF with UD shows that dynamic downscaling employing WRF not only improves CESM simulations but also improves ERA reanalysis.

Figure 5 shows scatter plots comparing CESM/WRF and ERA/WRF for annual average temperature (°C) for 20 y and

20-y average temperature (°C). CESM/WRF and ERA/WRF agree well for both annual average temperature for 20 y and 20-y average temperature with high CCs of 0.97 and 1.00, respectively.

Table 3 shows the root-mean-squared differences (RMSDs), mean differences (MDs), which is E[ERA/WRF − CESM/WRF], and CCs of CESM/WRF and ERA/WRF for simulated annual average temperature (°C) for 20 y and 20-y average temperature (°C) for land only, sea only, and both land and sea. CESM/WRF and ERA/WRF climatological temperature agree well for both land and sea. Overall RMSDs for annual average temperature for 20 y and 20-y average temperature are 0.44 and 0.11°C, respectively. MDs between CESM/WRF and ERA/WRF are almost zero. CCs between CESM/WRF and ERA/WRF for annual average temperature for 20 y and 20-y average temperature are 0.97 and 1.00, respectively.

3.3. Evaluation of Climatological Precipitation Simulated by CESM and CESM/WRF Using Observations and Reanalysis.
The accuracies of CESM- and CESM/WRF-simulated climatological precipitation are evaluated using observations and reanalysis. CESM/WRF in this section is the original 10 km resolution CESM/WRF convolved with a Gaussian function having full width at half maximum (FWHM) of 50 km. Figure 6 compares 20-y average annual precipitation (mm/y) simulated by CESM and CESM/WRF with CPC, CRU, GPCC, UD, and ERA and the average of CPC, CRU, GPCC, UD, and ERA. All are on the same CESM/WRF inner grid.

(a)

(b)

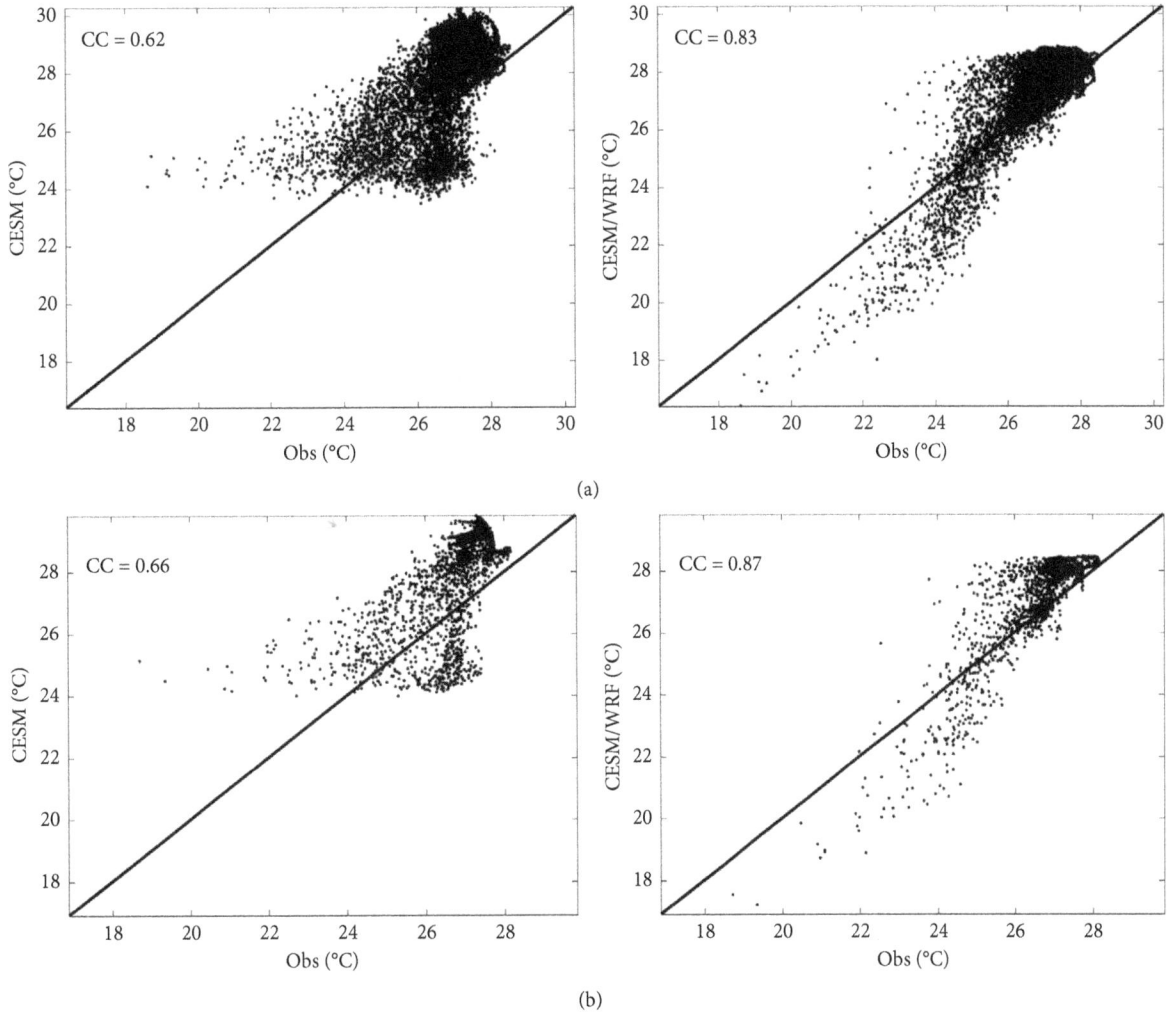

FIGURE 3: Scatter plots comparing CESM and CESM/WRF simulations with the average of CRU, UD, and ERA for (a) annual average temperature (°C) for 20 y and (b) 20-y average temperature (°C). CESM and ERA are at 1° and 0.75° resolutions, respectively. CRU and UD are at 0.5° resolution. CESM/WRF is at 50 km resolution.

TABLE 1: RMSEs, MEs (E[observations − simulations]), and CCs of simulations and observations for CESM and CESM/WRF-simulated annual average temperature (°C) for 20 y evaluated using the average of CRU, UD, and ERA for land only, sea only, and both land and sea, where CESM/WRF is at 50 km resolution.

Metric	CESM			CESM/WRF		
	Land	Sea	All	Land	Sea	All
RMSE	1.49	1.74	1.69	**0.96**	**1.05**	**1.03**
ME	**−0.12**	−1.51	−1.18	0.23	−0.82	−0.57
CC	0.31	**0.29**	0.62	**0.90**	0.23	**0.83**

Boldface highlights the model performing best for each performance metric.

TABLE 2: RMSEs, MEs (E[observations − simulations]), and CCs of simulations and observations for CESM and CESM/WRF-simulated 20-y average temperature (°C) evaluated using the average of CRU, UD, and ERA for land only, sea only, and both land and sea, where CESM/WRF is at 50 km resolution.

Metric	CESM			CESM/WRF		
	Land	Sea	All	Land	Sea	All
RMSE	1.45	1.67	1.62	**0.92**	**0.95**	**0.94**
ME	**−0.12**	−1.51	−1.18	0.23	−0.82	−0.57
CC	0.32	0.39	0.66	**0.92**	**0.44**	**0.87**

Boldface highlights the model performing best for each performance metric.

Comparison of the observation and reanalysis datasets shows some differences due to the sources and methods employed for different datasets and their spatial resolutions. Comparison of observations over Malaysia shows that CRU has high precipitation at the center of Malaysia and surrounding areas, CPC has high precipitation along the east coast, GPCC and UD have high precipitation along the east

and west coasts, and ERA has high precipitation along the west coast. Only ERA has precipitation data over the sea. Three main precipitation patterns consistent with most observation datasets include: (1) high precipitation over land along the west coasts of Sumatra and Java islands and lower precipitation over land along the east coasts, (2) high precipitation over the Indian Ocean to the west of Sumatra, and

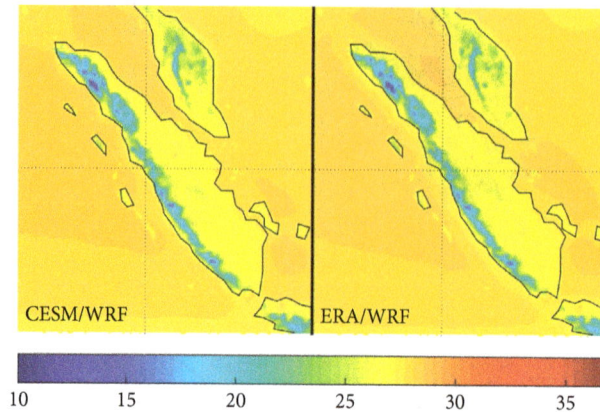

FIGURE 4: Comparisons of CESM/WRF and ERA/WRF 20-y average temperature (°C). Both CESM/WRF and ERA/WRF are at 10 km resolution.

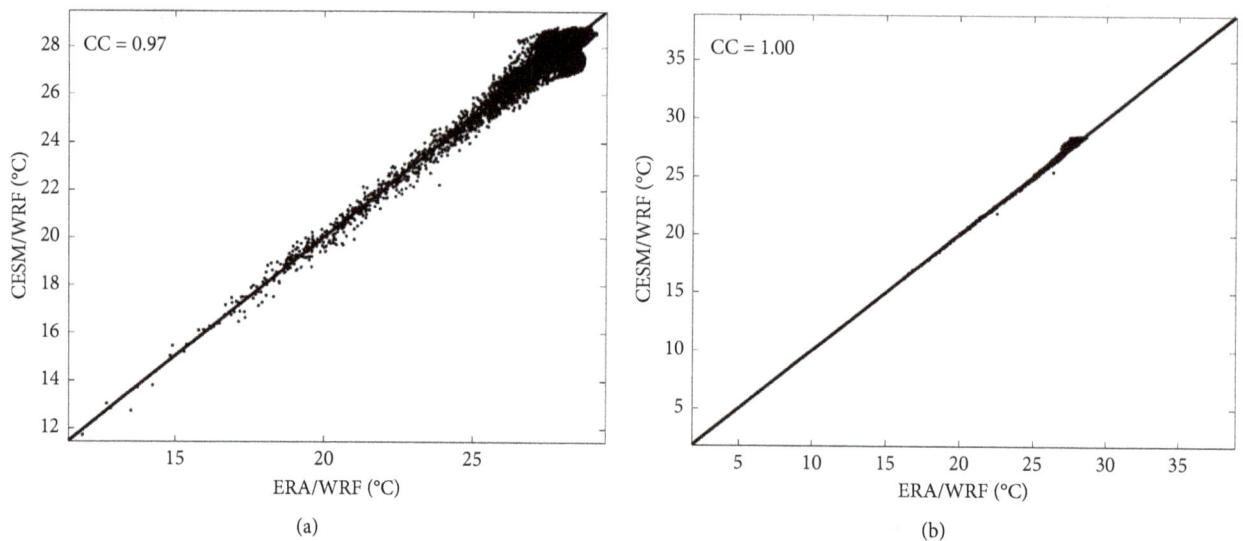

FIGURE 5: Scatter plots comparing CESM/WRF and ERA/WRF for (a) annual average temperature (°C) for 20 y and (b) 20-y average temperature (°C). Both CESM/WRF and ERA/WRF are at 10 km resolution.

TABLE 3: RMSDs, MDs (E[ERA/WRF − CESM/WRF]), and CCs of CESM/WRF and ERA/WRF for simulated annual average temperature (°C) for 20 y and 20-y average temperature (°C) for land only, sea only, and both land and sea, where both CESM/WRF and ERA/WRF are at 10 km resolution.

Metric	Annual average temperature for 20 y			20 y average temperature		
	Land	Sea	All	Land	Sea	All
RMSD	0.35	0.47	0.44	0.11	0.10	0.11
MD	0.06	0.01	0.02	0.06	0.01	0.02
CC	0.99	0.34	0.97	1.00	0.92	1.00

(3) low precipitation over the Pacific Ocean to the east of Sumatra and Malaysia.

The 20-y average annual precipitation simulated by CESM contradicts the observations and does not have any of the 3 main precipitation patterns shown in observations.

CESM has high precipitation over land in northwest Sumatra, all regions of Malaysia, southeast Sumatra, and along the east coast of Java. CESM has low precipitation over the Indian Ocean to the west of Sumatra and high precipitation over the Pacific Ocean to the east of Sumatra and Malaysia.

The pattern, intensity, and spatial resolution of 20-y average annual precipitation simulated by CESM and CESM/WRF are very different. The resolution and accuracies of simulated precipitation are significantly improved by the dynamically downscaling method employed in this study. CESM/WRF can simulate the 3 main precipitation patterns shown in the observations well. Its high precipitation at the center of Malaysia and surrounding areas is also consistent with that of CRU. The main difference between CESM/WRF and observations is the higher precipitation amount of CESM/WRF over high mountains along the west coasts of Sumatra and Java

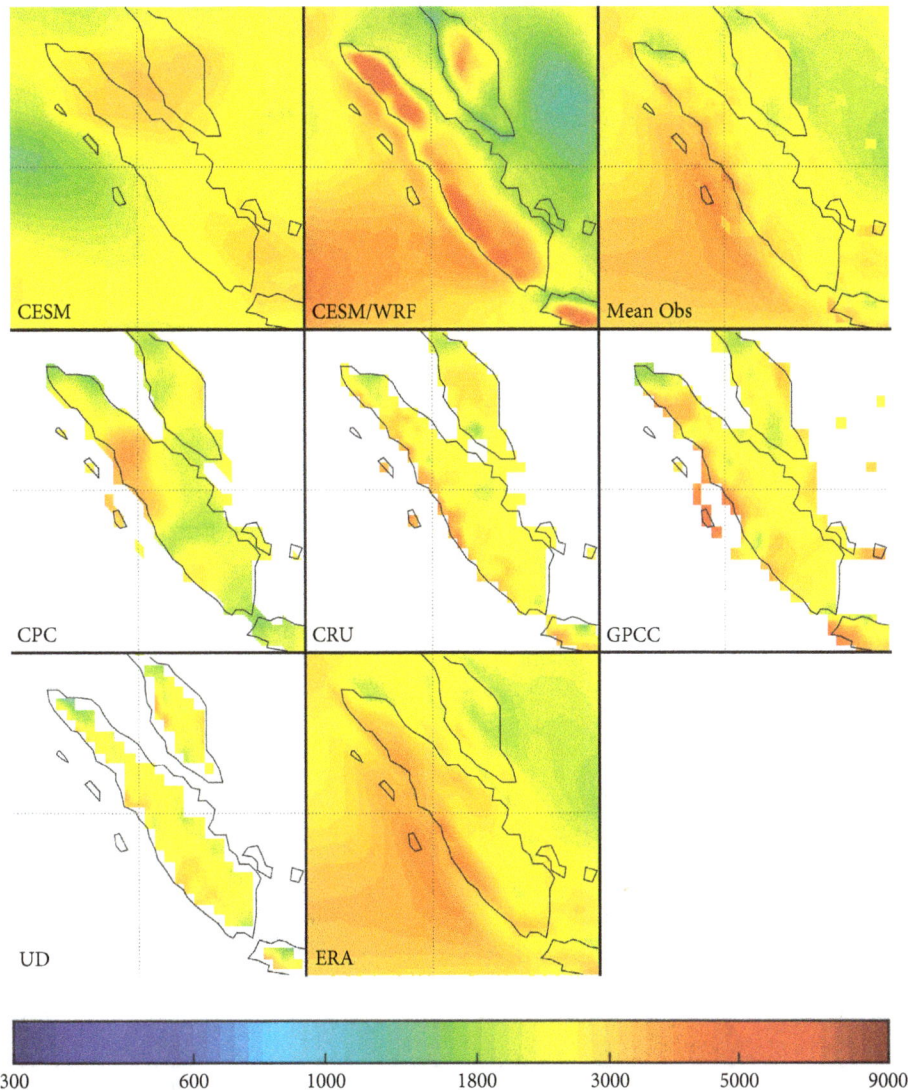

FIGURE 6: Comparisons of 20-y average annual precipitation (mm/y) simulated by CESM and CESM/WRF with CPC, CRU, GPCC, UD, and ERA and the average of CPC, CRU, GPCC, UD, and ERA. CESM and ERA are at 1° and 0.75° resolutions, respectively. CPC, CRU, GPCC, and UD are at 0.5° resolution. CESM/WRF is at 50 km resolution.

islands. This could be due to (1) the significantly higher resolution of CESM/WRF compared to those of all observations, (2) errors in CESM, and (3) errors in WRF physics.

Figure 7 compares CESM- and CESM/WRF-simulated 20-y average seasonal precipitation for the rainy season, i.e., from October to April, with CPC, CRU, GPCC, UD, and ERA and the average of CPC, CRU, GPCC, UD, and ERA. Figure 8 compares CESM- and CESM/WRF-simulated 20-y average seasonal precipitation for the dry season, i.e., from May to September, with CPC, CRU, GPCC, UD, and ERA and the average of CPC, CRU, GPCC, UD, and ERA. Results shown in Figures 7 and 8 are similar to those for the simulated 20 y annual precipitation shown in Figure 6, that is, (1) the observed precipitation for the two seasons has the 3 main precipitation patterns, (2) the precipitation pattern of the original CESM contradicts the observations, and (3) CESM/WRF can

simulate the 3 precipitation patterns well and performs significantly better than CESM. CESM/WRF-simulated precipitation for the two seasons over mountains along west coasts of Sumatra and Java islands are higher than observations.

The annual precipitation (mm/y) for individual years from 1980 to 1999 simulated by CESM and CESM/WRF is compared with the average of all five observation datasets using scatter plots in Figure 9(a). It is important to note that precipitation evaluation using scatter plots is difficult to get good agreement since different observation and reanalysis datasets also have differences among themselves. Downscaling CESM using WRF significantly improves the agreement of simulated annual precipitation with observations, where the correlation coefficient (CC) of simulations and observations improves from −0.14 for the original CESM to 0.47 for CESM/WRF. The range of CESM/WRF annual precipitation is larger than that of both CESM and

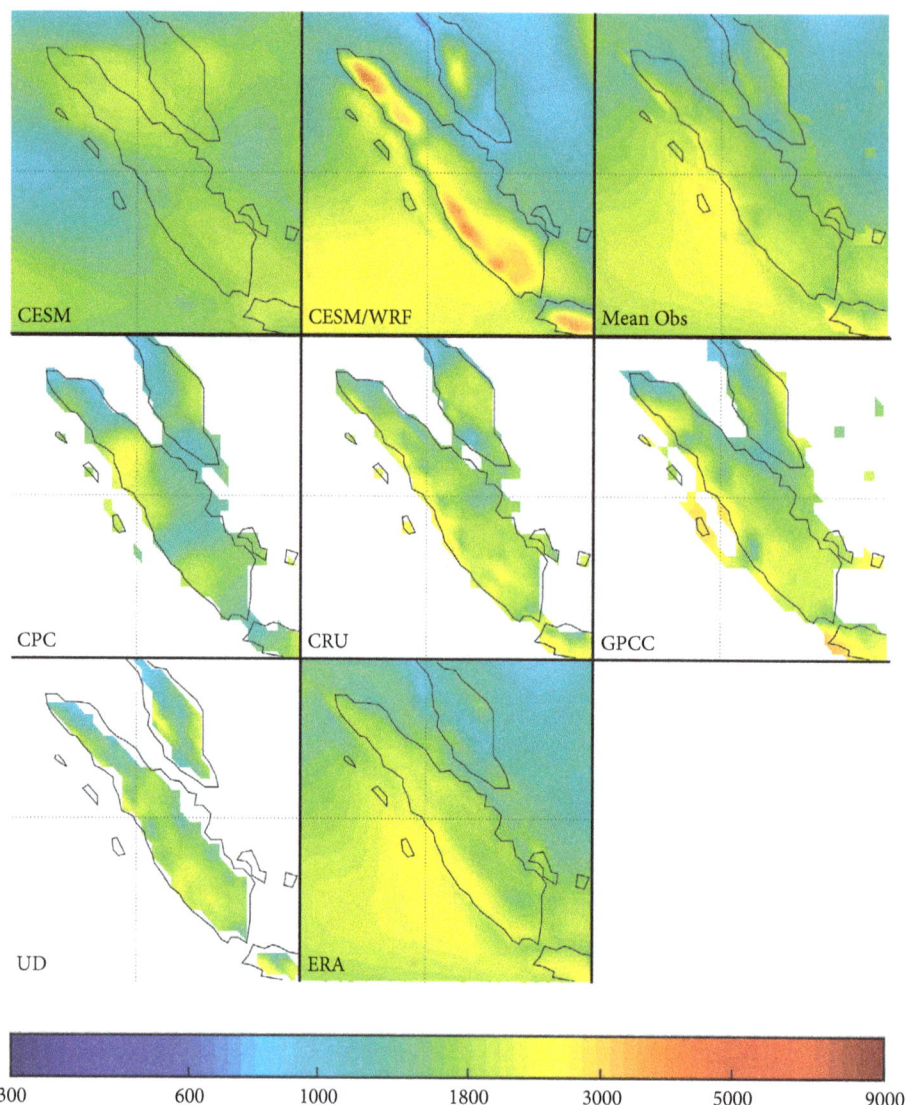

FIGURE 7: Comparisons of 20-y average seasonal precipitation for the rainy season (mm/season) simulated by CESM and CESM/WRF with CPC, CRU, GPCC, UD, and ERA and the average of CPC, CRU, GPCC, UD, and ERA. CESM and ERA are at 1° and 0.75° resolutions, respectively. CPC, CRU, GPCC, and UD are at 0.5° resolution. CESM/WRF is at 50 km resolution.

observations. The CESM/WRF's larger range is mainly due to its higher spatial resolution.

The 20-y average annual precipitation (mm/y) simulated by CESM and CESM/WRF is compared with the average of all five observation datasets in Figure 9(b). CESM/WRF is obviously more accurate than CESM. The dynamic downscaling significantly improves CC of simulations and observations from −0.38 for the original CESM to 0.69 for CESM/WRF. The increases of CCs in both Figures 9(a) and 9(b) are consistent with results shown in Figure 6, where the precipitation pattern of CESM/WRF agrees much better with observations.

Tables 4 and 5 show the performance metrics of annual precipitation for individual years of 1980–1999 and 20 y average annual precipitation, respectively, simulated by CESM and CESM/WRF evaluated using the average of CPC, CRU, ERA, GPCC, and UD for land only, sea only,

and both land and sea. Boldface highlights the model performing best for each performance metric. Accuracies for both annual precipitation for individual years and 20-y average annual precipitation are significantly improved by dynamical downscaling CESM using WRF. Overall CC significantly increases from −0.14 to 0.47 for annual precipitation for 20 y and from −0.38 to 0.69 for 20 y average annual precipitation. Overall RMSE improves from 922.32 to 904.15 mm/y for annual precipitation for 20 y and from 728.92 to 581.61 for 20-y average annual precipitation. Overall ME significantly reduces by ~78%. CESM/WRF performs worse than CESM in terms of RMSE and ME over land. This is consistent with CESM/WRF's higher precipitation amount over high mountains along the west coasts of Sumatra and Java islands in Figure 6. The performance of CESM/WRF to simulate climatological precipitation pattern over land is much better than that of

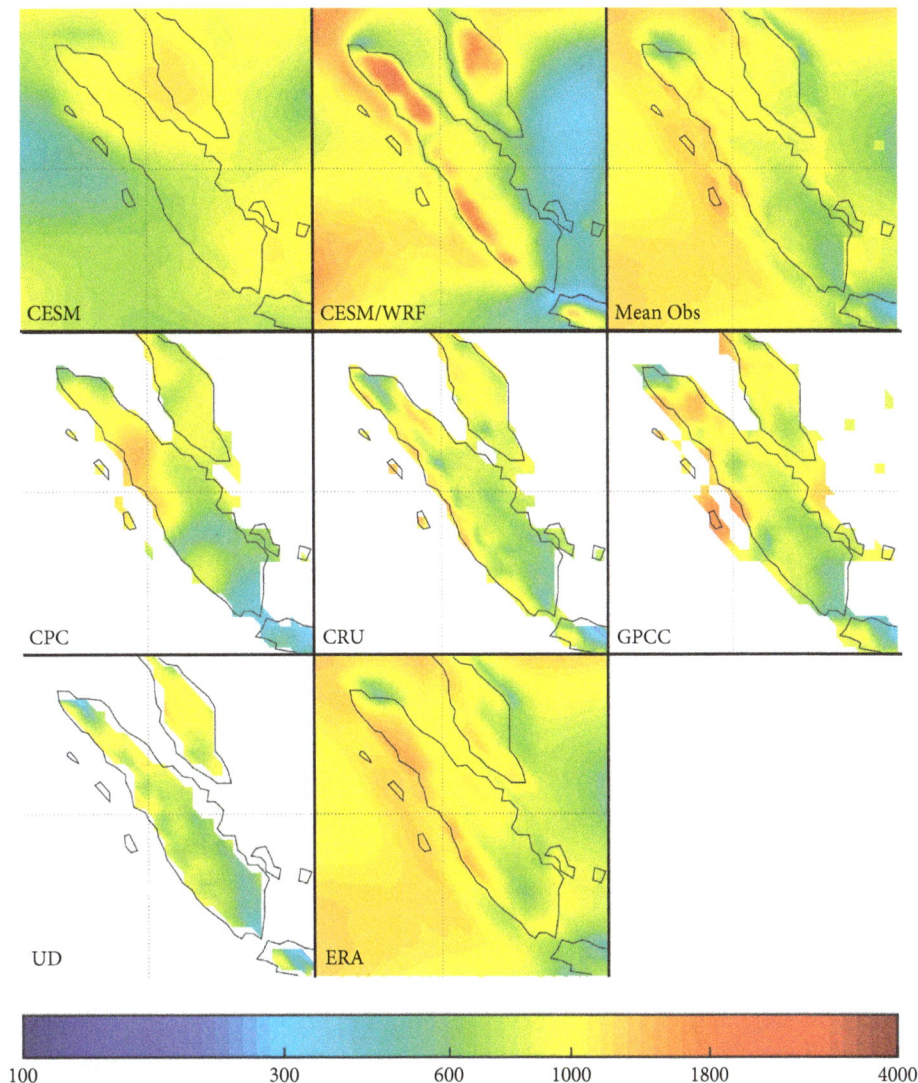

FIGURE 8: Comparisons of 20-y average seasonal precipitation for the dry season (mm/season) simulated by CESM and CESM/WRF with CPC, CRU, GPCC, UD, ERA, and the average of CPC, CRU, GPCC, UD, and ERA. CESM and ERA are at 1° and 0.75° resolutions, respectively. CPC, CRU, GPCC, and UD are at 0.5° resolution. CESM/WRF is at 50 km resolution.

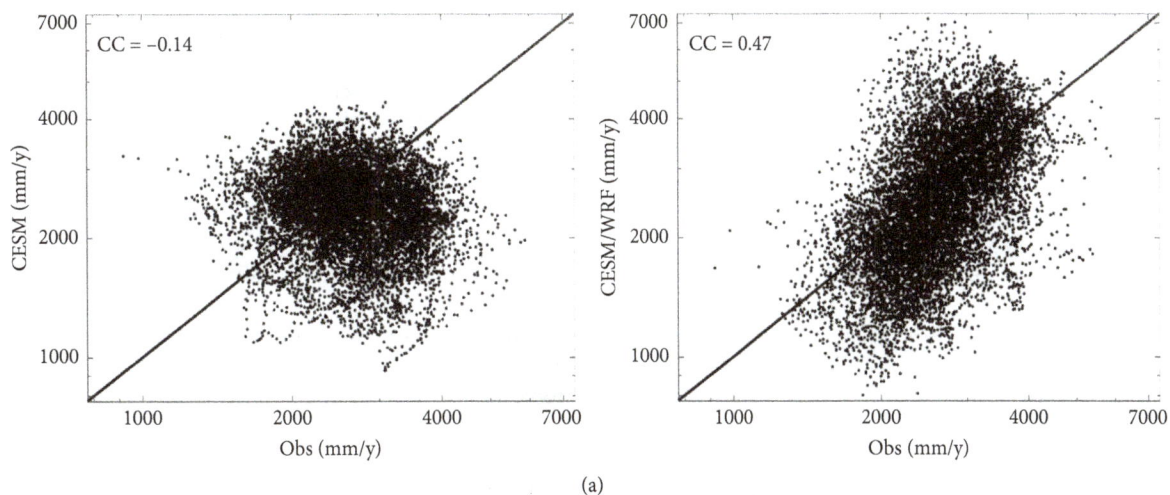

(a)

FIGURE 9: Continued.

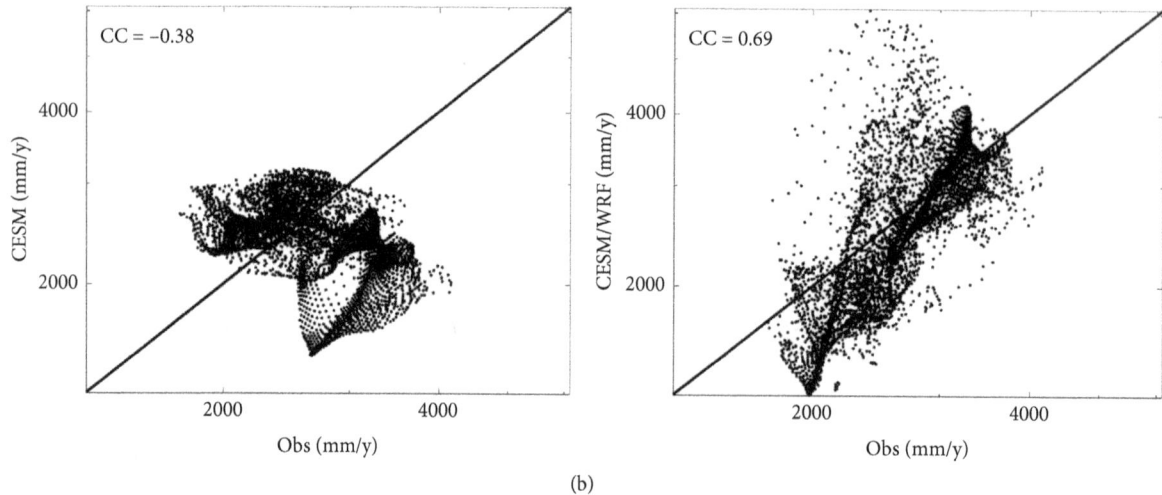

(b)

FIGURE 9: Scatter plots comparing CESM and CESM/WRF simulations with the average of CPC, CRU, GPCC, UD, and ERA for (a) annual precipitation (mm/y) for 20 y and (b) 20-y average annual precipitation (mm/y). CESM/WRF is at 50 km resolution.

TABLE 4: RMSEs, MEs (E[observations – simulations]), and CCs of simulations and observations for CESM and CESM/WRF-simulated annual precipitation (mm) for 20 y evaluated using the average of CPC, CRU, ERA, GPCC, and UD for land only, sea only, and both land and sea, where CESM/WRF is at 50 km resolution.

Metric	CESM			CESM/WRF		
	Land	Sea	All	Land	Sea	All
RMSE	**659.13**	990.53	922.32	1,061.87	**848.69**	**904.15**
ME	**−105.82**	371.76	257.78	−341.19	**182.19**	**57.28**
CC	−0.02	−0.14	−0.14	**0.29**	**0.55**	**0.47**

Boldface highlights the model performing best for each performance metric.

TABLE 5: RMSEs, MEs (E[observations – simulations]), and CCs of simulations and observations for CESM and CESM/WRF simulated 20-y average annual precipitation (mm) evaluated using the average of CPC, CRU, ERA, GPCC, and UD for land only, sea only, and both land and sea, where CESM/WRF is at 50 km resolution.

Metric	CESM			CESM/WRF		
	Land	Sea	All	Land	Sea	All
RMSE	**471.59**	792.57	728.92	940.19	**408.92**	**581.61**
ME	**−105.82**	371.76	257.78	−341.19	**182.19**	**57.28**
CC	−0.29	−0.36	−0.38	**0.35**	**0.88**	**0.69**

Boldface highlights the model performing best for each performance metric.

CESM. This can be seen from the improvement of CC from −0.29 for CESM to 0.35 for CESM/WRF.

3.4. Comparison of High-Resolution Climatological Precipitation Simulated by CESM/WRF and ERA/WRF. CESM/WRF-simulated climatological precipitation at 10 km resolution is compared with ERA/WRF at 10 km resolution. Top to bottom rows of Figure 10 compare CESM/WRF and ERA/WRF for 20-y average annual precipitation (mm/y) and 20-y average seasonal precipitation for rainy and dry seasons, respectively. CESM/WRF- and ERA/WRF-simulated 20-y average annual precipitation (mm/y) agree well for both precipitation pattern and intensity, i.e., precipitation is high along west coasts of Sumatra and Java islands, in the middle of Malaysia, and over the Indian Ocean to the west of Sumatra, and precipitation is low along the east coasts of Sumatra and Java islands, in the south of Malaysia, and over the Pacific Ocean to the east of Sumatra and Malaysia. The locations of CESM/WRF's precipitation peaks also agree well with those of ERA/WRF. Their main difference is over the Pacific Ocean to the east of Sumatra and Malaysia where ERA/WRF has lower precipitation. Seasonal precipitation of CESM/WRF and ERA/WRF agree well for both rainy and dry seasons. Their main difference is over the Indian Ocean to the west of Sumatra and over the Pacific Ocean to the east of Sumatra and Malaysia, where CESM/WRF has higher precipitation.

The scatter plots in Figure 11 compare CESM/WRF-simulated annual precipitation for 20 y and 20-y average annual precipitation with ERA/WRF. CESM/WRF-simulated annual precipitation and its long-term average agree well with ERA/WRF with CCs of 0.73 and 0.95, respectively. CESM/WRF-simulated 20-y average annual precipitation is biased higher than ERA/WRF for annual precipitation lower than ~2,000 mm/y and agrees well with ERA/WRF, otherwise particularly for high precipitation. This is consistent with results shown in the top row of Figure 10.

The scatter plots in Figure 12 compare CESM/WRF-simulated seasonal precipitation (mm) for years 1980–1999 and 20-y average seasonal precipitation for the rainy and dry seasons with ERA/WRF. Results are consistent with image comparisons in Figure 10. CESM/WRF simulations agree well with ERA/WRF for both seasons with high CCs. CCs between CESM/WRF and ERA/WRF 20-y average seasonal precipitation are as high as 0.96 and 0.95 for rainy and dry seasons, respectively. CESM/WRF 20-y average seasonal precipitation is biased higher than ERA/WRF for precipitation lower than ~2,000 mm.

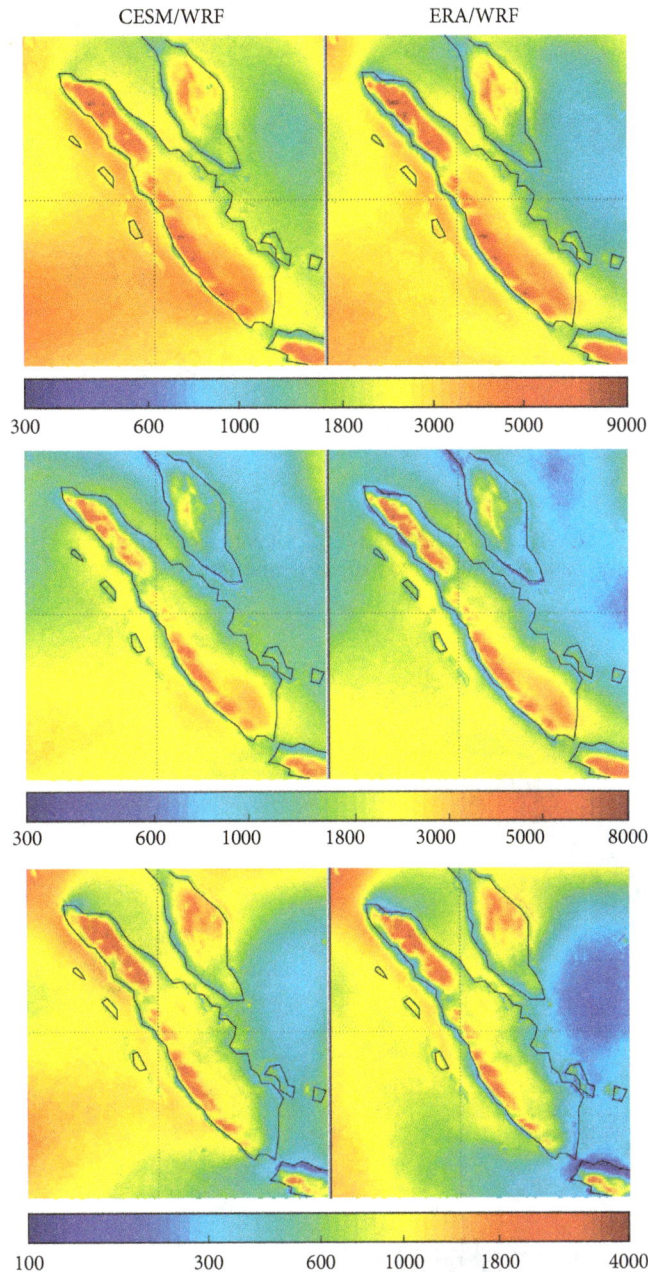

FIGURE 10: Top to bottom: comparisons of CESM/WRF and ERA/WRF for 20-y average annual precipitation (mm/y) and 20-y average seasonal precipitation (mm/season) for the rainy and dry seasons, respectively. Both CESM/WRF and ERA/WRF are at 10 km resolution.

Table 6 shows the root-mean-squared differences (RMSDs), mean differences (MDs), which is E[ERA/WRF − CESM/WRF], and CCs of CESM/WRF and ERA/WRF for simulated precipitation for 20 y and 20-y average precipitation for annual and rainy and dry seasons. CESM/WRF simulations are highly correlated with ERA/WRF for both annual and seasonal precipitation. CCs for 20 y annual and seasonal average precipitation are greater than or equal to 0.95. To get the idea about the size of RMSEs and MEs, numbers in the brackets in Table 6 are in percentages of the mean of ERA/WRF. RMSDs are 18.02, 17.08, and 24.94% of the mean of ERA/WRF for 20-y average annual precipitation and 20-y average seasonal precipitation for rainy and dry

seasons, respectively. CESM/WRF precipitation simulations are biased higher than ERA/WRF. MDs are −13.10, −10.32, and −19.06% of the mean of ERA/WRF for 20-y average annual precipitation and 20-y average seasonal precipitation for rainy and dry seasons, respectively.

4. Summary and Conclusion

The performance of a high-resolution climate simulation system CESM/WRF developed to be used for Sumatra and nearby regions is evaluated. CESM/WRF is composed of a mesoscale model WRF and outputs from the global climate model CESM used for initial and boundary conditions.

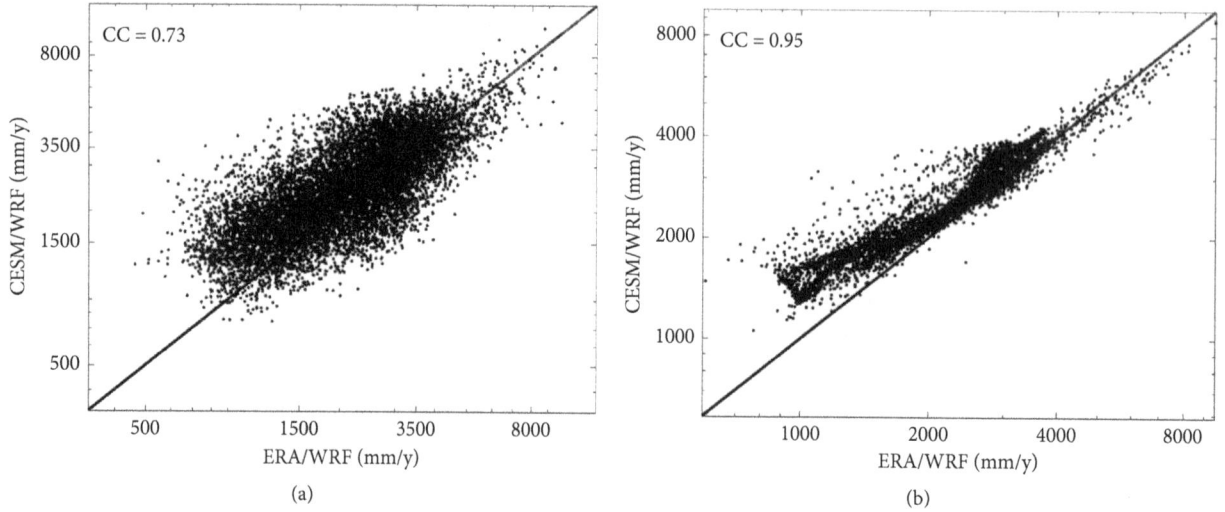

FIGURE 11: Scatter plots comparing CESM/WRF and ERA/WRF for (a) annual precipitation (mm/y) for 20 y and (b) 20-y average annual precipitation (mm/y). Both CESM/WRF and ERA/WRF are at 10 km resolution.

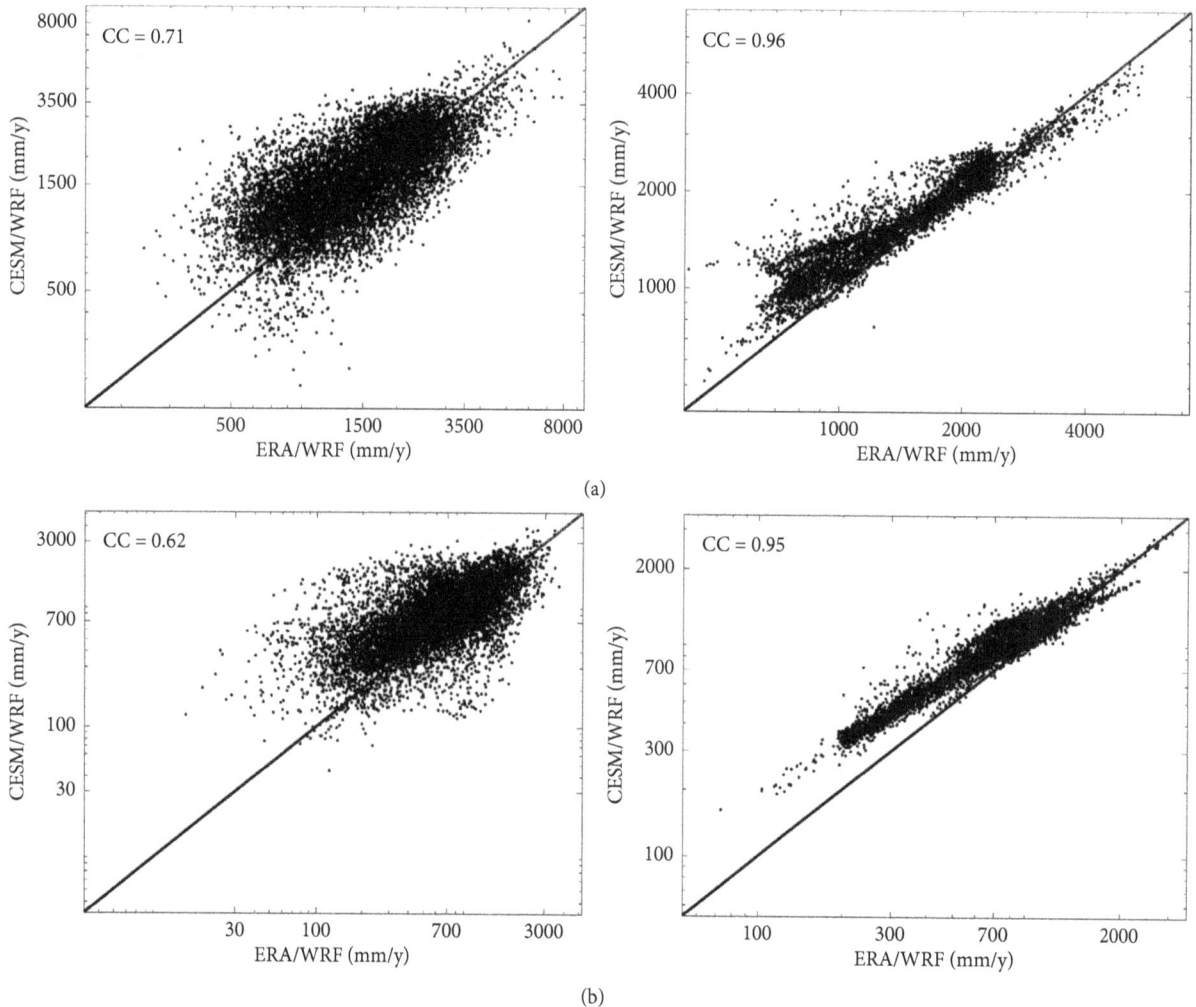

FIGURE 12: Scatter plots comparing CESM/WRF and ERA/WRF seasonal precipitation for 20 y (mm/season; left column) and 20-y average seasonal precipitation (mm/season; right column) for (a) rainy and (b) dry seasons. Both CESM/WRF and ERA/WRF are at 10 km resolution.

TABLE 6: RMSDs, MDs (E[ERA/WRF − CESM/WRF]), and CCs of CESM/WRF and ERA/WRF for simulated precipitation for 20 y and 20-y average precipitation for annual and rainy and dry seasons, where both CESM/WRF and ERA/WRF are at 10 km resolution.

Metric	Precipitation for 20 y			20 y average precipitation		
	Annual	Rain season	Dry season	Annual	Rain season	Dry season
RMSD	846.71 (36.41)	627.55 (39.54)	606.12 (58.61)	419.08 (18.02)	271.03 (17.08)	257.93 (24.94)
MD	−304.58 (−13.10)	−163.79 (−10.32)	−197.10 (−19.06)	−304.58 (−13.10)	−163.79 (−10.32)	−197.10 (−19.06)
CC	0.73	0.71	0.62	0.95	0.96	0.95

RMSDs and MDs in the brackets are in percentages of the mean of ERA/WRF.

CESM/WRF-simulated temperature and precipitation at 10 km resolution for Sumatra and nearby regions from 1980 to 1999 are evaluated using 4 observation datasets, including CPC, CRU, GPCC, UD, and the ERA reanalysis dataset treated in this study as an observation dataset. Since all observations have resolutions lower than 10 km and dense ground stations in Sumatra do not exist, the CESM/WRF-simulated climatological temperature and precipitation at 10 km resolution are compared with the downscaled reanalysis ERA/WRF at 10 km resolution.

Although different observation datasets have some differences among themselves for both climatological temperature and precipitation, there are patterns consistent for most observation datasets. While CESM contradicts all observation patterns for both temperature and precipitation, CESM/WRF can simulate all patterns well; CESM/WRF can simulate all patterns well. CESM/WRF can also resolve locations of very cold temperature at mountain peaks consistent with those of UD. Downscaling CESM using WRF significantly improves resolution and accuracies of the simulations. The main discrepancy between CESM/WRF-simulated precipitation and observations is CESM/WRF's higher precipitation over high mountains along the west coasts of Sumatra and Java islands and could be due to the significantly lower resolutions of observations, errors in CESM outputs, and errors in WRF physics.

Comparisons of CESM/WRF- and ERA/WRF-simulated climatological temperature and precipitation at 10 km resolution show that CESM/WRF simulations agree very well with ERA/WRF and there is no precipitation discrepancy over high mountains along the west coasts of Sumatra and Java islands. The high-resolution climate simulation system CESM/WRF developed in this study can provide useful simulated climatological temperature and precipitation for Sumatra and nearby regions. The good agreement between CESM/WRF and ERA/WRF also gains the confidence of employing CESM/WRF for climate projections for Sumatra and nearby regions.

Conflicts of Interest

The authors declare that they have no conflicts of interest.

Acknowledgments

This study was supported by the Interdisciplinary Graduate School of Earth System Science and Andaman Natural Disaster Management of the Prince of Songkla University, Phuket Campus, Thailand.

References

[1] T. F. Stocker, D. Qin, G.-K. Plattner et al., *Climate Change 2013: The Physical Science Basis. Contribution of Working Group I to the Fifth Assessment Report of the Intergovernmental Panel on Climate Change*, IPCC, Cambridge University Press, Cambridge, UK, 2013.

[2] C. B. Field, V. R. Barros, D. J. Dokken et al., *Climate Change 2014: Impacts, Adaptation, and Vulnerability. Part A: Global and Sectoral Aspects. Contribution of Working Group II to the Fifth Assessment Report of the Intergovernmental Panel on Climate Change*, IPCC, Cambridge University Press, Cambridge, UK, 2014.

[3] C. Chotamonsak, E. P. Salathé Jr., J. Kreasuwan, S. Chantara, and K. Siriwitayakom, "Projected climate change over Southeast Asia simulated using a WRF regional climate model," *Atmosphere Science Letter*, vol. 12, no. 2, pp. 161–239, 2011.

[4] J. Sentian and S. S. K. Kong, "High resolution climate change projection under SRES A2 scenario during summer and winter monsoons over Southeast Asia using PRECIS regional climate modeling system," *The SIJ Transactions on Computer Science Engineering & its Applications (CSEA)*, vol. 1, pp. 163–173, 2013, https://www.researchgate.net/publication/279455994.

[5] J. Katzfey, J. L. McGregor, K. Nguyen, and M. Thatcher, "Regional climate change projection development and interpretation for Indonesia," CSIRO Marine and Atmospheric Research Final Report for AusAID, CSIRO, Aspendale, VIC, Australia, 2010.

[6] M. S. Bukovsky and D. J. Karoly, "Precipitation simulation using WRF as a nested regional climate model," *Journal Applied Meteorology Climatology*, vol. 48, no. 10, pp. 2152–2159, 2009.

[7] J. Gula and W. R. Peltier, "Dynamical downscaling over the great lakes basin of North America using the WRF regional climate model: the impact of the great lakes system on regional greenhouse warming," *Journal of Climate*, vol. 25, no. 21, pp. 7723–7742, 2012.

[8] J. W. Hurrell, M. M. Holland, P. R. Gent et al., "The community Earth system model: a framework for collaborative research," *Bulletin of the American Meteorological Society*, vol. 94, no. 9, pp. 1339–1360, 2013.

[9] K. E. Taylor, R. J. Stouffer, and G. A. Meehl, "An overview of CMIP5 and the experiment design," *Bulletin of the American Meteorological Society*, vol. 93, no. 4, pp. 485–498, 2012.

[10] S. Levis, G. B. Bonan, E. Kluzek et al., "Interactive crop management in the Community Earth System Model (CESM1): seasonal influences on land–atmosphere fluxes," *Journal of Climate*, vol. 25, no. 14, pp. 4839–4859, 2012.

[11] J. E. Kay, C. Deser, A. Phillips et al., "The Community Earth System Model (CESM) large ensemble project: a community resource for studying climate change in the presence of internal climate variability," *Bulletin of the American Meteorological Society*, vol. 96, no. 8, pp. 1333–1349, 2015.

[12] F. Lehner, F. Joos, C. C. Raible et al., "Climate and carbon cycle dynamics in a CESM simulation from 850 to 2100 CE," *Earth System Dynamic*, vol. 6, no. 2, pp. 411–434, 2015.

[13] J. E. Kay, C. Wall, V. Yettella et al., "Global climate impact of fixing the southern ocean shortwave radiation bias in the community Earth system model (CESM)," *Journal of Climate*, vol. 29, no. 12, pp. 4617–4636, 2016.

[14] J. Dudhia, D. Gil, K. Manning, W. Wang, and C. Bruyere, "PSU/NCAR mesoscale modeling system tutorial class notes and users' guide (MM5 modeling system version 3)," 2015, http://www2.mmm.ucar.edu/mm5/documents/MM5_tut_Web_notes/tutorialTOC.htm.

[15] C. Surussavadee, "Evaluation of high-resolution tropical weather forecasting using satellite passive millimeter-wave observations," *IEEE Transaction on Geoscience and Remote Sensing*, vol. 52, no. 5, pp. 2780–2787, 2014.

[16] W. C. Skamarock and Coauthors, "A description of the advanced research WRF version 3," NCAR Technical Note NCAR/TN-475 + STR, NCAR, Boulder, CO, USA, 2008.

[17] C. Surussavadee, "Evaluation of WRF near-surface wind simulation in Tropics employing different planetary boundary layer scheme," in *Proceeding IEEE 8th International Renewable Energy Congress 2017*, pp. 1–4, Amman, Jordan, 2017.

[18] T. Afrizal and C. Surussavadee, "Evaluation of CESM/WRF climate simulations at high resolution over Sumatra," in *Proceeding IEEE 2nd International Conference on Information Technology, Information System and Electrical Engineering (ICITISEE)*, pp. 278–281, Institute of Electrical and Electronics Engineers, Yogyakarta, Indonesia, 2017.

[19] P. M. M. Soares, R. M. Cardoso, P. M. A. Miranda, J. d. Medeiros, M. Belo-Pereira, and F. Espirito-Santo, "WRF high resolution dynamical downscaling of ERA-Interim for Portugal," *Climate Dynamics*, vol. 39, no. 9-10, pp. 2497–2522, 2012.

[20] D. Huang and S. Gao, "Impact of different reanalysis data on WRF dynamical downscaling over China," *Atmospheric Research*, vol. 200, pp. 25–35, 2018.

[21] Y. Qiu, Q. Hu, and C. Zhang, "WRF simulation and downscaling of local climate in Central Asia," *International Journal of Climatology*, vol. 37, pp. 513–528, 2017.

[22] D. P. Dee, S. M. Uppala, A. J. Simmons et al., "The ERA-Interim reanalysis: configuration and performance of the data assimilation system," *Quarterly Journal of the Royal Meteorological Society*, vol. 137, no. 656, pp. 553–597, 2011.

[23] H. S. Lee, "General rainfall patterns in Indonesia and the potential impacts of local seas on rainfall intensity," *Water*, vol. 7, no. 12, pp. 1751–1768, 2015.

[24] K. Matsuura and C. J. Willmott, "Terrestrial air temperature 1900-2010 gridded monthly time series (version 3.01)," Janury 2016, http://climate.geog.udel.edu/~climate/html_pages/Global2011/README.GlobalTsT2011.html.

[25] I. Harris, P. D. Jones, T. J. Osborn, and D. H. Lister, "Updated high-resolution grids of monthly climatic observations—the CRU TS3.10 Dataset," *International Journal of Climatology*, vol. 34, no. 3, pp. 623–642, 2014.

[26] A. Becker, P. Finger, A. Meyer-Christoffer et al., "A description of the global land-surface precipitation data products of the Global Precipitation Climatology Centre with sample applications including centennial (trend) analysis from 1901-present," *Earth System Science Data*, vol. 5, no. 1, pp. 71–99, 2013.

[27] P. Xie, A. Yatagai, M. Chen et al., "A gauge-based analysis of daily precipitation over East Asia," *Journal of Hydrometeorology*, vol. 8, no. 3, pp. 607–626, 2007.

[28] G. Miguez-Macho, G. L. Stenchikov, and A. Robock, "Spectral nudging to eliminate the effects of domain position and geometry in regional climate model simulations," *Journal of Geophysical Research*, vol. 109, no. D13, pp. 1–14, 2004.

[29] C. Surussavadee and P. Aonchart, "Evaluation of WRF physics options for high-resolution weather forecasting in tropics using satellite passive millimeter-wave observations," in *Proceeding IEEE International on Geoscience and Remote Sensing Symposium (IGARSS 2013)*, pp. 2262–2265, Melbourne, VIC, Australia, July 2013.

[30] K. S. S. Lim and S. Y. Hong, "Development of an effective double moment cloud microphysics scheme with prognostic cloud condensation nuclei (CCN) for weather and climate model," *Monthly Weather Review*, vol. 138, no. 5, pp. 1587–1612, 2010.

[31] Z. I. Janjic, "The step-mountain eta coordinate model: further developments of the convection, viscous sublayer and turbulence closure schemes," *Monthly Weather Review*, vol. 122, no. 5, pp. 927–945, 1994.

[32] C. S. Bretherton and S. Park, "A new moist turbulence parameterization in the community atmosphere model," *Journal of Climate*, vol. 22, no. 12, pp. 3422–3448, 2008.

[33] A. S. Monin and A. M. Obukhov, "Basic laws of turbulent mixing in the surface layer of the atmosphere," *Trudy Geofizicheskogo Instituta*, Akademiya Nauk SSSR, vol. 24, pp. 163–187, 1954.

[34] F. Chen and J. Dudhia, "Coupling an advanced land-surface/hydrology model with the Penn State/NCAR MM5 modeling system. Part I: model description and implementation," *Monthly Weather Review*, vol. 129, no. 4, pp. 569–585, 2001.

9

Comparison of 3DVar and EnSRF Data Assimilation using Radar Observations for the Analysis and Prediction of an MCS

Shibo Gao(ID) **and Jinzhong Min**(ID)

Key Laboratory of Meteorological Disaster, Ministry of Education (KLME), Joint International Research Laboratory of Climate and Environment Change (ILCEC), Collaborative Innovation Center on Forecast and Evaluation of Meteorological Disaster (CICFEMD), Nanjing University of Information Science and Technology, Nanjing 210044, China

Correspondence should be addressed to Jinzhong Min; minjz@nuist.edu.cn

Academic Editor: Stefano Federico

Using radar observations, the performances of the ensemble square root filter (EnSRF) and an indirect three-dimensional variational (3DVar) data assimilation method were compared for a mesoscale convective system (MCS) that occurred in the Front Range of the Rocky Mountains, Colorado (USA). The results showed that the root mean square innovations (RMSIs) of EnSRF were lower than 3DVar for radar reflectivity and radial velocity and that the spread of EnSRF was generally consistent with its RMSIs. EnSRF substantially improved the analysis of the MCS compared with an experiment without radar data assimilation, and it produced a slight but noticeable improvement over 3DVar in terms of both coverage and intensity. Forecast results initiated from the final analysis revealed that EnSRF generally produced the best prediction of the MCS, with improved quantitative reflectivity and precipitation forecast skills. EnSRF also demonstrated better performance than 3DVar in the prediction of neighborhood probability for reflectivity at thresholds of 20 and 35 dBZ, which better matched the observed radar reflectivity in terms of both shape and extension. Additionally, the humidity, temperature, and wind fields were also improved by EnSRF; the largest error reduction was found in the water vapor field near the surface and at upper levels.

1. Introduction

Mesoscale convective systems (MCSs) are severe convective storms that can cause injury and damage property. However, accurate prediction of MCSs remains a challenge because of their rapid development and evolution. Doppler radars are platforms capable of observing MCS structure with high resolution and frequency, and such observations have been used for convective-scale data assimilation to improve the initial conditions of numerical weather prediction (NWP) models.

The three-dimensional variational (3DVar) data assimilation (DA) system, which includes mass continuity equations and other model equations as weak constraints, is efficient for storm-scale DA. Xiao et al. [1, 2] developed a Weather Research and Forecasting DA (WRFDA) 3DVar system to assimilate radial velocity and radar reflectivity by considering the total water vapor mixing ratio as a control variable, which improved quantitative precipitation forecasts for a hurricane. However, their system is limited to warm rain microphysics. Gao and Stensrud [3] used the background temperature to classify hydrometers to include ice and snow microphysics. Wang et al. [4] developed an approach for indirect radar reflectivity assimilation that assimilates retrieved rainwater and in-cloud water vapor estimated from radar reflectivity. An advantage of this new approach is that it avoids linearization errors attributable to the nonlinear relationship between reflectivity and microphysical variables. It has been used in several recent studies and it has demonstrated capability in improving short-term forecasts of convective storms [5–7]. However, it cannot overcome the inherent weakness of the 3DVar method that results from the neglect of the flow-dependent nature of the background error covariance (BEC). This problem is most severe in storm-scale

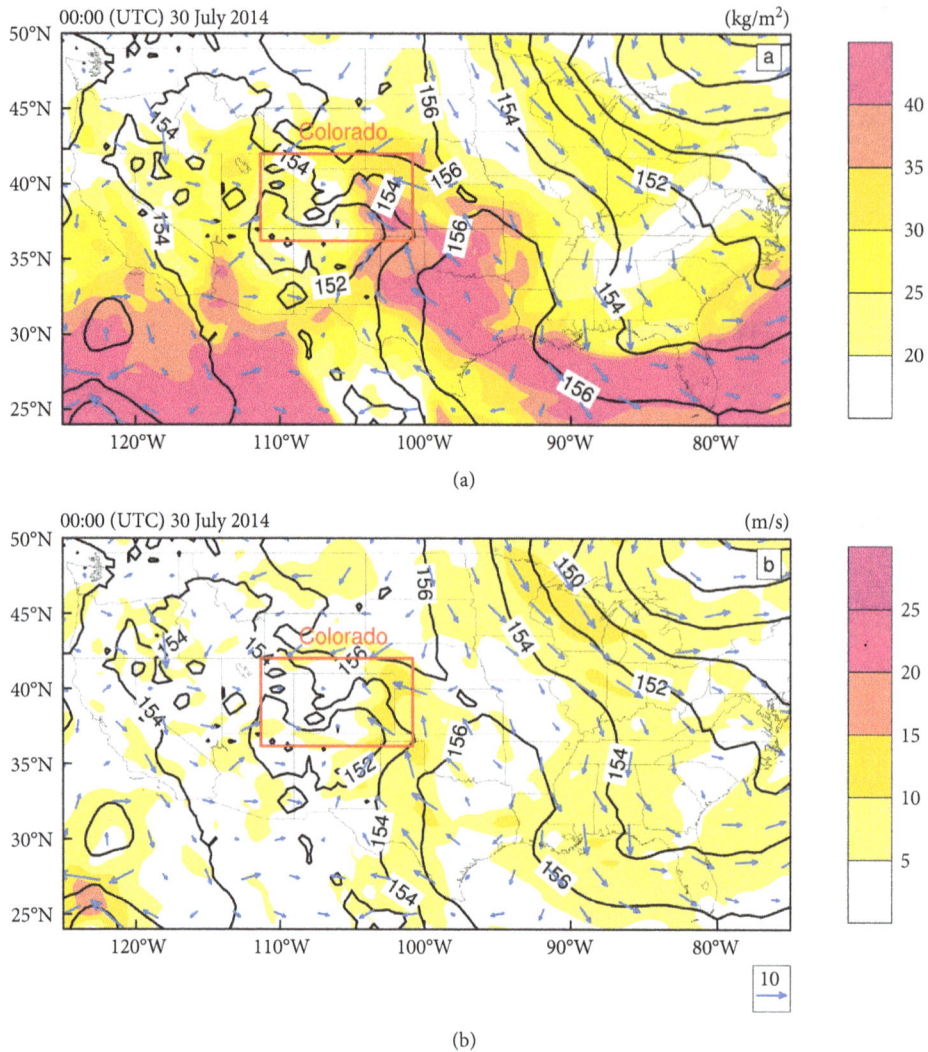

FIGURE 1: Environmental variables based on Global Forecast System (GFS) analysis data at 850 hPa at 0000 UTC 30 July 2014: geopotential height (black solid lines, unit: dagpm), wind vectors (blue vector, unit: m·s^{-1}). (a) Column-integrated precipitable water (shaded, unit: kg·m^{-2}); (b) wind speed (shaded, unit: m·s^{-1}).

DA because few state variables are observed directly and large-scale balance relationships are invalid.

Another advanced DA method for convective-scale NWP is the ensemble Kalman filter (EnKF), which estimates BEC using an ensemble of forecasts. EnKF is able to evolve the BEC dynamically throughout multiple assimilation cycles. Cross covariance produced by EnKF is especially important for convective-scale DA because state variables that are not observed directly can be updated. Snyder and Zhang [8] and Zhang et al. [9] first demonstrated the capability of EnKF for convective-scale DA using radar radial winds acquired during observing system simulation experiments. Subsequent studies by Xue et al. [10, 11] showed the performance of EnKF could be improved further when radar reflectivity was also assimilated into a compressible model with complex ice microphysics. The application of EnKF for assimilating real radar data has also produced successful results for several convective cases [12, 13]. It has been found that EnKF can handle complex and highly

nonlinear processes involved in DA, which makes it attractive for convective-scale DA [14–16].

Several studies have demonstrated that the performance of EnKF was better than 3DVar in global- to mesoscale DA [17–20]. Whitaker et al. [18] found that EnKF produced an improvement in analysis and forecasting relative to 3DVar when using the National Centers for Environmental Prediction (NCEP) Global DA system, especially in data-sparse regions. Meng and Zhang [20] showed that EnKF generally outperformed 3DVar by assimilating observations from either individual or multiple data platforms (e.g., soundings, and surface and wind profilers) for a mesoscale convective vortex event. However, few studies have compared the performances of 3DVar and EnKF for real convective storms. Johnson et al. [21] compared EnKF and 3DVar at multiple scales, and they found that the method of radar assimilation using EnKF could maintain the storm features throughout the entire forecast period, whereas a 3DVar forecast produced some deficits after the first hour.

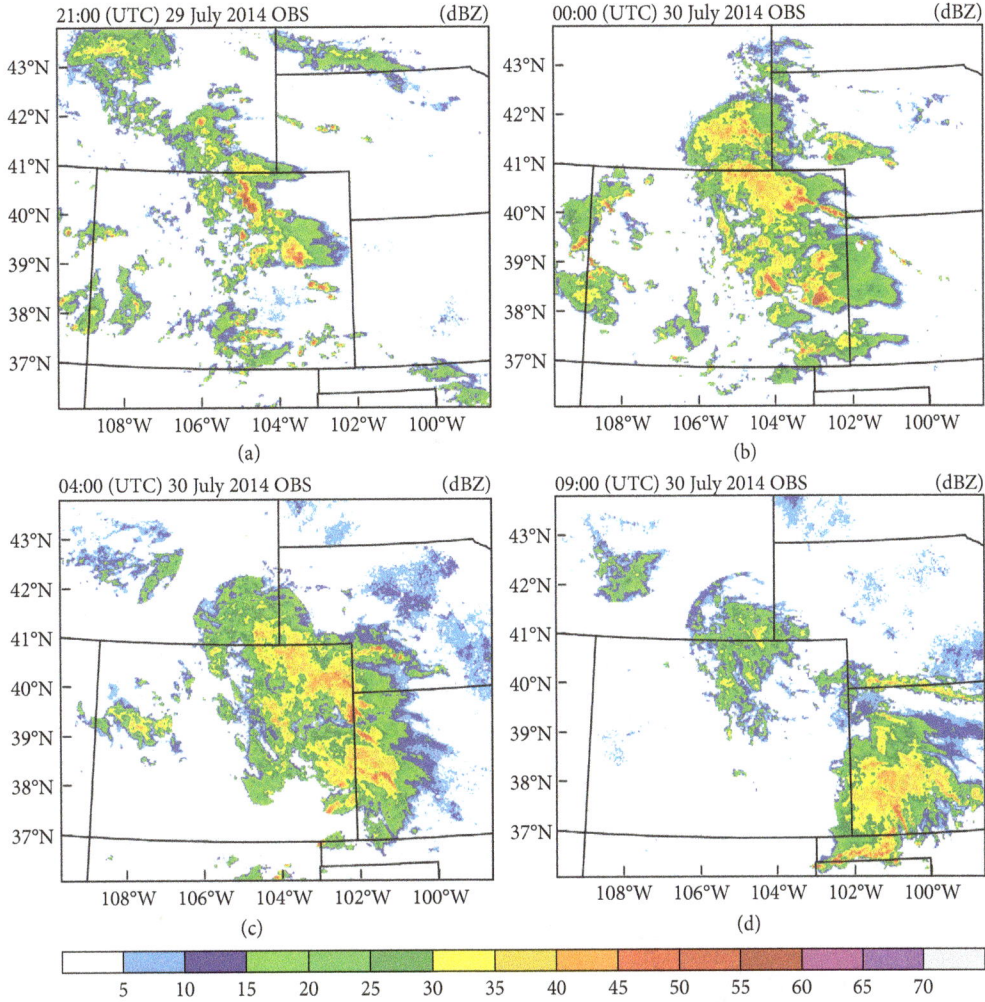

FIGURE 2: Observations of composite radar reflectivity (unit: dBZ) showing evolution of the mesoscale convective system (MCS) at 2100 UTC 29 July 2014 (a), 0000 UTC 30 July 2014 (b), 0400 UTC 30 July 2014 (c), and 0900 UTC 30 July 2014 (d).

In the context of convective-scale DA, the present study systematically compared the ensemble square root filter (EnSRF) and a newly developed indirect 3DVar method. The indirect 3DVar method used was that proposed by Wang et al. [4], which includes a procedure to assimilate an estimated in-cloud humidity field. A case study was conducted to investigate the impact of using different radar DA techniques on both the analysis and the subsequent prediction of an MCS that occurred over the Front Range of the Rocky Mountains in Colorado (USA) on 30 July 2014, which demonstrated the promising results of EnSRF over 3DVar in MCS analysis and forecasting. The remainder of this paper is arranged as follows. Section 2 presents the indirect 3DVar and EnSRF radar DA methods. The case description and the experimental setups are outlined in Section 3. Section 4 describes the analysis and forecast results, and the summary is presented in Section 5.

2. Method

2.1. WRFDA 3DVar Indirect Radar Data Assimilation Method.
The indirect reflectivity assimilation approach of

the WRFDA 3DVar system for convective-scale radar DA [4] was used in this study. The cost function in an updated system can be expressed as follows:

$$J = J_b + J_o, \tag{1}$$

where the subscript b represents the background and the subscript o represents the observation terms. To avoid the linearization error of the reflectivity–rainwater equation, two additional observation terms, corresponding to the rainwater mixing ratio and in-cloud water vapor (both estimated from reflectivity), are added in the cost function. Here, we extended the system by including snow and graupel in J_b and J_o, where

$$J_{qm} = \frac{1}{2}\left(\mathbf{q}_m - \mathbf{q}_m^b\right)^T \mathbf{B}_{qm}^{-1}\left(\mathbf{q}_m - \mathbf{q}_m^b\right)$$

$$+ \frac{1}{2}\left(\mathbf{q}_m - \mathbf{q}_m^o\right)^T \mathbf{R}_{qm}^{-1}\left(\mathbf{q}_m - \mathbf{q}_m^o\right), \tag{2}$$

$$J_{qv} = \frac{1}{2}\left(\mathbf{q}_v - \mathbf{q}_v^o\right)^T \mathbf{R}_{qv}^{-1}\left(\mathbf{q}_v - \mathbf{q}_v^o\right).$$

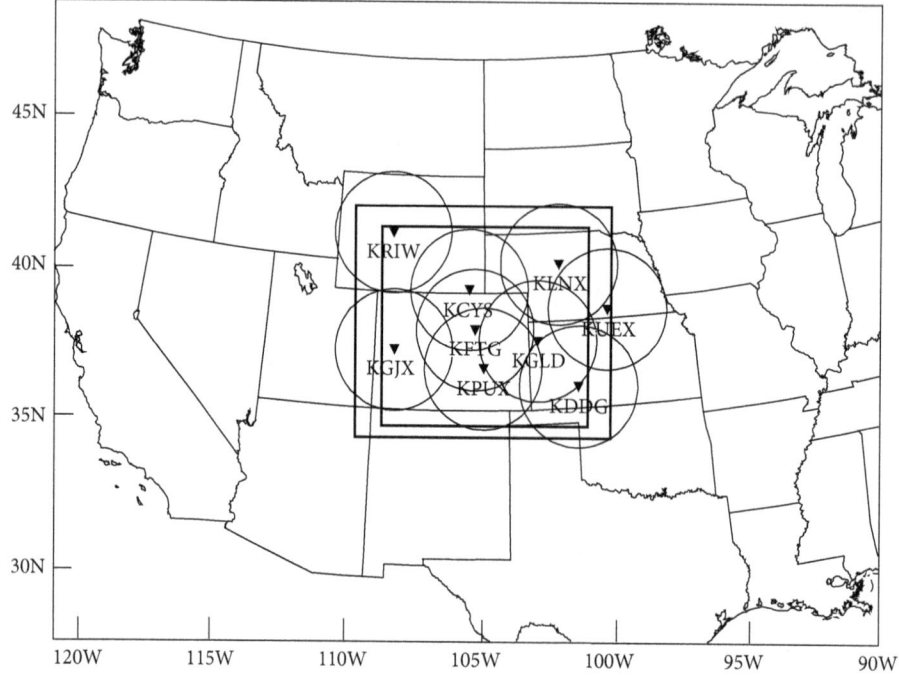

FIGURE 3: The three model domains and locations of the nine radar stations (inverted black triangles): KRIW (Riverton, WY), KLNX (North Platte, NE), KCYS (Cheyenne, WY), KUEX (Hastings, NE), KGJX (Grand Junction, CO), KFTG (Denver, CO), KGLD (Goodland, KS), KPUX (Pueblo, CO), and KDDC (Dodge City, KS). Maximum ranges of radars are shown by circles.

TABLE 1: List of experiments.

Experiments	Description
NoDA	No data assimilation
3DVar	3DVar assimilating both radial velocity and reflectivity
EnSRF	EnSRF assimilating both radial velocity and reflectivity

Here, q_m represents the hydrometeor mixing ratios (rainwater, snow, graupel–hail) of the atmospheric state, q_m^b are their background first guesses, q_m^o are their observations retrieved from radar reflectivity, and R_{qm} and B_{qm} are their related observational variances and background error matrix, respectively; q_v, q_v^o, and R_{qv} are the water vapor mixing ratios, their observations retrieved from radar reflectivity, and their related observational variances, respectively. The radar radial velocity assimilation followed the procedure in Xiao et al. [1]. In the assimilation of radial velocity, the operator includes the rainwater terminal velocity, and the terminal velocity can be calculated according to the rainwater mixing ratio.

2.2. EnSRF. For the standard EnKF, the update equations can be formulated as follows:

$$x_i^b(t) = \mathbf{M} x_i^a(t-1), \tag{3}$$

$$x_i^a = x_i^b + \mathbf{K}\left(y_i^o - H x_i^b\right), \tag{4}$$

$$\mathbf{K} = \mathbf{P}^b \mathbf{H}^T\left(\mathbf{H}\mathbf{P}^b\mathbf{H}^T + \mathbf{R}\right)^{-1},$$

$$\mathbf{H}\mathbf{P}^b\mathbf{H}^T \cong \frac{1}{N-1}\sum_i^N\left[H\left(x_i^b\right) - \overline{H(x^b)}\right]\left[H\left(x_i^b\right) - \overline{H(x^b)}\right]^T,$$

$$\mathbf{P}^b\mathbf{H}^T \cong \frac{1}{N-1}\sum_i^N\left(x_i^b - \overline{x^b}\right)\left[H\left(x_i^b\right) - \overline{H(x^b)}\right]^T,$$

$$\tag{5}$$

where x is the ensemble state vector, $\mathbf{K}$ is the Kalman gain matrix, H is the observation operator that projects state variables to observed quantities, and $\mathbf{H}$ is the linearized version of H. The covariance matrices for the observation and background errors are $\mathbf{R}$ and $\mathbf{P}$, respectively, $\mathbf{M}$ denotes the NWP model, and t is the current analysis time. Here, subscript i denotes the ensemble member order ranging from 1 to N, and N is the ensemble size. The superscripts b, a, and o denote the background, analysis, and observation, respectively, and the overbar $^-$ represents an ensemble mean.

To avoid introducing additional sampling errors to the ensemble, we used the EnSRF algorithm [22], which replaces (4) with the following equation:

$$\overline{x}^a = \overline{x}^b + \mathbf{K}\left(y^o - H\left(\overline{x}^b\right)\right),$$

$$x_i^{'a} = \beta(\mathbf{I} - \alpha\mathbf{K}\mathbf{H})x_i^{'b},$$

$$\alpha = \left[1 + \sqrt{\mathbf{R}\left(\mathbf{H}\mathbf{P}^b\mathbf{H}^T + \mathbf{R}\right)^{-1}}\right]^{-1}, \tag{6}$$

where α is a factor in the square root algorithm and β is the covariance inflation factor [23], which is used to deal with sampling and model errors. $\mathbf{I}$ is an identity matrix. The

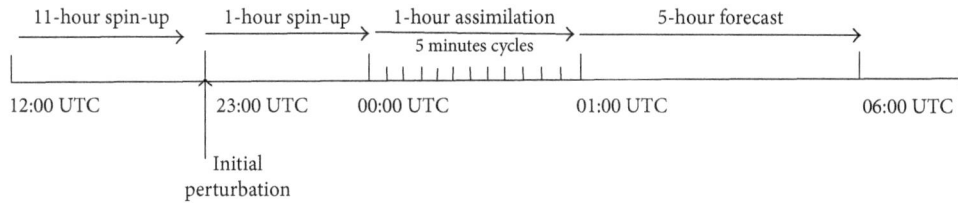

FIGURE 4: Diagram of initial spin-up forecast, data assimilation cycles, and subsequent forecast for the EnSRF experiment.

prime symbol (') in the equations represents the perturbation of the ensemble members.

3. Case Description and Experiment Design

An MCS that occurred over the Front Range of the Rocky Mountains in Colorado (USA) on 29–30 July 2014 was selected to investigate the differences between 3DVar and EnKF in storm-scale DA. The MCS originated in an environment associated with a cold front and a cyclone. The low-level precipitable water and wind field at 850 hPa 0000 UTC 30 July 2014 are shown in Figure 1. Convergence between the northerly cold dry air and southerly warm moist air caused intense convection over northern Colorado, where the precipitable water was >40 kg·m^{-2} (Figure 1(a)) and the maximum wind speed was up to 15 m·s^{-1} (Figure 1(b)). The system persisted for 13 h and it caused heavy rain, strong winds, and frequent lightning.

Figure 2 shows the evolution of this MCS in terms of radar composite reflectivity at different times. At 2100 UTC 29 July 2014 (Figure 2(a)), convective cells had become organized into a convective line over eastern Colorado, with maximum reflectivity of 55 dBZ. There were also some isolated convective cells in southern Wyoming and western Colorado. At 0000 UTC 30 July (Figure 2(b)), the MCS developed and became substantial. The northern part developed rapidly with a large area of reflectivity of 30–45 dBZ. At 0400 UTC 30 July (Figure 2(c)), the MCS was in the mature stage and its southern part became larger and intensified further. During the southeastward movement, some isolated convective cells became incorporated into the moving convective system. By 0900 UTC 30 July (Figure 2(d)), the area of intense radar echoes had contracted, weakened, and moved into Kansas.

The Advanced Research WRF (ARW) version 3.9 was adopted as the forecast model used in this study. The model domain was set using two-way nesting between three nested grids, which had horizontal grid spacings of 15, 3, and 1 km that generated 211×161, 321×281, and 802×751 horizontal grid points, respectively. All domains had 51 vertical grid levels from the surface up to 50 hPa.

The model domains and the locations of the radar stations are shown in Figure 3. The Kain–Fritsch (KF) cumulus parameterization scheme [24] was used only in domain 1. Other model physics schemes adopted included the Thompson microphysics scheme [25], Mellor–Yamada–Janjić (MYJ) planetary boundary layer model [26], Noah land surface model [27], and Rapid Radiative Transfer

Model for GCMs (RRTMG) for longwave and shortwave radiation [28].

As shown in Figure 3, nine high-resolution radars of the Weather Surveillance Radars-1988 Doppler network were used in the DA. Quality control that included despeckling, removal of ground clutter for reflectivity, and velocity de-aliasing (unfolding) [29] was conducted before the radar data were assimilated in the 3 km domain. For the overlapping radar observations, a data location check is conducted in both 3DVar and EnSRF DA systems to avoid counting the same radar observations. When the distance between radar data and radar site is longer than another data, the radar observation is not assimilated. Observations used for rainfall verification were quantitative precipitation estimates from the Multi-Radar/Multi-Sensor System developed by the National Severe Storms Laboratory [30].

Three experiments comprising NoDA, 3DVar, and EnSRF (Table 1) were designed to examine the impact of different radar DA methods on both the analysis and the prediction of the MCS. The NoDA experiment did not assimilate any data, and the initial and lateral boundary conditions were obtained from the $0.5° \times 0.5°$ NCEP operational Global Forecast System (GFS). For the 3DVar experiment, a 12 h spin-up forecast was performed starting from 1200 UTC 29 July 2014. The radar data were assimilated every 5 min for 1 h in the 3 km domain, and the National Meteorological Center (NMC) method [31] was used to estimate background error covariance for assimilation of radar observations. Then, a 5 h deterministic forecast was conducted from 0100 to 0600 UTC 30 July 2014 in the 1 km domain.

For the EnSRF experiment, as shown in Figure 4, 40 members were generated using the random-CV [32] method after an 11 h spin-up forecast initialized from the NCEP GFS analysis at 1200 UTC 29 July 2014. The amplitudes of the perturbations for the horizontal wind (u, v), potential temperature (θ), and water vapor mixing ratio (q_v) were approximately 2 m·s^{-1}, 2 K, and 1 g·kg^{-1}, respectively. Then, a 1 h ensemble forecast was conducted to develop the BEC for sampling the mesoscale environmental uncertainties. Radar data were assimilated every 5 min until 0100 UTC 30 July 2014 in the 3 km domain. The horizontal and vertical covariance localization radii were 12 and 4 km, respectively. A multiplicative covariance inflation following Xue et al. [11] with a factor of 1.25 was used to help maintain the ensemble spread. Additive noise [33] was also added to the analysis members with standard deviations of 0.5 m·s^{-1} for u and v, 0.5 K for θ, and 0.1 g·kg^{-1} for q_v.

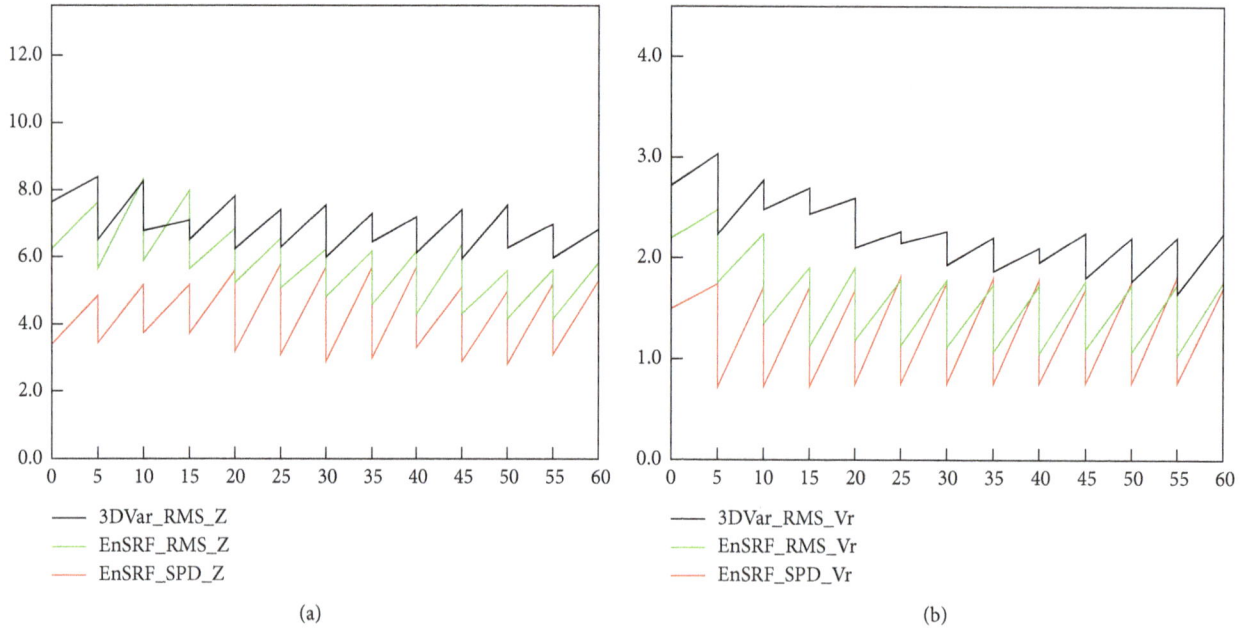

(a)   (b)

FIGURE 5: Root mean square innovations (RMSIs) and ensemble spreads for the experiments of 3DVar and EnSRF for reflectivity (unit: dBZ) (a) and radial velocity (unit: $m \cdot s^{-1}$) (b) calculated against the KFTG (Denver, CO) radar during the assimilation period.

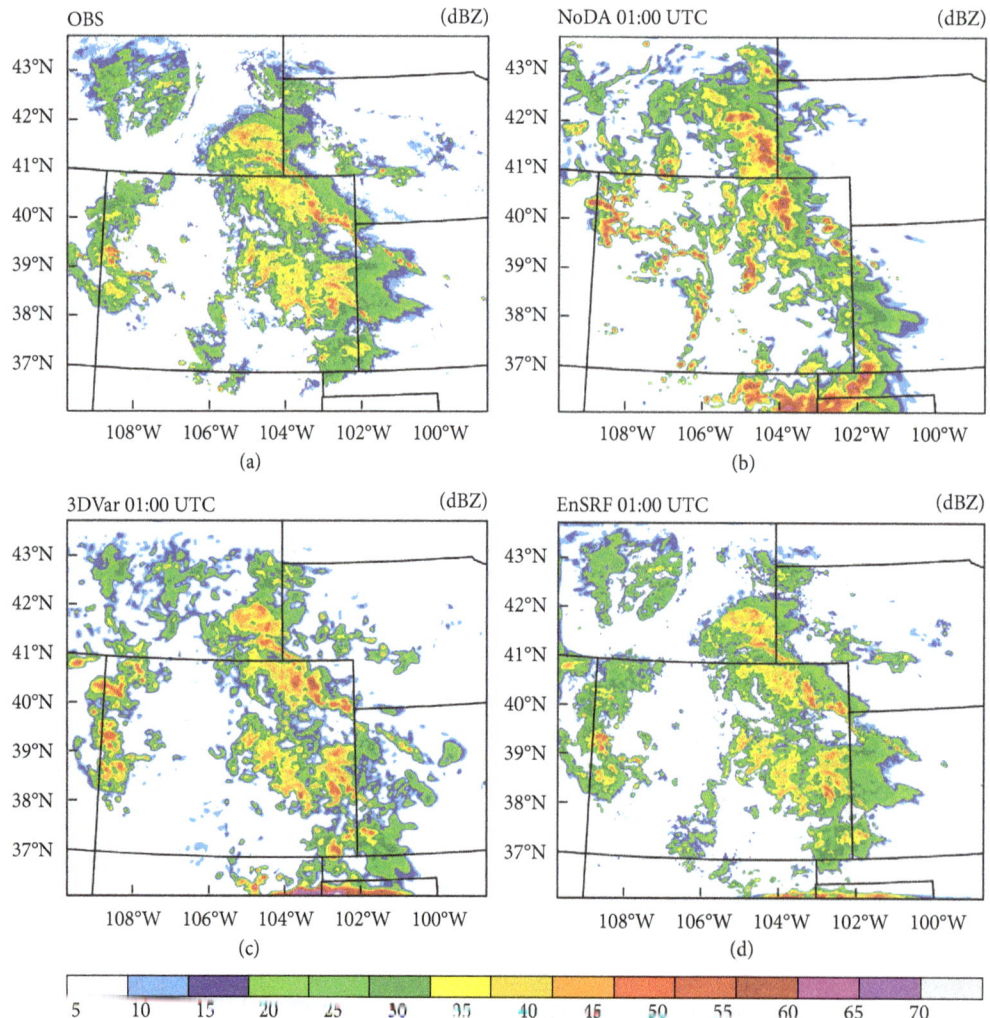

FIGURE 6: Composite radar reflectivity (unit: dBZ) of the mesoscale convective system (MCS) observation (OBS) (a) and results from the NoDA (b), 3DVar (c), and EnSRF (d) experiments at 0100 UTC 30 July 2014.

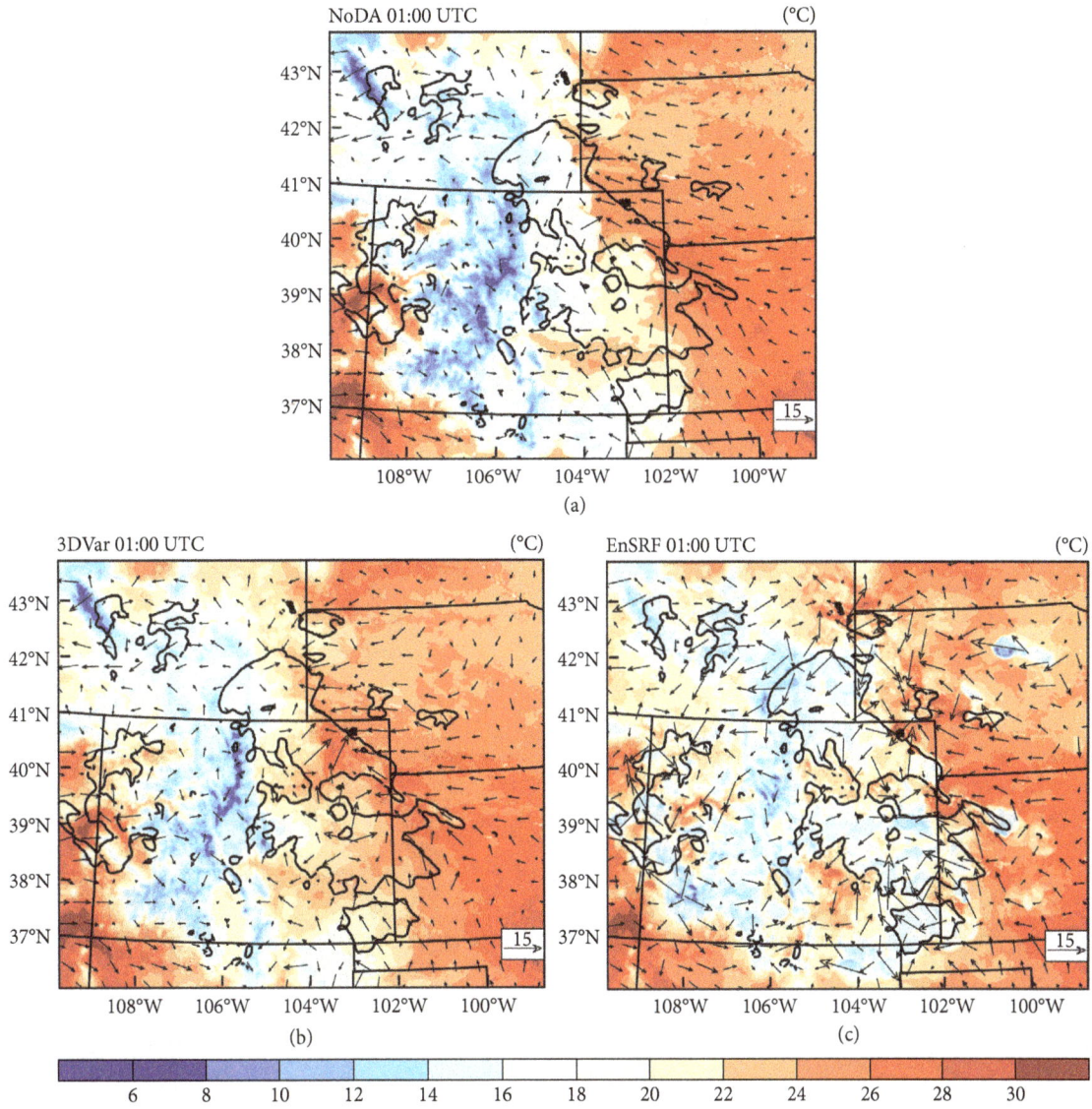

FIGURE 7: Surface temperature (unit:°C) and wind (unit: m·s^{-1}) from the NoDA (a), 3DVar (b), and EnSRF (c) experiments at 0100 UTC 30 July 2014. Solid black lines represent contours of observed reflectivity > 20 dBZ.

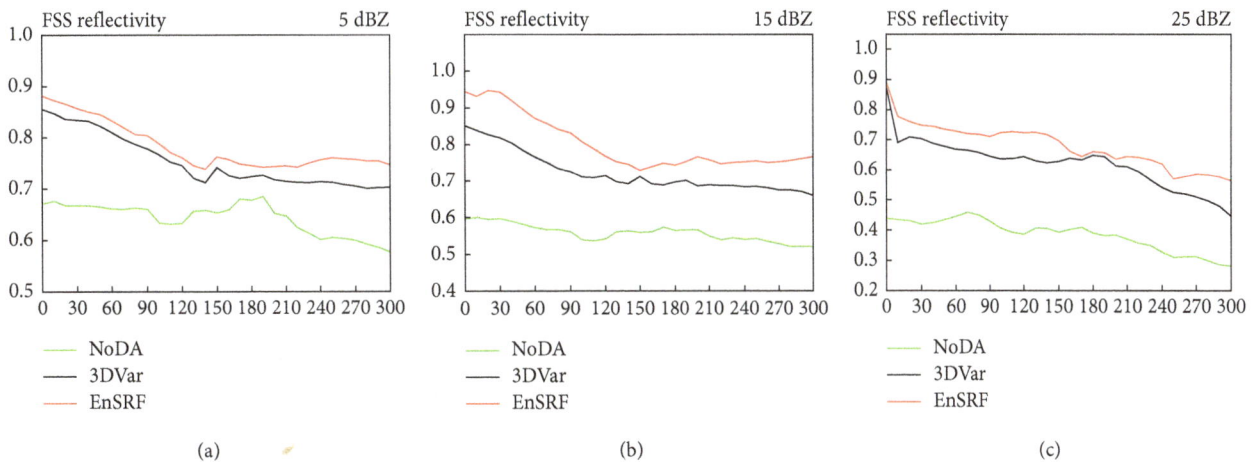

FIGURE 8: Continued.

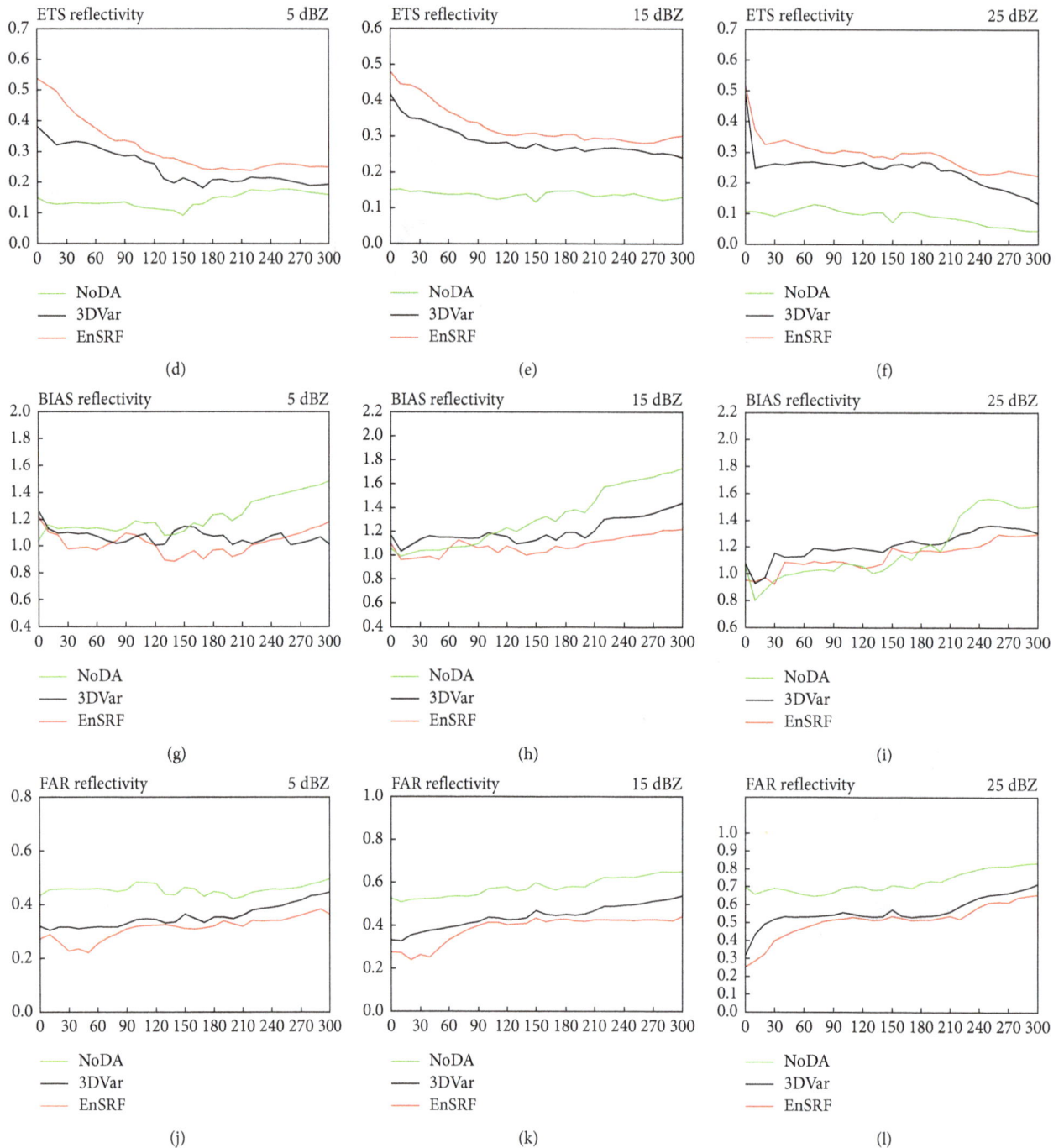

FIGURE 8: Fractions skill score (FSS) (a–c), equitable threat score (ETS) (d–f), BIAS (g–i), and false alarm ratio (FAR) (j–l) for reflectivity thresholds of 5 (a, d, g, j), 15 (b, e, h, k), and 25 dBZ (c, f, i, l) for the three experiments NoDA (green curve), 3DVar (blue curve), and EnSRF (red curve).

Finally, 5 h deterministic and ensemble forecasts were produced at 0100 UTC 30 July 2014 in the 1 km domain from the interpolated ensemble mean and the analysis members, respectively.

4. Results

4.1. Analysis Result. To evaluate the performances of the 3DVar and the EnSRF analyses quantitatively, the root mean square innovations (RMSIs) and the ensemble spread were calculated for radar reflectivity and radial velocity during the 1 h assimilation period (Figure 5). The RMSIs provide a measure of the overall fit of the model state to the observations, and the ensemble spread can be used to examine analysis uncertainty. The calculation was limited to regions where reflectivity was >15 dBZ. Generally, the RMSIs of radar reflectivity and radial velocity for both 3DVar and EnSRF tended to decrease with time. EnSRF

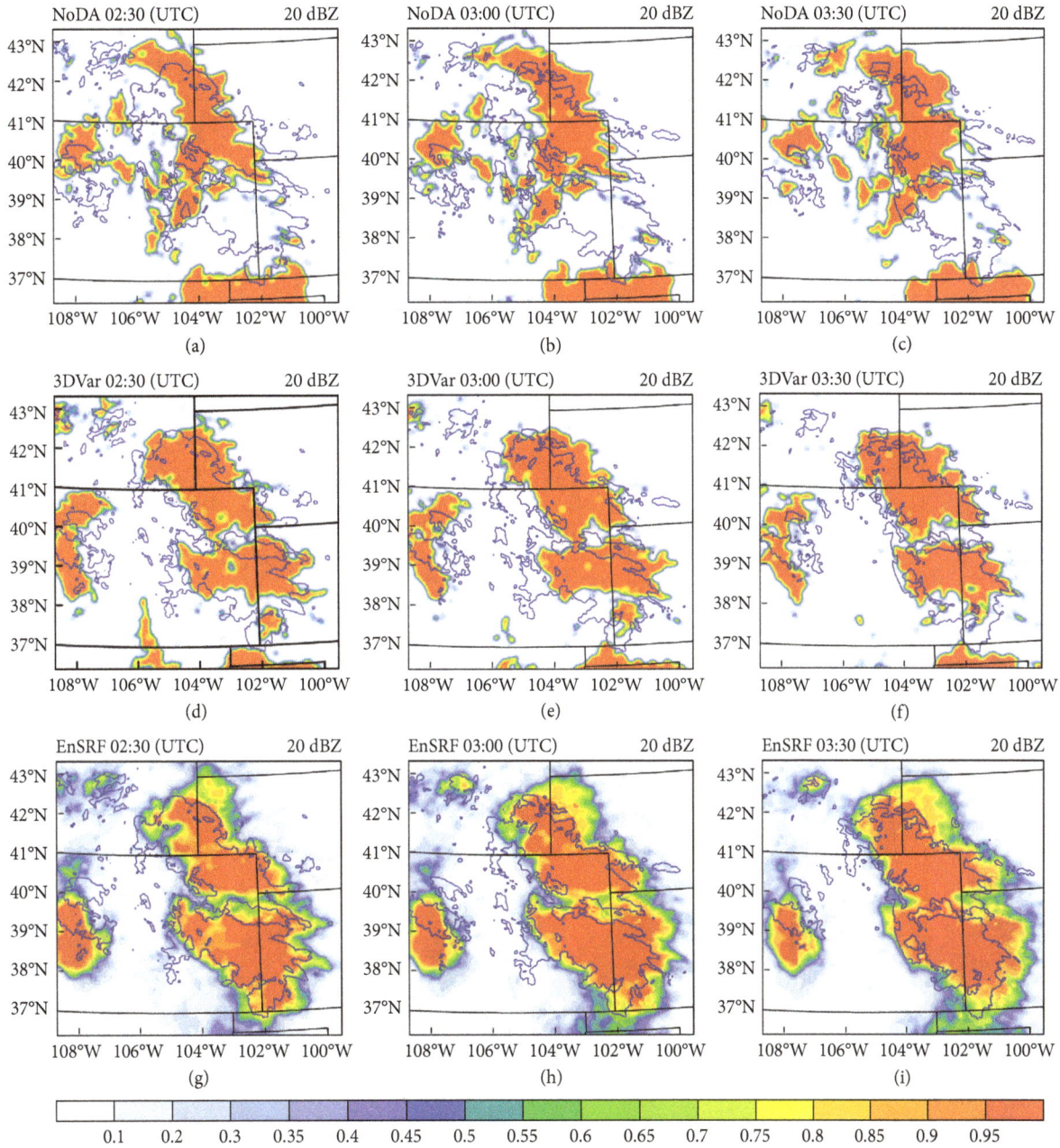

FIGURE 9: Neighborhood probabilities (NPs) of forecast radar reflectivity > 20 dBZ at 0230 UTC (a, d, g), 0300 UTC (b, e, h), and 0330 UTC (c, f, i) on 30 July 2014 from the NoDA (a–c), 3DVar (d–f), and EnSRF (g–i) experiments. Regions of reflectivity > 20 dBZ are outlined by the bold blue contour.

showed lower RMSIs than 3DVar during all analysis and forecast cycles, especially for the radial velocity at $t = 10$–30 min. This suggests EnSRF had smaller analysis error compared with 3DVar. The spread of EnSRF was slightly lower than the RMSIs for radar reflectivity and radial velocity in the first 20 min, suggesting underdispersion of the ensemble. Such underdispersion is a common problem in real radar DA at the convective scale [12, 34]. However, the ensemble spreads of radar reflectivity and radial velocity were consistent with the RMSI values in the following cycles, indicating the forecast error was representative of the ensemble spread.

Figure 6 shows the analyzed composite radar reflectivity of the NoDA, 3DVar, and EnSRF experiments against observations at 0100 UTC 30 July 2014. The observed reflectivity at this time showed a strong MCS covering eastern Colorado and southeastern Wyoming, with a convective line at the Colorado–Wyoming border. Some weak convective cells also formed behind the MCS during its southeastward movement (Figure 6(a)). The NoDA experiment clearly overestimated a northward shift of the northern MCS, and it produced only disorganized convection in the southern MCS (Figure 6(b)). Conversely, both 3DVar and EnSRF successfully captured the distribution and location of the

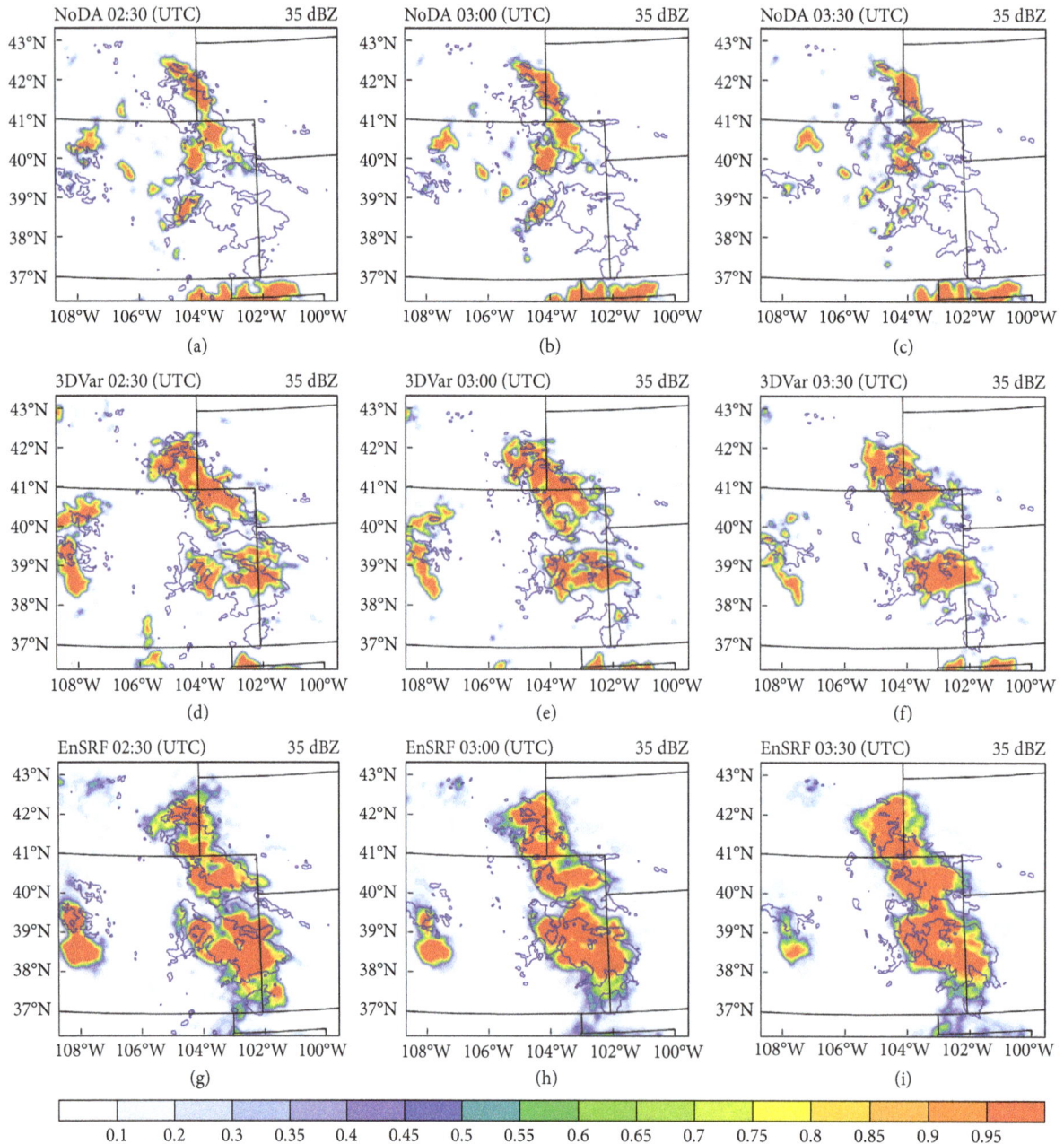

FIGURE 10: Neighborhood probabilities (NPs) of forecast radar reflectivity > 35 dBZ at 0230 UTC (a, d, g), 0300 UTC (b, e, h), and 0330 UTC (c, f, i) on 30 July 2014 from the NoDA (a–c), 3DVar (d–f), and EnSRF (g–i) experiments. Regions of reflectivity > 35 dBZ are outlined by the bold blue contour.

MCS, reflecting the positive effect of radar DA. However, 3DVar overpredicted the intensity of convection with reflectivity of 30–40 dBZ (Figure 6(c)), while EnSRF reduced these overestimations, producing results much closer to the observations (Figure 6(d)). The surrounding weaker convection with reflectivity of 20–30 dBZ was also analyzed better by EnSRF, particularly west of Kansas.

Surface temperature and wind field differences among the experiments at 0100 UTC were also pronounced (Figure 7). EnSRF presented the strongest cold pool with a broad area of low temperatures ranging from 10 to 18°C in the convective region, resulting from precipitation evaporation. Note that the

solid black line represents the contour of reflectivity > 20 dBZ. The temperatures in the NoDA and 3DVar experiments were 12–20°C and 16–22°C, respectively. The different features of the cold pool between the 3DVar and EnSRF experiments might be due to the lack of cross-variable correlations in the static BEC for hydrometeors in 3DVar [21]. Moreover, EnSRF adjusted the wind field better by producing the strongest wind, which plays an important role in maintaining convection.

4.2. Forecast Result.
To assess the overall forecast performances quantitatively, four verification metrics were

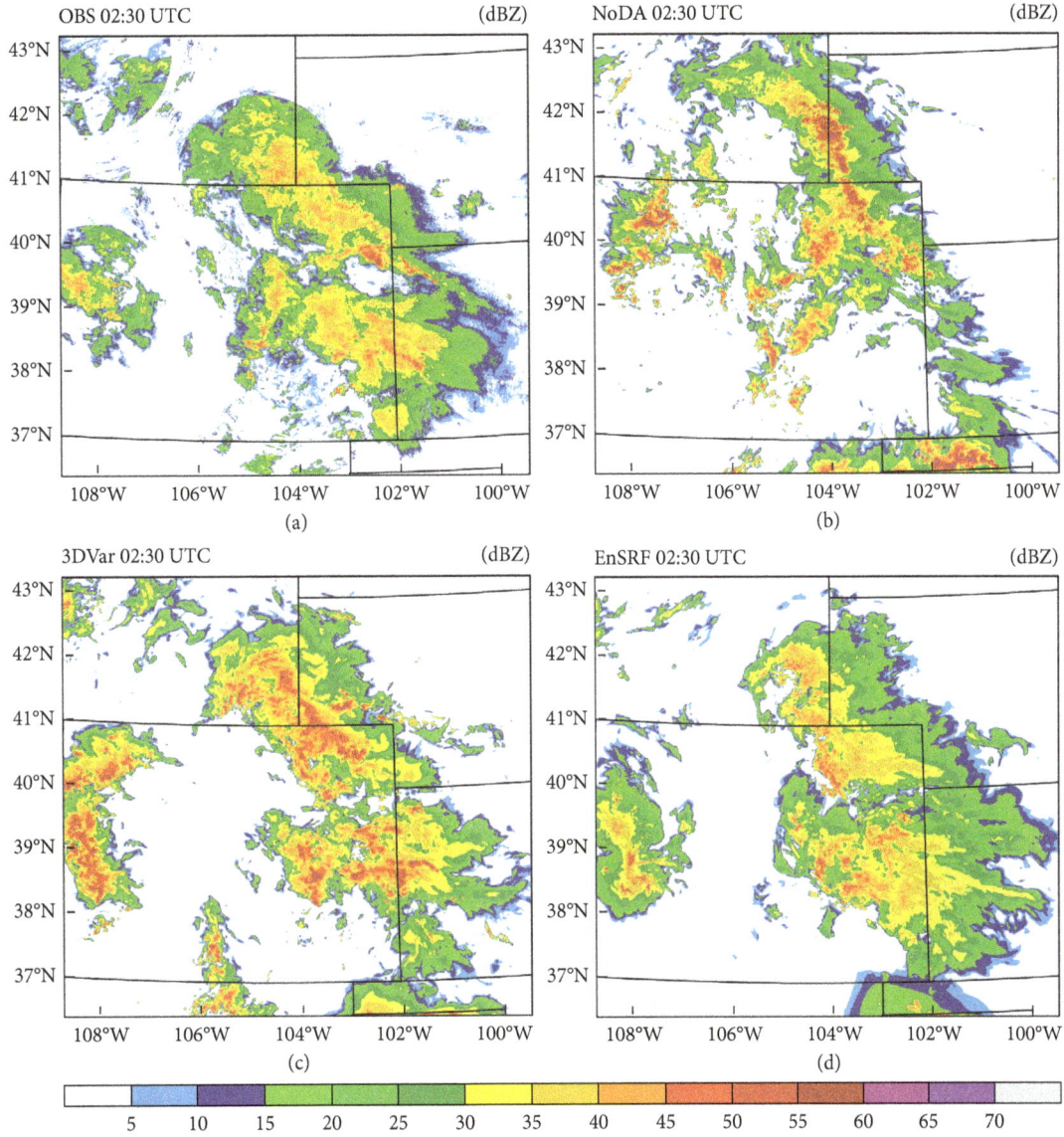

FIGURE 11: Composite reflectivity (unit: dBZ) of observations (a) and results from the NoDA (b), 3DVar (c), and EnSRF (d) experiments at 0230 UTC 30 July 2014.

adopted: the fractions skill score (FSS) [35], equitable threat score (ETS), BIAS, and false alarm ratio (FAR) [36]. The neighborhood-based FSS is defined as

$$FSS = 1 - \frac{(1/N)\sum_N \left(p_f - p_o\right)^2}{(1/N)\left(\sum_N p_f^2 + \sum_N p_o^2\right)},$$ (7)

where p_f and p_o are the forecast and observed fractional coverage of an elementary area by reflectivity or rainfall larger than a given threshold value, respectively, and N represents the number of grid points in the verification domain. The FSS was calculated at each forecast time and each grid based on neighborhood sizes of 6 km. It was then averaged over the domain every 10 min.

The ETS calculates the fraction of observed events predicted correctly, while the BIAS and FAR represent the

bias and false alarms of the reflectivity or precipitation forecast, respectively. They are defined as follows:

$$ETS = \frac{a - ch}{a + c - ch},$$

$$BIAS = \frac{a + b}{a + c},$$ (8)

$$FAR = \frac{b}{a + b},$$

where a, b, and c are the numbers of hits, false alarms, and missing grids, respectively, and ch is the number of hits expected through random chance. BIAS is greater (less) than 1.0 when the frequency of precipitation events, for a given threshold, is over-forecasted (under-forecasted). A perfect

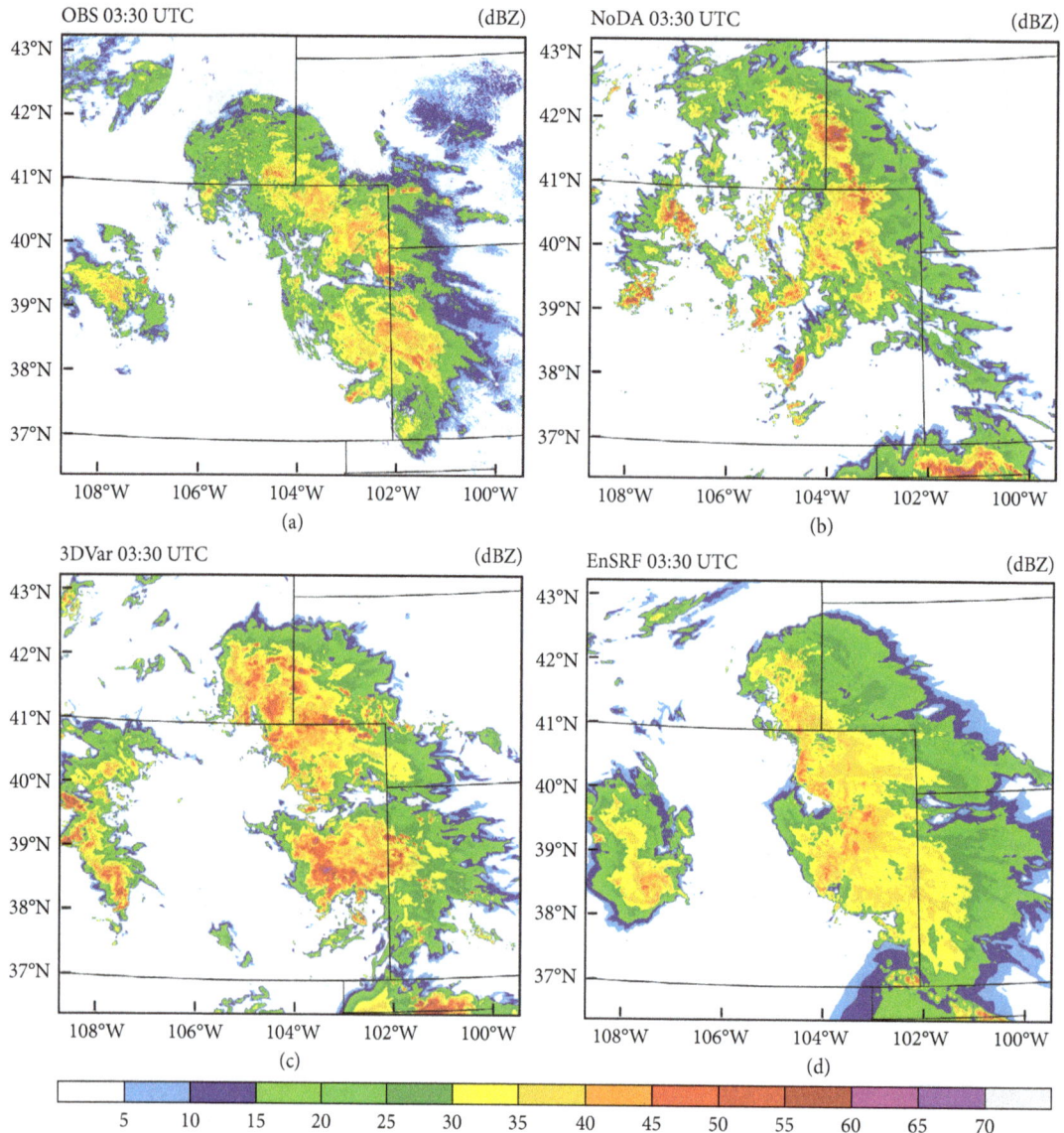

FIGURE 12: Composite reflectivity (unit: dBZ) of observations (a) and results from the NoDA (b), 3DVar (c), and EnSRF (d) experiments at 0330 UTC 30 July 2014.

forecast would be that the FSS, ETS, and BIAS are equal to 1.0 and FAR is equal to 0.0.

The results of FSS, ETS, BIAS, and FAR for the forecast reflectivity from the NoDA, 3DVar, and EnSRF experiments are compared in Figure 8 for reflectivity thresholds of 5, 15, and 25 dBZ. NoDA had the lowest FSS and ETS for all thresholds during the entire forecast period. 3DVar had substantially higher FSS and ETS than NoDA, but the scores decreased more quickly than EnSRF for all thresholds, except for FSS during the first 10 min and during 160–200 min for 25 dBZ (Figures 8(a)–8(f)). The BIAS shown in Figures 8(g)–8(i) above or below 1 indicates reflectivity higher or lower than the observation. NoDA generally overestimated reflectivity above 5 dBZ, whereas it was reduced by 3DVar and EnSRF, particularly for the period 160–300 min. This was also true for 15 dBZ at $t = 100$–300 min and for 25 dBZ at $t = 210$–300 min. It was

also found that EnSRF reduced the overestimation by 3DVar to some extent for all thresholds. 3DVar and EnSRF also showed consistent improvements over NoDA in terms of smaller FAR, while EnSRF showed slightly lower FAR than 3DVar (Figures 8(g)–8(l)).

A neighborhood probability (NP; Schwartz et al. [37]) with a radius of 6 km was also calculated to verify the forecast radar reflectivity of the three experiments. Figure 9 shows the NP of reflectivity (>20 dBZ) at 0230, 0300, and 0330 UTC 30 July 2014 for the NoDA, 3DVar, and EnSRF experiments. The NP forecast of reflectivity (>20 dBZ) of NoDA, for which no radar data were assimilated, was less skillful than 3DVar and EnSRF. It produced a large area of probabilities of 0.0 in the southern MCS, indicating convection was not predicted (Figures 9(a)–9(c)). The reflectivity forecast was generally improved in the 3DVar experiment; however, it contained some underpredictions in the southern MCS (Figures 9(d)–9(f)). EnSRF improved the

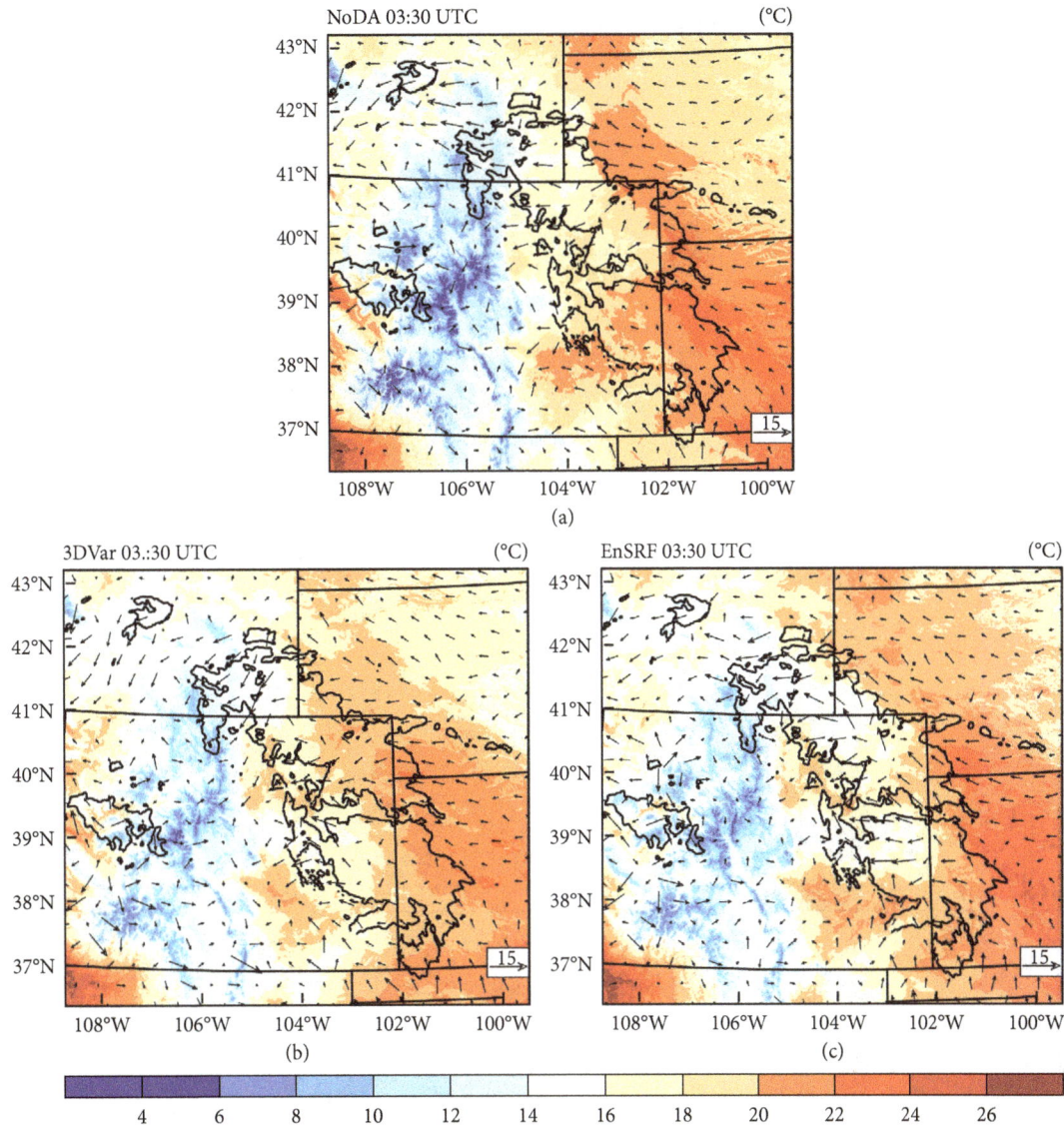

FIGURE 13: Surface temperature (unit: °C) and wind (unit: m·s⁻¹) from the NoDA (a), 3DVar (b), and EnSRF (c) experiments at 0330 UTC 30 July 2014. Solid black lines represent contours of observed reflectivity >20 dBZ.

forecast of the entire MCS compared with the other two experiments. It predicted broad areas of high probability that closely matched the observed reflectivity >20 dBZ in terms of shape, extent, and position (Figures 9(g)–9(i)). In addition, the spurious convection along the Oklahoma panhandle, seen in both NoDA and 3DVar, was suppressed by EnSRF. The higher forecast skill of EnSRF was also evident for the threshold of 35 dBZ, which is associated with stronger convection (Figure 10). Although NoDA and 3DVar exhibited greater forecast error than for reflectivity > 20 dBZ, EnSRF had large overlap of high probabilities in the observed reflectivity (>35 dBZ) during the 1 h forecast, particularly at 0330 UTC.

Figure 11 shows the observed composite radar reflectivity as well as the forecast results for the NoDA, 3DVar, and EnSRF experiments at 0230 UTC 30 July 2014. By this time, the observed MCS had developed into a well-defined

structure with an extensive area of stratiform precipitation. The convective center of the northern MCS in NoDA was highly overestimated with a northward displacement (Figure 11(b)). 3DVar produced a more realistic spatial pattern of the MCS, but the intensity of convection was overpredicted in the observed convective region (Figure 11(c)). EnSRF continued to present some improvements, including better representation of the stratiform precipitation and weaker reflectivity of 20–30 dBZ. The false storms produced over northwest Colorado by NoDA and 3DVar were also corrected (Figure 11(d)). At 0330 UTC (Figure 12), NoDA (panel b) presented a weaker MCS with a reduced area of high reflectivity (>55 dBZ); however, the main system remained displaced to the north. 3DVar (panel c) successfully predicted two parts of the MCS, and the degree of overestimation was reduced to some extent compared with 0230 UTC. The moderate stratiform

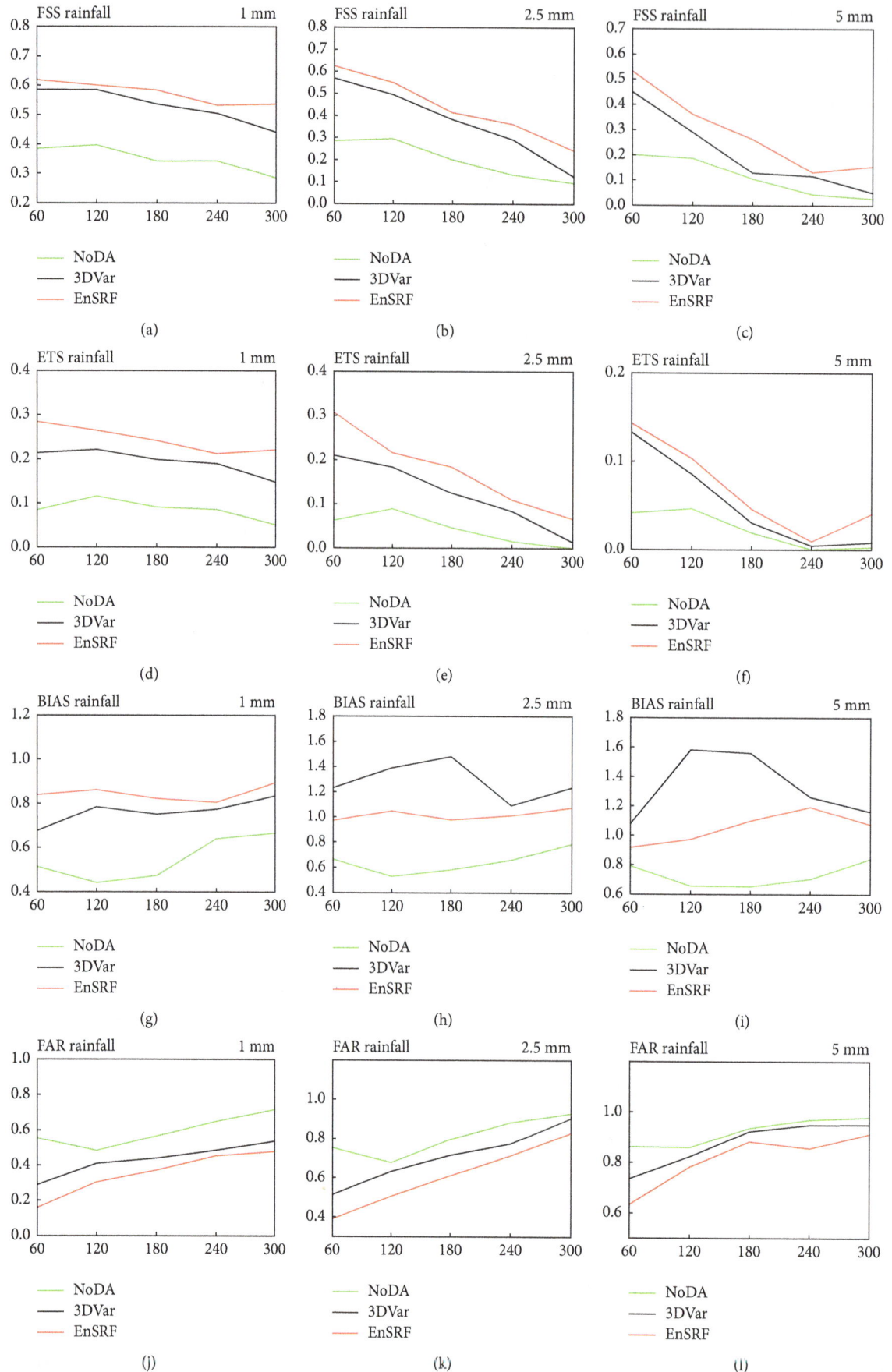

FIGURE 14: Fractions skill score (FSS) (a–c), equitable threat score (ETS) (d–f), BIAS (g–i), and false alarm ratio (FAR) (j–l) for precipitation thresholds of 1 (a, d, g, j), 2.5 (b, e, h, k), and 5 mm $(1\,\mathrm{h})^{-1}$ (c, f, i, l).

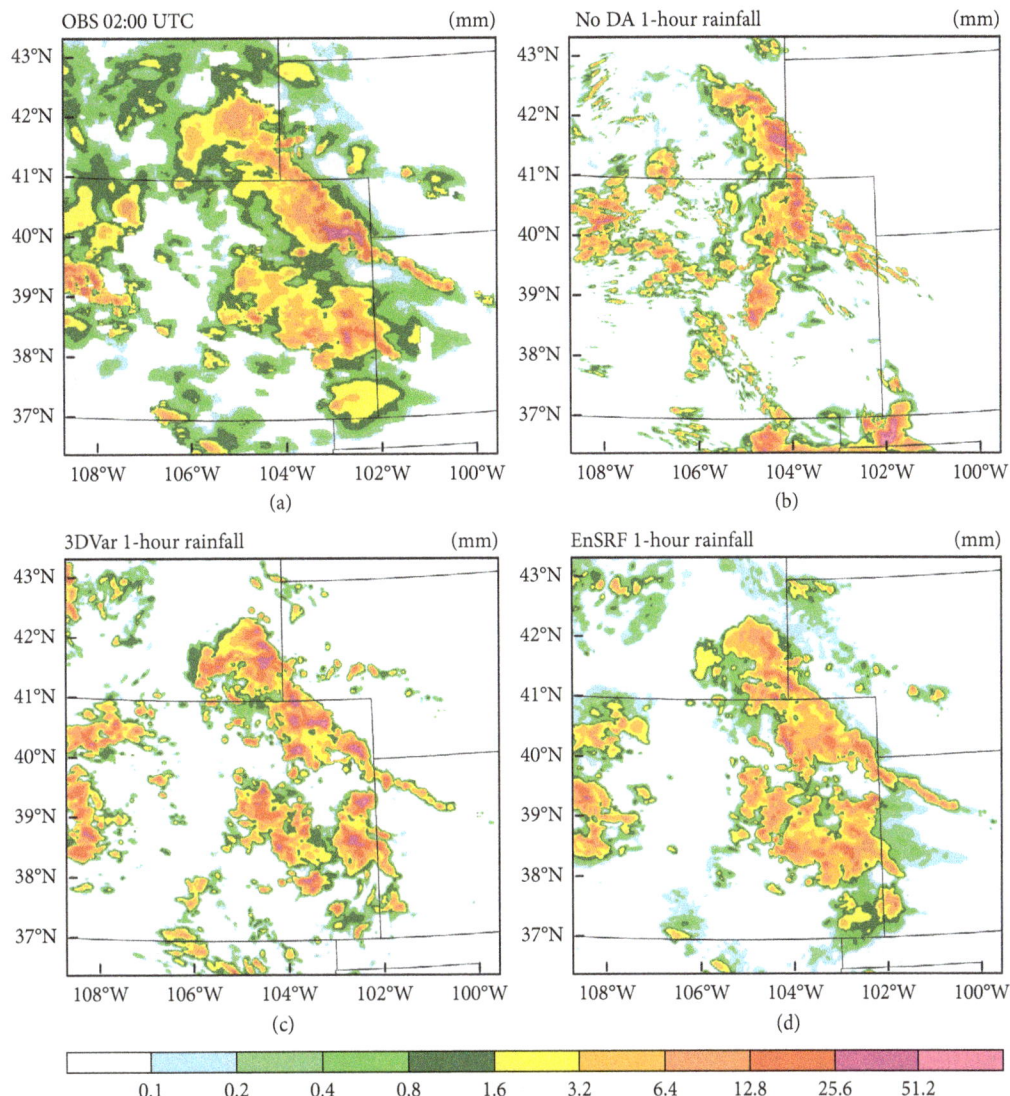

FIGURE 15: One-hour accumulated precipitation (unit: mm) of observations (a) and results from the NoDA (b), 3DVar (c), and EnSRF (d) experiments starting at 0100 UTC 30 July 2014.

precipitation was simulated better by EnSRF (panel d), especially for the southern MCS, although with a larger areal coverage than observed. Unlike the analysis results, the cold pool in 3DVar was stronger than NoDA at this time, while EnSRF still had the strongest cold pool and a near-surface wind that helped maintain a more realistic MCS structure (Figure 13).

Precipitation is another important variable in convective weather. Figure 14 compares the FSS, ETS, BIAS, and FAR for the 5 h forecast from all three experiments. It can be seen that the assimilation of radar data in 3DVar (blue curve) and EnSRF (red curve) resulted in much higher FSS and ETS than NoDA for the thresholds of 1.0, 2.5, and 5.0 mm $(1\,h)^{-1}$ during the entire forecast period. EnSRF showed relatively higher FSS and ETS than 3DVar, and the difference was larger in terms of FSS for the threshold of 5.0 mm $(1\,h)^{-1}$ at $t = 60$–180 min. EnSRF reduced the dry bias from 3DVar and NoDA for the threshold of 1.0 mm $(1\,h)^{-1}$, and the BIAS of EnSRF (which was close to 1) was between 3DVar and NoDA for the thresholds of 2.5 and 5.0 mm $(1\,h)^{-1}$. NoDA presented the largest FAR, followed in descending order by 3DVar and EnSRF.

Figure 15 compares the forecast 1 h accumulated precipitation of the NoDA, 3DVar, and EnSRF experiments. 3DVar and EnSRF were noticeably superior to NoDA in terms of spatial pattern and amount but with different strengths. 3DVar overpredicted the MCS and it produced a larger area of heavy precipitation (>25.6 mm); however, it also missed some moderate (1.6–12.8 mm) and weaker precipitation (0.2–1.6 mm). In contrast, EnSRF predicted a broader area of precipitation than the other experiments, which agreed better with the observations, particularly for precipitation in the range 0.2–1.6 mm. For the 3 h accumulated precipitation (Figure not shown), the observations produced a wider rainband with a larger area of heavy rain. 3DVar again produced too much heavy rain, whereas the EnSRF simulated precipitation had weaker amplitude but with wider coverage, as in the observations.

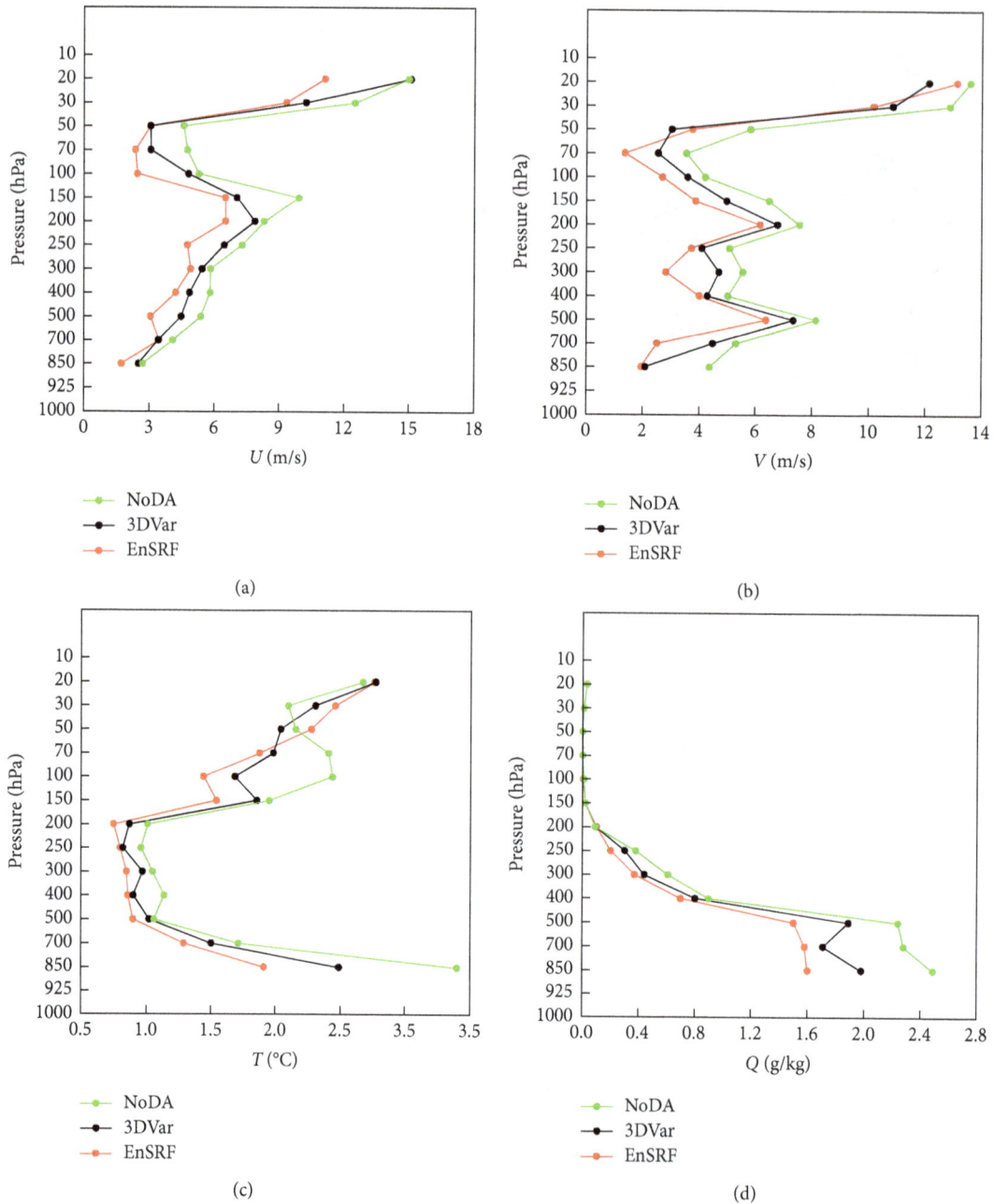

FIGURE 16: Vertical profiles of average root mean square errors (RMSEs) of forecasts against all radiosonde data in the 1 km domain for (a) the u component (unit: m s^{-1}), (b) v component (unit: m·s^{-1}), (c) T (unit: °C), and (d) Q (unit: g·kg^{-1}) at 0300 UTC 30 July 2014.

Figure 16 shows the vertical distribution of domain-averaged root mean square errors (RMSEs) of the horizontal wind components, temperature, and water vapor forecasts against all 10 radiosondes from the 1 km domain at 0300 UTC 30 July 2014. In comparison with NoDA and 3DVar, EnSRF (red curve) had slightly lower RMSEs for horizontal wind components and temperature below 70 hPa, and much smaller RMSEs for water vapor below 500 hPa. Figure 17 shows the RMSEs of horizontal wind components, temperature, and water vapor analyses against the data from all 781 surface METAR stations within the 1 km domain at 0300 UTC 30 July 2014. Evidently, EnSRF had the smallest RMSEs, while the errors of NoDA were the largest and significantly reduced by 3DVar and EnSRF for all variables.

5. Summary

In this study, the EnSRF and a recently developed indirect 3DVar method were used to assimilate radar data for an MCS that occurred over the Front Range of the Rocky Mountains in Colorado (USA) on 29–30 July 2014. Three experiments (NoDA, 3DVar, and EnSRF) were compared to investigate the impact of radar DA and different assimilation methods on both the analysis and the subsequent reflectivity

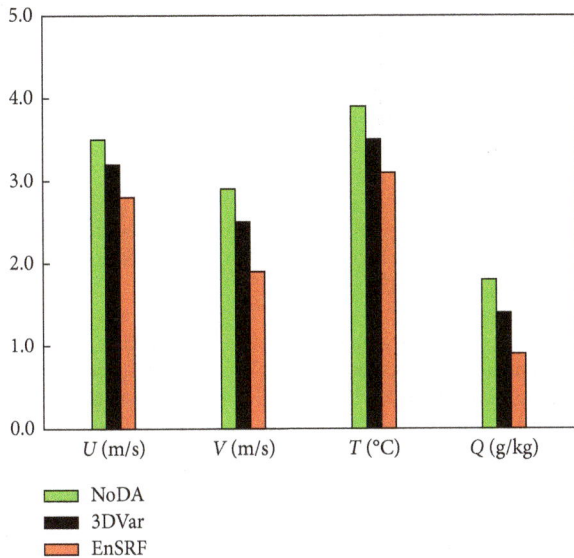

FIGURE 17: Average root mean square errors (RMSEs) of forecasts against all the data from METAR stations within the 1 km domain for the u component (unit: m·s^{-1}), v component (unit: m·s^{-1}), T (unit: °C), and Q (unit: g·kg^{-1}) at 0300 UTC 30 July 2014.

and precipitation forecasts. Both reflectivity and radial velocity data were assimilated from nine radars of the Weather Surveillance Radars-1988 Doppler network in the 3 km domain.

The analysis results showed that EnSRF reduced the RMSIs compared with 3DVar and that its ensemble spread and RMSI values were of comparable magnitude for both reflectivity and radial velocity in the analysis–forecast cycles. Assimilation of radar data resulted in considerable improvement of the analysis, and the analysis produced by EnSRF was more realistic than 3DVar in terms of intensity. EnSRF showed a stronger surface cold pool and wind outflow than 3DVar in the convective regions.

The positive impact of assimilating radar data was maintained for the forecast period, and EnSRF improved the quantitative reflectivity forecast skills measured by FSS, ETS, BIAS, and FAR over 3DVar. The NP of the reflectivity forecasts indicated that EnSRF produced the most skillful probabilistic forecast with high probabilities of the observed reflectivity with different thresholds that included moderate and strong convection. Although 3DVar improved the prediction of the extent of the MCS compared with NoDA, it was no better than EnSRF, especially for the southern MCS. The MCS structure was improved by both 3DVar and EnSRF in terms of location, intensity, and areal coverage, with EnSRF being the best. For precipitation, EnSRF also increased FSS and ETS and reduced FAR compared with NoDA and 3DVar, and it consistently produced the best BIAS for the different thresholds. The precipitation patterns and locations were represented well by 3DVar and EnSRF, although EnSRF reduced the overestimation of the heavy rainfall center compared with 3DVar. Verification against radiosonde sounding data in the 1 km domain suggested that the EnSRF reduced the RMSE of wind, temperature, and water vapor mixing ratio in comparison with NoDA and 3DVar.

The present study demonstrated the encouraging results of EnSRF over 3DVar in MCS analysis and forecasting; however, further research involving additional cases and longer forecast periods will be necessary to reach conclusions that are more reliable. Further studies on radar DA using more advanced methods such as 4DVar and hybrid DA are needed.

Conflicts of Interest

The authors declare no conflicts of interest.

Acknowledgments

This research was supported by the National Key Research and Development Program of China (Grant no. 2017YFC1502102) and the Research Innovation Program for College Graduates of Jiangsu Province (Grant no. KYLX_0829). The author Shibo Gao is supported by scholarship from the China Scholarship Council.

References

[1] Q. Xiao, Y.-H. Kuo, J. Sun et al., "Assimilation of Doppler radar observations with a regional 3DVAR system: impact of Doppler velocities on forecasts of a heavy rainfall case," *Journal of Applied Meteorology*, vol. 44, no. 6, pp. 768–788, 2005.

[2] Q. Xiao, Y. H. Kuo, J. Sun, W. C. Lee, D. M. Barker, and E. Lim, "An approach of radar reflectivity data assimilation and its assessment with the inland QPE of typhoon Rusa (2002) at landfall," *Journal of Applied Meteorology and Climatology*, vol. 46, no. 1, pp. 14–22, 2007.

[3] J. Gao and D. J. Stensrud, "Assimilation of reflectivity data in a convective-scale, cycled 3DVAR framework with hydrometeor classification," *Journal of the Atmospheric Sciences*, vol. 69, no. 3, pp. 1054–1065, 2012.

[4] H. Wang, J. Sun, S. Fan, and X. Y. Huang, "Indirect assimilation of radar reflectivity with WRF 3D-Var and its impact on prediction of four summertime convective events," *Journal of Applied Meteorology and Climatology*, vol. 52, no. 4, pp. 889–902, 2013.

[5] W. Tong, G. Li, J. Sun, X. Tang, and Y. Zhang, "Design strategies of an hourly update 3DVAR data assimilation system for improved convective forecasting," *Weather and Forecasting*, vol. 31, no. 5, pp. 1673–1695, 2016.

[6] E. P. Vendrasco, J. Sun, D. L. Herdies, and C. F. De Angelis, "Constraining a 3DVAR radar data assimilation system with large-scale analysis to improve short-range precipitation forecasts," *Journal of Applied Meteorology and Climatology*, vol. 55, no. 3, pp. 673–690, 2016.

[7] S. Gao, J. Sun, J. Min, Y. Zhang, and Z. Ying, "A scheme to assimilate "no rain" observations from Doppler radar," *Weather and Forecasting*, vol. 33, no. 1, pp. 71–88, 2018.

[8] C. Snyder and F. Zhang, "Assimilation of simulated Doppler radar observations with an ensemble Kalman filter," *Monthly Weather Review*, vol. 131, no. 8, pp. 1663–1677, 2003.

[9] F. Zhang, C. Snyder, and J. Sun, "Impacts of initial estimate and observation availability on convective-scale data assimilation with an ensemble Kalman filter," *Monthly Weather Review*, vol. 132, no. 5, pp. 1238–1253, 2004.

[10] M. Xue, M. Tong, and G. Zhang, "Simultaneous state estimation and attenuation correction for thunderstorms with

radar data using an Ensemble Kalman Filter: tests with simulated data," *Quarterly Journal of the Royal Meteorological Society*, vol. 135, no. 643, pp. 1409–1423, 2009.

[11] M. Xue, M. Tong, and K. K. Droegemeier, "An OSSE framework based on the ensemble square root Kalman filter for evaluating the impact of data from radar networks on thunderstorm analysis and forecasting," *Journal of Atmospheric and Oceanic Technology*, vol. 23, no. 1, pp. 46–66, 2006.

[12] F. Zhang, Y. Weng, J. F. Gamache, and F. D. Marks, "Performance of convection-permitting hurricane initialization and prediction during 2008-2010 with ensemble data assimilation of inner-core airborne Doppler radar observations," *Geophysical Research Letters*, vol. 38, no. 15, p. L15810, 2011.

[13] T. A. Jones, K. Knopfmeier, D. Wheatley, G. Creager, P. Minnis, and R. Palikonda, "Storm-scale data assimilation and ensemble forecasting with the nssl experimental warn-on-forecast system. Part II: combined radar and satellite data experiments," *Weather and Forecasting*, vol. 31, no. 1, pp. 297–327, 2016.

[14] M. Xue, Y. Jung, and G. Zhang, "State estimation of convective storms with a two-moment microphysics scheme and an ensemble Kalman filter: experiments with simulated radar data," *Quarterly Journal of the Royal Meteorological Society*, vol. 136, no. 648, pp. 685–700, 2010.

[15] N. Snook, Y. Jung, J. Brotzge, B. Putnam, and M. Xue, "Prediction and ensemble forecast verification of hail in the supercell storms of 20 May 2013," *Weather and Forecasting*, vol. 31, no. 3, pp. 811–825, 2016.

[16] D. T. Dawson, L. J. Wicker, E. R. Mansell, Y. Jung, and M. Xue, "Low-level polarimetric radar signatures in EnKF analyses and forecasts of the May 8, 2003 Oklahoma City tornadic supercell: impact of multimoment microphysics and comparisons with observation," *Advances in Meteorology* vol. 2013, Article ID 818394, 13 pages, 2013.

[17] Z. Pu, H. Zhang, and J. Anderson, "Ensemble Kalman filter assimilation of near-surface observations over complex terrain: comparison with 3DVAR for short-range forecasts," *Tellus, Series A: Dynamic Meteorology and Oceanography*, vol. 65, no. 1, p. 19620, 2013.

[18] J. S. Whitaker, G. P. Compo, and J.-N. Thépaut, "A comparison of variational and ensemble-based data assimilation systems for reanalysis of sparse observations," *Monthly Weather Review*, vol. 137, no. 6, pp. 1991–1999, 2009.

[19] X. Wang, "Application of the WRF hybrid ETKF–3DVAR data assimilation system for hurricane track forecasts," *Weather and Forecasting*, vol. 26, no. 6, pp. 868–884, 2011.

[20] Z. Meng and F. Zhang, "Limited-area ensemble-based data assimilation," *Monthly Weather Review*, vol. 139, no. 7, pp. 2025–2045, 2011.

[21] A. Johnson, X. Wang, J. R. Carley, L. J. Wicker, and C. Karstens, "A comparison of multiscale GSI-based EnKF and 3DVardata assimilation using radar and conventional observations for midlatitude convective-scale precipitation forecasts," *Monthly Weather Review*, vol. 143, no. 8, pp. 3087–3108, 2015.

[22] J. S. Whitaker and T. M. Hamill, "Ensemble data assimilation without perturbed observations," *Monthly Weather Review*, vol. 130, no. 7, pp. 1913–1924, 2002.

[23] J. L. Anderson, "An adaptive covariance inflation error correction algorithm for ensemble filters," *Tellus, Series A: Dynamic Meteorology and Oceanography*, vol. 59, no. 2, pp. 210–224, 2007.

[24] J. S. Kain and J. M. Fritsch, "A one-dimensional entraining/detraining plume model and its application in convective parameterization," *Journal of the Atmospheric Sciences*, vol. 47, no. 23, pp. 2784–2802, 1990.

[25] G. Thompson, R. M. Rasmussen, and K. Manning, "Explicit forecasts of winter precipitation using an improved bulk microphysics scheme. Part I: description and sensitivity analysis," *Monthly Weather Review*, vol. 132, no. 2, pp. 519–542, 2004.

[26] Z. I. Janjić, "The step-mountain Eta coordinate model: further developments of the convection, viscous sublayer, and turbulence closure schemes," *Monthly Weather Review*, vol. 122, no. 5, pp. 927–945, 1994.

[27] F. Chen and J. Dudhia, "Coupling an advanced land surface–hydrology model with the Penn State–NCAR MM5 modeling system. Part I: model implementation and sensitivity," *Monthly Weather Review*, vol. 129, no. 4, pp. 569–585, 2001.

[28] M. J. Iacono, J. S. Delamere, E. J. Mlawer, M. W. Shephard, S. A. Clough, and W. D. Collins, "Radiative forcing by long-lived greenhouse gases: calculations with the AER radiative transfer models," *Journal of Geophysical Research Atmospheres*, vol. 113, no. D13, 2008.

[29] K. Brewster, M. Hu, M. Xue, and J. Gao, "Efficient assimilation of radar data at high resolution for short range numerical weather prediction," in *Proceedings of World Weather Research Program Symposium and Nowcasting and Very Short-Range Forecasting*, pp. 1–14, Whistler, BC, Canada, August 2005.

[30] J. Zhang, K. Howard, C. Langston et al., "Multi-radar multi-sensor (MRMS) quantitative precipitation estimation: initial operating capabilities," *Bulletin of the American Meteorological Society*, vol. 97, no. 4, pp. 621–638, 2016.

[31] D. F. Parrish and J. C. Derber, "the national meteorological center's spectral statistical-interpolation analysis system," *Monthly Weather Review*, vol. 120, no. 8, pp. 1747–1763, 1992.

[32] D. M. Barker, "Southern high-latitude ensemble data assimilation in the Antarctic mesoscale prediction system," *Monthly Weather Review*, vol. 133, no. 12, pp. 3431–3449, 2005.

[33] D. C. Dowell and L. J. Wicker, "Additive noise for storm-scale ensemble data assimilation," *Journal of Atmospheric and Oceanic Technology*, vol. 26, no. 5, pp. 911–927, 2009.

[34] D. M. Wheatley, K. H. Knopfmeier, T. A. Jones, and G. J. Creager, "Storm-scale data assimilation and ensemble forecasting with the NSSL experimental warn-on-forecast system. Part I: radar data experiments," *Weather and Forecasting*, vol. 30, no. 6, pp. 272–294, 2015.

[35] M. Mittermaier and N. Roberts, "Intercomparison of spatial forecast verification methods: identifying skillful spatial scales using the fractions skill score," *Weather and Forecasting*, vol. 25, no. 1, pp. 343–354, 2010.

[36] F. Mashingia, F. Mtalo, and M. Bruen, "Validation of remotely sensed rainfall over major climatic regions in northeast Tanzania," *Physics and Chemistry of the Earth*, vol. 67–69, pp. 55–63, 2014.

[37] C. S. Schwartz, J. S. Kain, S. J. Weiss et al., "Toward improved convection-allowing ensembles: Model physics sensitivities and optimizing probabilistic guidance with small ensemble membership," *Weather and Forecasting*, vol. 25, no. 1, pp. 263–280, 2010.

10

Temporal-Spatial Characteristics of Drought in Guizhou Province, China, based on Multiple Drought Indices and Historical Disaster Records

Qingping Cheng,[1,2,3] Lu Gao (iD),[1,4,5,6] Ying Chen,[1,4,5,6] Meibing Liu,[1,4,5,6] Haijun Deng,[1,4,5,6] and Xingwei Chen[1,4,5,6]

[1]College of Geographical Science, Fujian Normal University, Fuzhou 350007, China
[2]Northwest Institute of Eco-Environmental and Resources Research, Chinese Academy of Sciences, Lanzhou 730000, China
[3]University of Chinese Academy of Sciences, Beijing 100049, China
[4]Institute of Geography, Fujian Normal University, Fuzhou 350007, China
[5]Fujian Provincial Engineering Research Center for Monitoring and Assessing Terrestrial Disasters, Fujian Normal University, Fuzhou 350007, China
[6]State Key Laboratory of Subtropical Mountain Ecology (Funded by Ministry of Science and Technology and Fujian Province), Fujian Normal University, Fuzhou 350007, China

Correspondence should be addressed to Lu Gao; l.gao@foxmail.com

Academic Editor: Stefano Dietrich

Guizhou Province, China, experienced several severe drought events over the period from 1960 to 2013, causing great economic loss and intractable conflicts over water. In this study, the spatial and temporal characteristics of droughts are analyzed with the standard precipitation index (SPI), comprehensive meteorological drought index (CI), and reconnaissance drought index (RDI). Meanwhile, historical drought records are used to test the performance of each index at identifying droughts. All three indices show decreasing annual and autumn trends, with the latter particularly prominent. 29, 30, and 32 drought events were identified during 1960–2013 by the SPI, CI, and RDI, respectively. Continuous drought is more frequent in winter–spring and summer–autumn. There is a significant increasing trend in drought event frequency, peak, and strength since the start of the 21st century. Drought duration indicated by CI shows longer durations in the higher-elevation region of central and western Guizhou. The corresponding drought severity is high in these regions. SPI and RDI indicate longer drought durations in the lower elevation central and eastern regions of Guizhou Province, where the corresponding drought severity is also very strong. SPI shows an increasing trend in drought duration and drought severity across most of the regions of Guizhou. In general, SPI and RDI show an increasing trend in the western Guizhou Province and a decreasing trend in central and eastern Guizhou. Comparing these three drought indices with historical records, the RDI is found to be more objective and reliable than the SPI and CI when identifying the periods of drought in Guizhou.

1. Introduction

Drought, a water shortage phenomenon caused by natural precipitation anomalies, is one of the most serious natural disasters, causing economic losses globally. The American Meteorological Society classified droughts into four types: meteorological drought, agricultural drought, hydrological drought, and socioeconomic drought [1]. Meteorological drought refers to water shortages caused by an imbalance in precipitation and evaporation. Drought disasters are a product of the coupling of the natural environmental and socioeconomic systems under specific time and space conditions [2]. Among different types of natural disasters, drought disasters are among those with the highest frequencies, widest ranges of influence, longest durations, and greatest losses; drought disasters lead not only to food production reduction, water shortages, and deterioration of ecosystems and the environment, but also to death and the change of dynasties, given that they are an important factor in restricting sustainable social development [3]. The factors influencing drought disasters are complex since there are great uncertainties relating to the occurrence and development of drought disasters in both time and space.

Drought is one of the most frequent and widespread natural disasters in China, where the total average area of land periodically influenced by droughts is 2.1×10^7 hm^2 (annual average value from 1950 to 2013), of which 9.4×10^6 hm^2 (annual average value from 1950 to 2013) suffers drought disasters in any particular year. Meteorological droughts, as described above, can develop into agricultural droughts [4, 5]. In China, droughts cause an annual average of 2.5×10^6 hm^2 of no-harvest area (annual average value from 1989 to 2013) and 1.62×10^{10} kg of grain loss (annual average value from 1950 to 2013); droughts also cause 2.7×10^7 people (annual average value from 1991 to 2013) and 2.0×10^7 livestock (annual average value from 1989 to 2013) to have difficulty finding sufficient drinking water; together, these factors contribute to an annual average direct economic loss of 1.0×10^{11} Chinese Yuan (annual average value from 1950 to 2013, http://www.mwr.gov.cn/sj/tjgb/zgshzhgb/201612/t20161222_776092.html) [6]. The abovementioned information indicates that China as a major agricultural country suffered severe meteorological droughts which caused great economic losses [7]. Drought is the hot spot of research for a long time. Zhai et al. [8] found that a significant dryness trend changes from the southwest to the northeast of China. In the early twenty-first century, the most severe droughts were located in the Southwest of China covering areas around 0.7 million km^2. Yu et al. [9] found that the severe and extreme droughts become more serious since late 1990s for the entire China via examining drought characteristics such as long-term trend and intensity duration. Meanwhile, the drought-prone regions in Northeast China, Southwest China, south China coastal region, and Northwest China were investigated by He et al. [7] and Ayantobo et al. [10]. Xu et al. [11] indicated that the three drought indices (SPI, RDI, and SPEI) have almost the same performances in the humid regions. However, SPI and RDI were more appropriate than SPEI in the arid regions. The Loess Plateau, Sichuan Basin, and Yunnan-Guizhou Plateau have significant dry trends, which is mainly caused by the significant decrease of precipitation. Liu et al. [12] found that the return periods of meteorological drought are longer, with an average of 42.1 years in China. Liu et al. [13] investigated the return period of concurrent drought events is 11 years in the water source area and the destination regions of water diversion project. The probability of concurrent drought events may significantly increase during 2020 to 2050. Shen et al. [14] revealed that the drought probability and intensity are rising and the affected areas of all degrees of drought have an increasing trend during the last 50 years based on the SPEI in Song-Liao River Basin.

The Southwest is one of the regions of China most frequently affected by drought disasters, with droughts of different degrees of severity occurring in this region almost every year, including a severe drought covering a large area every 5–10 years [15]. From 2009 to 2012, the five provinces of Southwest China (Yunnan Province, Sichuan Province, Chongqing City, Guizhou Province, and Guangxi Province) suffered a severe drought [16, 17]. This severe drought, which affected ~8.0×10^6 hm^2 of arable land, led not only to a large reduction in crop production but also caused drinking water shortages for 25 million people and 18 million livestock; meanwhile, the drought caused total direct economic losses of more than 40 billion Chinese Yuan [16, 17]. This drought was the worst in Southwest China since meteorological observations began [18]. The increasing frequency of severe droughts in the Southwest demonstrates that droughts are spreading from northern to southwestern China [19].

Sun et al. [20] assessed the contributions of decadal potential evapotranspiration (PET) anomalies to drought duration and intensity which could exceed those of precipitation in Southwest China. Li et al. [21] identified 87 drought events including 9 extreme events using the daily composite drought index (CI) at 101 stations in Southwest China. The droughts are more frequent from November to next April, and the frequency and intensity of drought increased with a significant decrease in precipitation and increase in temperature. Gao et al. [22] found that the significant soil drying trend happened in autumn, which can be sustained to the next spring. Han et al. [23] showed that the eastern part of southwestern China had an extremely high drought risk, which was greater in the north than south. Recently, several extreme drought disasters have hit Guizhou Province, such as that from September 2009 to March 2010, which caused drinking water shortages for 4.85 million people, with 7.01×10^5 hm^2 of crops suffering from drought, and direct economic losses of 2.3 billion Chinese Yuan. A subsequent extreme summer drought in 2011 caused drinking water shortages for 5.5 million people and 2.8 million livestock, with 1.763×10^6 hm^2 of crops affected, resulting in an economic loss of 15.76 billion Chinese Yuan. Only two years later, the extreme summer drought of 2013 caused drinking water shortages for 2.645 million people and 1.12 million livestock, with 1.763×10^6 hm^2 of crops affected, causing an economic loss of ~9.64 billion Chinese Yuan [24–26].

Droughts are typically measured and quantified using drought indices; a variety of indices for different applications have been developed [27, 28]. Based on World Meteorological Organization (WMO) statistics, there are 55 commonly used categories of drought indices. Among these, the comprehensive meteorological drought index (CI), standardized precipitation index (SPI), standardized precipitation evapotranspiration index (SPEI), and reconnaissance drought index (RDI) are widely used in various regions [21, 29–36]. At present, case studies of droughts in Guizhou Province are rare, with most such studies based on a single drought index [37, 38]. Furthermore, no study has validated these drought indices using historical disaster records, despite validation of the reliability of these indices being of great importance. This study aims at building a link between drought indices and real drought events in Guizhou Province, China.

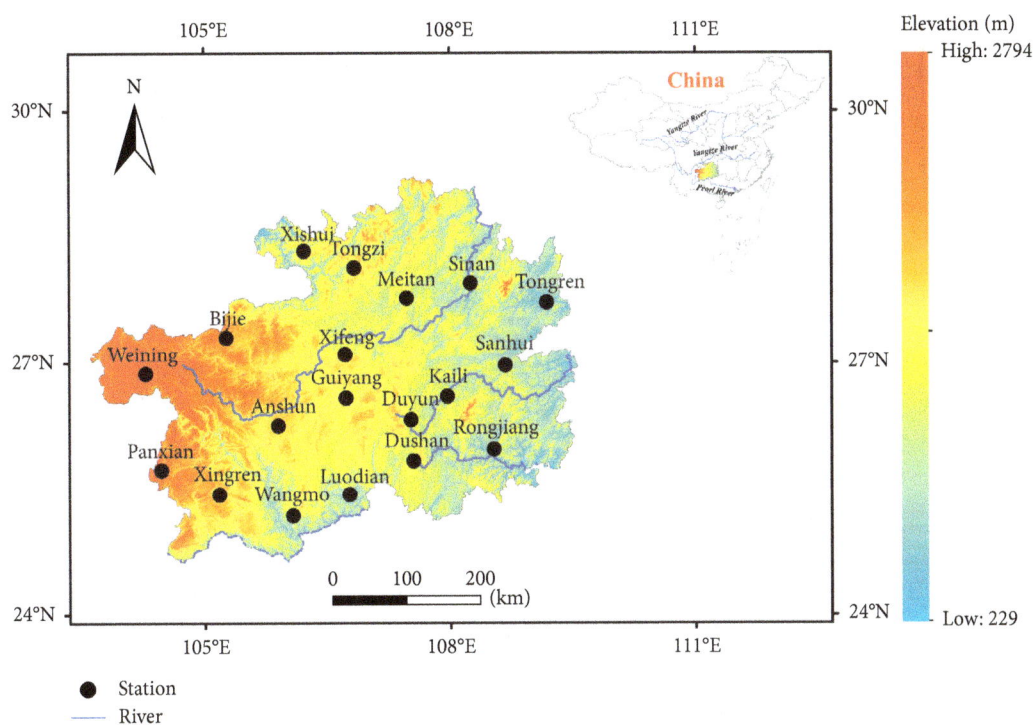

FIGURE 1: Location of meteorological stations in Guizhou Province.

2. Study Region and Data Resources

2.1. Study Region. Guizhou Province (103°36′–109°35′E; 24°37′–29°13′N), with an area of 176167 km², is located in the eastern Yunnan-Guizhou Plateau of China (Figure 1). The elevation of Guizhou Province ranges from 229 to 2794 m, higher in the west than that in the east of province [39]. The topography is dominated by plateau and mountains: carbonate rocks in the karst area are widespread and account for 62% of the total area of Guizhou Province. Guizhou has a humid subtropical monsoon climate with an annual mean temperature of 15°C and mean annual precipitation of 1400 mm. Over 70% of the annual rainfall occurs from May to September [39–41]. In general, the ecology and environment of Guizhou Province is extremely fragile, which causes frequent land-surface droughts, as illustrated by an old saying describing drought in Guizhou Province: "a drought every year, a mild drought every three years, a moderate drought every five years, a severe drought every decade."

2.2. Data Resources. The meteorological data for daily precipitation and pan evaporation (from January 1, 1959, to February 28, 2014) data set are used in this paper from the China Meteorological Data Sharing Service Network (http://data.cma.cn/) V3.0 version. Rigorous quality control had been conducted by China Meteorological Data Sharing Service Network before the data were released. The software used to detect and adjust shifts in the time series of daily precipitation and pan evaporation is RHtestsV3 and RHtests-dlyPrcp (http://etccdi.pacificclimate.org/software.shtml), respectively. Finally, 19 out of 32 national basic meteorological stations (no gaps exceeding two consecutive weeks) are selected for this

study. It should be noted that some evaporation data (since 2002) were recorded with E601B equipment. The E-601B-type evaporator was installed for meteorological stations in China from 1985. The E-601B-type evaporation evaporator is recommended by the World Meteorological Organization (WMO). This instrument has the advantages of corrosion-resistant and stable thermal effect, which made the measurements more close to nature [42]. In order to ensure the continuity, uniformity, and reliability of records, a linear regression is therefore applied to calibrate evaporation data collected by E601B (2002–2014) to 20 cm evaporating dish data (1998–2001), according to previous studies [43, 44]. The elevation data (DEM) are from the Shuttle Radar Topography Mission (SRTM) with a resolution of 90 m, derived from the Geospatial Data Cloud of China (http://www.gscloud.cn/). The historical disaster records are derived from China Meteorological Disaster Yearbook (Guizhou volume) [45–51]. The information such as drought duration, severity, and peaks was extracted from the yearbooks according to the disaster statistics which were originally recorded by the local meteorological department. Seasons are classified based on meteorological divisions: spring (March–May), summer (June–August), autumn (September–November), and winter (December–February), respectively. The distribution of meteorological stations, together with related information, is shown in Figure 1 and Table 1.

3. Methods

3.1. Drought Indices (SPI, CI, and RDI)

3.1.1. Standard Precipitation Index (SPI). The SPI was developed by McKee et al. [29]. Within a certain geographic

TABLE 1: Information of meteorological stations and average precipitation and 20 cm pan evaporation in 1960–2013.

Station	Longitude (°E)	Latitude (°N)	Elevation (m)	Pan precipitation (mm)	Evaporation (mm)	$P-C_V$	$E-C_V$
Anshun	105.9	26.25	1431.1	1310.1	1250.5	0.18	0.09
Bijie	105.27	27.3	1510.6	883.2	986.1	0.15	0.10
Duyun	107.52	26.32	969.1	1419.9	1239.1	0.17	0.08
Dushan	107.55	25.83	1013.3	1307.5	1201.4	0.14	0.09
Guiyang	106.73	26.58	1223.8	1099.6	1377.5	0.16	0.10
Kaili	107.97	26.6	720.3	1142.1	1347.8	0.18	0.09
Luodian	106.77	25.43	440.3	1142.1	1246.5	0.18	0.07
Meitan	107.47	27.77	792.2	1116.5	1052.6	0.15	0.10
Panxian	104.47	25.72	1800	1355	1599.1	0.18	0.12
Rongjiang	108.53	25.97	285.7	1181.6	1174.1	0.17	0.11
Sanhui	108.67	26.97	626.9	1102	1129.7	0.15	0.11
Sinan	108.25	27.95	416.8	1120.1	1148.4	0.19	0.10
Tongzi	106.83	28.13	972	1259.8	1198	0.14	0.09
Tongren	109.17	27.72	353.2	1008.4	1091.5	0.13	0.15
Wangmo	106.08	25.18	566.8	1222.2	1453.8	0.16	0.09
Weining	104.28	26.87	2237.5	889.7	1350.2	0.19	0.08
Xifeng	106.72	27.1	1112.1	1108.4	1282.5	0.15	0.12
Xishui	106.22	28.33	1180.2	1094.3	1002.3	0.14	0.10
Xingren	105.18	25.43	1378.5	1317.3	1517.1	0.15	0.09

Note. $P-C_V$, coefficient variation of precipitation; $E-C_V$, coefficient variation of pan evaporation.

TABLE 2: Classification of SPI, CI, and RDI.

SPI/CI/RDI value	Drought grades
Value ≤ -2.0	Extreme drought
$-2.0 <$ value ≤ -1.5	Severe drought
$-1.5 <$ value ≤ -1.0	Moderate drought
$-1.0 <$ value ≤ 0	Mild drought

area, the precipitation usually fluctuates regularly. If the precipitation is less than the average annual precipitation, a drought may therefore occur in this area. On the contrary, precipitation exceeding the annual average may induce flooding. The SPI has many advantages such as being dimensionless and standardized, working on multiple scales, and being easy to calculate. To calculate the SPI, a frequency distribution function is first constructed from a series of long-term precipitation observations. A gamma probability density function is then fitted to the series, and the cumulative probability of an observed precipitation is computed. The inverse normal (*Gaussian*) function, with a mean of 0 and a variance of 1, is then applied to transform the cumulative distribution to the standard normal distribution. Because the SPI is based on the cumulative probability of a given timescale, here the total amount of precipitation in the current month and previous i months ($i = 1, 2, 3, \ldots$) is used to calculate the SPI on a timescale of $i + 1$ month. Here, SPI_{12} (1–12 monthly cumulative precipitation) represents annual timescales, and SPI_3 (3 monthly cumulative precipitation) represents seasonal timescales. Drought classification is shown in Table 2.

3.1.2. Comprehensive Meteorological Drought Index (CI).
The comprehensive meteorological drought index (CI) is effective for meteorological drought monitoring and assessment [52]. Both 30 day (month scale) and 90 day

(seasonal scale) standardized precipitation indices, combined with a 30 day relative humidity index, can be used to calculate a comprehensive meteorological drought index. Since the CI can indicate precipitation climate anomalies on both short (months) and long timescales (seasons) [52], this index is therefore suitable for meteorological drought monitoring and historical drought assessment. The first step of the calculation is as follows:

$$\text{MI} = \frac{\left(P_{ij} - \text{PET}_{ij}\right)}{\text{PET}_{ij}}, \tag{1}$$

where MI is the relative moisture index in the recent 30 days, P_{ij} refers to the total amount of precipitation in the recent 30 days (unit: mm), and PET_{ij} is the total potential evapotranspiration in the recent 30 days (mm; here we use evaporation of a 20 cm evaporating dish). The CI is then calculated as follows:

$$\text{CI} = a\text{SPI}_{30} + b\text{SPI}_{90} + c\text{M}_{30}, \tag{2}$$

where SPI_{30} and SPI_{90} are the standardized precipitation indexes for 30 d and 90 d periods, respectively. M_{30} refers to the MI of 30 days. a, b, and c are set as 0.4, 0.4, and 0.8. In theory, the weight coefficients a, b, and c are from the average values above light drought levels of SPI_{30}, SPI_{90}, and MI_{30} divided by the smallest history of SPI_{30}, SPI_{90}, and MI_{30}, respectively (GBT 20481-2006 meteorological drought level) [52]. The drought classification scheme is displayed in Table 2.

3.1.3. Reconnaissance Drought Index (RDI).
The drought detection index was proposed by Tsakiris et al. [31, 32] and takes into account the effects of precipitation and evapotranspiration on drought. The RDI has three modes of expression: the initial value RDI (α_0) is presented in an aggregated form using a monthly time step and calculated

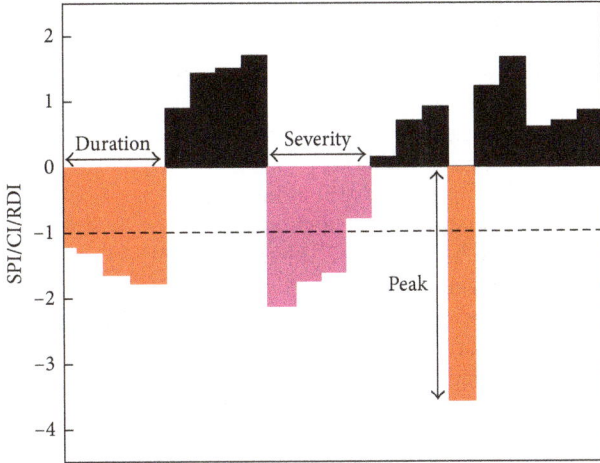

FIGURE 2: Definition of drought characteristics for SPI, CI, and RDI based on Run Theory.

for each month of a hydrological year or a complete year. The second expression is normalized RDI (RDI_n), and the third expression is standardized RDI (RDI_{st}). The initial value α_0 can be calculated with the following formula:

$$a_0^i = \frac{\sum_{i=1}^{12} P_{ij}}{\sum_{j=1}^{12} PET_{ij}}, \quad i = 1, \ldots, N \text{ and } j = 1, \ldots, 12, \quad (3)$$

where P_{ij} and PET_{ij} are precipitation and potential evapotranspiration (we use evaporation of the 20 cm evaporating dish) in jth month of ith hydrological year, respectively. N is the total number of years. Equation (3) can calculate the RDI for any period of the year.

The normalized RDI, RDI_n, is calculated using the following equation for each year, in which it is evident that the parameter $\bar{a}_0$ is the arithmetic mean of a_0 values calculated for the N years of data:

$$RDI_n^{(i)} = \frac{a_0^{(i)}}{\bar{a}_0} - 1. \quad (4)$$

The standard RDI (RDI_{st}) is similar to the standard precipitation index (SPI) and is calculated as follows:

$$RDI_{st}k^{(i)} = \frac{y_k^{(i)} - \overline{y_k}}{\sigma_{yk}}, \quad (5)$$

where $y_k^{(i)} = \ln(a_0^{(i)})$, $\overline{y_k}$ is the arithmetic mean of y_k, and σ_{yk} is the standard deviation of $y_k^{(i)}$. The drought classification scheme is shown in Table 2.

3.2. Drought Variables. According to McKee et al. [29] and Spinoni et al. [53], a drought event is defined as being when SPI, CI, and RDI values are lower than −1 (included in this month) to positive value (excluding this month), with at least two consecutive such months used to define drought events from 1960 to 2013 in this study. Drought duration-severity-area-intensity/frequency is widely used in drought research [8, 10, 11, 54]. The derived drought variables [54, 55] based on the Run Theory follow the definitions

(Figure 2). Drought duration is defined as the number of months from the first month in which the indicator goes lower than −1 to the last month with a negative value before the indicator returns to positive values. Drought intensity is defined as the number of months in which the drought indicator remains lower than −1. Drought severity is defined as the sum of the monthly absolute values of the index when the index is ≤−1 over the period 1960–2013. Drought peak refers to the month in the "drought event" with the lowest value of the indicator [36].

3.3. Mann–Kendall Test. The Mann–Kendall (M-K) nonparametric statistical test method, proposed by Mann [56] and Kendall [57] and recommended by the World Meteorological Organization (WMO). The M-K test does not require samples to follow a certain distribution nor is affected by a few abnormal values. It is widely used in the data of nonnormal distribution of hydrology and meteorology due to its simplicity. Here, the M-K test is applied to analyze the temporal characteristics of SPI, CI, and RDI. For a time series, $X = \{x_1, x_2, \ldots, x_n\}$, where $n > 10$. The test statistic Z_{mk} is calculated as follows:

$$Z_{mk} = \begin{cases} \dfrac{S-1}{\sqrt{\mathrm{Var}(S)}}, & S > 0 \\ 0, & S = 0 \\ \dfrac{S+1}{\sqrt{\mathrm{Var}(S)}}, & S < 0 \end{cases},$$

where, $S = \sum_{i=1}^{n-1} \sum_{k=i+1}^{n} \mathrm{sgn}(x_k - x_i)$,

$$\mathrm{Var}(S) = \frac{[n(n-1)(2n+5) - \sum_{i=1}^{m} t_i(t_i-1)(2t_i+5)]}{18}, \quad (6)$$

where $\mathrm{Var}(S)$ is the variance of the statistic S; x_k and x_i are the sequential data values; m is the number of tied groups; t_i denotes the number of data points in the ith group; n is the length of the data set; and $\mathrm{sgn}(x_k - x_i)$ is the sign function, determined as

$$\mathrm{sgn}(x_k - x_i) = \begin{cases} +1, & (x_k - x_i) > 0 \\ 0, & (x_k - x_i) = 0 \\ -1, & (x_k - x_i) < 0 \end{cases}. \quad (7)$$

For the statistic Z_{mk} value, $Z_{mk} > 0$ indicates that the time series has a rising (increasing) trend, while time series with $Z_{mk} < 0$ has a falling (decreasing) trend. Absolute values of $Z_{mk} \geq 1.65$, 1.96, and 2.58 are adopted, respectively, indicating significance levels of $\alpha = 0.1$, 0.05, and 0.01.

When the M-K test is further used to test the sequence mutation, the test statistic is different from the above Z_{mk}, by constructing a rank sequence:

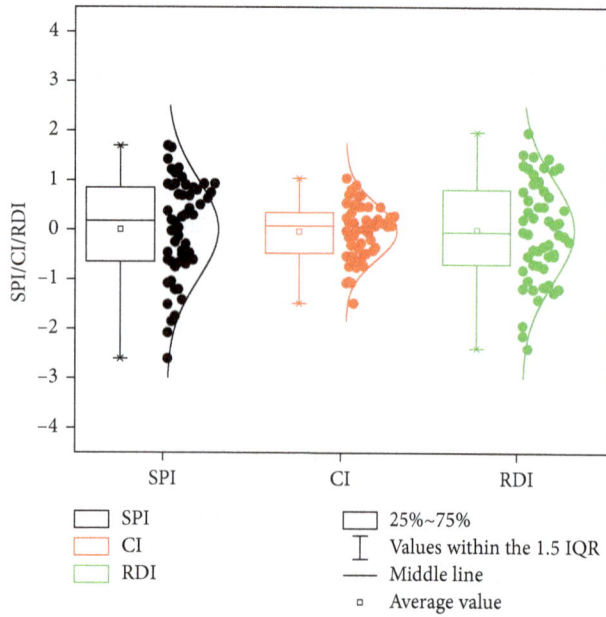

SPI | CI | RDI
SPI | 25%~75%
CI | Values within the 1.5 IQR
RDI | Middle line
Average value

FIGURE 3: Box-plot line and normal distribution curve for SPI, CI, and RDI.

$$S_k = \sum_{i=1}^{i} \sum_{j}^{i-1} \alpha_{ij} \quad (k = 2, \ 3, \ 4, \ldots, n),$$

$$\alpha_{ij} = \begin{cases} 1 & x_i > x_j, \\ 0 & x_i < x_j, \end{cases} \quad 1 \le j \le i,$$

$$\mathrm{UF}_k = \frac{[S_k - E(S_k)]}{\sqrt{\mathrm{Var}(S_k)}} \quad (k = 1, \ 2, \ldots, n), \tag{8}$$

$$E(S_k) = \frac{k(k+1)}{4},$$

$$\mathrm{Var}(S_k) = \frac{k(k-1)(2k+5)}{72},$$

where UF_k is a standard normal distribution and a significant level α is given. If there is a significant trend change, the time series x is arranged in reverse order and then is calculated according to the formula:

$$\begin{aligned} \mathrm{UB}_k &= -\mathrm{UF}_k \\ k &= n + 1 - k \quad (k = 1, \ 2, \ldots, n), \end{aligned} \tag{9}$$

where UF_k is a positive sequence and UB_k is a reverse sequence. If UF_k exceeds 0, the sequence shows a rising trend, and a value of <0 indicates a falling trend. The rising or falling trend is significant when these parameters exceed the critical line. If the UF_k and UB_k curves intersect and the intersection is between the critical straight lines, the corresponding moment of intersection is defined as the moment when the mutation begins.

4. Results

4.1. Temporal Variability. Figure 3 shows the box-plot line and normal distribution curve for CI, SPI, and RDI. All three indices conform to normal distribution, and the distribution of drought indices is also very similar in the box-plot for SPI and RDI. The normal distribution of the CI is concentrated, and the box-plot reflects drought ranks' relative light. Figure 4 shows annual and seasonal SPI trends and the M-K test in Guizhou Province. Annual and seasonal Z values were, respectively, −2.33, −1.99, −0.39, −2.30, and −0.72, and all showed a decreasing trend. Annual, spring, and autumn trends were significant at the 0.05 significance level. The magnitude of the decreasing trend for the annual and autumn trends is larger, at −0.020/10a and −0.023/10a, respectively. As illustrated in Figure 4(a), the annual decreasing trend is significant in 1980–1990 and 2000–2013 at the 0.05 significance level. UF and UB intersect in 2006 and break through the boundary line in 2012-2013. In spring, UF and UB intersect in 1984 and break the boundary line in 1998–2001, 2007, and 2010–2013. In autumn, UF and UB intersect in 1986 and break the boundary line in 2003–2013, indicating a significant abrupt decrease in the trend. However, summer and winter mutations are not significant. According to the drought index, annual severe droughts or extreme droughts are found in 1966, 2009, 1989, and 2013. For seasons, severe or extreme droughts in spring are more often in 1979, 1986, 1988, 1991, and 2011, while in summer they are found in 1972, 1981, 2011, and 2013. For autumn, the severe or extreme droughts are found in 1969, 1978, 1992, 2002, and 2006. For winter, the years with severe or extreme winter droughts are 1978, 1985, 2009, and 2012.

Z values for annual and seasonal droughts were, respectively, −2.26, −0.66, −0.24, −2.69, and −1.51; all showed a decreasing trend, with the trend for annual and autumn timescales significant at the 0.05 significance level (Figure 5). The rate at which the trend decreases for annual and autumn timescales is larger, at −0.012/10a and −0.018/10a, respectively. As illustrated in Figure 5(a), the decreasing trend of annual UF is significant in 1980–1990 and 2000–2013 at the 0.05 significance level. UF and UB intersect in 2006 and break through the boundary line in 2012-2013. In autumn, UF and UB intersect in 1992 and break through the boundary line in 2005–2013, indicating a significant abrupt decrease in the trend. However, the trends in spring, summer, and winter are not significant. From the drought index, the only year with an annual severe drought is 2011. Years with severe or extreme droughts in the spring are 1987, 1988, 2010, and 2011. Years with severe or extreme droughts in the summer are 1972, 2011, and 2013. Years with severe or extreme droughts in the autumn are 1992 and 2009. Years with severe or extreme droughts in the winter are 1962 and 2009.

Figure 6 shows annual and seasonal trends in the RDI alongside an M-K test for Guizhou Province. The annual, spring, summer, autumn, and winter Z values were, respectively, −1.25, −1.24, −0.12, −2.34, and −0.98, and all showed a decreasing trend for autumn at the 0.05 significance level. The rate at which the trend decreases on annual and autumn timescales is larger, at −0.013/10a and −0.022/10a,

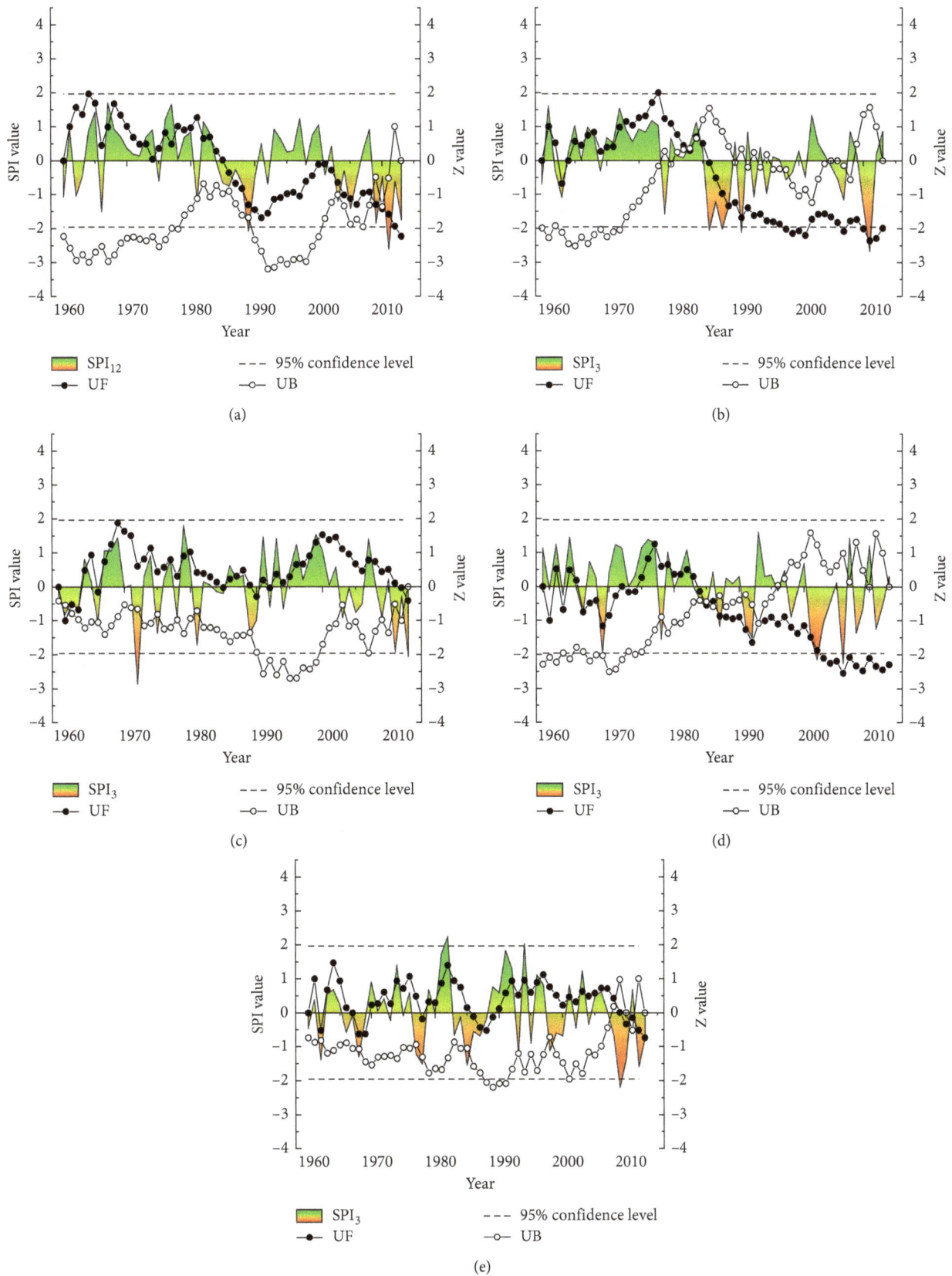

FIGURE 4: M-K trend test of the SPI (if the UF value > 0, the sequence shows a rising trend and indicating wet; UF value < 0 shows a falling trend and indicating drought): (a) annual, (b) spring, (c) summer, (d) autumn, and (e) winter.

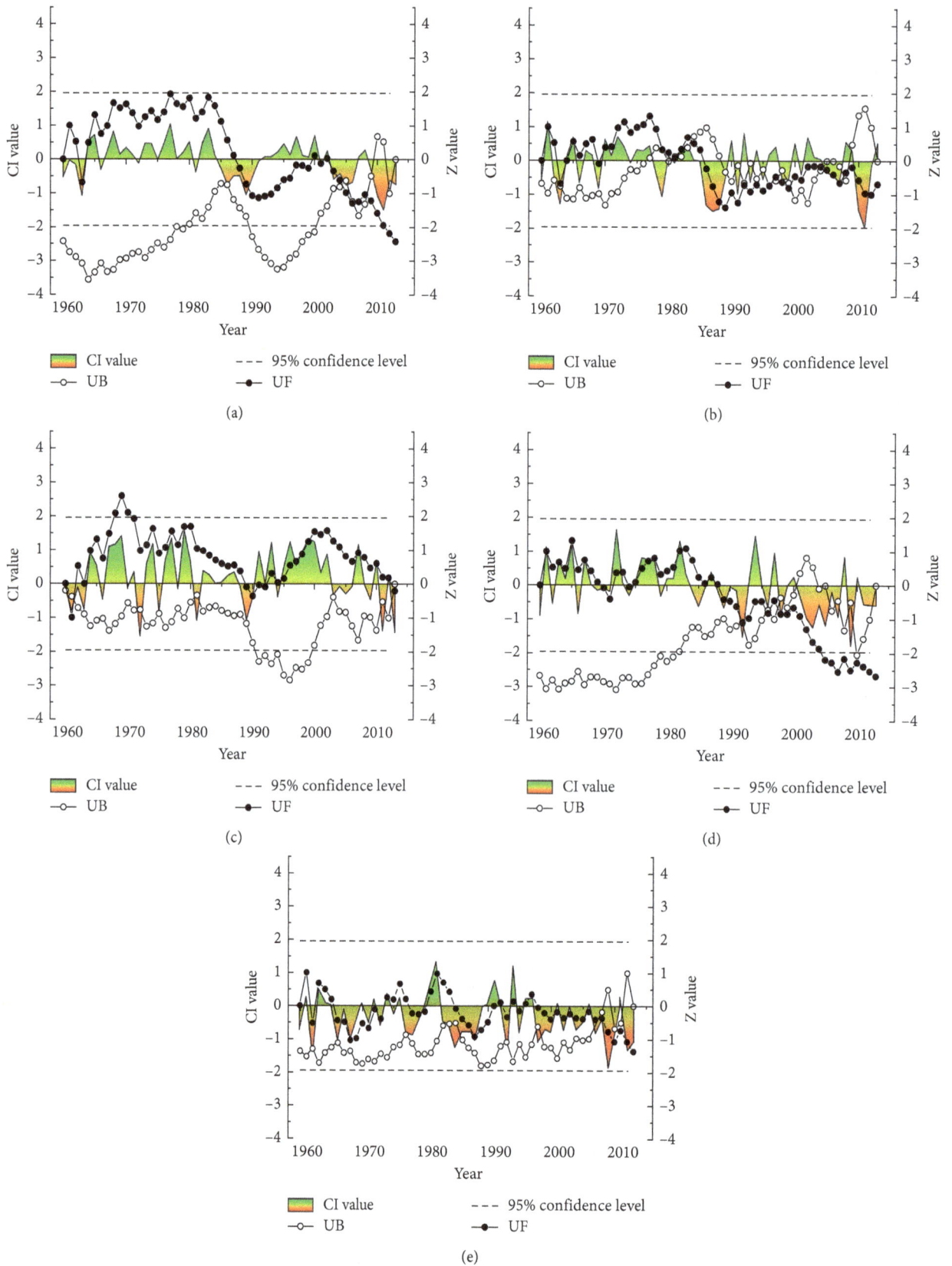

FIGURE 5: M-K trend test of the CI: (a) annual, (b) spring, (c) summer, (d) autumn, and (e) winter.

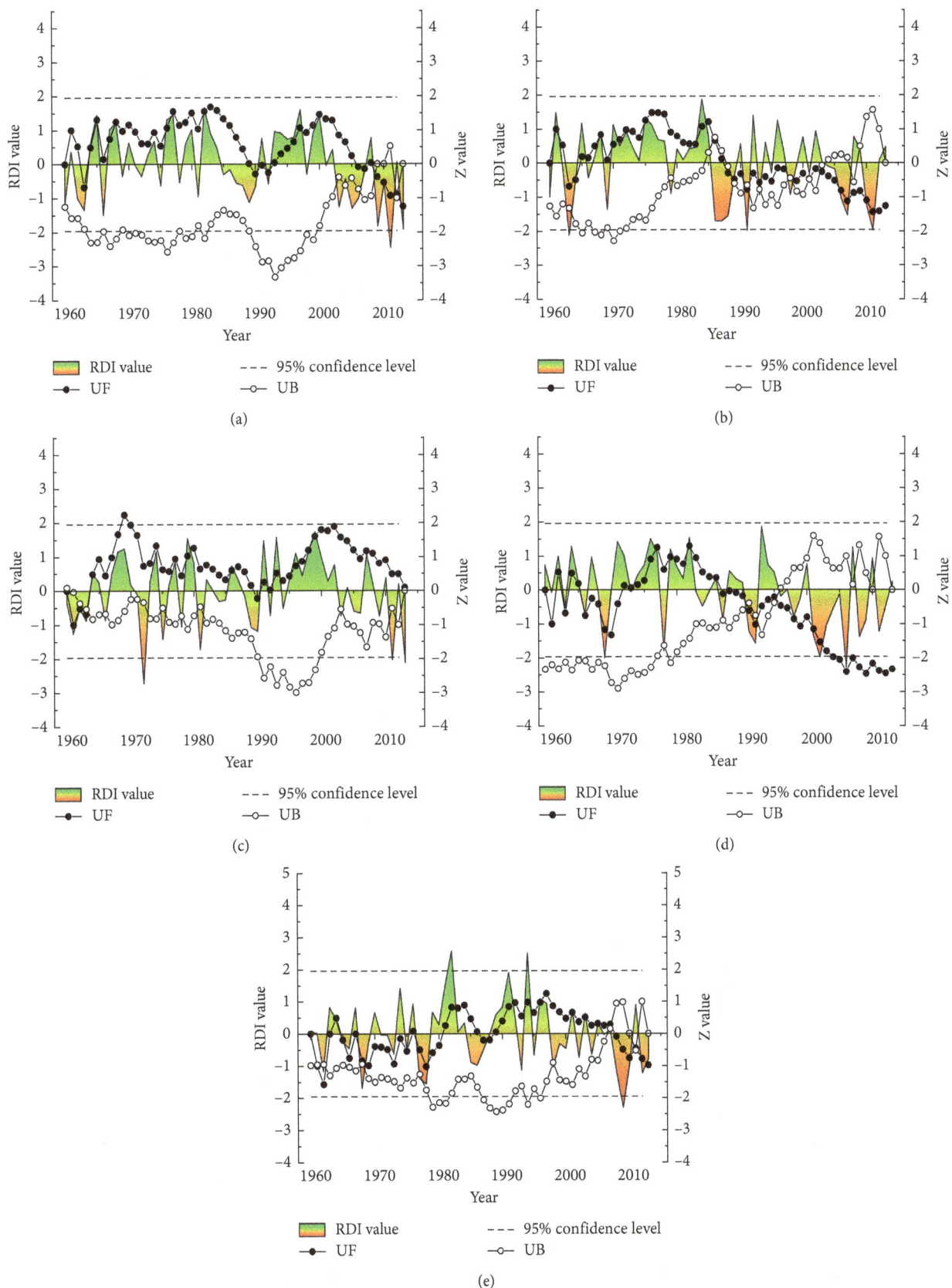

FIGURE 6: M-K trend test of the RDI: (a) annual, (b) spring, (c) summer, (d) autumn, and (e) winter.

TABLE 3: Identification of typical drought events by SPI, CI, and RDI in 1960–2013.

Event	SPI Peak	SPI Intensity	CI Event	CI Peak	CI Intensity	RDI Event	RDI Peak	RDI Intensity
1960 (4-6/8-10)	**1960-1/8**	3	**1960 (1-4/8-12)**	**1960-1/10**	5	**1960 (4-6/8-12)**	**1960-4/8**	5
1963 (3-6/8-9)	**1963-6/8**	5	**1963 (3-6/8-9)**	**1963-6/8**	4	**1963 (3-6/8-9)**	**1963-6/8**	7
1966 (8-9)	**1966-9**	3	**1966 (1-4/8-9/11-12)**	**1966-3/9/11**	4	**1966 (8-9)**	**1966-9**	2
1966 1967 (11-1)	1967-1	2	1969 (2-5)	1969-2	3	1966/1967 (11-1)	1967-1	2
1969 (2-4)	1969-2	1	1978 (2-4/7-9)	1978-2/7	3	1969 (2-5/9-10)	1969-2/9	3
1978 (2-4/7-9)	**1978-2/7**	2	**1979 (1-5)**	**1979-3**	3	**1978 (2-4/7-9)**	**1978-2/7**	3
1978 1979 (12-5)	**1978-12**	3	**1979/1980 (11-1)**	**1979-11**	2	**1978/1979 (12-4)**	**1978-12**	4
1985 1986 (12-6)	**1986-5**	4	**1985/1986 (12-6)**	**1986-5**	5	**1985/1986 (10-6)**	**1986-5**	4
1987 (3-4)	**1987-3**	2	**1987 (3-5)**	**1987-3**	2	**1987/1988 (3-5/12-1)**	**1987-3/12**	3
1987 1988 (12-1/3-5)	**1987-12,1988-5**	4	**1988 (1-7/11-12)**	**1988-3/11**	6	**1988/1989 (3-7/11-1)**	**1988-5/11**	4
1988 1989 (11-1/5-8)	**1988-11,1989-5**	2	**1989 (5-12)**	**1989-11**	2	**1989 (7-8)**	**1989-8**	1
1992 (8-12)	**1992-8**	3	**1992/1993 (8-2)**	**1992-8**	4	**1992 (8-12)**	**1992-8**	4
1993,1994 (3-5/10-2)	1993-2/12	2	1995 (2-5)	1995-2	2	1993/1994 (3-6/12-2)	1993-3/12	2
1998 (9-12)	1998-9	1	1998/1999 (11-3)	1999-3	3	1998 (4-5/9-12)	1998-4/9	2
2002 (9-12)	2002-9	2	2002/2003 (11-3)	2002-11	2	2002 (9-12)	2002-9	2
2003 (8-11)	2003-8	2	2003/2004 (8-1)	2003-8	3	2003 (2-3/8-12)	2003-2/8	3
2005 (9-11)	2005-9	1	2005/2006 (9-1)	2005-9	3	2005 (9-12)	2005-9	1
2009 2010 (8-5)	**2010-2**	7	**2009/2010 (8-5)**	**2010-2**	9	**2009/2010 (8-5)**	**2010-2**	7
2011 (2-5/7-8)	**2011-4/7**	6	**2011 (2-5/7-9)**	**2011-4/7**	7	**2011 (4-5/7-9)**	**2011-4/7**	5
2013-2014 (1-2/7-8/10-2)	**2013-1/7/12**	5	**2012 (1-4/10-12)**	**2012-1/12**	5	**2013-2014 (1-2/6-8/10-2)**	**2013-1/7/12**	6
			2013/2014 (1-2/7-2)	**2013-1/7**	6			

Note. The bold values mean that they are consistent with the historical records.

(a)

(b)

(c)

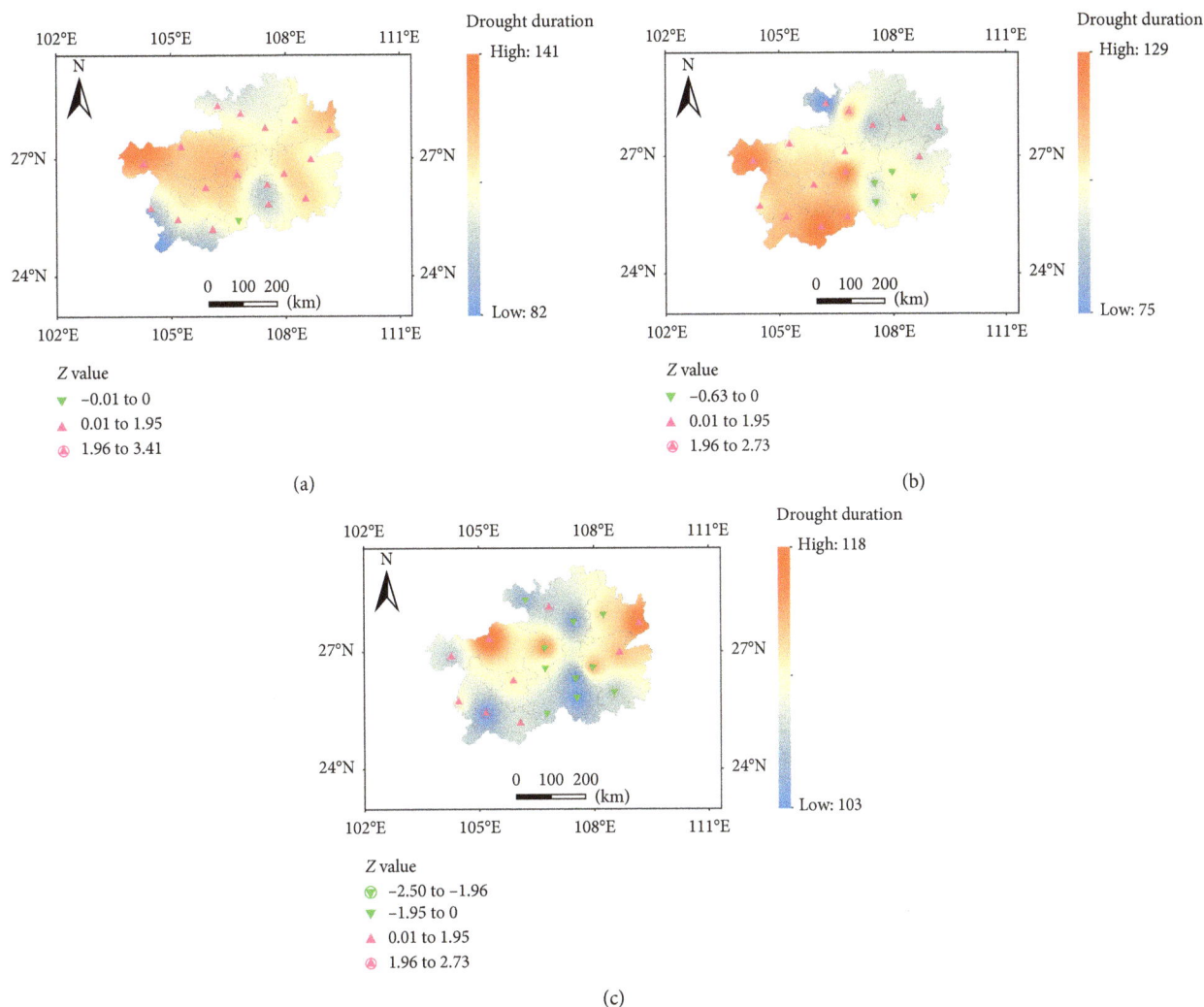

FIGURE 7: Trend distribution of drought duration: (a) SPI, (b) CI, and (c) RDI. The red dots indicate a significant positive trend ($Z > 1.96$) and the purple points indicate the insignificant positive trend ($0 < Z > 1.96$). The green points indicate the significant negative trend ($Z < -1.95$).

respectively. Figure 6(d) shows that UF and UB intersect in 1995 for autumn, breaking the boundary line in 2004–2013. However, the mutation in spring, summer, and winter was not significant. From the RDI, years with annual severe drought or extreme drought are 1966 and 2013, and 2009 and 2011, respectively. Years with severe or extreme droughts in the spring are 1986, 1987, 1988, 1991, and 2007, and 1963, 1991, and 2011, respectively. Years with severe or extreme droughts in the summer are 1981, and 1972, 2011, and 2013, respectively. Years with severe or extreme droughts in the autumn are 1978, 1992, and 1969, and 2002 and 2011, respectively. Years with severe or extreme droughts in the winter are 1968 and 1978, and 2009, respectively.

As shown in Table 3, 29, 30, and 32 drought events were identified from the SPI, CI, and RDI indices, respectively. The performances of the three indices are close with small differences on month scales. Identification of drought events in 1963, 1966, 1978-1979, 1985-1986, 1987-1988, 1988-1989, 1992, 2009-2010, 2011, and 2013-2014 is consistent for all three indices. We note that there were more droughts in the 1960s, 1980s, and 2000s, with a particular rise since the

beginning of the 21st century. The drought peak also increased significantly since the beginning of the 21st century. Droughts classified as severe occurred in 1963, 1985-1986, 1987-1988, 1992, 2009-2010, 2011, and 2013-2014. In addition, as shown in Table 3, drought events took place in all seasons, especially in winter–spring and summer–autumn. There was a persistent drought in summer–autumn–winter–spring 2009-2010, a persistent drought in spring–summer 2011, and a persistent drought in winter–spring–summer–autumn 2013-2014.

4.2. Interannual Variability

4.2.1. Spatial Distribution and Trends of Drought Duration. The spatial distribution of drought durations and trends for the three indices is shown in Figure 7. Drought duration is longer in the northwest and relatively short in the southwest of Guizhou Province. In terms of the trend, only one station (Luodian station) shows a decreasing trend (i.e., a tendency to be wet). All other stations showed an increasing trend.

(a)

(b)

(c)

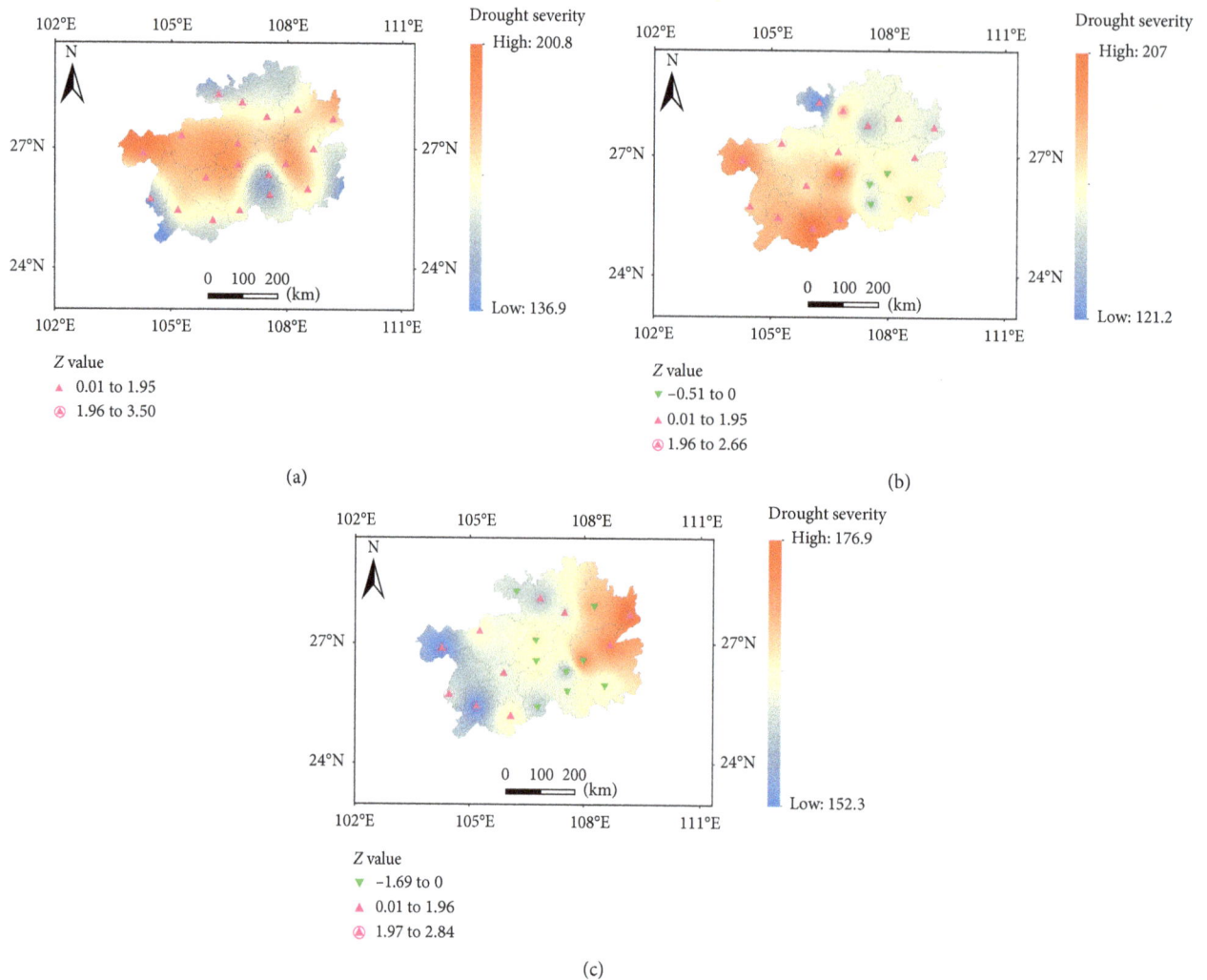

FIGURE 8: Trend distribution of drought severity: (a) SPI, (b) CI, and (c) RDI. The red dots indicate the significant positive trend ($Z > 1.96$) and the purple points indicate the insignificant positive trend ($0 < Z > 1.96$). The green points indicate the significant negative trend ($Z < -1.95$).

Among them, five stations (Weining, Guiyang, Xifeng, Xishui, and Tongzi stations) showed a significant increasing trend; these are mainly located in the northwest of Guizhou Province. The CI shows that droughts lasted for more and less time in western and northeast Guizhou Province, respectively. In terms of changes in the trend, four stations (Weining, Bijie, Tongzi, and Xingren stations) showed significant increasing trends; three of these stations are located in the west. Meanwhile, four stations (Kaili, Duyun, Dushan, and Rongjiang stations) showed non-significant decreasing trends in the southeast. The RDI suggests that drought duration is longer in northwest and northeast regions and shorter in southern Guizhou Province. Nine stations located in western Guizhou Province increased significantly. Furthermore, ten stations in central and eastern Guizhou Province had a decreasing trend. Among these was the one station in southeastern Guizhou (Rongjiang station) with a significant decreasing trend.

4.2.2. Spatial Distribution and Trends of Drought Severity. Figure 8 shows the spatial distribution and trends of drought severity. The spatial distribution of drought severity is almost consistent with that of drought duration. However, more severity droughts are typically found in the northwest of Guizhou Province, where all stations show an increasing trend. Stations with significant increasing trends are mainly distributed in the northwest and northeast of Guizhou Province. The drought severity determined by the CI is also consistent with drought duration. Drought intensity is of higher magnitude in western Guizhou Province. Among the four stations with significant increasing trends (Weining, Bijie, Panxian, and Tongzi stations), three (Weining, Bijie, and Panxian stations) are located in the west of the province, while drought duration showed a decreasing trend in southeast Guizhou Province. The drought intensity is also consistent with drought duration based on the RDI. Severe droughts are more frequent in eastern Guizhou. However, stations with significant increasing trends are primarily

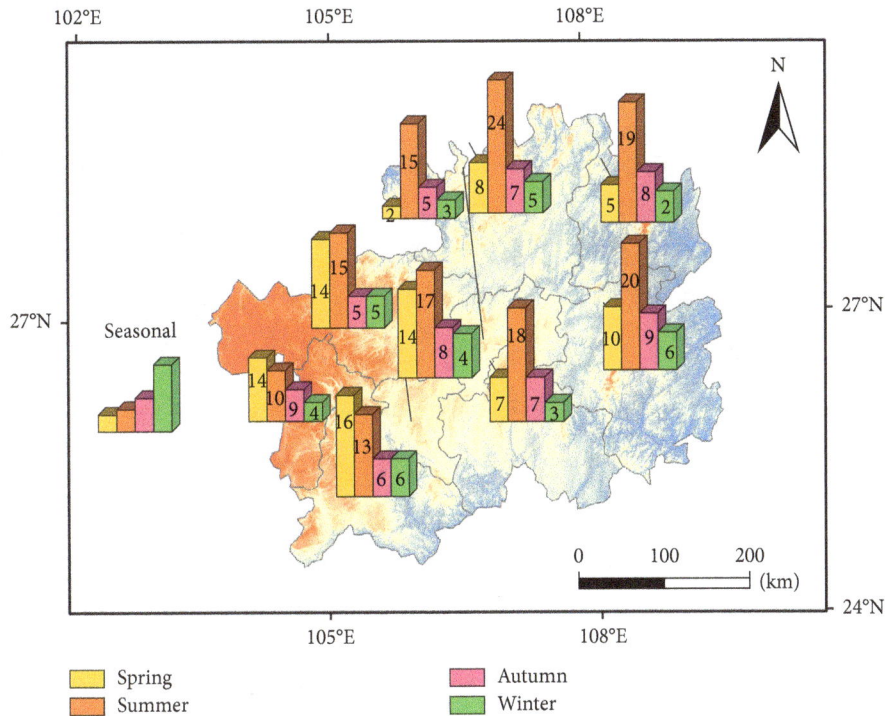

FIGURE 9: Statistics of seasonal drought frequency based on historical records in Guizhou Province in 1960–2013.

located in western Guizhou, while stations with both increasing and decreasing trends are located in northeast Guizhou, with stations with decreasing trends located in central and eastern Guizhou.

4.3. Validation of Three Drought Indices Based on Historical Disaster Records. Drought frequency in different seasons from 1960 to 2013 in Guizhou Province is shown in Figure 9, based on statistics of Chinese meteorological disasters [45–59]. More drought events are shown to have happened in spring and summer in Guizhou Province. Spring droughts are more frequent in the central and west of the Province, where Anshun City, Bijie City, and Qianxinan City are located. Summer droughts are more frequent in the central and east of Guizhou Province, home to Zunyi City, Tongren City, and Qiandongnan City. Moderate, severe, and extreme droughts are more frequent in spring in western Guizhou and summer in eastern Guizhou Province (Figure 10). Moderate and extreme droughts are more frequent in autumn and winter and mainly affect eastern Guizhou Province.

Figure 11 shows that droughts are more frequent in spring, summer, and autumn based on the SPI. However, droughts are less frequent in spring and summer than the historical records (Figure 9), while droughts in autumn are more frequent than the historical records. Winter droughts are highly consistent with historical records based on SPI. The CI suggests that drought occurrence increased in winter and spring. But the historical records show fewer droughts in winter. The CI is relatively close to historical records in spring, followed by autumn and summer. Autumn droughts occurred more frequently than the historical records. Fewer droughts in summer were found in the historical records.

Drought predictions from the RDI are close to the historical records in spring and summer. However, this index suggests more droughts in autumn and winter, particularly in winter.

The mild and moderate seasonal droughts identified by the SPI are more frequent than those found in historical records. However, the severe and extreme seasonal droughts are identified less frequently than the historical records, especially in spring and summer (Figures 10 and 12).

The CI identifies more frequent mild and moderate seasonal droughts than the historical records, while it identifies fewer severe and extreme droughts than the historical record (Figures 10 and 13).

The RDI identifies more mild droughts than historical records indicate. However, the moderate, severe, and extreme droughts identified by the RDI are close to the historical records (Figures 10 and 14).

The drought frequency analysis (Figures 9 and 14) for SPI, CI, and RDI compared to the historical records shows that the mild and moderate droughts in winter are more than the historical records. The historical records describe the severity of the crop yield reduction. However, the drought indices do not take this into account. Thus, the drought statistics by indices are possible more frequently than the historical records. Overall, the severe and extreme droughts are less frequent than the historical records, especially CI. The RDI is closer to the historical records compared to the SPI and CI.

Figure 15 shows variation of the three drought indices in the area historically affected by droughts; among these data are the typical drought years shown in Table 4. The three drought indices in the drought-affected area were highest in 2011. However, the three drought indices in the affected area, particularly CI, are inconsistent with historical records in 2010, 1992, 1990, and 1988. Together with Figures 9–14,

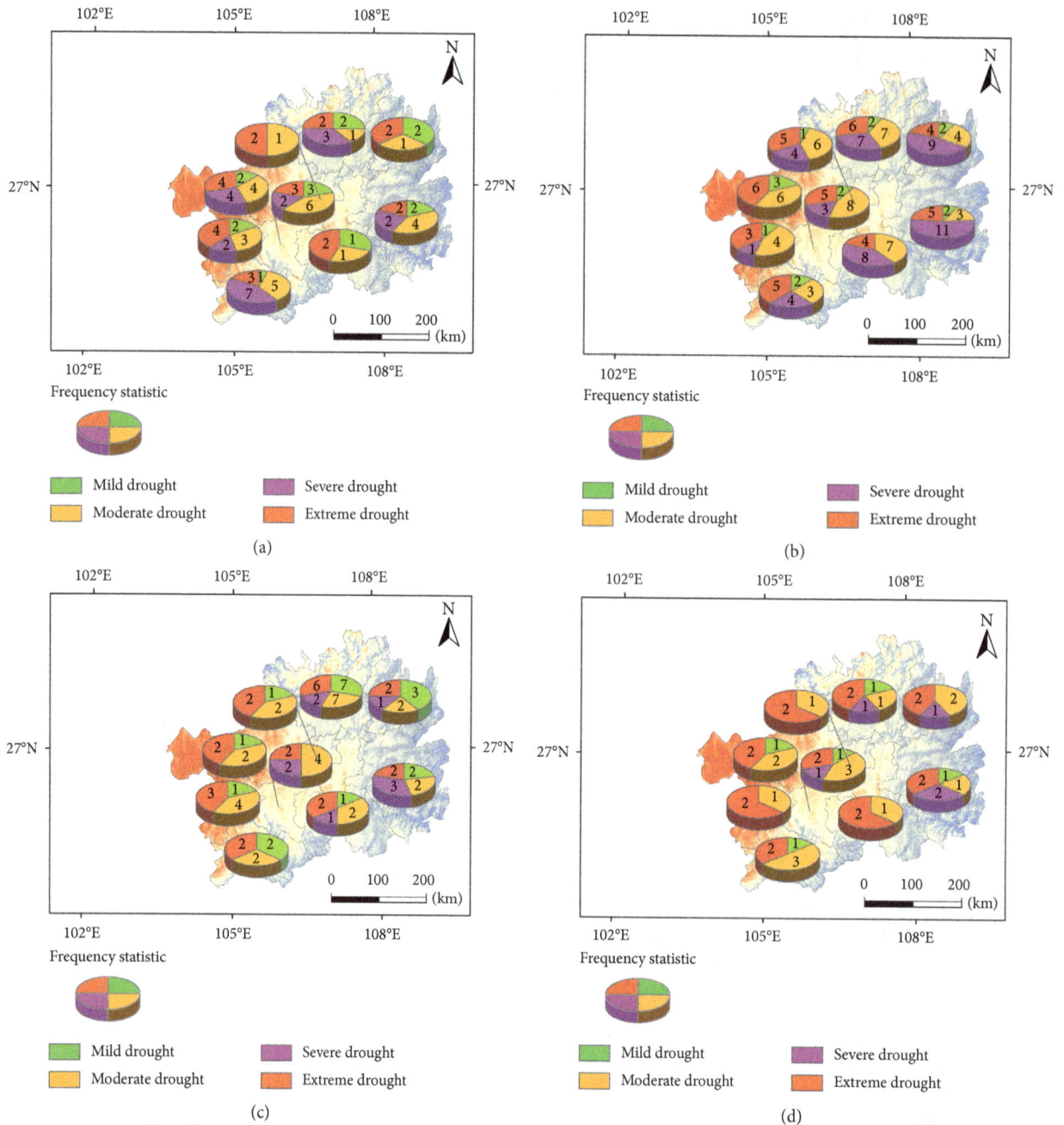

FIGURE 10: Statistics of seasonal drought frequency in different drought grades based on historical records in Guizhou Province in 1960–2013: (a) spring, (b) summer, (c) autumn, and (d) winter.

it is therefore shown that the RDI is more objective and reliable at indicating drought than the CI and SPI (the SPI_{12} value is shown here). Therefore, the abovementioned analysis indicated that the relationship between the historical records and drought index still needs to be further quantified in the future.

5. Discussion and Conclusions

All three drought indices showed decreasing trends in annual and seasonal in the past 54 years. The results are consistent with Zhai et al., Xu et al., Dai, and Milly and

Dunne [8, 11, 60, 61]. However, Sheffield et al. [62] discovered that a little change in global drought for the period of 1948–2008 based on the Palmer drought severity index. Further, the presented results demonstrated a significant drought trend in autumn for the three drought indices, which are consistent with Li et al. [21] and Gao et al. [22]. These results also show that the drought in spring and summer are dominant from the historical records, which are increasing [43–51]. The autumn drought also shows a significant increasing trend in Guizhou Province, which may have a great impact on autumn crops. Gao et al. [22] found that autumn soil moisture anomaly is helpful to further

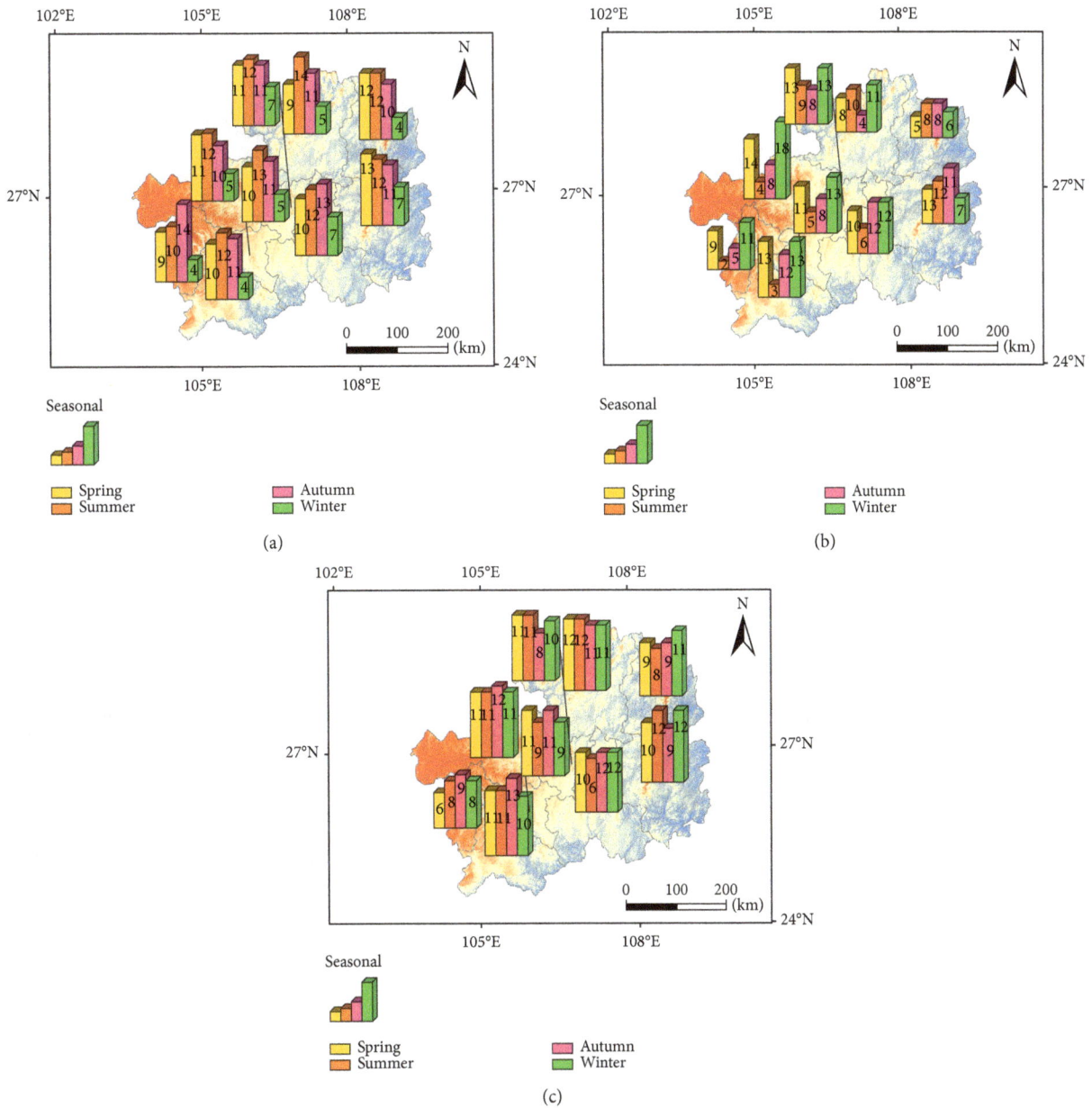

FIGURE 11: Statistics of seasonal drought frequency based on (a) SPI, (b) CI, and (c) RDI in Guizhou Province in 1960–2013.

understand the nature of the drought in Southwest China and may provide a clue for drought monitoring and risk management. The SPI, CI, and RDI identified 29, 30, and 32 drought events, respectively. Winter–spring and summer–autumn droughts have become more frequent since the beginning of the 21st century. The increase in frequency and strengthening trends of drought frequency, duration, peak, and intensity is significant over the period 1960–2013. These results are also consistent with Zhai et al., Yu et al., Xu et al., Li et al., and Gao et al. [8, 9, 11, 21, 22].

In terms of drought duration, the spatial distribution of the SPI is close with the RDI during 1960–2013. However, the spatial distribution of the CI is inconsistent with those of the SPI and RDI. As Section 3.1.2 mentioned that the CI index is composed of SPI and MI; however, some scholars point out that SPI and MI have certain defects. For instance,

the SPI only utilizes precipitation information, without considering other meteorological variables that may play an important role for drought. In addition, the weight coefficients are relatively artificial and random, which may affect the ability of the CI [63–65]. Thus, it is possible to be the main reason for the disagreement with the distribution of RDI and SPI. For drought severity, the spatial distributions of the three drought indices are also inconsistent. In the present study, the drought severity is based on annual statistics. However, the seasonal statistics show that SPI and CI account for a large proportion in spring, while RDI accounts for a large proportion in summer. Therefore, SPI and CI show higher drought severity in the western province. The RDI shows higher drought severity in the eastern which is consistent with the historical records. Moreover, Xu et al. [11] also revealed that the spatial distribution of

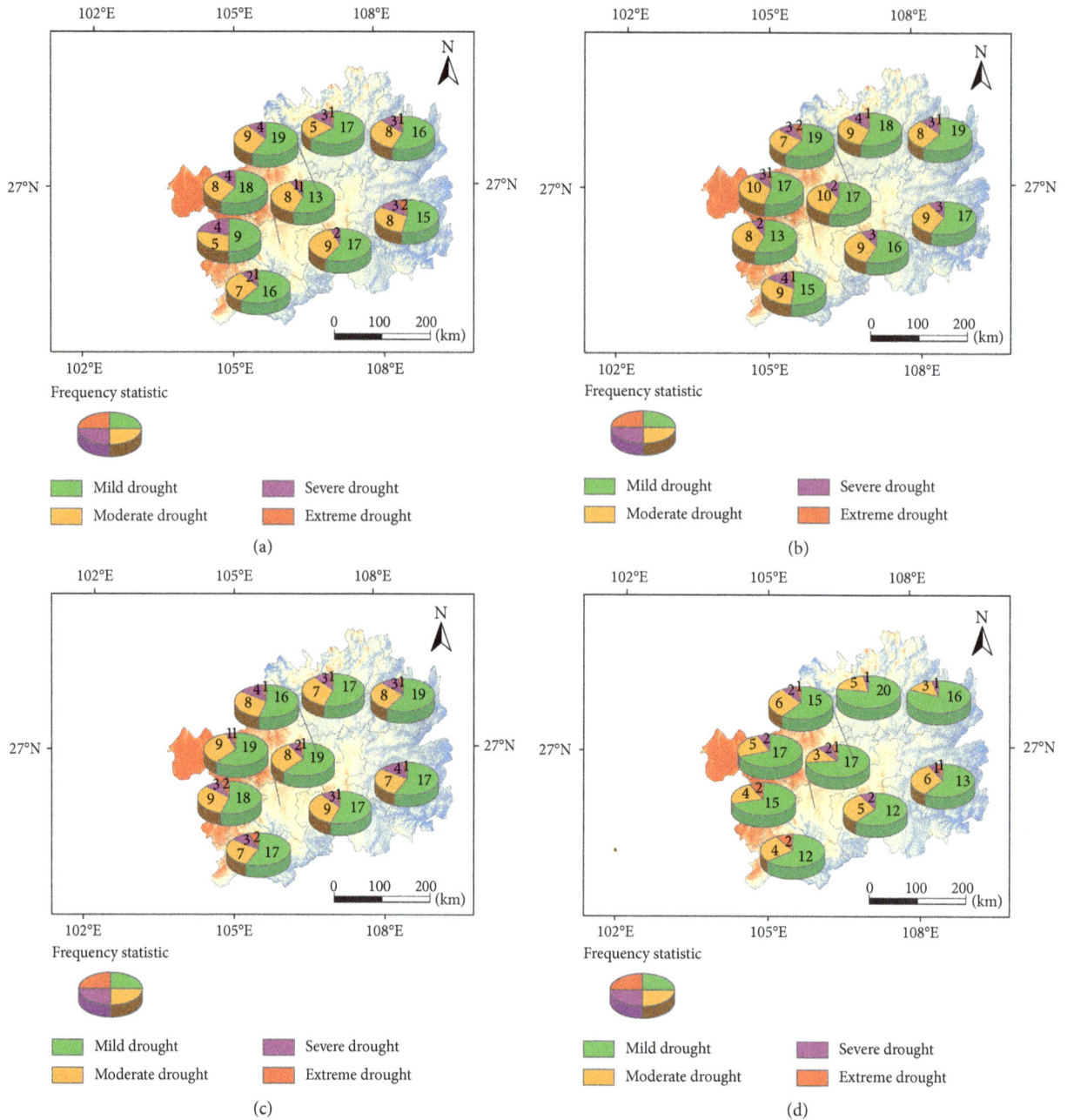

FIGURE 12: Statistics of seasonal drought frequency in different drought grades based on the SPI in Guizhou Province in 1960–2013: (a) spring, (b) summer, (c) autumn, and (d) winter.

drought severity using RDI$_3$ (3 months reconnaissance drought index) is almost the same as that using SPI$_3$ (3 months standardized precipitation index). However, the distributions of SPEI$_3$ (3 months standardized precipitation evapotranspiration index) are quite different with SPI$_3$ and RDI$_3$ as well as the trends. Based on the above analysis and the historical records (Table 3) of disasters in the drought-affected area that consider seasonal drought frequency and magnitude, the RDI performs more objectively and reliably than SPI and CI. However, the SPI, CI, and RDI all indicate drought frequencies and durations less or more than those indicated by the historical records. This may be related to the defects of the SPI, CI, and RDI. Previous studies have

revealed that meteorological droughts are the water shortages caused by an imbalance precipitation and evaporation [66]. The most of the drought indices are mainly based on the precipitation and evaporation calculation. Therefore, they play a vital role in the capture of drought characteristics [11, 62]. Evaporation is always the focus of drought research. However, compared to precipitation, there are still many uncertainties in evaporation measurement. Therefore, different evaporation models may not get the same results. Previous studies applied PDSI, SPI, RDI, and SPEI [11, 60–62, 67], which mainly adopted the Thornthwaite and Penman–Monteith or other regimes to calculate the reference evapotranspiration (ETo). Therefore, different drought

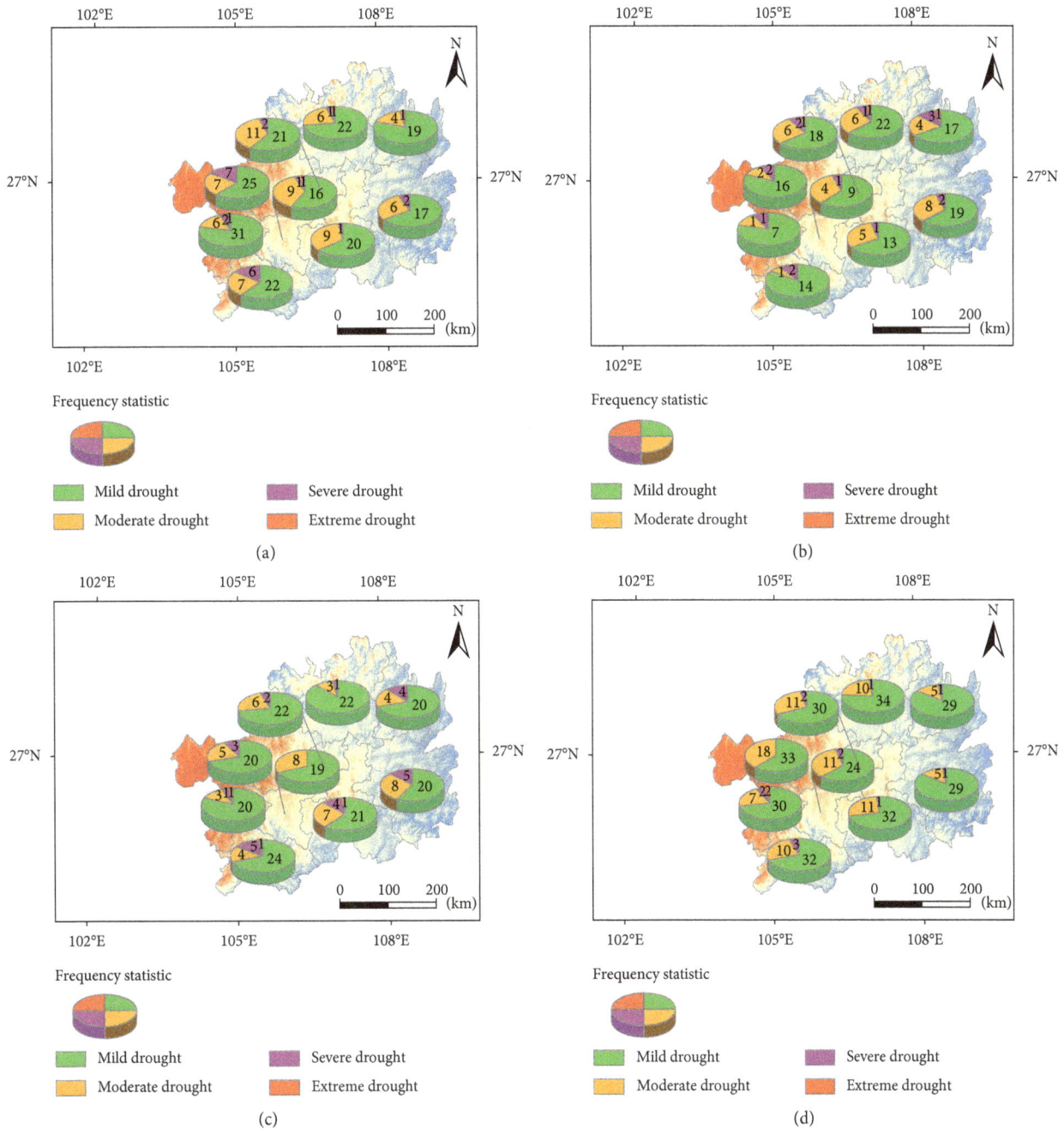

FIGURE 13: Statistics of seasonal drought frequency in different drought grades based on the CI in Guizhou Province in 1960–2013: (a) spring, (b) summer, (c) autumn, and (d) winter.

trends were obtained; for example, Dai [60] demonstrated that the observed global aridity changes are consistent with model predictions up to 2010, which suggest more severe and widespread droughts in the next 30–90 years caused by decreased precipitation or increased evaporation. Meanwhile, Milly and Dunne [61] also found that the historical and future tendencies are towards continental drying. However, Sheffield et al. [62] indicated that the previous reported increase in global drought is overestimated, and there was little change in drought over the period of 1948–2008. In addition, the results based on different drought indices are also inconsistent. For example, Zarch et al. [67] showed that the percentage of

drought-prone areas estimated by the SPI is higher than that by the RDI for the period prior to 1998, while it is the converse for the period after 1998. Xu et al. [11] indicated that SPEI and RDI are sensitive to ETo. The RDI based on the Thornthwaite equation overestimates the influence of air temperature. Thus, it overestimates the grade of drought. Besides, Vicente-Serrano et al. [68] pointed out that SPI, PDSI, SPDI, and SPEI are sensitive to precipitation and ETo. The results may be quite different with respect to different indices.

All three drought indices indicate that mild droughts occurred more frequently than what is shown in the historical records, across different seasons and levels of

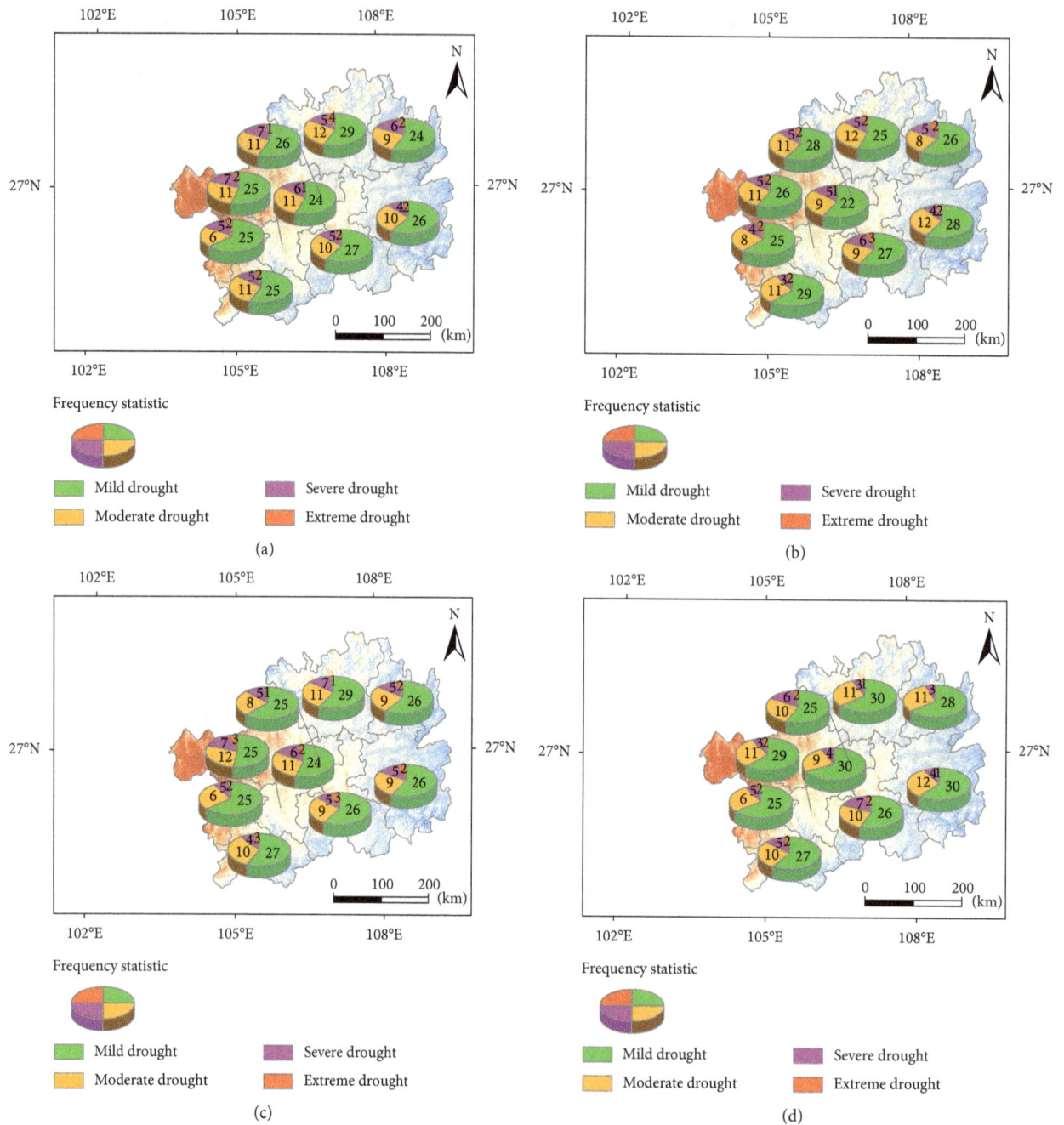

FIGURE 14: Statistics of seasonal drought frequency in different drought grades based on the RDI in Guizhou Province in 1960–2013: (a) spring, (b) summer, (c) autumn, and (d) winter.

drought. This may be related to different statistical analysis methods. In this paper, any interval when the indices are between −1 and 0 is classified as an occurrence of mild drought. However, it is necessary for a drought to cause agricultural and socioeconomic damage in order for it to be noted in historical records. We also point out that the drought-affected area was highest in 2011, consistent with RDI and CI, but not with SPI. The density of meteorological stations may also play a role. In this study, only data from 19 stations are considered. However, the records of drought-affected areas are based on statistics covering over 88 counties in the entire province. Thus, a higher density of weather stations may overcome the index-historical data mismatch.

Previous studies [69–73] have stated that the occurrence of droughts in the southwestern region of Guizhou Province is close to related atmospheric circulation anomalies and special topography [69–75]. In addition, the significant decrease in precipitation [11, 21] is an important factor for drought. Meanwhile, the change of potential evaporation is also a critical factor [20]. Chen et al. [41] pointed out that the number of continuous wet days (CWD) was decreasing

(a)

(b)

(c)

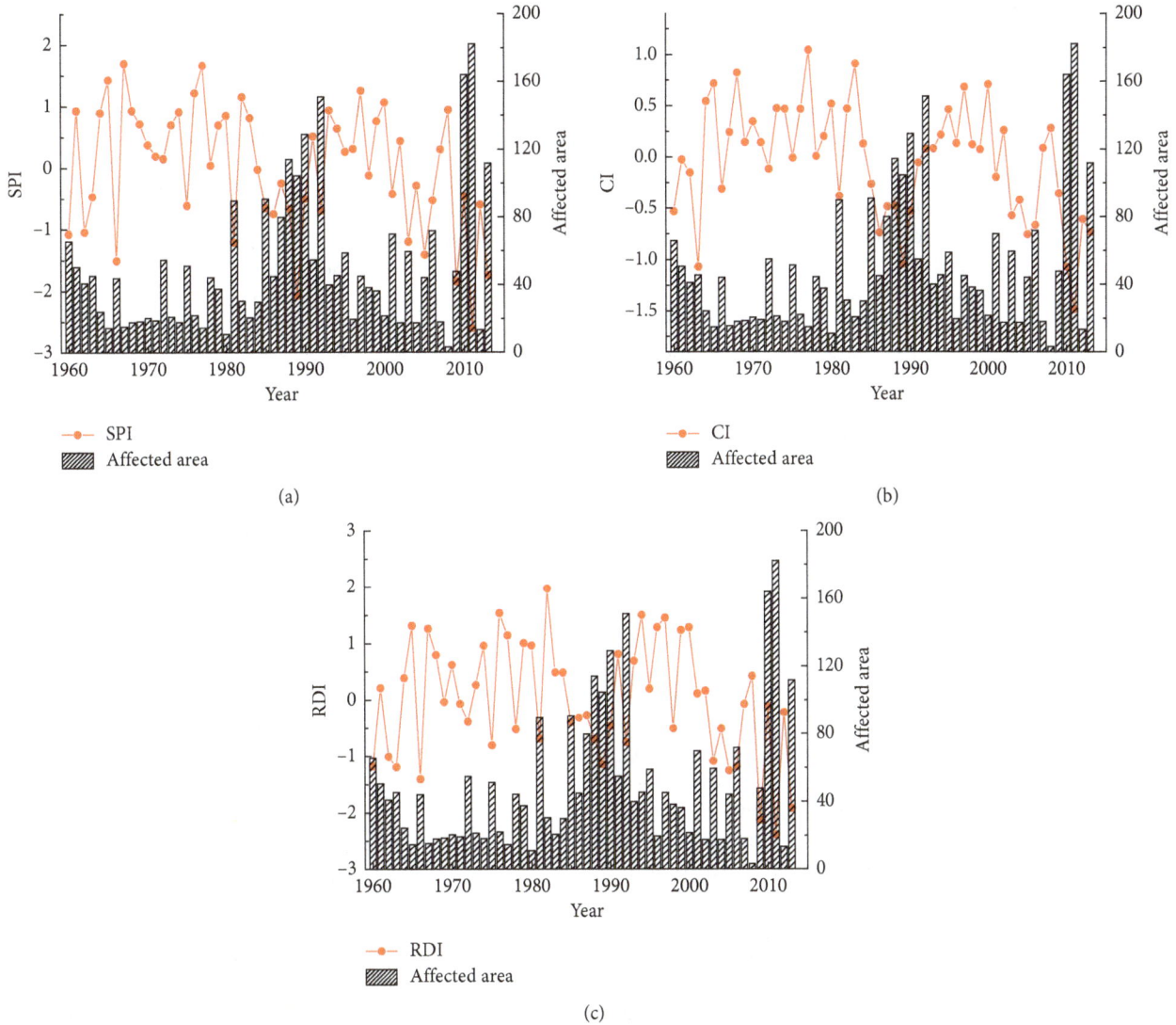

FIGURE 15: Comparison of drought-affected areas based on (a) SPI, (b) CI, and (c) RDI.

TABLE 4: Comparison of historical affected area in typical drought years based on the SPI, CI, and RDI.

Affected area ($\times 10^4$ hm^2)	1988	1989	1990	1992	2010	2011	2013
Historical	114.07	104.6	128.93	150.93	163.9	182.25	111.78
SPI historical	−0.7	−2.1	−0.5	−0.7	−0.5	−2.6	−1.7
CI historical	−0.5	−1.0	−0.5	−0.1	−1.1	−1.5	−0.7
RDI historical	−0.7	−1.1	−0.5	−0.7	−0.1	−2.4	−1.9

significantly while the largest 5 days of rainfall (RX5 day), strong precipitation (R95), and strongest rainy day (R20mm) measures did not have significant decreasing trends in response to the decreasing trend of the three indices (when considering Guizhou Province). In terms of drought distribution, all three drought indices indicated more frequent spring droughts in western Guizhou, and more frequent summer droughts in eastern Guizhou. Shen et al. [75] pointed out that drought characteristics are mainly the result of uneven spatiotemporal distribution of water resources in Guizhou Province. The spring drought is the most severe in Bijie City and Liupanshui City in western

Guizhou Province. The rainy season in western Guizhou starts in June; when these rains are late, a spring drought is triggered. Zunyi, Tongren, and Qiannan Cities in eastern Guizhou Province are prone to summer droughts. This may be the result of the rainy season starting early (April) in the area. A precipitation decrease will likely cause a summer drought. Moreover, Milly and Dunne [61] and Sheffield et al. [62] stated that other factors such as runoff, relative humidity, wind speed, and other physical mechanisms should also be taken into account. The relationship between global drought and climate change can be assessed more accurately by combining physical hydrological models and large quantities of measured and satellite remote-sensing data. Furthermore, the influence of human activities is also an important factor that cannot be ignored.

The karst landform is also an important factor for the drought in Guizhou Province [72]. The karst topography is widely distributed in Guizhou Province, and the arable land is mainly located in the high mountains [73, 74]. However, the water source for irrigation is located at the bottom of the valley. Due to the widespread karst, the soil layer is infertile with a poor water storage capacity. Further, water permeability is strong, and water moves quickly through the rocks. Therefore, in such a region, once drought occurs, it will have an important impact on agricultural production and domestic water. The historical records indicate that spring and summer droughts have begun to occur more frequently in Guizhou Province. Based on analysis of the three indices considered in this paper, the annual and seasonal drought trends, especially for autumn droughts, are more significant. These results demonstrate that the government of Guizhou Province should focus on monitoring and damage prevention not only for the spring and summer droughts but also for the autumn drought.

In this study, three indices are used to describe the spatiotemporal characteristics of Guizhou Province during 1960–2013. The comparison analysis shows that the RDI is much closer to the historical records than CI and SPI. The RDI may be more reliable for drought monitoring in this region. However, this study is limited in some aspects. The historical records are more often qualitative descriptions. When we extracted the drought information, the disaster loss (such as crop loss), duration, and other information were comprehensively considered. However, the drought index is a quantitative indicator that is more sensitive to weather conditions. It does not identify the crop loss information. Thus, the drought index tends to identify more mild droughts. Definitely, it is a great challenge to match the qualitative description for the quantitative indicator. However, we believe that this study still can provide a useful reference for drought monitoring and assessment. This issue will be further improved in the future work. In addition, the atmosphere or other meteorological variables are not investigated. The mechanisms responsible for the drought in Guizhou Province need to be further explored. Improving and modifying the drought index is also the topic of ongoing work and future research.

Conflicts of Interest

The authors declare that there are no conflicts of interest in this paper.

Acknowledgments

This study was supported by the National Natural Science Foundation of China (Grant no. 41501106) and the Scientific Research Foundation for Returned Scholars, Ministry of Education of the People's Republic of China (Grant no. 2014-1685). The meteorological data have been provided by China Meteorological Data Sharing Service System of National Meteorological Information Center (http://data.cma.cn/).

References

[1] A. K. Mishra and V. P. Singh, "Review of drought concepts," *Journal of Hydrology*, vol. 91, no. 1-2, pp. 202–216, 2010.

[2] T. Stocker, D. G. Qin, G. Plattner, M. Tignor, S. Allen, and J. Boschung, *IPCC, 2013: Climate Change 2013: The Physical Science Basis. Contribution of Working Group I to the Fifth Assessment Report of the Intergovernmental Panel on Climate Change*, Cambridge University Press, Cambridge, UK, 2013.

[3] Y. Wang, S. Sha, and S. P. Wang, "Assessment of drought disaster risk in southern China," *Acta Prataculturae Sinica*, vol. 24, no. 5, pp. 12–24, 2015.

[4] D. Alexander, *Confronting Catastrophe*, Oxford University Press, Oxford, UK, 2000.

[5] R. Mechler, *Natural Disaster Risk Management and Financing Disaster Losses in Developing Countries*, Ph. D. dissertation, Verlag fuer Versicherungswissenschaft, Karlsruhe, Germany, 2004.

[6] N. Liu, X. Z. T. Zhang, Y. T. Tian, S. F. Kuang, and X. F. Liu, "The State Flood Control and Drought Relief Headquarters Ministry of water resources of People's Republic of China," in *Bulletin of Drought and Flood Disaster China in 2013*, pp. 35–38, Water Conservancy and Hydropower Press, Beijing, 2014, in Chinese.

[7] J. He, X. H. Yang, Z. Li, X. J. Zhang, and Q. H. Tang, "Spatiotemporal variations of meteorological droughts in China during 1961–2014: an investigation based on multi-threshold identification," *International Journal of Disaster Risk Science*, vol. 7, no. 1, pp. 63–76, 2016.

[8] J. Q. Zhai, J. L. Huang, B. D. Su et al., "Intensity–area–duration analysis of droughts in China 1960–2013," *Climate Dynamics*, vol. 48, no. 1-2, pp. 151–168, 2017.

[9] M. X. Yu, Q. F. Li, M. J. Hayes, M. D. Svoboda, and R. R. Heim, "Are droughts becoming more frequent or severe in China based on the standardized precipitation evapotranspiration index: 1951–2010?," *International Journal of Climatology*, vol. 34, no. 3, pp. 545–558, 2013.

[10] O. O. Ayantobo, Y. Li, S. B. Song, and N. Yao, "Spatial comparability of drought characteristics and related return periods in mainland China over 1961–2013," *Journal of Hydrology*, vol. 550, pp. 549–567, 2017.

[11] K. Xu, D. W. Yang, H. B. Yang, Z. Li, Y. Qin, and Y. Shen, "Spatio-temporal variation of drought in China during 1961–2012: a climatic perspective," *Journal of Hydrology*, vol. 526, pp. 253–264, 2015.

[12] X. F. Liu, S. X. Wang, Y. Zhou, F. T. Wang, G. Yang, and W. L. Liu, "Spatial analysis of meteorological drought return periods in china using copulas," *Natural Hazards*, vol. 80, no. 1, pp. 367–388, 2016.

[13] X. M. Liu, Y. Z. Luo, T. T. Yang, K. Liang, M. H. Zhang, and C. M. Liu, "Investigation of the probability of concurrent drought events between the water source and destination regions of china's water diversion project," *Geophysical Research Letters*, vol. 42, no. 20, pp. 8424–8431, 2016.

[14] X. J. Shen, X. Wu, X. M. Xie, Z. Z. Ma, and M. J. Yang, "Spatiotemporal analysis of drought characteristics in Song-Liao river basin in China," *Advances in Meteorology*, vol. 2017, Article ID 3484363, 13 pages, 2017.

[15] Q. Zhang, X. B. Pan, and Z. G. Ma, *2009 Drought*, China Meteorological Press, Beijing, China, 2009, in Chinese.

[16] D. Barriopedro, C. M. Gouveia, R. M. Trigo, and L. Wang, "The 2009/10 drought in China: possible causes and impacts on vegetation," *Journal of Hydrometeorology*, vol. 13, no. 4, pp. 1251–1267, 2012.

[17] L. Zhang, J. Xiao, J. Li, K. Wang, L. P. Lei, and H. D. Guo, "The 2010 spring drought reduced primary productivity in southwestern China," *Environmental Research Letters*, vol. 7, no. 4, pp. 1–10, 2012.

[18] J. H. Yang, Q. Zhang, J. S. Wang, J. Y. Jia, and J. Wang, "Spring persistent droughts anomaly characteristics of over the southwest China in recent 60 years," *Arid Land Geography*, vol. 38, no. 2, pp. 215–222, 2014, in Chinese.

[19] J. Y. He, M. J. Zhang, P. Wang, S. J. Wang, and X. M. Wang, "Climate characteristics of the extreme drought events in Southwest China during recent 50 Years," *Acta Geographica Sinica*, vol. 66, no. 9, pp. 1179–1190, 2011.

[20] S. L. Sun, H. S. Chen, W. M. Ju et al., "On the coupling between precipitation and potential evapotranspiration: contributions to decadal drought anomalies in the Southwest China," *Climate Dynamics*, vol. 48, no. 11-12, pp. 3779–3797, 2016.

[21] Y. J. Li, F. M. Ren, Y. P. Li, P. L. Wang, and H. M. Yan, "Characteristics of the regional meteorological drought events in Southwest China during 1960–2010," *Journal of Meteorological Research*, vol. 283, no. 3, pp. 381–392, 2014.

[22] C. J. Gao, H. S. Chen, S. L. Sun et al., "A potential predictor of multi-season droughts in Southwest China: soil moisture and its memory," *Natural Hazards*, vol. 91, no. 2, pp. 553–566, 2017.

[23] L. Y. Han, Q. Zhang, P. L. Ma, J. Y. Jia, and J. S. Wang, "The spatial distribution characteristics of a comprehensive drought risk index in southwestern China and underlying causes," *Theoretical and Applied Climatology*, vol. 124, no. 3-4, pp. 517–528, 2015.

[24] Z. X. Chi, Z. J. Du, Z. M. Chen et al., "Analyses on meteorological elements and general circulation of drought in Guizhou Province in autumn-winter-spring from 2009 to 2010," *Plateau Meteorology*, vol. 31, no. 1, pp. 176–184, 2012, in Chinese.

[25] X. J. Wang, H. Bai, W. Y. Zhou, Z. H. Chen, and D. H. Zhang, "Comparative analysis of drought in Guizhou from July to August between the year of 2011 and 2013 and the effect on agriculture," *Tianjin Agricultural Sciences*, vol. 20, no. 11, pp. 118–124, 2014, in Chinese.

[26] Z. H. Wu, W. Y. Zhou, and D. H. Zhang, "Analysis and evaluation severe summer drought in 2013 of Guizhou Province," *Journal of Guizhou Province Meteorology*, vol. 38, no. 4, pp. 39–42, 2015, in Chinese.

[27] J. A. Dracup, K. S. Lee, and E. G. Paulson, "On the definition of droughts," *Water Resources Research*, vol. 16, no. 2, pp. 297–303, 1980.

[28] D. A. Wilhite and M. H. Glantz, "Understanding the drought phenomenon: the role of definitions," *Water International*, vol. 10, no. 3, pp. 111–120, 1985.

[29] T. B. McKee, N. J. Doeskin, and J. Kleist, "The relationship of drought frequency and duration to time scales," in *Proceedings of the 8th Conference on Applied Climatology*, pp. 179–184, American Meteorological Society, Boston, MA, USA, 1993.

[30] G. Tsakiris, D. Pangalou, and H. Vangelis, "Regional drought assessment based on the reconnaissance drought index (RDI)," *Water Resources Management*, vol. 21, no. 5, pp. 821–833, 2007.

[31] G. Tsakiris and H. Vangelis, "Establishing a drought index incorporating evapotranspiration," *European Water*, vol. 9, no. 10, pp. 3–11, 2005.

[32] S. M. Vicente-Serrano, J. I. Lopez-moreno, J. Lorenzo-lacruz, A. E. Kenawy, C. Azorin-Molina, and E. Morán-Tejeda, "The NAO impact on droughts in the Mediterranean region," *Advances in Global Change Research*, vol. 46, pp. 23–40, 2011.

[33] M. R. Kousari, M. T. Dastorani, Y. Niazi, E. Soheili, M. Hayatzadeh, and J. Chezgi, "Trend detection of drought in arid and semi-arid regions of Iran based on implementation of reconnaissance drought index (RDI) and application of non-parametrical statistical method," *Water Resources Management*, vol. 28, no. 7, pp. 1857–1872, 2014.

[34] J. Spinoni, G. Naumann, J. Vogt, and P. Barbosa, "European drought climatologies and trends based on a multi-indicator approach," *Global and Planetary Change*, vol. 127, no. 6, pp. 50–57, 2015.

[35] A. R. Zarei, M. M. Moghimi, and M. R. Mahmoudi, "Analysis of changes in spatial pattern of drought using RDI index in south of Iran," *Water Resources Management*, vol. 30, no. 11, pp. 3723–3743, 2016.

[36] S. Mitra and P. Srivastava, "Spatiotemporal variability of meteorological droughts in southeastern USA," *Natural Hazards*, vol. 86, no. 3, pp. 1007–1038, 2017.

[37] Z. H. Wu, P. G. Zhan, Z. H. Chen, H. Bai, and J. S. Wang, "The nearly 40-year drought evolution characteristics of Anshun Municipality assessed by CI and K drought indexes," *Journal of Glaciology and Geocrylogy*, vol. 35, no. 4, pp. 1044–1055, 2013, in Chinese.

[38] Y. Feng, N. B. Cui, Y. M. Xu, Z. P. Zhang, and J. P. Wang, "Temporal and spatial distribution characteristics of meteorological drought in Guizhou Province," *Journal of Arid Land Resources and Environment*, vol. 29, no. 8, pp. 82–86, 2015, in Chinese.

[39] X. Y. Bai, S. J. Wang, and K. N. Xiong, "Assessing spatial-temporal evolution processes of karst rocky desertification and indications for restoration strategies," *Land Degradation and Development*, vol. 24, no. 81, pp. 47–56, 2013.

[40] Y. C. Tian, X. Y. Bai, S. J. Wang, L. Y. Qin, and Y. Li, "Spatial-temporal changes of vegetation cover in Guizhou Province, Southern China," *Chinese Geographical Science*, vol. 27, no. 1, pp. 25–38, 2017.

[41] X. K. Chen, J. X. Xu, J. P. Hu, Z. Z. Zhang, and X. Huang, "Feature of spatial and temporal variation of extreme precipitation in Guizhou Province from 1961 to 2012," *Journal of Water Resources and Water Engineering*, vol. 26, no. 4, pp. 50–61, 2015, in Chinese.

[42] S. L. Song and B. L. Wang, "Principle and performance comparison of different evaporation sensors," *Meteorological Science and Technology*, vol. 38, no. 1, pp. 111–113, 2010, in Chinese.

[43] D. L. Qi, H. B. Xiao, and X. D. Li, "Variation characteristics of pan evaporation in different ecological function areas of Qinghai Province during 1964-2013," *Journal of Arid Meteorology*, vol. 34, no. 2, pp. 234–242, 2016, in Chinese.

[44] D. L. Qi, X. D. Li, H. B. Xiao et al., "Study on changing characteristics and impact factor of evaporation over the three-river source area in recent 50 years," *Resources and Environment in the Yangze Basin*, vol. 24, no. 9, pp. 1613–1620, 2015, in Chinese.

[45] K. G. Wen, *China Meteorological Disaster (Guizhou Volume)*, Meteorological Press, Beijing, China, 2006, in Chinese.

[46] X. F. Xu, Y. Yu, and C. Z. Wang, *China Meteorological Disaster Yearbook*, China Meteorological Press, Beijing, China, 2006, in Chinese.

[47] X. F. Xu, Y. Yu, and C. Z. Wang, *China Meteorological Disaster Yearbook*, China Meteorological Press, Beijing, China, 2007, in Chinese.

[48] X. F. Xu, Y. Yu, and C. Z. Wang, *China Meteorological Disaster Yearbook*, China Meteorological Press, Beijing, China, 2008, in Chinese.

[49] X. F. Xu, Y. Yu, and C. Z. Wang, *China Meteorological Disaster Yearbook*, China Meteorological Press, Beijing, China, 2009, in Chinese.

[50] X. F. Xu, Y. Yu, and C. Z. Wang, *China Meteorological Disaster Yearbook*, China Meteorological Press, Beijing, China, 2010, in Chinese.

[51] X. F. Xu, Y. Yu, and C. Z. Wang, *China Meteorological Disaster Yearbook*, China Meteorological Press, Beijing, China, 2011, in Chinese.

[52] People's Republic of China General Administration of quality supervision, Inspection and Quarantine, *China National Standardization Management Committee, GB/20481-2006 Meteorological Drought Level*, China Standard Press, Beijing, China, 2006, in Chinese.

[53] J. Spinoni, G. Naumann, H. Carrao, P. Barbosa, and J. Vogt, "World drought frequency, duration, and severity for 1951-2010," *International Journal of Climatology*, vol. 34, no. 8, pp. 2792-2804, 2014.

[54] D. Halwatura, A. M. Lechner, and S. Arnold, "Drought severity-duration-frequency curves: a foundation for risk assessment and planning tool for ecosystem establishment in post-mining landscapes," *Hydrology and Earth System Sciences*, vol. 19, no. 7230, pp. 1069-1091, 2015.

[55] V. M. Yevjevich, *An Objective Approach to Definitions and Investigations of Continental Hydrologic Droughts. Hydrology Paper No.23*, Colorado State University, Fort Collins, CO, USA, 1967.

[56] H. B. Mann, "Nonparametric tests against trend," *Econometrica*, vol. 13, no. 3, pp. 245-259, 1945.

[57] M. G. Kendall, *Rank Correlation Method//Rank Correlation Methods*, Griffin, London, UK, 1955.

[58] X. F. Xu, Y. Yu, and C. Z. Wang, *China Meteorological Disaster Yearbook*, China Meteorological Press, Beijing, China, 2012, in Chinese.

[59] X. F. Xu, Y. Yu, and C. Z. Wang, *China Meteorological Disaster Yearbook*, China Meteorological Press, Beijing, China, 2013, in Chinese.

[60] A. G. Dai, "Increasing drought under global warming in observations and model," *Nature Climate Change*, vol. 3, no. 1, pp. 52-58, 2013.

[61] P. C. D. Milly and K. A. Dunne, "Potential evapotranspiration and continental drying," *Nature Climate Change*, vol. 6, no. 10, pp. 1-6, 2016.

[62] J. Sheffield, E. F. Wood, and M. L. Roderick, "Little change in global drought over the past 60 years," *Nature*, vol. 491, no. 7424, pp. 435-438, 2012.

[63] A. J. Teuling, A. F. V. Loon, S. I. Seneviratne et al., "Evapotranspiration amplifies European summer drought," *Geophysical Research Letters*, vol. 40, no. 10, pp. 2071-2075, 2013.

[64] C. L. Wang, J. Guo, L. F. Xue, and L. J. Ding, "An improved comprehensive meteorological drought index CI_(new) and its applicability analysis," *Chinese Journal of Agrometeorology*, vol. 32, no. 24, pp. 621-626, 2011, in Chinese.

[65] L. H. Yang, J. Y. Gao, S. U. Ru-Bo, and X. F. Lin, "Analysis on the suitability of improved comprehensive meteorological drought index in Fujian Province," *Chinese Journal of Agrometeorology*, vol. 33, no. 4, pp. 603-608, 2013, in Chinese.

[66] M. Hayes, M. Svoboda, N. Wall, and M. Widhalm, "The lincoln declaration on drought indices: universal meteorological drought index recommended," *Bulletin of the American Meteorological Society*, vol. 92, no. 4, pp. 485-488, 2011.

[67] M. A. A. Zarch, B. Sivakumar, and A. Sharma, "Droughts in a warming climate: a global assessment of standardized precipitation index (SPI) and reconnaissance drought index (RDI)," *Journal of Hydrology*, vol. 526, pp. 183-195, 2015.

[68] S. M. Vicente-Serrano, G. V. D. Schrier, S. Beguería, C. Azorin-Molina, and J. I. Lopez-Moreno, "Contribution of precipitation and reference evapotranspiration to drought indices under different climates," *Journal of Hydrology*, vol. 526, pp. 42-54, 2015.

[69] W. Zhang, F. F. Jin, J. X. Zhao, L. Qi, and H. L. Ren, "The possible influence of a nonconventional El Niño on the severe autumn drought of 2009 in southwest China," *Journal of Climate*, vol. 26, no. 21, pp. 8392-8405, 2013.

[70] W. Zhang, F. F. Jin, and A. Turner, "Increasing autumn drought over southern china associated with ENSO regime shift," *Geophysical Research Letters*, vol. 41, no. 11, pp. 4020-4026, 2014.

[71] L. Feng, T. Li, and W. Yu, "Cause of severe droughts in southwest China during 1951-2010," *Climate Dynamics*, vol. 43, no. 7-8, pp. 2033-2042, 2014.

[72] B. J. Liu, C. C. Chen, Y. Q. Lian, J. F. Chen, and X. H. Chen, "Long-term change of wet and dry climatic conditions in the southwest karst area of China," *Global and Planetary Change*, vol. 127, pp. 1-11, 2015.

[73] L. H. Wu, S. J. Wang, X. Y. Bai et al., "Quantitative assessment of the impacts of climate change and human activities on runoff change in a typical karst watershed, SW China," *Science of the Total Environment*, vol. 601-602, pp. 1449-1465, 2017.

[74] Z. W. Li, X. L. Xu, C. H. Xu, M. X. Liu, and K. L. Wang, "Dam construction impacts on multiscale characterization of sediment discharge in two typical karst watersheds of southwest China," *Journal of Hydrology*, vol. 558, pp. 42-54, 2018.

[75] D. F. Shen, C. J. Shang, X. Y. Fang, and J. X. Xu, "Spatio-temporal characteristics of drought duration and drought severity in Guizhou," *Journal of Arid Land Resources and Environment*, vol. 30, no. 7, pp. 138-143, 2016, in Chinese.

11

Nonlocal Inadvertent Weather Modification Associated with Wind Farms in the Central United States

Matthew J. Lauridsen and **Brian C. Ancell**

Texas Tech University, Lubbock, TX, USA

Correspondence should be addressed to Matthew J. Lauridsen; matthew.j.lauridsen@gmail.com

Academic Editor: Anthony R. Lupo

Local effects of inadvertent weather changes within and near wind farms have been well documented by a number of modeling studies and observational campaigns; however, the broader nonlocal atmospheric effects of wind farms are much less clear. The goal of this study is to determine whether wind farm-induced perturbations are able to evolve over periods of days, and over areas of thousands of square kilometers, to modify specific atmospheric features that have large impacts on society and the environment, specifically midlatitude and tropical cyclones. Here, an ensemble modeling approach is utilized with a wind farm parameterization to quantify the sensitivity of meteorological variables to the presence of wind farms. The results show that perturbations to nonlocal midlatitude cyclones caused by a wind farm are statistically significant, with magnitudes of roughly 1 hPa for mean sea-level pressure, 4 m/s for surface wind speed, and 15 mm for maximum 30-minute accumulated precipitation. Cyclone perturbation magnitude is also found to be dependent on wind farm size and location relative to the midlatitude cyclone genesis region and track.

1. Introduction

Amid growing emphasis on domestic, renewable, and environmentally friendly energy sources, several renewable energy industries have experienced large growth in the United States and across the globe. One such prominent industry expanding in the Central United States is wind power. To produce power, wind turbines extract kinetic energy from the mean flow, which produces drag on the mean wind and also enhances turbulent kinetic energy (TKE) downwind of the turbine. In turn, an important potential result of the creation of wind farms is the inadvertent modification of weather. Local modification of the atmospheric state near wind farms has been well documented by a number of observational campaigns and modeling studies. Observations show surface temperature is increased at night [1, 2] and hub-height wind speed is decreased within and immediately downstream of wind farms [3]. Modeling studies have concluded a decrease of wind speed at hub height [4–9] but a wind speed increase at the

surface due to momentum transfer in enhanced vertical mixing [4, 7]. Surface water vapor mixing ratio has been shown to decrease due to this vertical mixing [4] since the surface is the source of moisture. Additionally, potential temperature has been found to increase under stable atmospheric conditions and at times decrease when the boundary layer is unstable [4, 8–12].

In terms of how local atmospheric modifications may evolve, many studies suggest that relatively small perturbations can grow rapidly to modify the atmosphere nonlocally, potentially with regard to high-impact weather events. Lorenz [13] first reported on the fundamental nature underlying such perturbation growth, a phenomenon known as chaos. In recent studies more relevant to the full atmospheric state, Zhang et al. [14] showed that smaller initial condition perturbations grew faster, in both magnitude and spatial size, than larger perturbations introduced to 36-hour forecasts with the MM5 mesoscale model. Other studies have explored high-impact weather features such as tropical cyclones [15, 16] and midlatitude cyclones [17–20] that exhibit a large sensitivity to initial conditions. Highly sensitive regions involved with tropical cyclone evolution are found to be collocated with the flow toward the cyclone at upper levels [15], and these maxima in sensitivity can occur as far as 4000 km from the cyclone [16], often to the northwest of the storm. Midlatitude cyclogenesis can be triggered by initial perturbations as shown by an adjoint model [19], and midlatitude cyclones have been shown to be sensitive to low-level temperature advection [17, 18, 20], specifically during cyclogenesis. Additionally, low-level moisture [17] and upper-level potential vorticity [20] perturbations to the initial state are shown to have a large impact on developing midlatitude cyclones.

Since wind farms have been shown to directly modify the local atmospheric state and large sensitivity of high-impact weather features to the lower atmosphere has been demonstrated, it is reasonable to expect wind farm-induced perturbations may evolve rapidly downstream to affect the weather in a significant way. However, the broader nonlocal atmospheric effects of wind farms on timescales of days are much less clear than those within and surrounding wind farms. Therefore, there is a need to establish whether wind farms are capable of producing modifications to the atmosphere on large temporal and spatial scales which is the goal of this study. This research goal is particularly aimed at addressing the knowledge gap with regard to nonlocal inadvertent weather modification pointed out by the American Meteorological Society, which stated that anthropogenic forcing which can be derived from wind farms could "modify atmospheric circulation and weather patterns on all scales, including synoptic storm tracks, in ways that are just beginning to be explored" [21].

In this study, we examine the sensitivity of midlatitude and tropical cyclones to wind farm-induced perturbations, as these events can significantly affect weather over large areas over periods of days. The key tool we use here to investigate the potential link between wind farms and the nonlocal atmospheric state is the Weather Research and Forecasting (WRF) mesoscale model with a wind farm parameterization introduced in [7]. Rather than assessing wind farm effects on forecast skill, we aim to fundamentally understand the sensitivity of the atmosphere to wind farm effects on the timescale of hours to days. While an adjoint approach [22] is one way to examine the sensitivity of wind farms to nonlocal weather features, it is limited by computational cost and nonlinearity over the simulation window of our experiments (96 hours). Thus, here we use an ensemble-like approach, designed to incorporate wind farm size and location variability, to determine the relationship between nonlocal midlatitude and tropical cyclones to wind farms. The experimental setup is described in the next section. The results are then presented in Section 3. A summary and conclusions of results are provided in Section 4.

2. Methodology

Here, we use the Weather Research and Forecasting-Advanced Research WRF (WRF-ARW) mesoscale model version 3.5.1 [23] and simulate wind farms using the wind farm parameterization outlined in [7]. The single domain utilized is 5500 km by 3700 km as shown in Figure 1, with a horizontal grid spacing of 10 km and 40 vertical levels. Wind turbines being simulated in the model are based off of the Siemens SWT-2.3-108 turbine. The turbines have a hub height of 100 m, a rotor diameter of 108 m, cut-in and cut-out speeds of 3.0 and 25.0 ms^{-1}, respectively, a power output of 2.3 MW, and a standing thrust coefficient of 0.158. The Fitch scheme to parameterize a wind farm is run under the idealized configuration, with the turbines being placed equidistant from each other in a grid pattern, and the generic thrust and power coefficient curves that are approximate to an actual wind turbine are used. In each horizontal grid cell that makes up the wind farm, there are 324 wind turbines. This equates to a turbine spacing of 555 5/9 m, or rotor diameter of 5.14 m. GFS data are used for initialization and lateral boundary conditions, which are updated every 6 hours, and the time step is 60 seconds. With the resolution used here (10 km), this ends up being about a 3 : 1 dynamical downscaling ratio.

Other parameterizations used here include the Thompson scheme [24] for microphysics, the rapid radiative transfer model scheme [25] for longwave radiation and Dudhia scheme [26] for shortwave radiation with 30 minutes between radiation physics calls, the Monin–Obukhov (Janjic) Eta similarity scheme [27] for surface-layer physics, the unified Noah land-surface model [28] for land-surface physics, the Mellor–Yamada–Nakanishi–Niino (MYNN) 2.5-level TKE scheme [29, 30] for boundary-layer physics, and the Kain–Fritsch scheme [31] for cumulus physics with 5 minutes between cumulus physics calls. It should be noted that the wind farm parameterization is only functional when used in conjunction with the MYNN 2.5-level TKE scheme, and this boundary-layer physics parameterization was chosen by the wind farm parameterization developers due to its more reliable prediction of TKE. The TKE calculation by the MYNN 2.5-level TKE scheme is tuned to match results from large eddy simulation (LES).

Ten total cases of midlatitude cyclogenesis (six cases) and tropical cyclogenesis (four cases) are included in this study (listed in Table 1). The model is initialized at 00Z about 24 hours before cyclone formation for the cases of midlatitude cyclogenesis in an attempt to capture any interaction between the wind farm perturbations and the localized sensitivity field associated with the cyclones demonstrated through prior studies. While it is possible that sensitivity would be different within a natural state that included the wind farm configurations in this study, we assume such differences would be small enough with regard to the climatology of the storm track such that our experiments still reveal valid results regarding short-term cyclone evolution. For the tropical cyclogenesis cases, the 96-hour forecast is initialized at 00Z when the cyclone was already within the model domain but before the cyclone strengthened. All midlatitude cyclones in this study form along the Rocky Mountains, deepen, and propagate east or northeast across the Central United States. Two of the four tropical cyclones originate in the Gulf of Mexico and make landfall along the

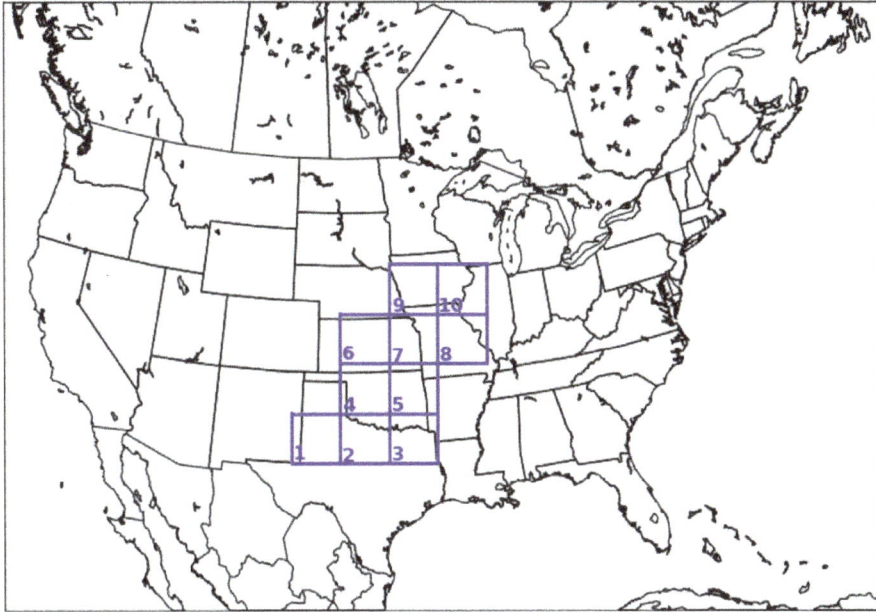

FIGURE 1: Members, numbered, of the wind farm location ensemble (in outlined boxes).

TABLE 1: Cases used in the study, the type of cyclone observed in model simulation, and the location ensemble with the largest perturbation.

Cases	Type of cyclone	Member
24 August 2011 00Z–28 August 2011 00Z	Tropical	4
26 August 2012 00Z–30 August 2012 00Z	Tropical	1
28 May 2013 00Z–1 June 2013 00Z	Midlatitude	4
3 October 2013 00Z–7 October 2013 00Z	Tropical	8
10 October 2013 00Z–14 October 2013 00Z	Midlatitude	6
13 October 2013 00Z–17 October 2013 00Z	Midlatitude	1
26 April 2014 00Z–30 April 2014 00Z	Midlatitude	9
6 May 2014 00Z–10 May 2014 00Z	Midlatitude	6
1 July 2014 00Z–5 July 2014 00Z	Tropical	4
12 October 2014 00Z–16 October 2014 00Z	Midlatitude	4

Gulf Coast of the United States, while the other two are located in the Atlantic Ocean and follow a path that tracks just offshore of the southeastern United States. In turn, unlike the midlatitude cyclones that track much closer to the prescribed wind farms, the tropical cyclones examined here must take advantage of perturbations that have evolved over long distances if they are to be modified.

Predetermined ensembles of wind farm locations and sizes are utilized for each case. First, an ensemble of fixed locations is run for each case, consisting of 10 wind farm locations in the Central United States as seen in Figure 1. The wind farms in this ensemble of locations are 300 km by 300 km, or 90,000 km^2. While this wind farm size is very large, the purpose of this location ensemble is to capture the

most sensitive areas with regard to the evolution of the cyclones, providing an appropriate place to execute an ensemble of wind farm sizes. The member of the ensemble that produces the most significant changes to the midlatitude cyclone or tropical cyclone is then chosen as the location to run the fixed ensemble of wind farm sizes. To determine which wind farm location produces the largest perturbations, a control run is utilized in which the wind farm parameterization is turned off. The members of the location ensemble are then differenced with the control run (location member minus control; examples are shown in Figures 2 and 3), and the location that produces the largest perturbations to minimum cyclone pressure and maximum cyclone 10-meter wind speed is then selected as the location for the size ensemble. Wind farm sizes included in the size ensemble are shown in Figure 4, which in this example are located over region 6. Some members in the wind farm size ensemble are much larger than wind farms in operation today. However, they are included in the ensemble to have an estimation of how large wind farms must be to produce a significant change to nonlocal meteorology. Additionally, wind farms larger than current wind farms can be seen as an aggregation of many wind farms in the same region.

After an initial examination of results, it was determined that some perturbations (found by differencing the wind farm run with the control run) are not directly created by the wind farm in the model but are possibly created by the spurious propagation of noise that can then evolve throughout the forecast. These "unrealistic" perturbations occur in regions where convective precipitation and non-convective precipitation are occurring, possibly due in part to the Boolean response of the convection parameterization as described in [14, 32, 33]. However, turning off the cumulus parameterization with a finer grid spacing was ineffective in removing the unrealistic noise here. Both studies [34] and [35] found evidence of these unrealistic processes

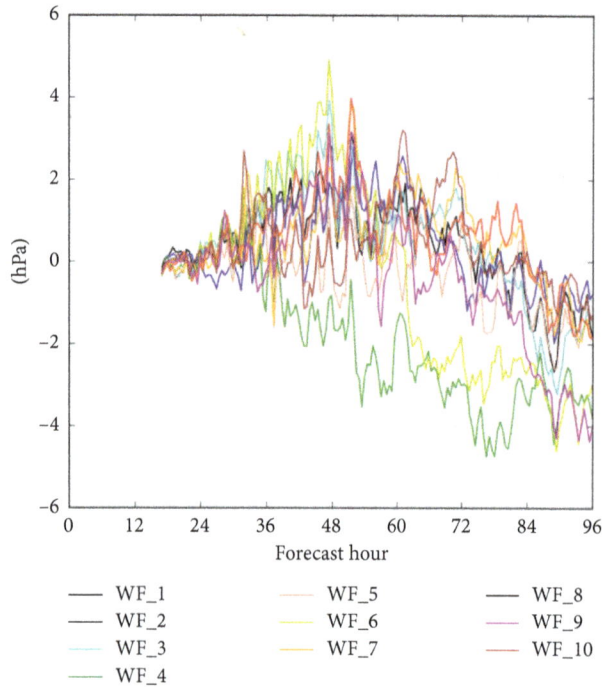

FIGURE 2: Example time series of cyclone minimum sea-level pressure perturbation from the control run. The case shown here represents the tropical cyclone case initialized on 24 August 2011 00Z.

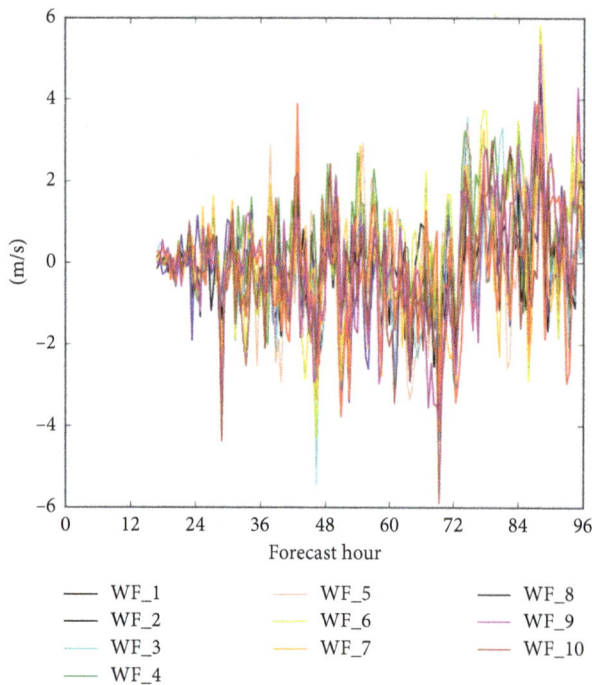

FIGURE 3: Example time series of cyclone maximum 10 m wind speed perturbation from the control run. The case shown here represents the tropical cyclone case initialized on 24 August 2011 00Z.

influencing perturbation experiments, effectively contaminating the results. In turn, it is crucial in our experiments here to address this issue if we are to understand the true relationships between wind farms and the evolution of the

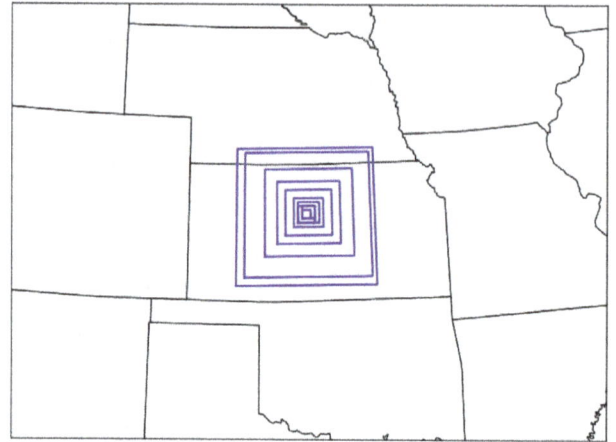

FIGURE 4: Wind farms (in outlined boxes) of the size ensemble shown for location 6 from Figure 1.

atmosphere at the time and spatial scales examined. An example of these unrealistic perturbations can be found in Figure 5. Here, the perturbations in northern Missouri at forecast hour 05 are approximately 800 km from the wind farm and were present in the beginning at half an hour after initialization, but at a much smaller magnitude. If realistic, the perturbations would have traveled at 1600 kilometers per hour, far too fast for the advection of realistic perturbations via gravity or acoustic waves. Several other model runs have also shown these unrealistic perturbations occurring at locations unexplainable by gravity or acoustic waves. In any case, experimental measures must be taken to ensure that no misinterpretations of downstream effects are being made associated with this discovered unrealistic propagation of noise.

To show an example of the unrealistic propagation of noise discussed above, the model is run with no wind farm but instead a small potential temperature perturbation across one grid cell in the stratosphere in the northeastern part of the domain noted by the green marker in Figure 6. This area is selected due to the average zonal flow across the area from west to east; therefore, the stratospheric perturbation should likely advect out of the domain and not impact in a realistic way the features of importance in the domain such as the tropical and midlatitude cyclones. Instead, the same unrealistic perturbations observed with perturbations induced by wind farms again occur. An image of the unrealistic perturbations due to a stratospheric temperature perturbation can be found in Figure 6, and an image of the simulated reflectivity for the control run is shown in Figure 7. It can be seen that these perturbations in question are indeed occurring in and near a region of precipitation where moist physics is playing a crucial role. These moist processes are seen to be associated with rapid initial growth of noise through nonlinear chaos [36]. Other various model setups including altering the parameterizations used, changing the number of wind turbines per cell, varying the upper level damping layer and vertical velocity damping, and utilizing a nested domain with greater horizontal resolution over the wind farm were all unable to remove these unrealistic effects.

Surface pressure (mb) [mem1 – ctrl] Valid: 2011-08-24_05:00Z f05

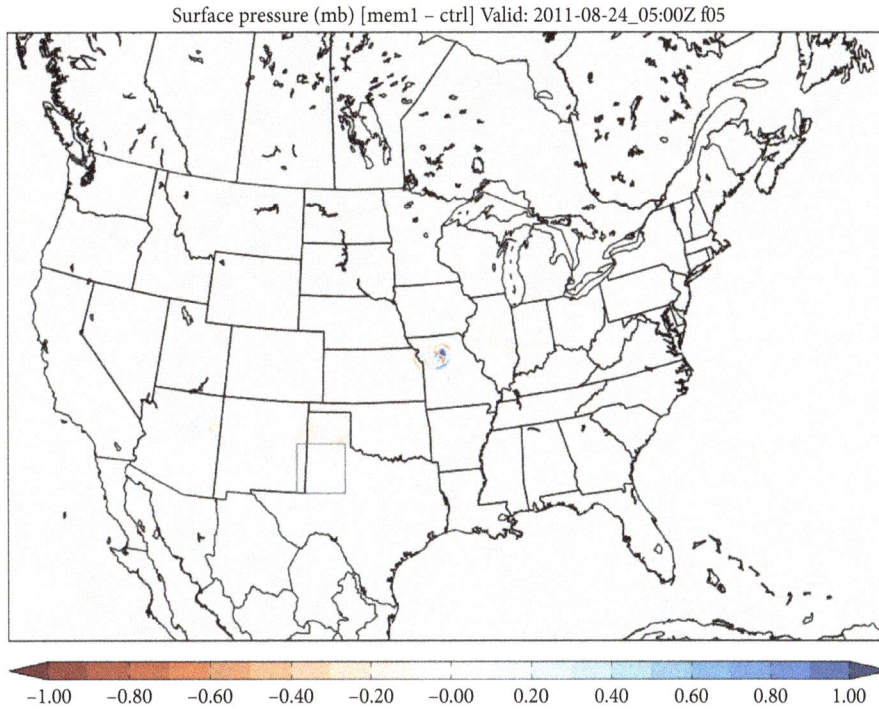

FIGURE 5: Differences in surface pressure at forecast hour 05 between a wind farm run and a control run without any wind farms. Wind farms in box are centered over western Texas.

Surface pressure (mb) [memalt2.0 – ctrl] Valid: 2011-08-24_05:00Z f05

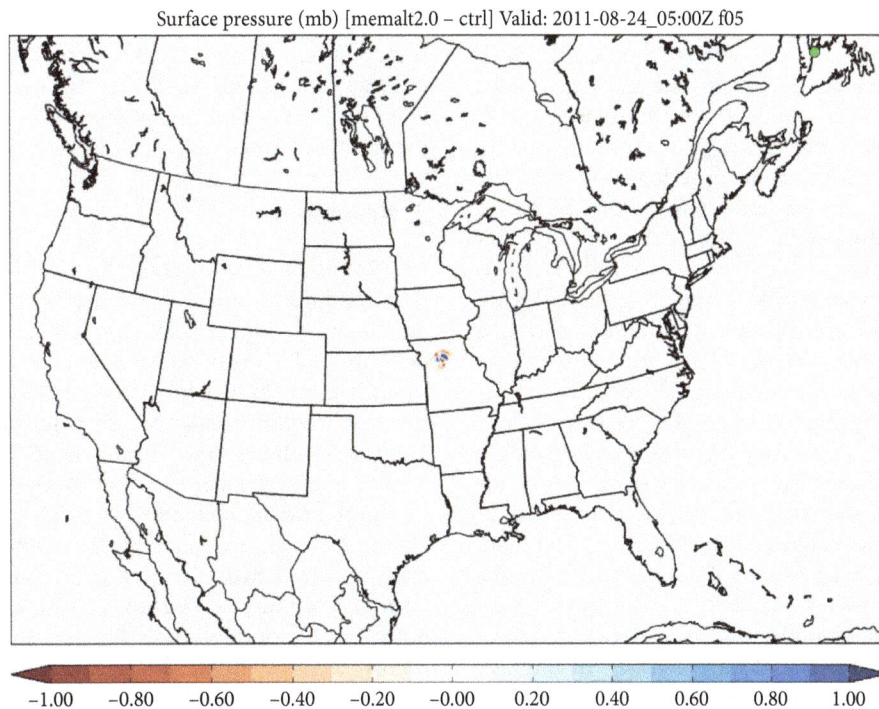

FIGURE 6: Differences in surface pressure at forecast hour 05 between a run with a 2°C stratospheric potential temperature perturbation and a control run. Location of temperature perturbation is shown by a green marker in the upper right part of the image.

To determine whether the results obtained from the wind farm runs are realistic and are produced by the wind farm parameterization rather than the unrealistic, rapid propagation of noise, an ensemble of ten members with varying stratospheric potential temperature perturbations is run to characterize the effects of the unrealistic perturbations. For situations where the effects from this potential temperature perturbation are not statistically different from the wind

Simulated radar reflectivity (dBZ) [ctrl] Valid: 2011-08-24_05:00Z f05

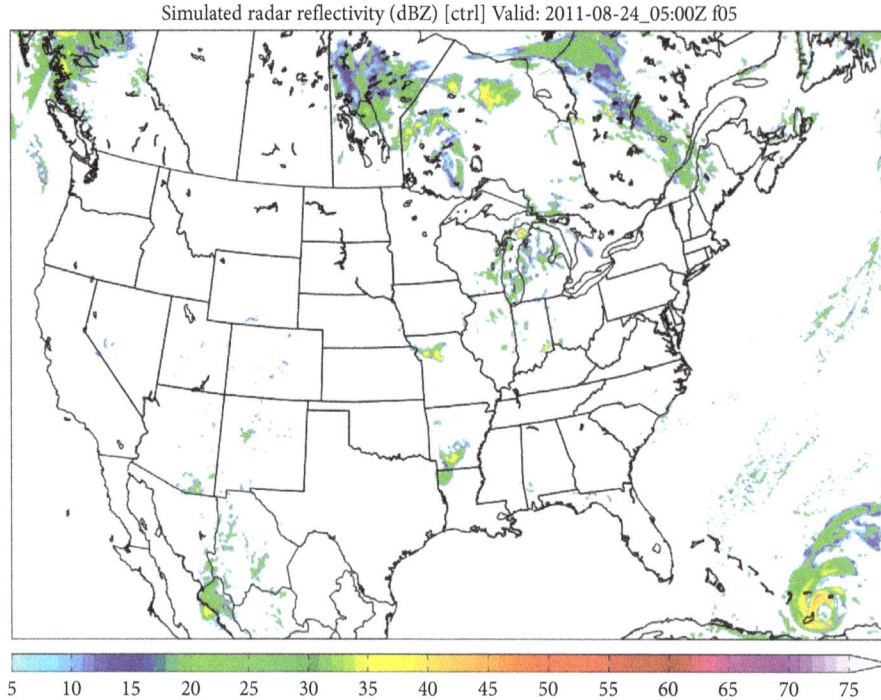

5   10   15   20   25   30   35   40   45   50   55   60   65   70   75

FIGURE 7: Simulated reflectivity at forecast hour 05 for the control run.

farm runs, we will be unable to conclude that any simulated perturbations are realistically generated from the wind farm. This technique to ensure significance results are not false positives parallel to a strategy in Lorenz et al. [37, 38], which utilized field significance testing to account for spatial correlation of results and multiplicity, and is also in line with [36] which notes there is no avoiding chaos seeding in WRF and one must account for it in the experimental setup. The ensemble here consists of potential temperature increases from the control between 0.2°C and 2.0°C in 0.2°C increments and will serve as the unrealistic benchmark ensemble throughout the study. All temperature increases are made in the stratosphere at an eta level of 0.240. By testing against this group of model runs with the stratospheric warming and associated unrealistic model effects, this effectively shows significance in terms of what can be concluded as realistic processes (an important aspect of the experiments as recommended in Lorenz et al. [38]). Additionally, Ancell [36] showed the unrealistic processes have no correlation with the magnitude of the perturbation in an ensemble of unrealistic runs. Thus, by testing whether the realistic ensemble shows any correlation of the metric to the initial perturbation magnitudes (and thus a sensitivity to the said magnitude), one would have shown real physical relationships beyond the more randomized noise.

The list of metrics used in this study includes both cyclone minimum sea-level pressure and maximum 10-meter wind speed, as well as cyclone maximum 30-minute accumulated precipitation, maximum and minimum 2-meter temperature, maximum and minimum 2-meter potential temperature, maximum boundary layer height, and maximum 2-meter water vapor mixing ratio. To determine whether the wind farm-induced perturbations are more

significant than the unrealistic effects, linear regressions are composed between the cyclone perturbation magnitude and size of the wind farm or stratospheric potential temperature perturbation. These linear regressions will determine whether the perturbations to the cyclone show any relationship to the wind farms beyond the unrealistic effects of the noise described above, a technique also used in [36].

3. Results

3.1. Tropical Cyclone Example. Selected surface pressure perturbations for the 3 October 2013 tropical cyclone case are shown in Figure 8, and Figure 9 shows the sea-level pressure and 10 m wind for the control run. Beginning at forecast hour 05, perturbations across the domain are restricted to within and near the wind farm in Oklahoma. Here, a southerly flow in the wind farm region led to a positive pressure perturbation along the southern edge of the wind farm and extending through the wind farm due to propagation of the perturbation. Downwind, a negative pressure perturbation existed at the northern edge of the farm and propagated northward with the flow. By forecast hour 24, the southerly flow has moved pressure perturbations created by the wind farm as far north as Canada. The pressure perturbations at this time were not well organized; but by forecast hour 42, it can be seen that the perturbations were beginning to organize.

At forecast hour 64, the pressure perturbation dipole suggests a positional change of the tropical cyclone between the control run and model run with the wind farm. These dipole pressure perturbations are common throughout many cases in the forecast hour range of 48–96 and may be a systematic effect of wind farms on cyclones. However,

Surface pressure (mb) [mem4 – ctrl] Valid: 2013-10-03_05:00Z f05

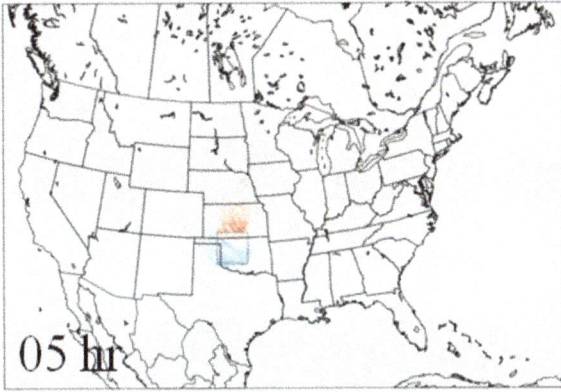

−1.00 −0.80 −0.60 −0.40 −0.20 −0.00 0.20 0.40 0.60 0.80 1.00

(a)

Surface pressure (mb) [mem4 – ctrl] Valid: 2013-10-04_00:00Z f24

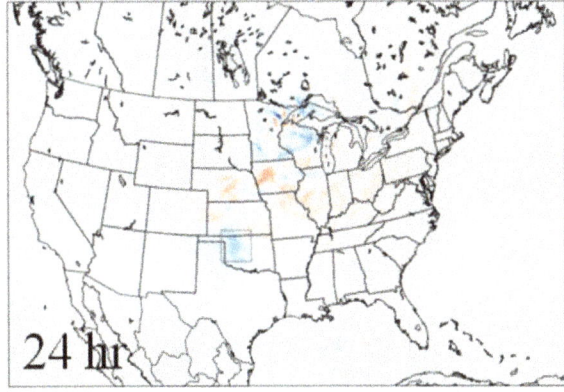

−1.00 −0.80 −0.60 −0.40 −0.20 −0.00 0.20 0.40 0.60 0.80 1.00

(b)

Surface pressure (mb) [mem4 – ctrl] Valid: 2013-10-04_18:00Z f42

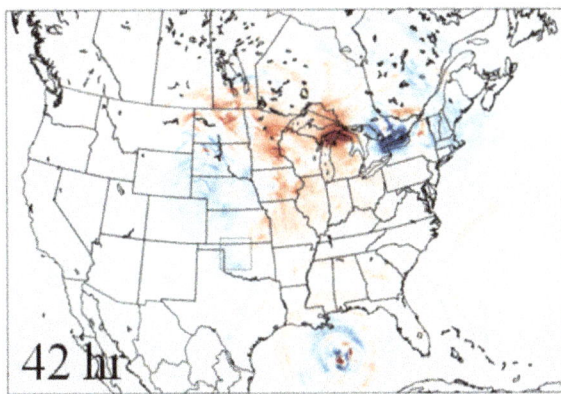

−1.00 −0.80 −0.60 −0.40 −0.20 −0.00 0.20 0.40 0.60 0.80 1.00

(c)

Surface pressure (mb) [mem4 – ctrl] Valid: 2013-10-05_16:00Z f64

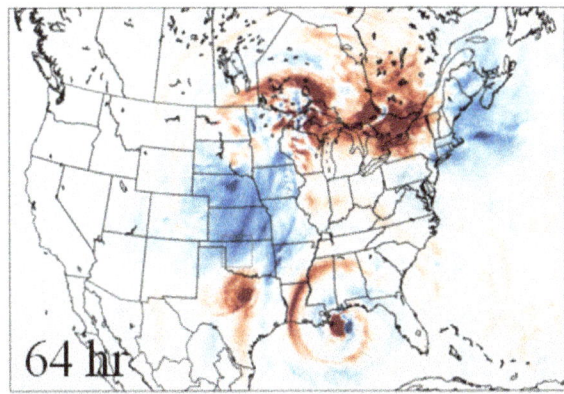

−1.00 −0.80 −0.60 −0.40 −0.20 −0.00 0.20 0.40 0.60 0.80 1.00

(d)

Surface pressure (mb) [mem4 – ctrl] Valid: 2013-10-06_10:00Z f82

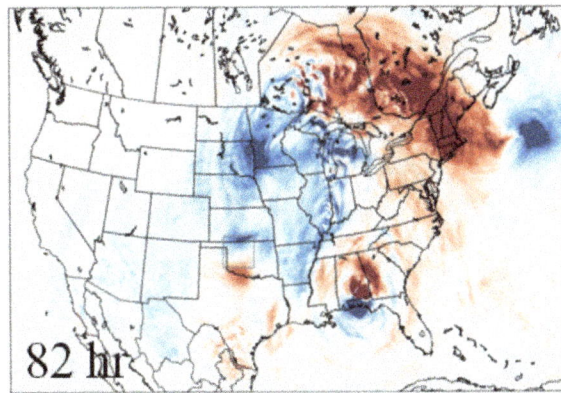

−1.00 −0.80 −0.60 −0.40 −0.20 −0.00 0.20 0.40 0.60 0.80 1.00

(e)

Surface pressure (mb) [mem4 – ctrl] Valid: 2013-10-07_00:00Z f96

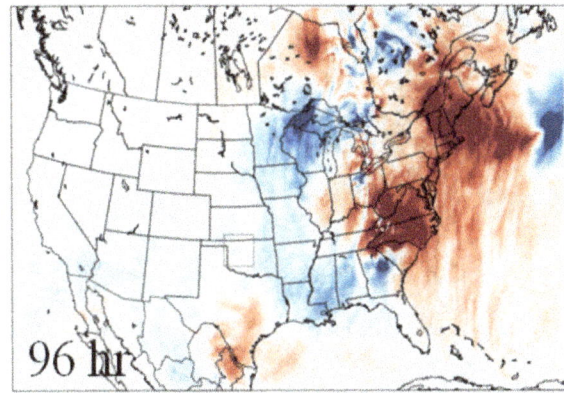

−1.00 −0.80 −0.60 −0.40 −0.20 −0.00 0.20 0.40 0.60 0.80 1.00

(f)

FIGURE 8: Surface pressure perturbations at forecast hours 05 (a), 24 (b), 42 (c), 64 (d), 82 (e), and 96 (f) for the 3 October 2013 case due to a wind farm in box shown in Oklahoma.

Sea-level pressure (mb), 10 m wind (kts) [ctrl]
Valid: 2013-10-03_05:00Z f05

Sea-level pressure (mb), 10 m wind (kts) [ctrl]
Valid: 2013-10-04_00:00Z f24

(a)

(b)

Sea-level pressure (mb), 10 m wind (kts) [ctrl]
Valid: 2013-10-04_18:00Z f42

Sea-level pressure (mb), 10 m wind (kts) [ctrl]
Valid: 2013-10-05_16:00Z f64

(c)

(d)

Sea-level pressure (mb), 10 m wind (kts) [ctrl]
Valid: 2013-10-06_10:00Z f82

Sea-level pressure (mb), 10 m wind (kts) [ctrl]
Valid: 2013-10-07_00:00Z f96

(e)

(f)

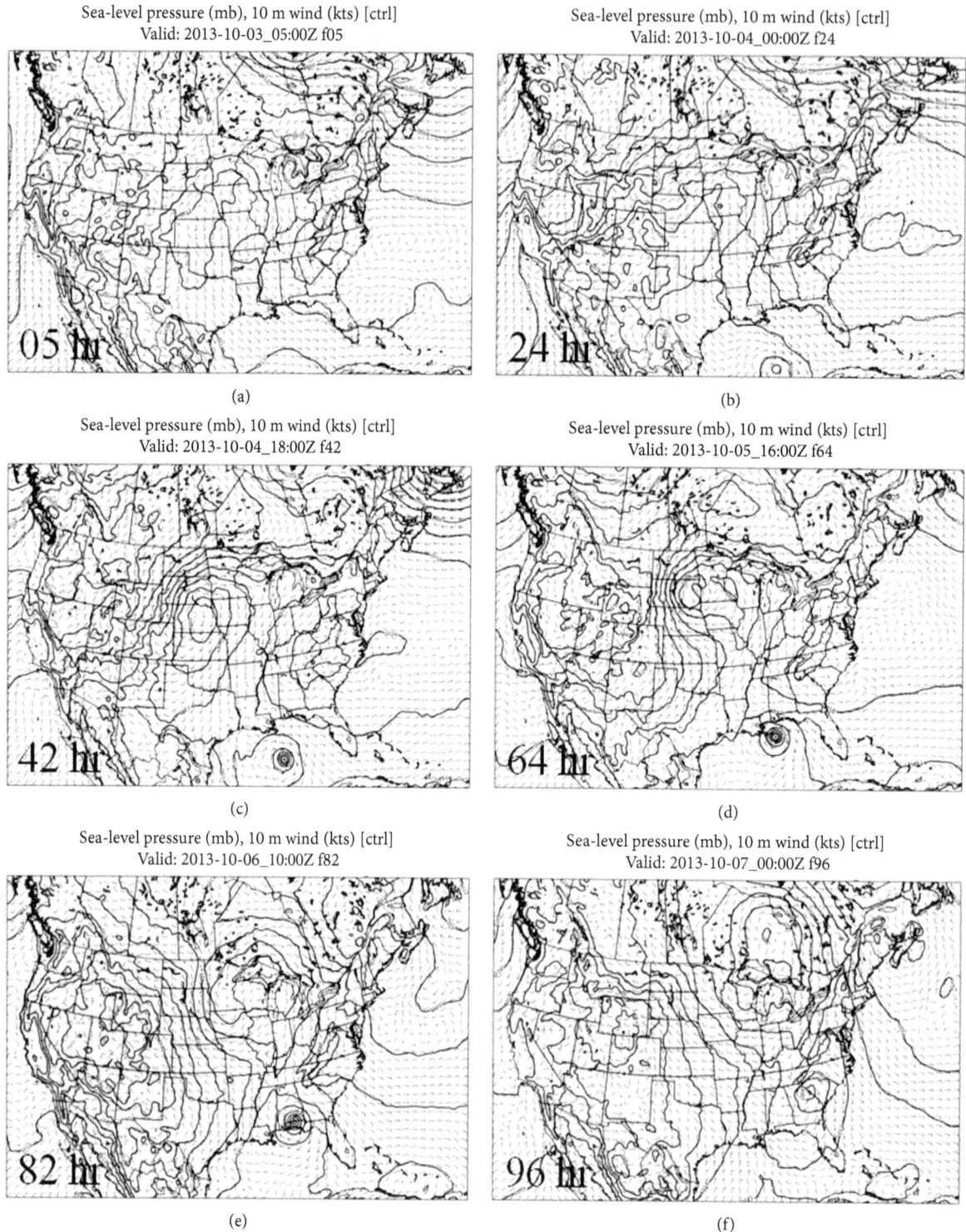

FIGURE 9: Surface sea-level pressure and 10 m wind speed at forecast hours 05 (a), 24 (b), 42 (c), 64 (d), 82 (e), and 96 (f) for the 3 October 2013 control case.

these systematic effects are not consistent in each case, as the cyclone track change is dependent on many factors including wind farm location, cyclone location, and cyclone intensity. Also, the waves emanating from the tropical cyclone evident in the figure are regarded as realistic perturbations, likely due to gravity waves. These are not the

unrealistic model effects that occur instantaneously at model initiation throughout the entire domain discussed previously. By the end of the forecast, the tropical cyclone pressure perturbations have interacted with perturbations in the larger synoptic flow, presenting a negative perturbation to pressure. It is also worth noting that the perturbations are almost entirely limited to along the longitude of the wind farm and eastward, due to the west-to-east synoptic zonal flow across the United States.

3.2. Midlatitude Cyclone Example.

Examining the surface pressure perturbations for the 26 April 2014 midlatitude cyclone case also reveals similar features as the tropical cyclone case above. An area of increased pressure formed on the upwind side of the wind farm in Kansas, in this case the southern side, and a resulting low pressure perturbation occurred on the northern side of the wind farm. One explanation for this is with more mass accumulating on the upwind side of wind farms due to interaction with turbines, a decreased pressure forms on the downwind side due to conservation of mass, as seen at forecast hour 05 in Figure 10. Another possible mechanism for the pressure perturbation signature is a gravity wave caused by the divergence and convergence pattern around the wind farm [12]. For reference, the sea-level pressure and 10 m wind for the control run are shown in Figure 11. The decreased pressure perturbation advanced with the flow to the northeast and began to interact with unrealistic perturbations by forecast hour 16. Cyclogenesis began in northeastern Colorado shortly afterward, and the area of increased pressure stayed in place.

By forecast hour 40, the cyclone that developed in Colorado had moved eastward and became modified by the wind farm-induced perturbations. The cyclone pressure was increased due to the presence of the increased pressure perturbation in the area. Throughout the remainder of the forecast, the midlatitude cyclone continued to exhibit an increased pressure due to the wind farm in Kansas. A feature of decreased pressure in northern Wisconsin interacted with the cyclone in forecast hours 80 through 96 as the cyclone continued to progress northeastward.

The positioning of the wind farm is likely very important for midlatitude cyclone perturbations, as midlatitude cyclogenesis occurs near the wind farms. The sign and magnitude of perturbations to the cyclone are dependent on the flow direction, as the positive and negative pressure perturbations are formed on the upwind and downwind sides of the wind farm, respectively. The advancement of these perturbations to the area of cyclogenesis or to the cyclone's path appears to result in the precise sign of pressure perturbation realized by the cyclone.

3.3. Sensitivity to Wind Farm Location.

The perturbations caused by wind farms appear to have an impact on nonlocal atmospheric features downwind of the farm and are able to modify high-impact features as seen in Figures 8 and 10. However, simply examining differences between the control and perturbed runs may be misleading due to the effects of unrealistic noise in the model. Thus, the cyclones in these simulations are analyzed objectively to determine how they are affected by wind farms. First, time-series plots were utilized to determine which wind farm locations are in regions most sensitive to cyclogenesis, cyclone propagation, and cyclolysis. The member of the ensemble that produced the most significant changes to cyclone is chosen as the location in which the ensemble of wind farm sizes is run. The two metrics chosen due to their close association with cyclone intensity include the cyclone minimum mean sea-level pressure and maximum 10 m wind speed. The time-series plots of one tropical cyclone and one midlatitude cyclone for these variables are shown in Figures 12 and 13, showing model runs with the ten different wind farm locations included in the location ensemble and how that location influences the cyclone perturbations. The location that produces the largest perturbations over the extended forecast period is chosen as the location to run the ensemble of wind farm sizes. Another impact of wind farms on cyclones considered by this study is the cyclone track itself. Figure 14 shows the cyclone track for all ten wind farm location runs as well as the control for the cyclones considered in Figures 12 and 13. Little change in the cyclone track was noticed for all cyclones in this study, and further analysis focuses on the intensity-based metrics introduced above, as well as other metrics that are described later.

3.4. Sensitivity to Wind Farm Size.

With the wind farm locations determined to have the largest impacts on the cyclone development, the ensemble of wind farm sizes is then run to determine the dependency of nonlocal perturbation magnitude on the wind farm size. Locations where the wind farm size ensemble is run for each case are shown in Table 1. Table 2 shows the sizes of the wind farms used for the size ensemble. The largest perturbations over the forecast for each of the ten wind farm sizes are compared to determine the trend between perturbation magnitude and sign and the wind farm size. Statistical analysis in the form of the coefficient of determination, r^2, is utilized to determine how well the perturbations with varying wind farm sizes fit a linear regression line. The slope of the linear regression is also analyzed, and a p value is calculated to determine the statistical significance of that slope relative to the null hypothesis that there exists no relationship between the chosen metrics and the wind farm size.

Arguably, the most important metric to determine the strength of cyclones is the minimum sea-level pressure. When comparing wind farm size and the perturbation to cyclone minimum sea-level pressure, unique trends appear. Generally for tropical cyclones, the wind farm decreases the central cyclone pressure. Also, three of four tropical cyclone cases show that the minimum pressure perturbation becomes more negative with increasing wind farm size, while one case showed the opposite trend.

On the contrary, when it comes to midlatitude cyclones, the opposite is usually true. In five of six cases, most wind farm sizes increase the minimum sea-level pressure in the cyclone, thus weakening the midlatitude cyclone. Four of the six cases also show that the minimum pressure perturbation becomes more positive with increasing wind farm size. Shown in Figures 15 and 16 are the largest minimum sea-level pressure perturbations that occur with the ten wind farm sizes, the linear regression for these perturbations, and statistical values for the tropical and midlatitude cyclones

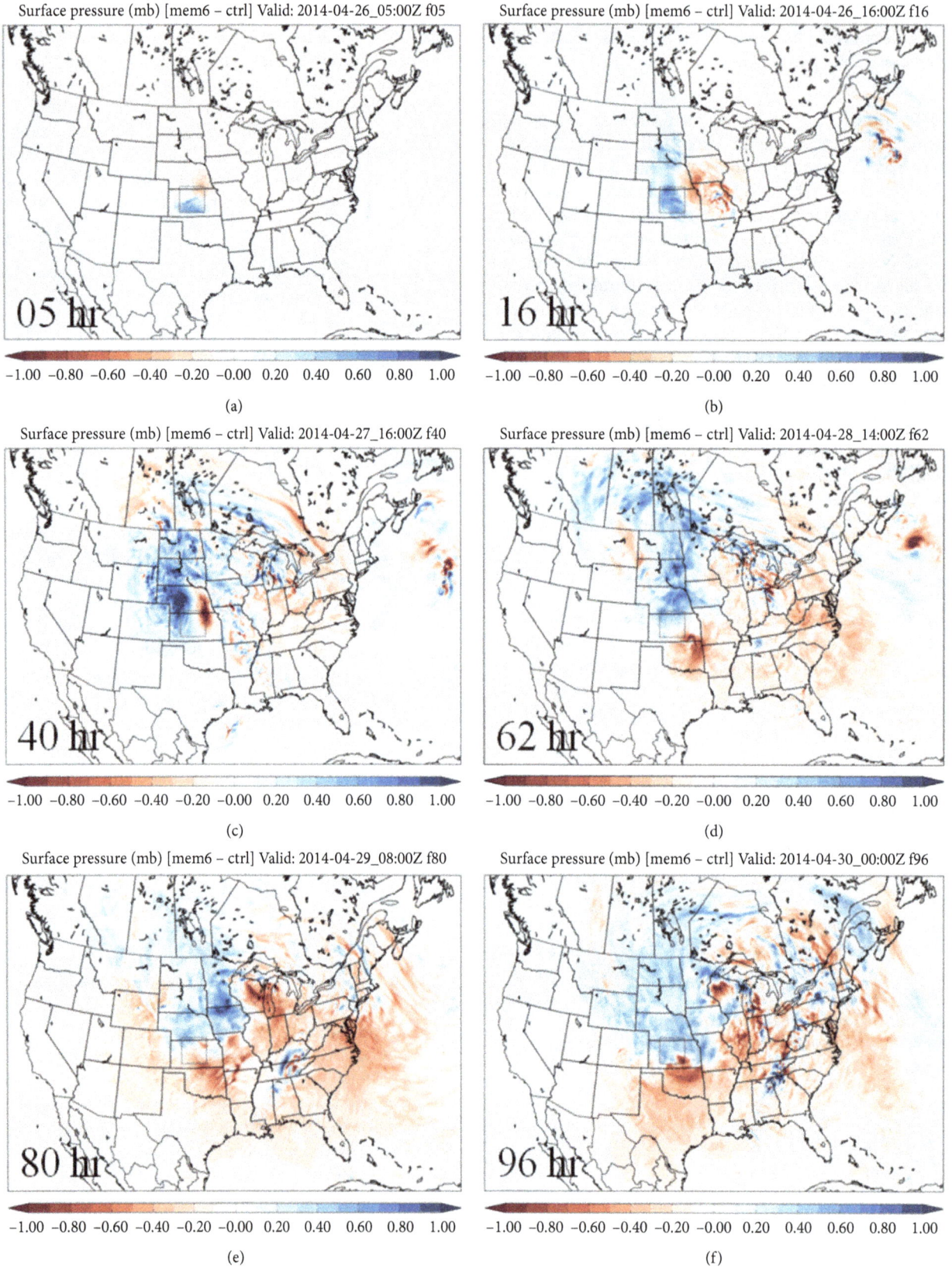

Surface pressure (mb) [mem6 – ctrl] Valid: 2014-04-26_05:00Z f05

−1.00 −0.80 −0.60 −0.40 −0.20 −0.00 0.20 0.40 0.60 0.80 1.00

(a)

Surface pressure (mb) [mem6 – ctrl] Valid: 2014-04-26_16:00Z f16

−1.00 −0.80 −0.60 −0.40 −0.20 −0.00 0.20 0.40 0.60 0.80 1.00

(b)

Surface pressure (mb) [mem6 – ctrl] Valid: 2014-04-27_16:00Z f40

−1.00 −0.80 −0.60 −0.40 −0.20 −0.00 0.20 0.40 0.60 0.80 1.00

(c)

Surface pressure (mb) [mem6 – ctrl] Valid: 2014-04-28_14:00Z f62

−1.00 −0.80 −0.60 −0.40 −0.20 −0.00 0.20 0.40 0.60 0.80 1.00

(d)

Surface pressure (mb) [mem6 – ctrl] Valid: 2014-04-29_08:00Z f80

−1.00 −0.80 −0.60 −0.40 −0.20 −0.00 0.20 0.40 0.60 0.80 1.00

(e)

Surface pressure (mb) [mem6 – ctrl] Valid: 2014-04-30_00:00Z f96

−1.00 −0.80 −0.60 −0.40 −0.20 −0.00 0.20 0.40 0.60 0.80 1.00

(f)

Figure 10: Surface pressure perturbations at forecast hours 05 (a), 16 (b), 40 (c), 62 (d), 80 (e), and 96 (f) for the 26 April 2014 case due to a wind farm in box shown in Kansas.

Sea-level pressure (mb), 10 m wind (kts) [ctrl]
Valid: 2014-04-26_05:00Z f05

Sea-level pressure (mb), 10 m wind (kts) [ctrl]
Valid: 2014-04-26_16:00Z f16

(a)

(b)

Sea-level pressure (mb), 10 m wind (kts) [ctrl]
Valid: 2014-04-27_16:00Z f40

Sea-level pressure (mb), 10 m wind (kts) [ctrl]
Valid: 2014-04-28_14:00Z f62

(c)

(d)

Sea-level pressure (mb), 10 m wind (kts) [ctrl]
Valid: 2014-04-29_08:00Z f80

Sea-level pressure (mb), 10 m wind (kts) [ctrl]
Valid: 2014-04-30_00:00Z f96

(e)

(f)

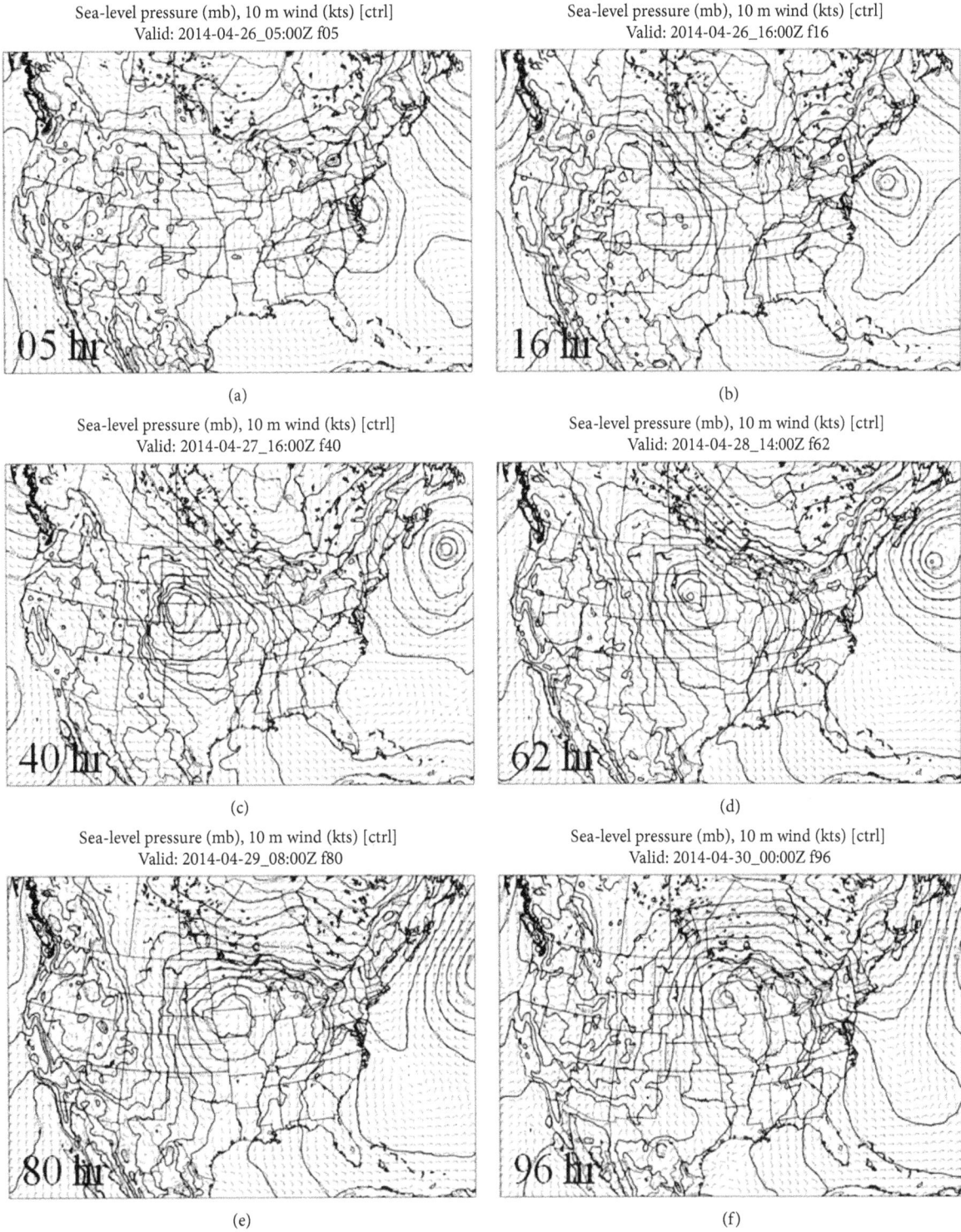

FIGURE 11: Surface sea-level pressure and 10 m wind speed at forecast hours 05 (a), 16 (b), 40 (c), 62 (d), 80 (e), and 96 (f) for the 26 April 2014 control case.

covered earlier. Each dot shows one model run and how the wind farm size (or magnitude of unrealistic temperature perturbation) corresponds to the magnitude of perturbation to the cyclone. The linear relationship shown in the mid-latitude cyclone case (Figure 16) with varying wind farm sizes shows the relationship between wind farm size and

cyclone central pressure is significant, unlike the tropical cyclone case shown in Figure 15 or either of the unrealistic ensembles. In this case, the five largest-sized wind farms produced a noticeable change to the midlatitude cyclone.

Multiple cases show a relationship between the magnitude of cyclone minimum sea-level pressure and the wind

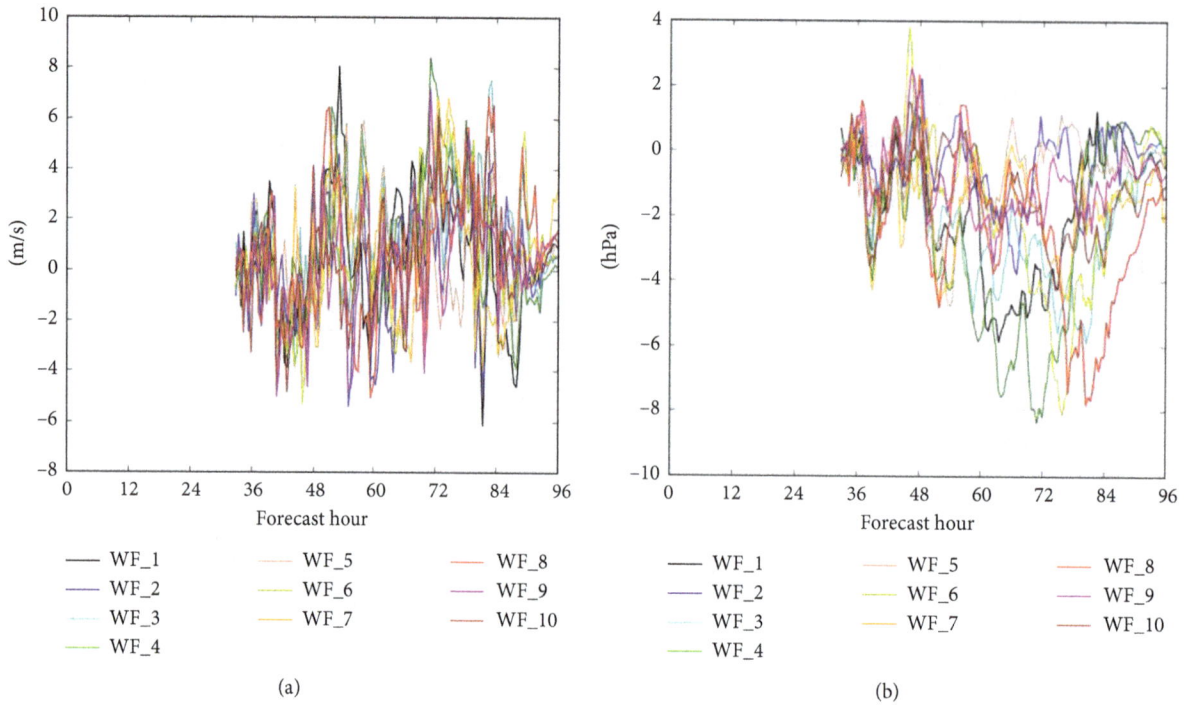

Figure 12: Cyclone maximum 10 m wind speed (a) and minimum sea-level pressure perturbation (b) for the 3 October 2013 tropical cyclone case.

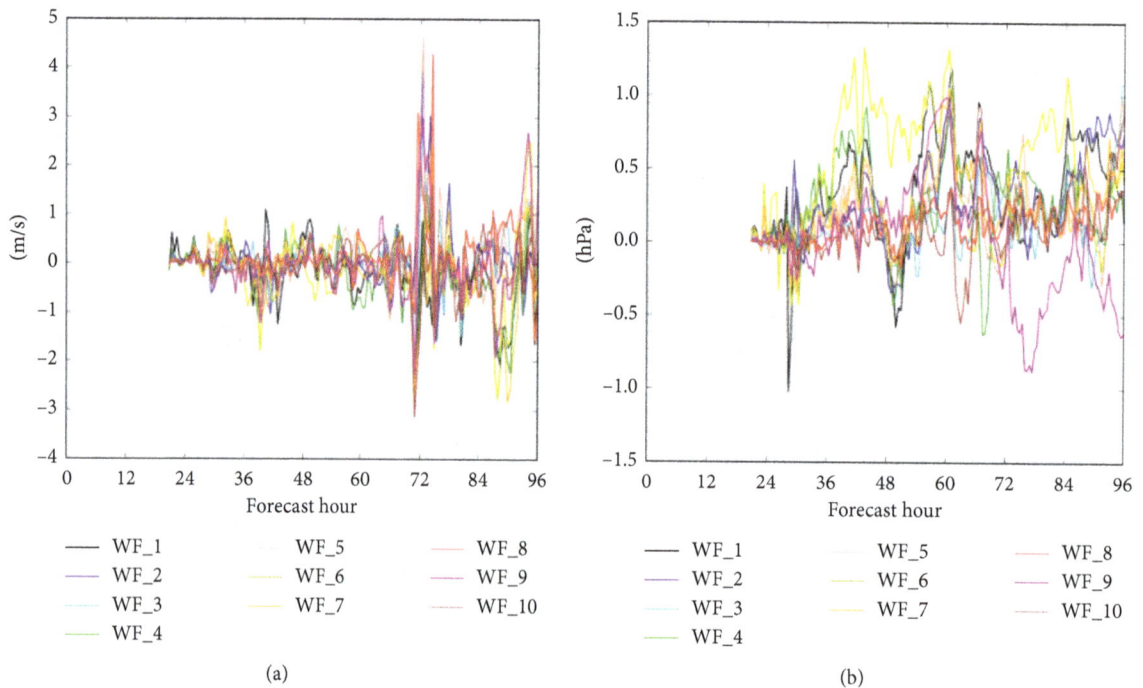

Figure 13: Cyclone maximum 10 m wind speed (a) and minimum sea-level pressure perturbation (b) for the 26 April 2014 midlatitude cyclone case.

farm size—five of the six midlatitude cases showed a significant p value (equal to or less than 0.05), while none of the tropical cyclone cases did. The average p value for midlatitude cyclones is 0.1043, while the average p value for tropical cyclones is 0.5229, shown in Tables 3 and 4.

Additionally, the coefficient of determination (r^2) shows many cases where the data fit a linear regression well. The average r^2 value for midlatitude cyclones is 0.5900, while average r^2 value for tropical cyclones is 0.0961, also shown in Tables 3 and 4. Again, these statistics are calculated for each

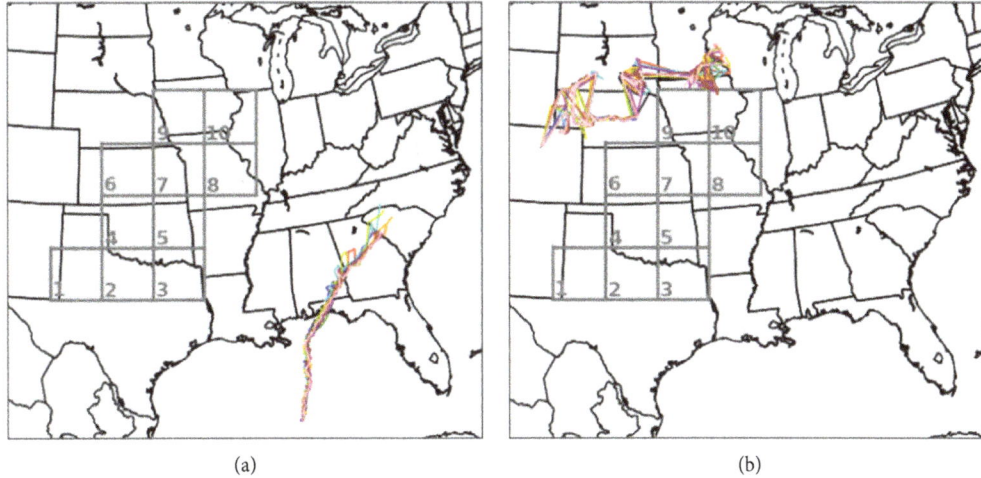

FIGURE 14: Cyclone tracks for the control (black) and wind farm location ensemble members (color) for the 3 October 2013 tropical cyclone (a) and 26 April 2014 midlatitude cyclone (b). Wind farm locations are outlined in gray.

case (six midlatitude and four tropical), among all ten members of the wind farm size ensemble, and averaged among all cases. The r^2 value is simply a linear regression and therefore is a measure of how consistently the perturbation magnitude is changed as the wind farm size changes. The p value here indicates how significant this linear regression is compared to no change at all. These statistical tests are also applied to all other metrics included in the study and are also summarized in Tables 3 and 4.

Another important metric for determining cyclone strength, especially tropical cyclones, is wind speed. The maximum 10 m wind speed increased in three of the four tropical cyclone cases due to a wind farm in the Central United States. Among these four cases, increasing the size of the wind farm increased wind speeds in two cases and decreased the maximum wind speed in the other two cases. Four of the six midlatitude cyclones observed increases in maximum wind speed with a wind farm in the model. Also, five of the six midlatitude cyclones realized greater maximum wind speed with increasing wind farm size. The average r^2 value for midlatitude cyclones is 0.2984, while the average r^2 value for tropical cyclones is 0.0803. The average p value for midlatitude cyclones is 0.2566, while the average p value for tropical cyclones is 0.3487.

When considering weather that has a large impact on the safety and protection of life and property, rainfall and flooding rank among the most powerful ones. Here, we analyze the maximum 30-minute accumulated precipitation at the same grid point within the cyclone. The maximum 30-minute accumulated precipitation is seen to increase in two tropical cyclone cases and additionally in another only for the largest wind farms. Among these four cases, increasing the size of the wind farm increased maximum 30-minute accumulated precipitation in two cases, and the other two cases saw a decrease in maximum 30-minute accumulated precipitation with larger wind farms. Three of the six midlatitude cyclones observed increases in maximum 30-minute accumulated precipitation with a wind farm in the model, and the other three cases realized increases to

TABLE 2: Wind farm size ensemble member sizes.

Dimensions	Number of turbines
20 km × 20 km	1,296
30 km × 30 km	2,916
40 km × 40 km	5,184
50 km × 50 km	8,100
70 km × 70 km	15,876
110 km × 110 km	39,204
150 km × 150 km	72,900
210 km × 210 km	142,884
300 km × 300 km	291,600
330 km × 330 km	352,836

maximum precipitation only for the largest wind farms. Also, five of the six midlatitude cyclones realized greater maximum 30-minute accumulated precipitation with increasing wind farm size. The average r^2 value for midlatitude cyclones is 0.3713 (Table 3), while the average r^2 value for tropical cyclones is 0.0408 for the maximum 30-minute accumulated precipitation metric (Table 4). The average p value for midlatitude cyclones is 0.1000, while the average p value for tropical cyclones is 0.7128.

It is important to demonstrate the significance of the perturbation magnitudes found in this study. The perturbations found here exhibit magnitudes of nearly 8 hPa for sea-level pressure and roughly 8 m/s for surface wind speed. The typical variability of sea-level pressure and surface wind speed, either temporally or spatially, exhibits perhaps a 50 hPa and 20 m/s range, respectively. Thus, the perturbations modeled here show a substantial fraction (greater than 10%) of the natural range of variability of these variables and can certainly be considered significant as they achieved these magnitudes over a 4-day simulation window.

3.5. Unrealistic Perturbations to Nonlocal cyclones. To determine whether the results above are due to the wind farm and not the unrealistic effects observed in the model, the

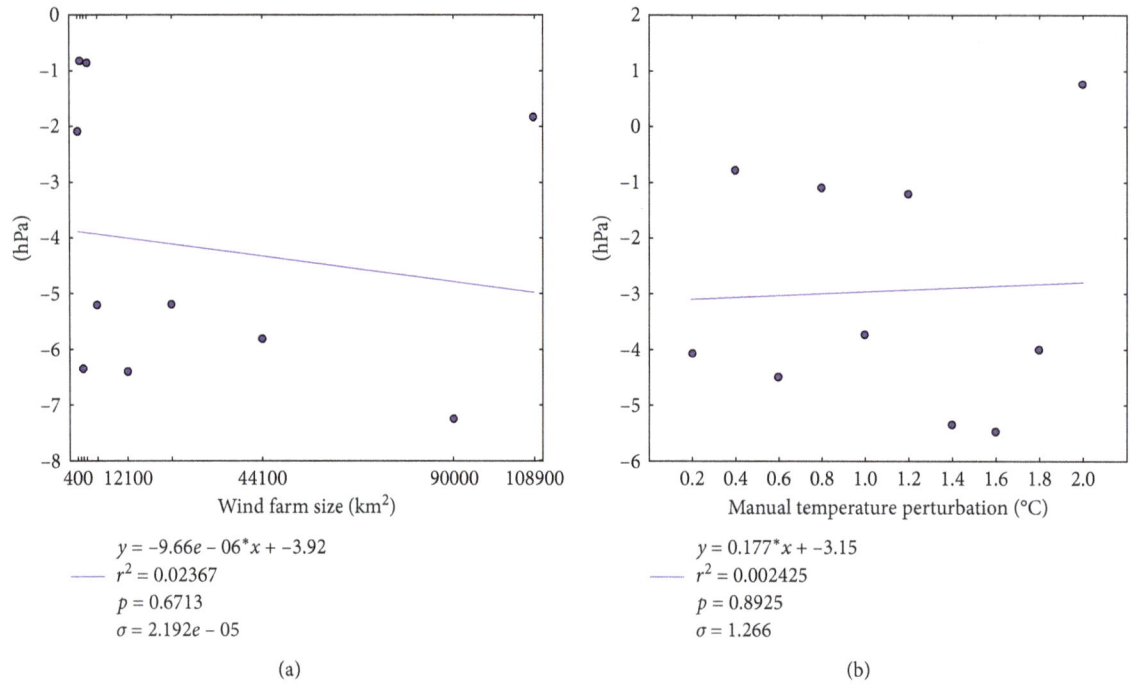

$y = -9.66e - 06^*x + -3.92$
$r^2 = 0.02367$
$p = 0.6713$
$\sigma = 2.192e - 05$

(a)

$y = 0.177^*x + -3.15$
$r^2 = 0.002425$
$p = 0.8925$
$\sigma = 1.266$

(b)

FIGURE 15: Largest perturbation of cyclone minimum sea-level pressure for the ten wind farm sizes (a) and unrealistic effects ensemble (b) for the 3 October 2013 tropical cyclone case.

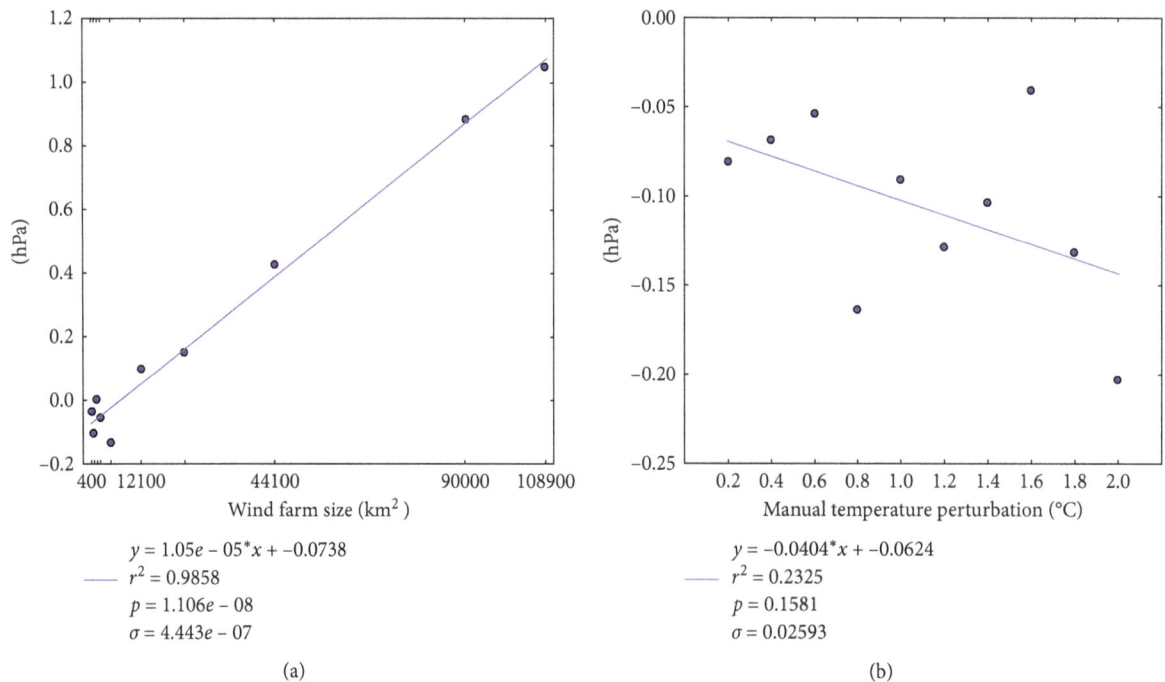

$y = 1.05e - 05^*x + -0.0738$
$r^2 = 0.9858$
$p = 1.106e - 08$
$\sigma = 4.443e - 07$

(a)

$y = -0.0404^*x + -0.0624$
$r^2 = 0.2325$
$p = 0.1581$
$\sigma = 0.02593$

(b)

FIGURE 16: Largest perturbation of cyclone minimum sea-level pressure for the ten wind farm sizes (a) and unrealistic effects ensemble (b) for the 26 April 2014 midlatitude cyclone case.

same analysis as shown above is also applied to the ensemble of ten stratospheric temperature perturbations in the northeastern part of the domain. If the coefficient of determination (r^2) is lower and p value is higher for this ensemble when compared to the wind farm size ensemble results, we can determine that, in fact the results are due to

the wind farm in the model. This is because the same relative unrealistic perturbations that occur throughout the domain with the wind farms included in the model, including the cyclones in question, are also observed when a perturbation is made to a grid cell in the stratosphere. This technique has been used in previous works such as [36].

TABLE 3: Average coefficient of determination (r^2), average p value with the ratio of significant cases (95% confidence), and largest perturbation to cyclone among midlatitude cyclones.

Metric	Average r^2	Average p value (ratio of significance)	Largest perturbation
Minimum mean sea-level pressure	0.5900	0.1043 (5/6)	+1.047 hPa
Maximum 10 m wind speed	0.2984	0.2566 (2/6)	+4.1922 m/s
Maximum 30-minute accumulated precipitation	0.3713	0.1000 (2/6)	+15.3757 mm
Maximum 2 m temperature	0.4898	0.1513 (3/6)	+0.520°C
Minimum 2 m temperature	0.1533	0.4678 (1/6)	−0.273°C
Maximum 2 m potential temperature	0.4328	0.2110 (2/6)	−0.139°C
Minimum 2 m potential temperature	0.2172	0.2976 (1/6)	+1.082°C
Maximum boundary layer height	0.2054	0.3974 (1/6)	+44.93 m
Maximum 2 m water vapor mixing ratio	0.3372	0.2459 (3/6)	+0.5536 g/kg

TABLE 4: Average coefficient of determination (r^2), average p value with the ratio of significant cases (95% confidence), and largest perturbation to cyclone among tropical cyclones.

Metric	Average r^2	Average p value (ratio of significance)	Largest perturbation
Minimum mean sea-level pressure	0.0961	0.5229 (0/4)	−7.269 hPa
Maximum 10 m wind speed	0.0803	0.6077 (0/4)	+4.5865 m/s
Maximum 30-minute accumulated precipitation	0.0408	0.7128 (0/4)	+12.415 mm
Maximum 2 m temperature	0.1953	0.3960 (1/4)	+0.6643°C
Minimum 2 m temperature	0.1672	0.3175 (0/4)	+0.9814°C
Maximum 2 m potential temperature	0.2299	0.2544 (1/4)	−0.698°C
Minimum 2 m potential temperature	0.2312	0.3884 (1/4)	+1.225°C
Maximum boundary layer height	0.1056	0.4518 (0/4)	−625.67 m
Maximum 2 m water vapor mixing ratio	0.1794	0.3826 (1/4)	+0.2981 g/kg

By comparison of the above results, Tables 5 and 6 display the difference in results between the wind farm size ensembles and stratospheric temperature perturbation ensembles. The greatest minimum sea-level pressure perturbations that occur with the ten potential temperature perturbations, the linear regression for these perturbations, and statistical values for all ten cases are included in Figures 15 and 16 for comparison to the earlier plots for the wind farm size ensembles.

For eight of the nine metrics regarding midlatitude cyclones, perturbations to the cyclone by the wind farms prove to be more significant, both in terms of the average coefficient of determination and the average p value, than the unrealistic perturbations. The only metric that this does not apply to is the cyclone minimum 2 m temperature. Additionally, all eight of these metrics saw more significant p values for the six midlatitude cyclone cases with the wind farm size ensemble. Observing the greatest perturbations to the nine metrics alone, the wind farm ensemble produced larger perturbations to all except for minimum sea-level pressure and minimum 2 m temperature.

Comparison of statistics between the two ensembles for tropical cyclones reveals that, for six of the nine metrics, perturbations to the cyclone by the wind farms prove to be more significant for both the average coefficient of determination and the average p value. The three metrics that were not more significant for the wind farm size ensemble are minimum sea-level pressure, maximum 30-minute accumulated precipitation, and maximum boundary layer height. However, the only metric that had cases with more significant p values is the maximum 2 m water vapor mixing ratio. There was no change in the number of significant

p values for six metrics, and the remaining two metrics of maximum 30-minute accumulated precipitation and maximum boundary layer height had more cases with significant p values for the unrealistic perturbation ensemble versus the wind farm size ensemble. Examining the greatest perturbations shows larger perturbations from wind farms to all of the metrics except for maximum 10 m wind speed and maximum 30-minute accumulated precipitation.

Generally, higher coefficients of determination and lower p values occur for the metric differences between the wind farm size ensemble and the control run than for the metric differences between the stratospheric temperature perturbations ensemble and the control run. Therefore, the observed perturbations to the cyclones are more dependent on the size of the wind farm than the magnitude of temperature perturbation made in the stratosphere far away from the cyclone. The metrics where we cannot make this claim are the minimum 2 m temperature for midlatitude cyclones and minimum sea-level pressure, maximum 30-minute accumulated precipitation, and maximum boundary layer height for tropical cyclones.

4. Summary and Conclusions

With the increase in the number of and spatial extent of wind farms in the United States, an important potential consequence is an inadvertent change in weather. The main goal of this study is to determine whether wind farm-induced perturbations can evolve over periods of days, and over areas of thousands of square kilometers, to modify specific atmospheric features that have large impacts on society and the environment. This research used an

TABLE 5: Difference between the wind farm ensemble and stratospheric temperature perturbation ensemble in the average coefficient of determination (r^2), average p value, and the ratio of significant cases for perturbations to midlatitude cyclones.

Metric	Average r^2	Average p value (ratio of significance)
Minimum mean sea-level pressure	+0.4758	−0.2797 (+5/6)
Maximum 10 m wind speed	+0.1385	−0.2593 (+1/6)
Maximum 30-minute accumulated precipitation	+0.2502	−0.2995 (+2/6)
Maximum 2 m temperature	+0.3932	−0.2858 (+3/6)
Minimum 2 m temperature	−0.0547	+0.1472 (0/6)
Maximum 2 m potential temperature	+0.3641	−0.2758 (+2/6)
Minimum 2 m potential temperature	+0.0730	−0.1464 (+1/6)
Maximum boundary layer height	+0.1489	−0.2304 (+1/6)
Maximum 2 m water vapor mixing ratio	+0.2140	−0.2205 (+2/6)

TABLE 6: Difference between the wind farm ensemble and stratospheric temperature perturbation ensemble in the average coefficient of determination (r^2), average p value, and the ratio of significant cases for perturbations to tropical cyclones.

Metric	Average r^2	Average p value (ratio of significance)
Minimum mean sea-level pressure	−0.0170	+0.0545 (0/4)
Maximum 10 m wind speed	+0.0701	−0.1867 (0/4)
Maximum 30-minute accumulated precipitation	−0.0731	+0.1809 (−1/4)
Maximum 2 m temperature	+0.0539	−0.1004 (0/4)
Minimum 2 m temperature	+0.0715	−0.1750 (0/4)
Maximum 2 m potential temperature	+0.0199	−0.1099 (0/4)
Minimum 2 m potential temperature	+0.017	−0.0044 (0/4)
Maximum boundary layer height	−0.1276	+0.0057 (−1/4)
Maximum 2 m water vapor mixing ratio	+0.0696	−0.0963 (+1/4)

ensemble approach with the WRF mesoscale model and the Fitch wind farm parameterization to quantify the sensitivity of meteorological variables to the presence of wind farms.

The results show that wind farms located in the Central United States were in fact capable of significantly altering midlatitude cyclones, but not tropical cyclones, and can cause large perturbations up to hundreds of kilometers away, including perturbations to the minimum cyclone pressure of 1 hPa, maximum 2 m temperature of 0.5°C, maximum 30-minute accumulated precipitation of 15 mm, and maximum 2 m wind speed of 4 m/s. Wind farms were shown to both increase and decrease the strength of midlatitude cyclones, and the outcome is likely due to the positioning of the wind farm with respect to the location of cyclogenesis. The majority of midlatitude cyclone cases show that wind farms placed nearer to or more upstream of the area of cyclogenesis lead to negative pressure perturbations and positive wind speed perturbations. Wind farms placed downstream tend to result in positive pressure perturbations. This implies that wind farms upstream of the region of cyclogenesis increase the strength of the cyclone, while wind farms downstream of cyclogenesis decrease the cyclone strength. Comparing the results of midlatitude cyclone perturbations by wind farms to the perturbations by unrealistic effects, it can be seen that the wind farm-induced perturbations are substantially more statistically significant than the unrealistic effects in general. This gives us confidence that results presented are truly due to the wind farm and not a result of the unrealistic noise.

As this work answers the question of whether wind farms can modify synoptic scales at timescales of days, it is impossible to conclude that such a systematic evolution is true more generally given the small sample size used here. Future work planned given the motivation gained in this study will aim to discover whether such systematic changes are robust.

The average coefficient of determination and the average p value among midlatitude cyclones were more significant for all metrics except for cyclone minimum 2 m temperature. Additionally, those metrics saw more significant p values when comparing wind farm size ensemble runs versus the unrealistic perturbation runs. The average coefficient of determination and the average p value among tropical cyclones were more significant for six of nine metrics, however by a relatively small amount. Also, the only metric that saw significant p values was the maximum 2 m water vapor mixing ratio.

When modeling nonlocal wind farm impacts on tropical cyclones, confidence in perturbation results is low. Tropical cyclones inherently have more moisture and convection than midlatitude cyclones. With it being shown that unrealistic effects were occurring in the model due to the rapid growth of noise in the moist physics, the ensemble of unrealistic perturbations showed that these wind farm modifications were not more statistically significant than the unrealistic perturbations to the tropical cyclone. Therefore, the signal obtained from tropical cyclone modifications primarily lies within the noise created by the model. More generally, the primary mechanism driving nonlocal inadvertent weather modification demonstrated here involves an initial local modification, followed by subsequent growth and evolution of those perturbations downstream similar to the processes examined in prior sensitivity studies of cyclones. This further explains the significance behind the relationship between midlatitude cyclones, which evolved near the wind farm locations, and the lack of such a relationship with regard to tropical cyclones, which did not.

It has been determined here that wind farms can significantly modify nonlocal midlatitude cyclones. In turn, this motivates future studies to better define the degree of nonlocal inadvertent modification of cyclones and whether these are systematic changes, and this is a planned next step in this research. The approach taken here can also be applied to other weather features, such as fronts or convective initiation, to obtain a greater understanding of how wind farms are altering the broader atmospheric state. This understanding is very applicable more broadly to forecasting such events, as large changes due to wind farms would suggest that forecast skill can depend on simulating the effects of wind farms by numerical weather prediction models. Additionally, the roles of different wind farm/turbine configurations in inadvertent weather modification are another avenue of future research. The size of future wind farms is relatively unknown, though, as it is dependent on public opinion of wind energy and the willingness of property owners to lease their land to utility companies for the placement of wind turbines. Moreover, our results may not apply generally to a range of coarser turbine spacings, which should be studied in future research. Finally, adjoint or ensemble sensitivity techniques can be utilized to better estimate the largest perturbations to high-impact weather features by first locating regions of high sensitivity and then placing an ensemble of wind farm sizes in that location. This approach would effectively determine "worst-case scenarios" for how wind farms might impact not only weather features but society and the environment as well, toward the responsible management of these effects in the future.

Disclosure

This paper was presented at the Workshop on Sensitivity Analysis and Data Assimilation in Meteorology and Oceanography (2015) and the Conference on Planned and Inadvertent Weather Modification (2015). Matthew J. Lauridsen is currently at Fleet Numerical Meteorology and Oceanography Center, Monterey, CA.

Conflicts of Interest

The authors declare that they have no conflicts of interest.

References

[1] L. Zhou, Y. Tian, S. Baidya Roy, C. Thorncroft, L. F. Bosart, and Y. Hu, "Impacts of wind farms on land surface temperature," *Nature Climate Change*, vol. 2, no. 7, pp. 539–543, 2012.

[2] L. Zhou, Y. Tian, H. Chen, Y. Dai, and R. A. Harris, "Effects of topography on assessing wind farm impacts using MODIS data," *Earth Interactions*, vol. 17, no. 13, pp. 1–18, 2013.

[3] M. B. Christiansen and C. B. Hasager, "Wake effects of large offshore wind farms identified from satellite SAR," *Remote Sensing of Environment*, vol. 98, no. 2-3, pp. 251–268, 2005.

[4] S. Baidya Roy, S. W. Pacala, and R. L. Walko, "Can large wind farms affect local meteorology?," *Journal of Geophysical Research*, vol. 109, p. 6, 2004.

[5] D. B. Barrie and D. B. Kirk-Davidoff, "Weather response to a large wind turbine array," *Atmospheric Chemistry and Physics*, vol. 10, no. 2, pp. 769–775, 2010.

[6] B. H. Fiedler and M. S. Bukovsky, "The effect of a giant wind farm on precipitation in a regional climate model," *Environmental Research Letters*, vol. 6, article 045101, 2011.

[7] A. C. Fitch, J. B. Olson, J. K. Lundquist et al., "Local and mesoscale impacts of wind farms as parameterized in a mesoscale NWP model," *Monthly Weather Review*, vol. 140, no. 9, pp. 3017–3038, 2012.

[8] D. W. Keith, J. F. Decarolis, D. C. Denkenberger et al., "The influence of large-scale wind power on global climate," *Proceedings of the National Academy of Sciences of the United States of America*, vol. 101, no. 46, pp. 16115–16120, 2004.

[9] D. B. Kirk-Davidoff and D. W. Keith, "On the climate impact of surface roughness anomalies," *Journal of the Atmospheric Sciences*, vol. 65, no. 7, pp. 2215–2234, 2008.

[10] C. Wang and R. G. Prinn, "Potential climatic impacts and reliability of very large-scale wind farms," *Atmospheric Chemistry and Physics*, vol. 10, no. 4, pp. 2053–2061, 2010.

[11] S Baidya Roy, "Simulating impacts of wind farms on local hydrometeorology," *Journal of Wind Engineering and Industrial Aerodynamic*, vol. 99, no. 4, pp. 491–498, 2011.

[12] A. C. Fitch, J. B. Olson, and J. K. Lundquist, "Parameterization of wind farms in climate models," *Journal of Climate*, vol. 26, no. 17, pp. 6439–6458, 2013.

[13] E. N. Lorenz, "Deterministic nonperiodic flow," *Journal of the Atmospheric Sciences*, vol. 20, no. 2, pp. 130–141, 1963.

[14] F. Zhang, C. Snyder, and R. Rotunno, "Effects of moist convection on mesoscale predictability," *Journal of the Atmospheric Sciences*, vol. 60, no. 9, pp. 1173–1185, 2003.

[15] M. S. Peng and C. A. Reynolds, "Sensitivity of tropical cyclone forecasts as revealed by singular vectors," *Journal of the Atmospheric Sciences*, vol. 63, no. 10, pp. 2508–2528, 2006.

[16] C. A. Reynolds, M. S. Peng, and J.-H. Chen, "Recurving tropical cyclones: singular vector sensitivity and downstream impacts," *Monthly Weather Review*, vol. 137, no. 4, pp. 1320–1337, 2009.

[17] B. C. Ancell and C. F. Mass, "Structure, growth rates, and tangent linear accuracy of adjoint sensitivities with respect to horizontal and vertical resolution," *Monthly Weather Review*, vol. 134, no. 10, pp. 2971–2988, 2006.

[18] R. H. Langland, R. L. Elsberry, and R. M. Errico, "Evaluation of physical processes in an idealized extratropical cyclone using adjoint sensitivity," *Quarterly Journal of the Royal Meteorological Society*, vol. 121, no. 526, pp. 1349–1386, 1995.

[19] T. Vukicevic and K. Raeder, "Use of an adjoint model for finding triggers for Alpine lee cyclogenesis," *Monthly Weather Review*, vol. 123, no. 3, pp. 800–816, 1995.

[20] X. Zou, Y.-H. Kuo, and S. Low-Nam, "Medium-range prediction of an extratropical oceanic cyclone: impact of initial state," *Monthly Weather Review*, vol. 126, no. 11, pp. 2737–2763, 1998.

[21] American Meteorological Society Council, "Inadvertent weather modification: an information statement of the American Meteorological Society, AMS," 2010, https://www.ametsoc.org/ams/index.cfm/about-ams/ams-statements/statements-of-the-ams-in-force/inadvertent-weather-modification/.

[22] R. M. Errico, "What is an adjoint model?," *Bulletin of the American Meteorological Society*, vol. 78, no. 11, pp. 2577–2591, 1997.

[23] W. C. Skamarock and Coauthors, "A description of the ad-

vanced research WRF version 3," NCAR Tech. Note NCAR/
TN-475+STR, 2008.

[24] G. Thompson, P. R. Field, R. M. Rasmussen, and W. D. Hall,
"Explicit forecasts of winter precipitation using an improved
bulk microphysics scheme. Part II: implementation of a new
snow parameterization," *Monthly Weather Review*, vol. 136,
no. 12, pp. 5095–5115, 2008.

[25] E. J. Mlawer, S. J. Taubman, P. D. Brown, M. J. Iacono, and
S. A. Clough, "Radiative transfer for inhomogeneous atmospheres: RRTM, a validated correlated-k model for the
longwave," *Journal of Geophysical Research: Atmospheres*,
vol. 102, no. D14, pp. 16663–16682, 1997.

[26] J. Dudhia, "Numerical study of convection observed during
the Winter Monsoon Experiment using a mesoscale two-
dimensional model," *Journal of the Atmospheric Sciences*,
vol. 46, no. 20, pp. 3077–3107, 1989.

[27] Z. Janjic, "The step-mountain eta coordinate model: further
developments of the convection, viscous sublayer, and turbulence closure schemes," *Monthly Weather Review*, vol. 122,
no. 5, pp. 927–945, 1994.

[28] F. Chen and J. Dudhia, "Coupling an advanced land
surface–hydrology model with the Penn State–NCAR MM5
modeling system. Part II: preliminary model validation,"
Monthly Weather Review, vol. 129, no. 4, pp. 587–604, 2001.

[29] M. Nakanishi and H. Niino, "An improved Mellor-Yamada
level-3 model: its numerical stability and application to
a regional prediction of advection fog," *Boundary-Layer
Meteor*, vol. 119, no. 2, pp. 397–407, 2006.

[30] M. Nakanishi and H. Niino, "Development of an improved
turbulence closure model for the atmospheric boundary
layer," *Journal of the Meteorological Society of Japan*, vol. 87,
no. 5, pp. 895–912, 2009.

[31] J. S. Kain, "The Kain–Fritsch convective parameterization: an
update," *Journal of Applied Meteorology*, vol. 43, no. 1,
pp. 170–181, 2004.

[32] F. Zhang, A. M. Odins, and J. W. Neilsen-Gammon, "Mesoscale predictability of an extreme warm-season precipitation event," *Weather and Forecasting*, vol. 21, no. 2,
pp. 149–166, 2006.

[33] J. P. Hacker, "Spatial and temporal scales of boundary layer
wind predictability in response to small-amplitude land
surface uncertainty," *Journal of the Atmospheric Sciences*,
vol. 67, no. 1, pp. 217–233, 2010.

[34] C. Hohenegger and C. Schar, "Predictability and error growth
dynamics in cloud-resolving models," *Journal of the Atmospheric Sciences*, vol. 64, no. 12, pp. 4467–4478, 2007.

[35] D. Hodyss and S. Majumdar, "The contamination of 'data
impact' in global models by rapidly growing mesoscale in-
stabilities," *Quarterly Journal of the Royal Meteorological
Society*, vol. 133, pp. 1865–1875, 2007.

[36] B. C. Ancell, A. Bogusz, M. J. Lauridsen, and C. J. Nauert,
"Seeding chaos: the dire consequences of numerical noise in
NWP perturbation experiments," *Bulletin of the American
Meteorological Society*, vol. 99, no. 3, pp. 615–628, 2018.

[37] M. J. Lauridsen, "Nonlocal inadvertent weather modification
associated with wind farms in the Central United States,"
Masters Thesis, Texas Tech University, 2015.

[38] R. Lorenz, A. J. Pitman, and S. A. Sisson, "Does Amazonian
deforestation cause global effects; can we be sure?," *Journal of
Geophysical Research: Atmospheres*, vol. 121, no. 10, pp. 5567–
5584, 2016.

12

Case Study of Ground-Based Glaciogenic Seeding of Clouds over the Pyeongchang Region

Ha-Young Yang⬡,[1] Ki-Ho Chang⬡,[1] Sanghee Chae⬡,[1] Eunsil Jung,[2]
Seongkyu Seo⬡,[1] Jin-Yim Jeong⬡,[1] Jung-Ho Lee,[1] Yonghun Ro,[1] and Baek-Jo Kim[1]

[1]*Applied Meteorology Research Division, National Institute of Meteorological Sciences, Seogwipo 63568, Republic of Korea*
[2]*School of Constructional Disaster Prevention and Environmental Engineering, Kyungpook National University,
Sangju 37224, Republic of Korea*

Correspondence should be addressed to Ki-Ho Chang; khchang@korea.kr

Academic Editor: István Geresdi

Ground-based glaciogenic seeding experiments were conducted at the Daegwallyeong Cloud Physics Observation Site (CPOS) from 2012 to 2015 for the target area Yongpyeong, which lies 9 km away. The preseeding (NOSEED) and seeding (SEED) periods were defined based on the simulation results of AgI concentration ($>10 \, \text{L}^{-1}$) in the Weather Research and Forecast (WRF) model with the modified Morrison scheme in microphysics. It was difficult to determine whether snow enhancement via seeding occurred over the entire target area due to uncertainties associated with limitations such as observations and numerical model based on only two points (seeding and target sites). However, in three of four cases, the vertical reflectivity from micro rain radar, total concentration, and average size of snow particles observed at PARSIVEL and precipitation increased in the seeding effect time. In two of four cases, the simulated increased precipitation during the seeding effect time was also observed. In one case that did not show changes after seeding, it is analyzed that a sufficient cloud depth was not supplied to the seeding region due to the blocking effect of the Taebaek Mountains.

1. Introduction

Clouds in the atmosphere are potential water resources; however, only a small proportion of clouds fall as precipitation. To secure water resources, cloud seeding can be used to enhance precipitation and accelerate precipitation efficiency by injecting artificial ice nuclei (IN) or cloud condensation nuclei into targeted clouds [1]. Many studies have been conducted since the first successful field experiment by Schaefer [2]. Experiments on winter snow enhancement through glaciogenic seeding from the upwind side of mountainous areas have shown a high probability of success [3, 4].

It is difficult to distinguish between the orographic increase in precipitation and precipitation increased by seeding. Microphysical studies through direct observations using airborne or surface observation may be essential for the verification of the seeding effect. Super and Heimbach

[5] and Super and Boe [6] performed aircraft sampling above the target area after ground-based inserting IN into mountain clouds during winter in the Rocky Mountains. They have shown that ice particle concentration increased in seeded zones compared to control zones. Direct evidences for the changes in precipitation rate, ice crystal habits, and concentration after seeding were confirmed on the ground.

In the silver iodide (AgI) Seeding Cloud Impact Investigation (ASCII) pilot project, Geerts et al. [7] confirmed an increase in radar reflectivity at the boundary layer (0.5–1.0 km) of mountainous areas during seeding. Pokharel et al. [8] separated upstream (control) and downstream (target) regions based on the location of the AgI generator and defined the preseeding (NOSEED) and seeding (SEED) periods. They confirmed an increase in the radar reflectivity during SEED compared with the NOSEED period.

Ground-based snow enhancement experiments at the Daegwallyeong Cloud Physics Observation Site (CPOS) in

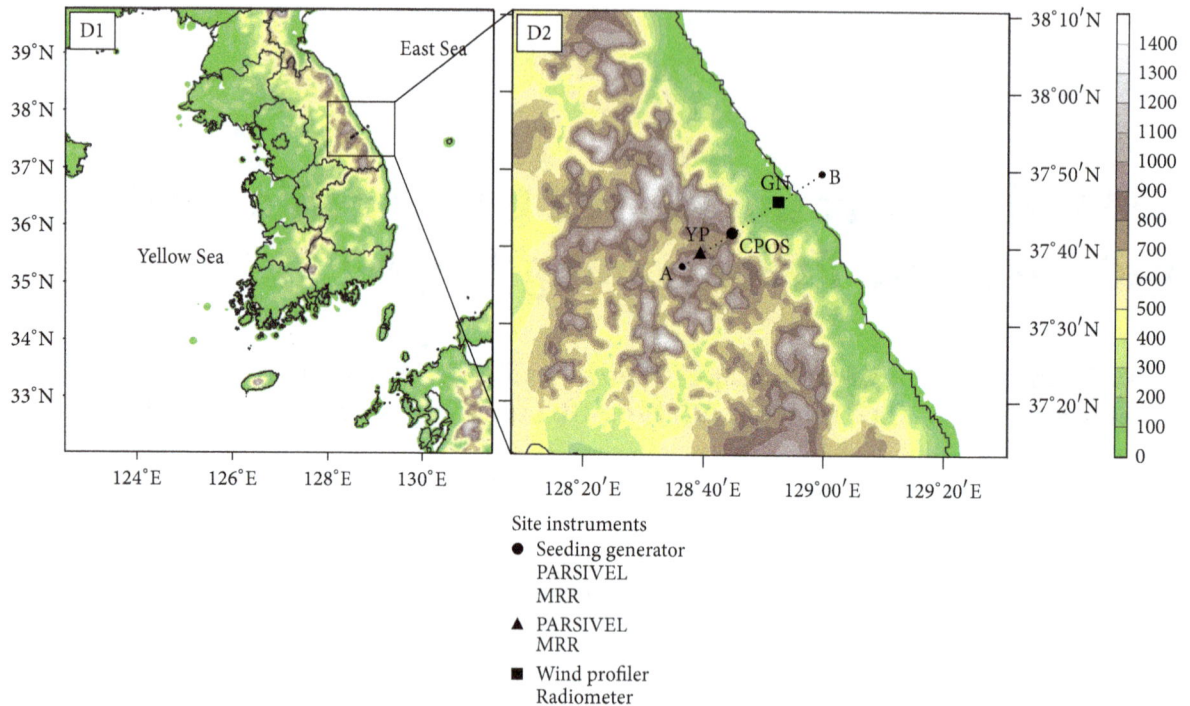

FIGURE 1: The model domain and topography of the experimental region. The triangle, circle, and square indicate the locations of the Yongpyeong observatory (YP), Cloud Physics Observation Site (CPOS), and Gangwon Regional Office of Meteorology (GN), respectively. The dashed line from A to B represents the horizontal cross section passing through YP, CPOS, and GN.

South Korea have been conducted continuously since 2006, targeting Yongpyeong, where the alpine skiing competition of the 2018 Winter Olympics was held. Lee et al. [9] verified an increase in snowfall at CPOS after ground-based AgI seeding in 2006. However, they examined the change in snowfall via rapidly forced condensation processes (~1 min) at the seeding site CPOS and did not analyze the seeding effects in the downwind target area to show the extension of the seeding effect.

In this study, we examined changes in clouds and precipitation in the downwind target area after ground-based AgI seeding, following the methods of Geerts et al. [7] and Pokharel et al. [8]. But, by the modified WRF (Weather, Research and Forecast) model, the simulated AgI dispersion was used to distinguish the periods unaffected (NOSEED) and affected (SEED) by seeding in the target region. For four seeding experiments, the observed and simulated changes of cloud and surface precipitation during the period between NOSEED and SEED were examined.

2. Methods

2.1. Experimental Design. Figure 1 shows the topography of the experimental area and the model domain. The domain for numerical simulations was set to the Korean Peninsula (domain 1, D1) and Gangwon (domain 2, D2), using a two-way grid nesting procedure to simulate D2 more precisely. The CPOS (37°41′N, 128°45′E, 843 m MSL), which was used as the seeding site, is located on the ridge of the Taebaek Mountain Ranges about 13 km southwest from the Gangwon

Regional Office of Meteorology (GN, 37°45′N, 128°53′E, 26 m MSL). The Yongpyeong observatory (YP, 37°38′N, 128°40′E, 772 m MSL) is about 9 km southwest from the CPOS. A-B in model domain 2 (D2) stands for a horizontal path of cross section, passing YP, CPOS, and GN. To minimize the possibility of orographic cloud and precipitation enhancement, the target area at YP was designed to be lower than the seeding point CPOS with an altitude difference of ~70 m.

Ground-based AgI seeding experiments were performed at CPOS from 2012 to 2015 by burning the 2% $AgI-NH_4I$-acetone solution, converting supercooled liquid water into ice crystals. In general, AgI is activated as IN at temperatures of −5°C or lower [10]. However, Blair [11] suggested that the $AgI-NH_4I$-acetone solution is activated as IN at warmer temperatures (−3°C) through a cloud chamber-wind tunnel experiment. Four experimental cases were selected in which, at a low temperature (<−3.5°C) and under the influence of northeasterly wind at CPOS, the AgI seeding amount was 0.63 g min^{-1}, the average liquid water content (1 h before experiment until the end of the experiment) was above 0.1 g m^{-3} (0.5–1 km) at GN. The observed meteorological information is listed in Table 1. Except for the missing EXP1 data, the average liquid water content at GN (1 h before seeding to the start of seeding) was 0.10–0.29 g m^{-3}. The height peak CPOS is 843 m MSL, and thus the LWC of incoming cloud is estimated in the vertical range of 500–1000 m. Stratiform clouds with sufficient liquid water content moved toward CPOS. Easterly winds of 5 m s^{-1} or lower flowed into CPOS during the seeding period, and AgI particles injected

TABLE 1: Summary of the ground-based glaciogenic seeding experiments during 2012–2015.

Exp. number	Location Date	GN LWC[1] (g m⁻³)	Cloud type[2]	Burning period (LST)	CPOS Temp.[3] (°C)	WS[3] (m s⁻¹)	WD[3] (deg)	Visibility[3] (m)	YP[4] NOSEED (LST)	SEED (LST)	Post-SEED (LST)	Enhanced rainfall[5] (mm)	Results MRR[6]	PARSIVEL[7] Total conc. (L⁻¹)	Diameter (mm)
EXP1	2012.02.25	N/A	StNs	1425–1455	−5	3.6	52	233	1345–1435	1435–1525	1525–1655	Yes (0.5)	Yes	N/A	N/A
EXP2	2013.03.13	0.29	StNs	1800–1900	−5.3	1.3	55	186	1705–1815	1815–1925	1925–2015	Yes (0.5)	Yes	2.3 → 4.5	0.63 → 0.72
EXP3	2014.01.04	0.28	St	1530–1600	−4.8	1	45	151	1455–1550	1550–1645	1645–1735	-	No	N/A	N/A
EXP4	2015.04.07	0.10	StAs	1510–1610	−3.7	N/A	64	123	1405–1520	1520–1635	1635–1725	Yes (0.5)	Yes	6.9 → 8.3	0.61 → 0.70

[1] Average liquid water content between 500 m and 1000 m MSL during the period from 1 h before turning the generator on to turning it off. [2] Cloud type based on human-eye observations during the period from 1 h before turning the generator on to turning it off. [3] Average weather conditions during the burning period. [4] SEED is the period during which more than 10⁻¹ L of AgI in the WRF simulation exist on the surface at YP. The NOSEED and SEED periods are equally long, whereas the post-SEED is the 50 min period after SEED as estimated by the SNPS at CPOS. [5] The enhanced precipitation by seeding is calculated as the difference between NOSEED and Tot-SEED (SEED + post-SEED). [6] "Yes" represents MRR reflectivity increased during Tot-SEED compared with that during NOSEED. [7] Variation in the average total concentration and diameter during NOSEED and Tot-SEED.

TABLE 2: Summary of model configurations.

	EXP1	EXP2	EXP3	EXP4
Date	25 Feb. 2012	13 Mar. 2013	4 Jan. 2014	7 Apr. 2015
Period (LST)	0900-2100	1200-0000	0900-2100	0900-2100
Domain	Domain 1 (D1)		Domain 2 (D2)	
Horizontal spacing	5 km		1 km	
Dimension (x, y, z)	$170 \times 170 \times 40$		$121 \times 106 \times 40$	
Time step (s)	15		5	
Model top	50 hPa			
Grid nesting	Two-way			
Initial & boundary layer	KLAPS (1 hour, 5 km $\times$ 5 km)			
Microphysics	Morrison scheme			
Planetary boundary layer	YSU PBL			
Land-surface model	Thermal diffusion scheme			
Land use data	USGS (e.g., topo_30s, soiltype_top_30s)			
Longwave radiation scheme	Rapid Radiative Transfer Model (RRTM)			
Shortwave radiation scheme	Goddard shortwave			

directly in the lower clouds (cold fog) with a visibility of ≤200 m at CPOS.

Numerical simulations of the four cases were conducted using the WRF mesoscale model, version 3.4, which is a fully compressible, nonhydrostatic, and primitive equation model [12]. The model configuration is listed in Table 2. The domain D2, used for the analyses of all the experiments, consists of 121 × 106 horizontal grid points with a grid spacing of 1 km and 40 vertical layers. Initial and boundary conditions for hindcast were set using 1 h Korea Local Analysis and Prediction System (KLAPS) data with a 5 km resolution. In this model, the Morrison microphysics scheme was modified to include the actual AgI release at a chosen surface point and ice nucleation process. A detailed description of the ice nucleation process is provided by Kim et al. [13] and Chae et al. [14].

2.2. Instruments. Several instruments were used both upwind and downwind of CPOS. Lee et al. [9] mentioned the importance of humidity in the upwind area as a factor influencing the seeding effect at CPOS. To examine the upwind weather conditions before the seeding, wind profiler (PCL1300, Degreane Horizon) and microwave radiometer (MWR; RPG-HATPRO, Radiometer Physics GmbH) data at GN were used. The wind profiler observes the upper-air wind speed and direction with high vertical resolutions by performing observations every 10 min [15]. The MWR receives microwave signals in 14 frequency bands from the atmosphere and represents measurements of brightness temperature. It may produce the sounding of temperature, humidity, water vapor amount, and liquid water content data.

At the seeding point CPOS, changes in weather conditions and aerosol due to seeding were observed with an automatic weather station (AWS, Vaisala) at 15 m above ground level and a scanning nanoparticle spectrometer

(SNPS; HCTm), respectively, throughout the experiment. The SNPS is connected to a differential mobility analyzer and condensation particle counter and measures the size distribution of 7–282 nm particles. At YP, which is in the downwind area, observation data from the micro rain radar (MRR; Metek) and a PARSIVEL disdrometer (PARSIVEL; OTT Hydromet) were used. Snow depth was not observed during the experimental period; therefore, the precipitation data obtained using a 0.5 mm tipping-bucket rain gauge (JY100097-2; Jinyang) were used for verification. The rain gauge for measuring the snowfall has a time delay of around 20 min due to internal heating (communication with GN). The PARSIVEL is a laser-based optical sensor used for the simultaneous measurement of liquid and solid precipitation particles [16, 17]. The MRR retrieves quantitative rain rate, drop size distribution, radar reflectivity, fall velocity of hydrometeors, and other rain parameters simultaneously on vertical profiles up to several kilometers above the radar [18]. To use the MRR data for snow observations, Maahn and Kollias [19] regarded the main signals from the Doppler power spectrum of raw data as the Doppler speed by snow and calculated the equivalent reflectivity after noise removal and dealiasing. In this study, we analyzed the reproduced data following their proposed MRR processing method.

3. Results

3.1. Description of Cases. Figure 2 shows the synoptic weather situation of the four cases from the Korea Meteorological Administration (KMA) website. At CPOS, air parcels in the seeding experiments move from the East Sea and frequent low clouds and fog occur because it is close to the east coast and there are steep slopes to the east [20]. When the synoptic chart over the Korean Peninsula is of a northern high

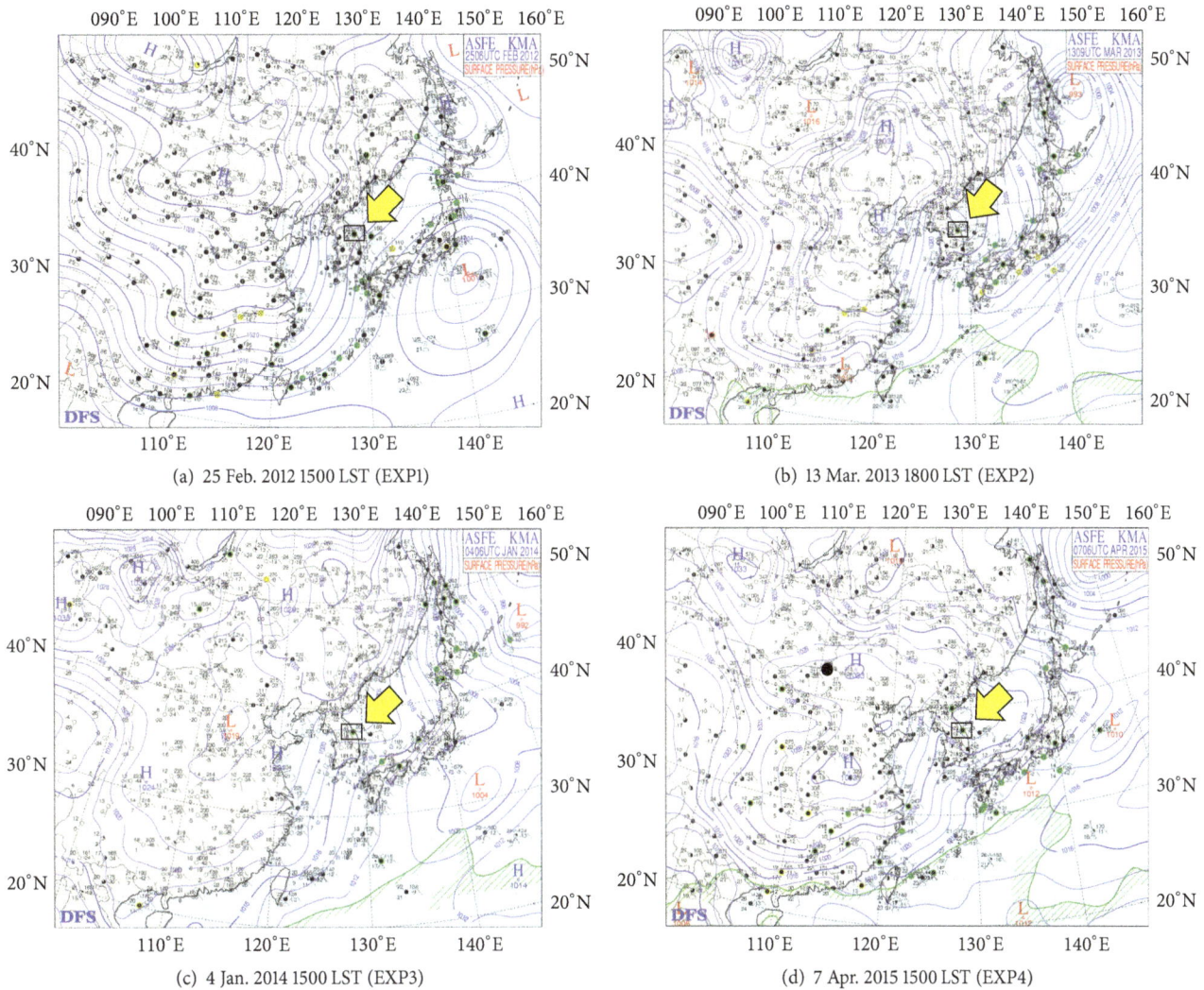

(a) 25 Feb. 2012 1500 LST (EXP1)

(b) 13 Mar. 2013 1800 LST (EXP2)

(c) 4 Jan. 2014 1500 LST (EXP3)

(d) 7 Apr. 2015 1500 LST (EXP4)

FIGURE 2: Synoptic weather charts for (a) EXP1, (b) EXP2, (c) EXP3, and (d) EXP4. The red box indicates the experimental region shown in Figure 1.

and southern low type, the long period of fog (low clouds) often occurs in the Daegwallyeong area due to the northeast airflow. This fog is effective for experiments of direct seeding into clouds because it moves upward via advection along the steep mountain of the Taebaek Mountain Ranges and flows into Daegwallyeong in a supercooled condition [21–23].

The EXP1 case exhibits a cold advection pattern (CA) with a typical high-west-low-east pressure pattern due to the effects of the migratory anticyclone in China and the trough in the Yeongdong region, South Korea, and northeasterly wind flows into the Yeongdong region (Figure 2(a)). Lee [24] confirmed that, in CA, the cold advection via northeasterly winds results in precipitation in the Yeongdong region in general. The EXP2, EXP3, and EXP4 cases exhibit the low-pressure patterns (LPs) that passed along the Japanese islands resulting in easterly airflow into the Taebaek mountains (Figures 2(b)–2(d)). In the LPs, an easterly wind with high

humidity lasting for 1-2 days produced easterly orographic low clouds in the Daegwallyeong region.

3.2. Simulated Results. We compared the simulated and observed precipitation to verify the performance of the WRF model. Figure 3 illustrates the 12 h accumulated simulated and observed precipitation over the simulation period of each case. In most cases, the simulated precipitation is slightly underestimated in the coastal area, overestimated in the mountainous area. However, the simulated precipitation pattern was reasonably consistent with the observations.

Figure 4 shows vertical cross sections of the AgI particle concentration, cloud water mixing ratio, temperature, and wind vectors at the 30 min after the AgI generator was turned on. AgI diffuses toward YP in all simulations. Cloud water mixing ratio more than $0.1\,g\,kg^{-1}$ was simulated above CPOS for all the cases, and northeasterly winds of $7–10\,m\,s^{-1}$ from CPOS to 1.5–2.0 km MSL were simulated from the CPOS

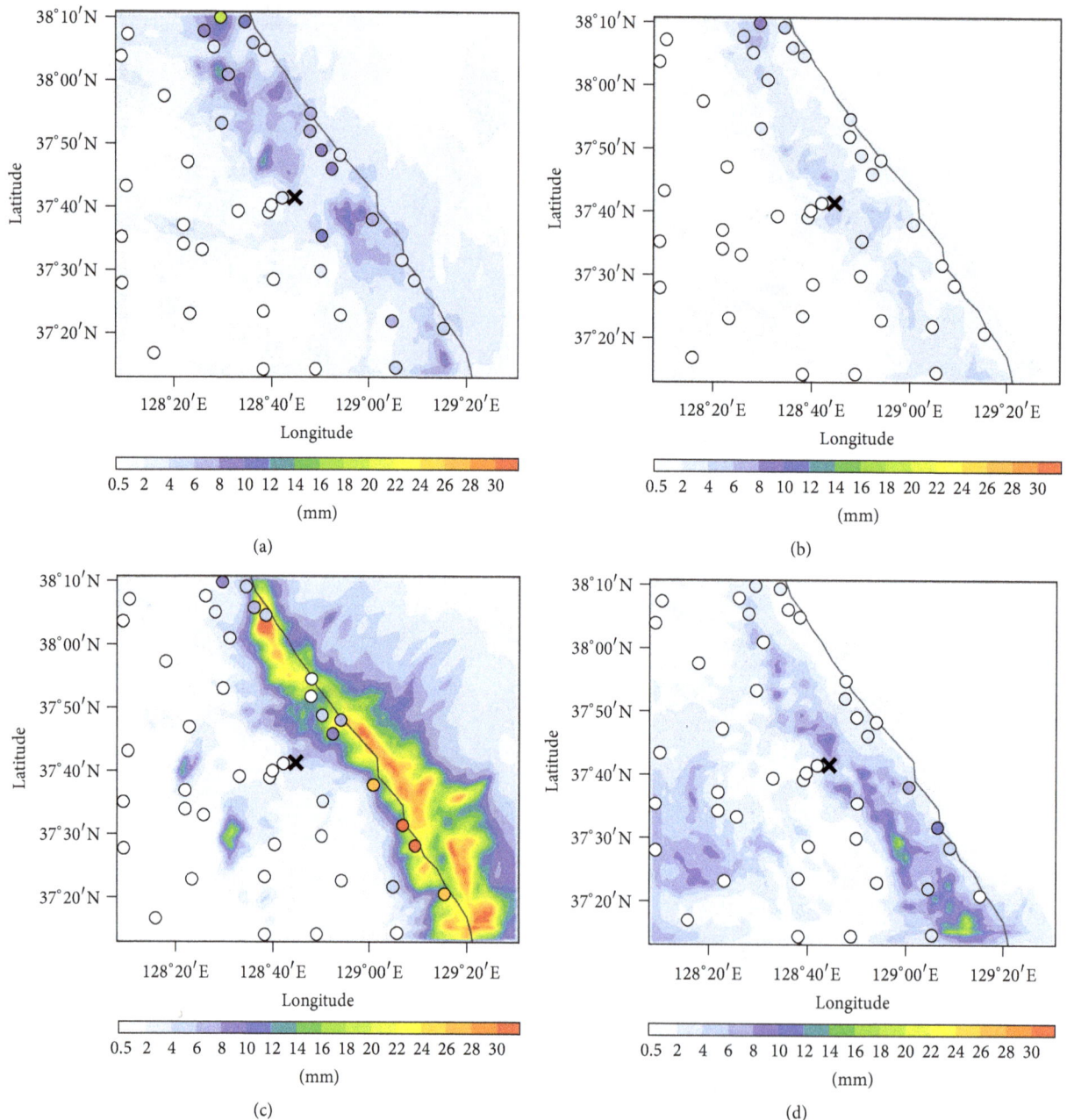

FIGURE 3: The 12 h accumulated precipitation (mm) of nonseeded simulation and rain gauges for (a) EXP1, (b) EXP2, (c) EXP3, and (d) EXP4. The simulated precipitation is color shaded, overlaid by the rain gauge precipitation in color-filled circles. The cross indicates the location of CPOS.

to YP in EXP1, EXP2, and EXP4. In EXP3, the relatively low wind speed of $5\,\mathrm{m\,s^{-1}}$ was simulated. The 9 h simulated vertical wind by this model for 1.5–2.0 km MSL at CPOS showed agreement with the observed wind by radiosonde at 1800 LST, 13 March 2013 [20]. Compared with the simulated mean wind speed for 1.5–2.0 km MSL at CPOS, the observed surface wind speed appears to be relatively very weak (Table 1) due to the roughness of mountain forests.

Under natural conditions, IN exists less than $1\,\mathrm{L^{-1}}$ [7]. The SEED period was defined in accordance with the simulation

results of the AgI concentration ($>10\,\mathrm{L^{-1}}$) at the surface of YP. This value was arbitrarily chosen value in the seeding simulation to induce some difference from the natural value. A period of the same duration as the SEED period before seeding was defined as NOSEED. Figure 5 shows the temporal behavior and residue of aerosol particles after seeding at CPOS. During the background measurements at CPOS, the accumulation mode was predominant, with a mean particle size of 110 nm. After the AgI generator was turned on at 1510 LST, aerosols of 10–100 nm were generated, and the total

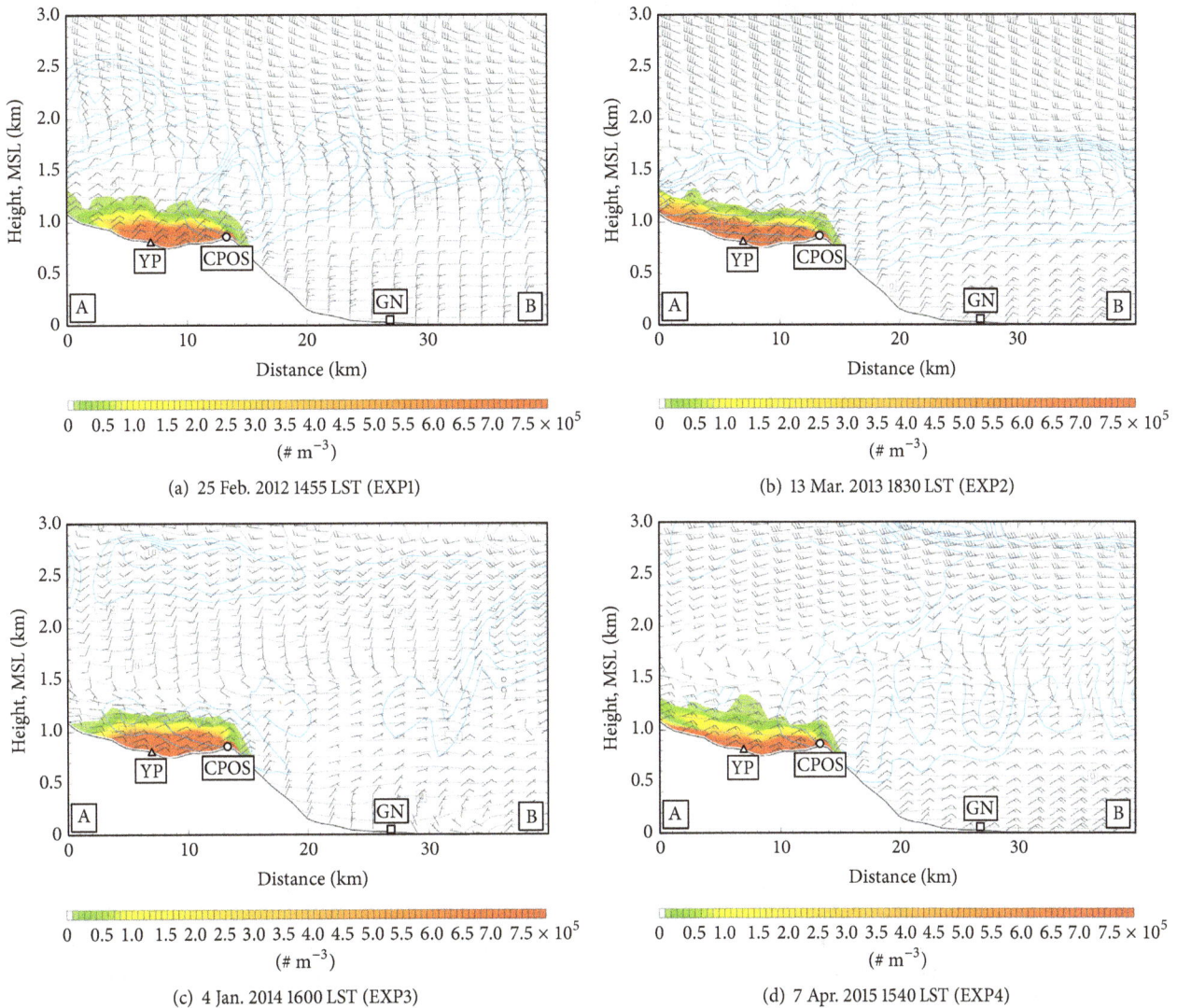

(a) 25 Feb. 2012 1455 LST (EXP1)

(b) 13 Mar. 2013 1830 LST (EXP2)

(c) 4 Jan. 2014 1600 LST (EXP3)

(d) 7 Apr. 2015 1540 LST (EXP4)

FIGURE 4: Vertical cross sections of the AgI particle number concentration (color shaded, m^{-3}), cloud water mixing ratio (blue solid line, $g\,kg^{-1}$), temperature (grey solid line, °C), and wind vectors at 30 min after seeding simulated by the WRF model. The A-B line is the same as shown in Figure 1.

concentration of aerosols increased to 22,000 cm^{-3}, reaching a maximum for particles around 40 nm. According to the results of Vali et al. [25], the dry AgI-NH_4I solution has a size range of 0.01–10 μm, and most aerosols exist in the range of 5–50 nm. Therefore, most of the aerosols observed after seeding seems to be the AgI particles. The particle concentrations steadily decreased after the generator was turned off and then returned to a similar state as the preseeding state at approximately 50 min after the end of seeding. The residue time (post-SEED) after the end of seeding was assumed as 50 min for all the experimental cases. It was assumed that the air mass moving toward YP was affected because the seeding material at CPOS exerted an influence until 50 min after the end of seeding (Figure 5). Therefore, 50 min after the SEED period was defined as the post-SEED period, and the total section affected by seeding was defined as Tot-SEED (SEED + post-SEED).

3.3. Vertical Reflectivity and Precipitation.

Figure 6 shows the 10 min accumulated precipitation from the 0.5 mm tipping-bucket rain gauge and the vertical reflectivity from the MRR at YP. The quality control method proposed by Maahn and Kollias [19] removes noise up to the second observation bin (400 m). In Figure 6, the increase in vertical reflectivity was well shown in Tot-SEED compared with that in NOSEED especially for EXP1, EXP2, and EXP4. During NOSEED (~1400 LST), the reflectivity of 0–5 dBZ appeared intermittently at altitudes below 1 km (Figure 6(a)). However, in Tot-SEED, the reflectivity of 5–15 dBZ appeared homogeneously from the ground up to around 1.5 km. In the cases of EXP2 and EXP4, no reflectivity was detected via MRR for NOSEED, but reflectivity was observed for Tot-SEED.

In the cases of EXP1 and EXP4, a 0.5 mm increase in precipitation was observed during post-SEED. However, in the case of EXP3, vertical reflectivity showed no significant

(a)

········ geometric mean

——— total concentration

(b)

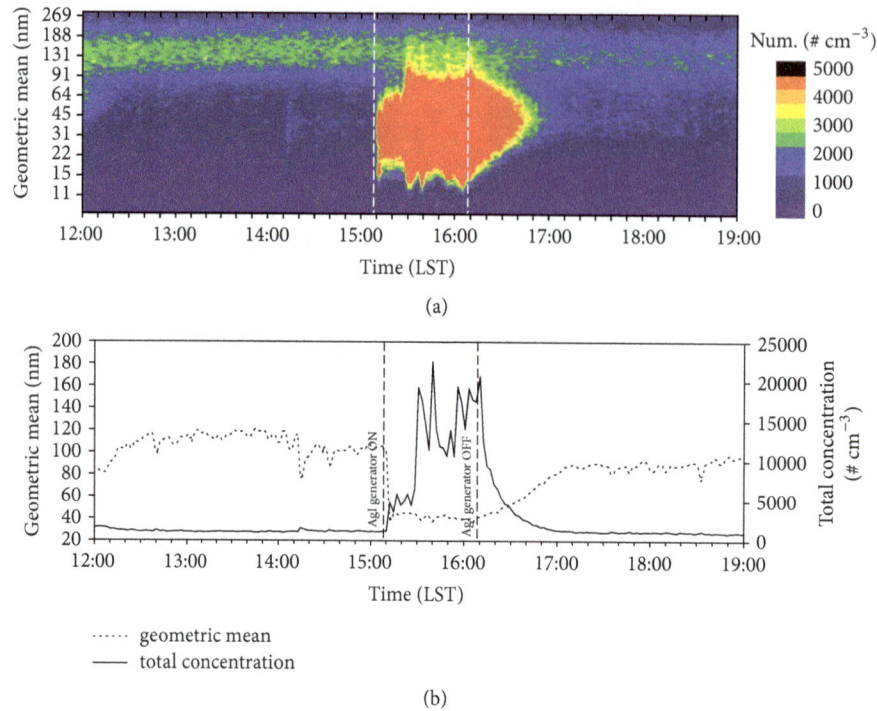

FIGURE 5: Distribution of the (a) particle number concentrations with geometric size and (b) time series of the geometric mean (dotted) and total particle number concentrations (solid) of aerosols measured by SNPS at CPOS on 7 Apr. 2015. The dashed lines represent the seeding period.

changes and no precipitation was observed in NOSEED and Tot-SEED. In addition, there was a difference in the reaction time of the seeding effect at YP that corresponded to the downwind side depending on the wind speed at CPOS (Figure 6). For example, the surface wind speeds at CPOS for EXP1 and EXP2 were $3.6 \, \text{m s}^{-1}$ and $1.3 \, \text{m s}^{-1}$, respectively. Therefore, EXP1 is believed to show more rapid precipitation enhancement more rapidly than the other cases. We assumed that the starting time of snowfall at YP was earlier than the detection of precipitation due to the delay in the snow melting time (estimated as about 20 min) in the rain gauge. In Figures 6(a) and 6(d), we believe that the precipitation appears later than the reflectivity due to this delay. On the other hand, in the case of EXP3, the vertical reflectivity showed no significant changes and no precipitation was observed in NOSEED or Tot-SEED (Figure 6(c)).

In the cases of EXP2, the precipitation of 0.5 mm occurred during post-SEED, but the increased reflective was smaller than other cases. For the EXP1 and EXP2, the difference of the seeded and nonseeded simulated precipitations was well appeared near the YP target area (Figure 7). In case of EXP4, an increase in reflectivity and precipitation was observed, but the increase in precipitation in the model by seeding was not shown. Figure 4(d) shows that the simulated temperature near the minimum optimal concentration for ice nucleation ($\sim 10^{-5} \, \text{m}^{-3}$), proposed by Xue et al. [26], in the middle between YP and CPOS is about $-2.5°C$, which is lower than the minimum temperature for ice nucleation ($-4°C$).

Therefore, we guess that the condition for the ice nucleation is insufficient in the simulation for EXP4 (Figure 7(d)).

To examine the vertical cloud changes before and after seeding, the difference in the frequency-by-altitude diagram (FAD) was calculated following the methods of Yuter and Houze [27] and Pokharel et al. [28]. These diagrams show the normalized frequency of vertical reflectivity by height. The difference FAD for the MRR vertical reflectivity composited during Tot-SEED and NOSEED is shown in Figure 8. The FAD difference analysis was not performed for EXP2 and EXP4 because no reflectivity was observed in NOSEED. In the case of EXP1, the reflectivity increased up to around 1.5 km and about 2 dBZ on the ground (Figure 8(a)). However, in the case of EXP3, the average reflectivity change between NOSEED and Tot-SEED is very small, lower than 0.5 dBZ, indicating that seeding had no significant effect (Figure 8(b)).

Figure 9 shows the vertical wind field observed by the wind profiler at GN. The dashed line indicates the height of wind change from east (0–135°) to west (225–360°). The altitudes at which the wind field changed one hour before the start of seeding in EXP1, EXP2, EXP3, and EXP4 were 1.6 km, 1.3 km, 0.9 km, and 1.2 km, respectively. In EXP3 (Figure 9(c)), in which there was no increase in precipitation at YP, the thickness of the cloud layer flowing to the seeding point appeared to be relatively thinner because the mountain (the height of peak is about 840 m) blocked the supply of sufficient liquid water (by simple calculation, the easterly cloud

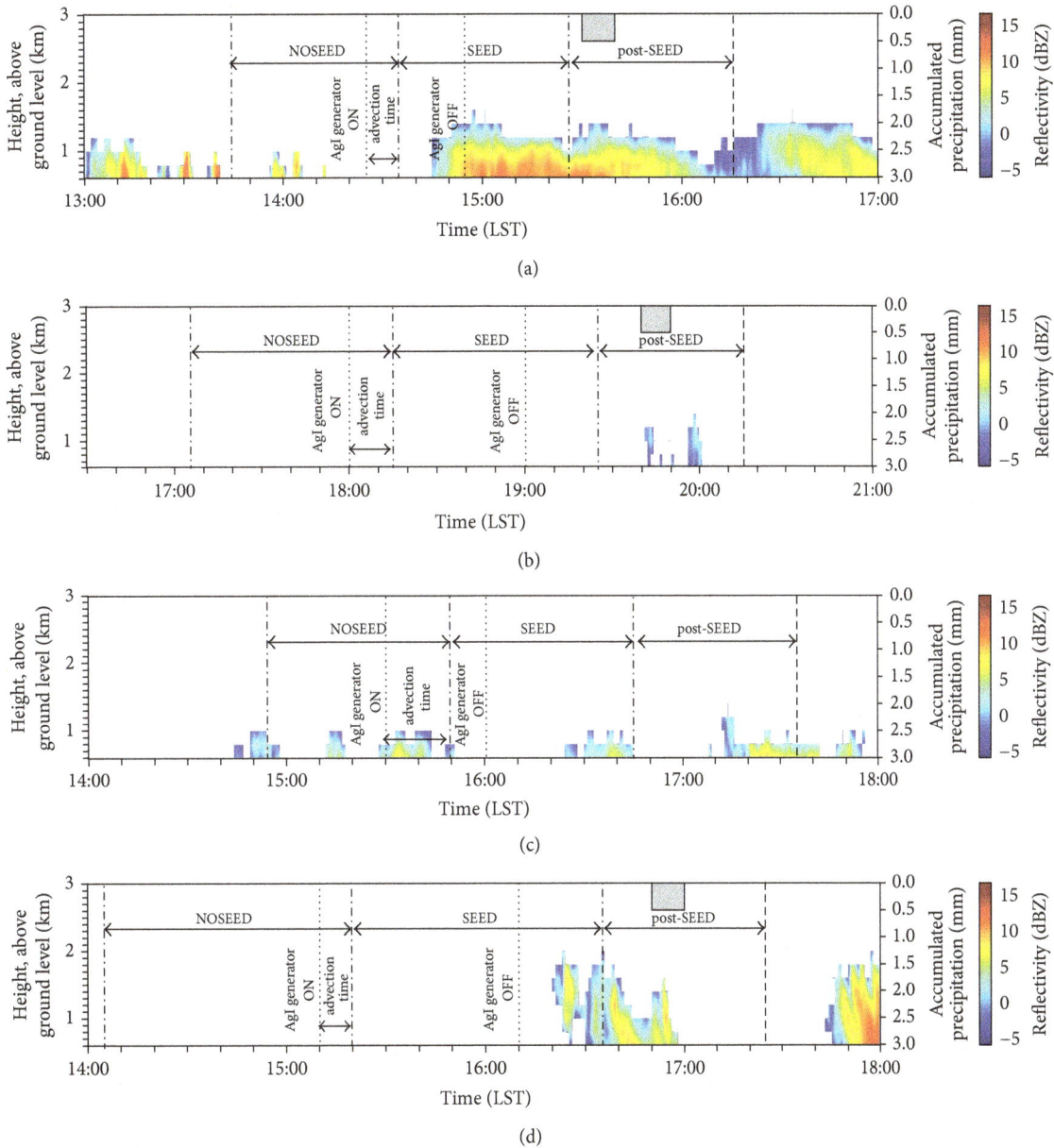

FIGURE 6: Time series of vertical reflectivity measured by micro rain radar (MRR, color shaded) and 10 min accumulated precipitation (grey bar) from the rain gauge at YP for (a) EXP1, (b) EXP2, (c) EXP3, and (d) EXP4.

thickness may be estimated as 60 m, the difference between the height of mountain peak and changed wind field).

To investigate whether the inflow of natural precipitation into YP was the cause of the increased reflectivity of Tot-SEED, 1 h averaged radar reflectivity of PPI0 after the start of seeding was examined (Figure 10). Radar reflectivity was not shown in the target area during the Tot-SEED period for EXP2, EXP 3, and EXP4. For the EXP1, the reflectivity at YP is small (−2–0 dBZ).

3.4. Changes in Snow Particles. We investigated microphysical changes such as number concentration and particle size before and after seeding using PARSIVEL data as in the ASCII study [28–30]. PARSIVEL observes snow particles between 0.062 mm and 25.5 mm with 32 size bins. Figure 11 shows the total snow concentration and mean diameter for EXP2 and EXP4 from PARSIVEL at YP. EXP1 and EXP3 were excluded from this analysis because of missing measurements. In EXP2 (Figure 11(a)), the snow particle number concentration rapidly increased from 1810 LST to 1830 LST. Although the average snow particle number concentration observed by PARSIVEL from 1630 LST to 1750 LST at CPOS (the seeding point) was below 5 L^{-1}, it increased sharply to over 15 L^{-1} from 1750 LST to 1820 LST before the start of seeding (not

(a) 1630 LST (EXP1)

(b) 2030 LST (EXP2)

(c) 1730 LST (EXP3)

(d) 1730 LST (EXP4)

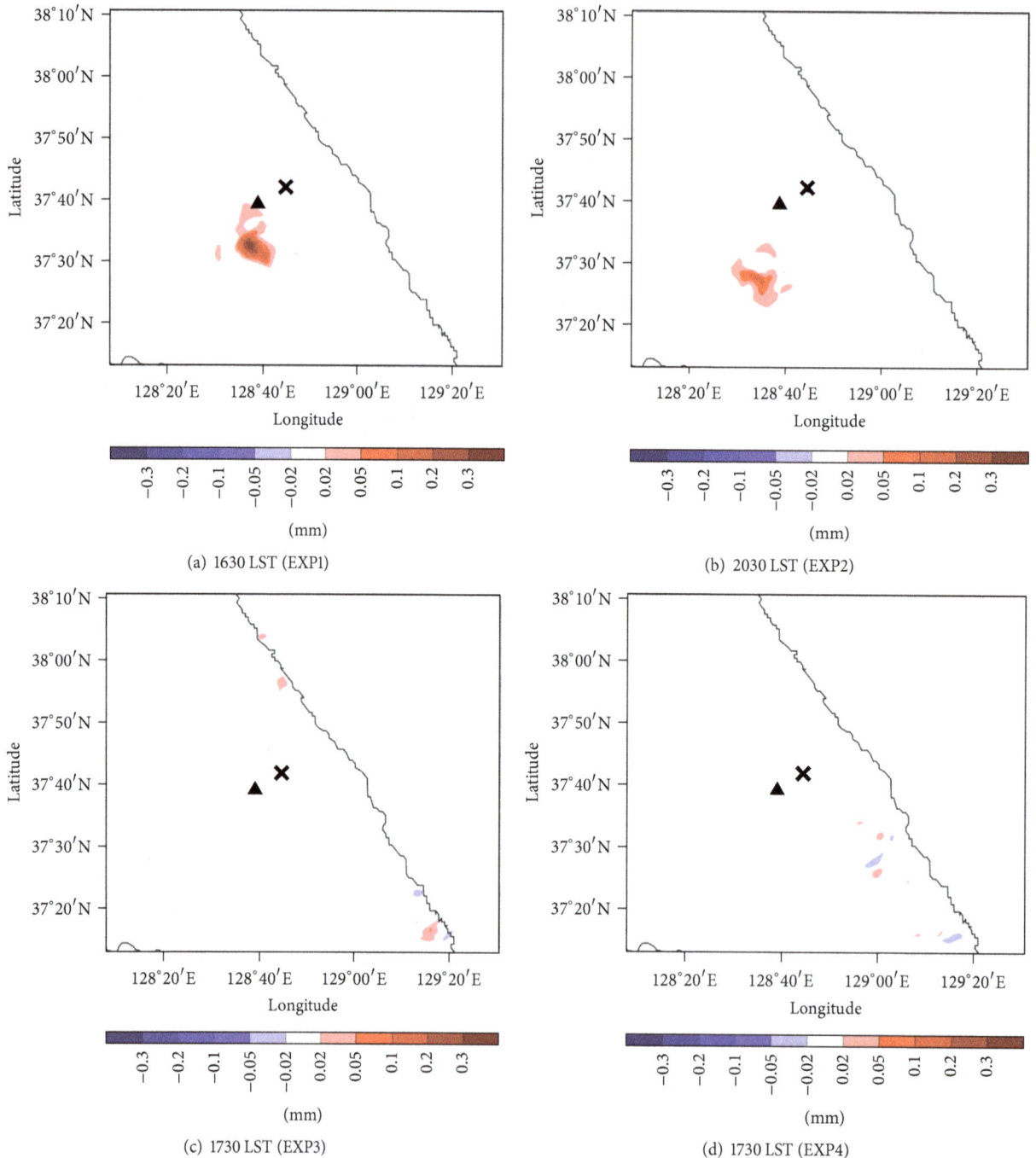

FIGURE 7: Differences between the accumulated seeded and nonseeded precipitation (mm) during the period from the start of seeding to the end of post-SEED. The triangle and cross indicate the locations of YP and CPOS, respectively.

shown). Therefore, the period from 1810 LST to 1830 LST when the number concentration at YP increased sharply was excluded from this analysis, considering that the change was not caused by seeding but by a temporary inflow of natural precipitation from CPOS. In the case of EXP2 (Table 1), the mean diameter of Tot-SEED increased from 0.63 mm to 0.72 mm compared to NOSEED, and the total snow particle concentration increased by 1.9 times from $2.3 \, L^{-1}$ to $4.5 \, L^{-1}$. In EXP4 (Figure 11(b)), the average particle size

increased from 0.61 mm to 0.70 mm and the particle number concentration increased by 1.2 times from $6.9 \, L^{-1}$ to $8.3 \, L^{-1}$ in Tot-SEED compared to NOSEED.

In the ASCII project, the total snow concentration was below $0.5 \, L^{-1}$ during NOSEED but increased to $2–15 \, L^{-1}$ during SEED, and a high particle number concentration was maintained even in the post-SEED period. The project suggested that the continuous increase in snow particles during the post-SEED period was caused by a delayed impact

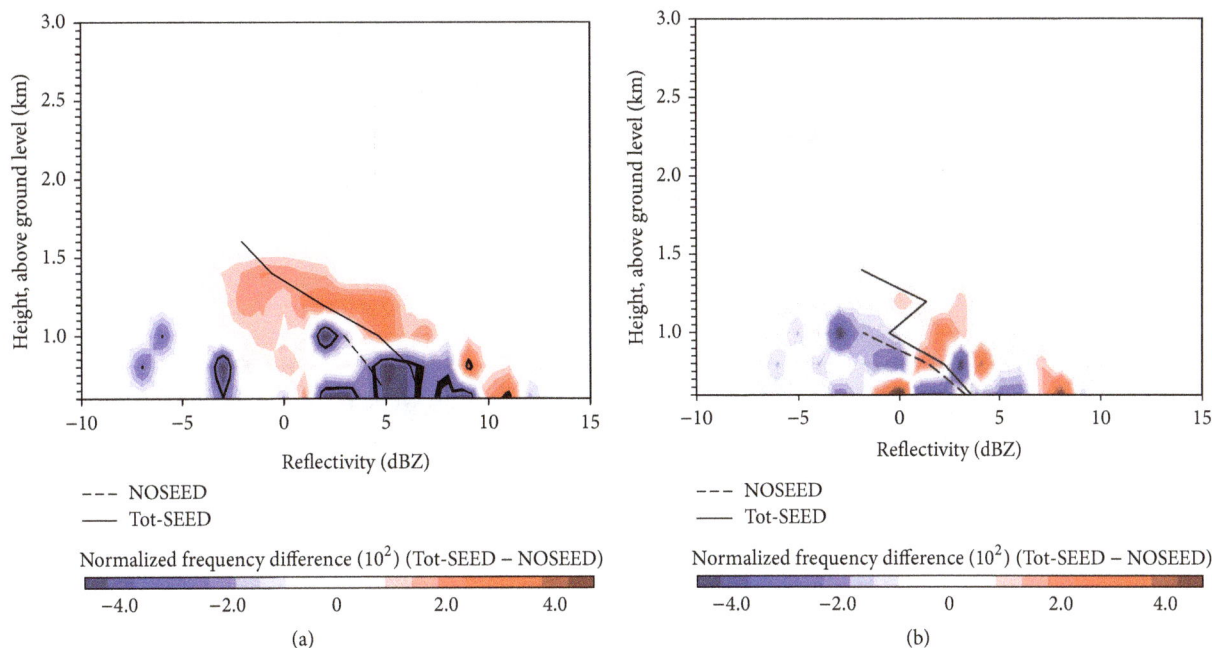

FIGURE 8: Normalized frequency difference from the MRR between NOSEED (dashed) and Tot-SEED (solid) at YP for (a) EXP1 and (b) EXP3.

due to the slow mixing of AgI nuclei in the clouds, the low transport speed, and the slow nucleation rate following the AgI nuclei seeding at the 300 m cloud base. This study also showed that the mean diameter and particle number concentration increased during post-SEED compared to SEED. The simulated AgI dispersion was faster than the wind speed at the same altitude observed with the wind profiler at GN. The relatively large increase of snow particles during the post-SEED appears to be partially due to the earlier simulated end time of SEED.

The particle size distribution data from PARSIVEL are grouped into two parts corresponding with the NOSEED and Tot-SEED. Figure 12 shows differences in the normalized frequency-by-diameter displays for NOSEED and Tot-SEED at YP based on the method of Pokharel et al. [28]. Compared with NOSEED, both large and small particles increased markedly during Tot-SEED. This indicates the effect of seeding, similar to the results of previous studies [30–32].

4. Conclusions and Discussions

In this study, experimental cases from 2012 to 2015 were analyzed to investigate the effects of ground-based AgI seeding on clouds over Pyeongchang region. At YP, the target area downwind of the seeding site (i.e., CPOS), the vertical reflectivity based on MRR increased at altitudes below 1 km during the Tot-SEED period, when the cloud thickness was sufficient. In those cases, the total snow particle number concentration and mean size measured by PARSIVEL also increased after seeding. In all the cases, the 1 h averaged composite reflectivity of grounded scanning radar was absent

or negligible after seeding, and thus the inflow of natural precipitation into YP appeared to be low. As a result, the precipitation increase in Tot-SEED could be attributed to the seeding effect in EXP1, EXP2, and EXP4. The precipitation enhancement is well shown in the simulation by the WRF with the modified Morrison scheme in microphysics seeding effect, especially for EXP1 and EXP2. In the case (EXP3) without any change after seeding, the cloud layer was thinner than the other cases, sufficient liquid water content was not supplied to the seeding area. This is due to the small vertical extent of the easterly wind field, thus showing almost no seeding effect.

There were uncertainties associated with the results. First, the simulated dispersion time of AgI used to define the NOSEED and SEED periods was slightly (below 30 min) faster than the time estimated by the advection with the observed wind speed from wind profiler at GN. However, because we used the same numerical model as that used by Kim et al. [13], it was assumed that the dispersion time of the seeding material was valid in this study.

Second, the inflow of natural precipitation was uncertain. Only the 1 h averaged PPI0 of the composite reflectivity data was used to determine the inflow of natural precipitation. Therefore, the inflow of natural precipitation could not be excluded completely in each case, especially below 1.5 km. Consequently, the increases in reflectivity, precipitation particle size, and particle number concentration that appeared after seeding could not be clearly attributed to the seeding effect. Furthermore, it is difficult to assert that the snow enhancement due to seeding existed in the entire Yongpyeong area due to uncertainties in the observations and numerical model. However, there were many seeding effects such as the

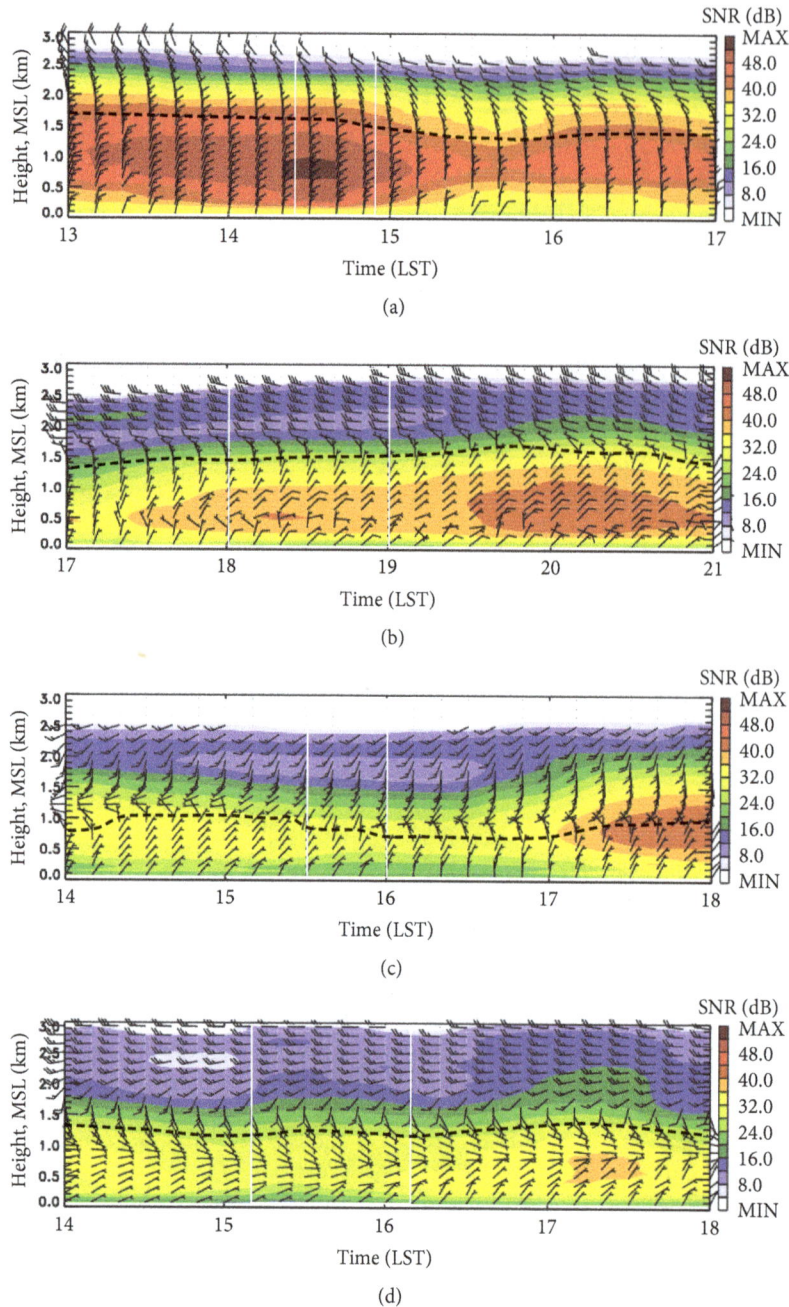

FIGURE 9: Time series of horizontal wind and the signal-to-noise ratio (SNR) of the wind profiler at GN for (a) EXP1, (b) EXP2, (c) EXP3, and (d) EXP4. The vertical solid lines mark the period of AgI generator operation at CPOS. The dashed line indicates the height of wind change. The details of each experiment are listed in Table 1.

microphysical and precipitation difference between NOSEED and Tot-SEED at YP after seeding. Finally, there was no snow depth measurement sensor that could be used to observe the actual snow accumulation before and after seeding at YP. The 0.5 mm tipping-bucket rain gauge used in this study only measured the snow melted by heating, and so the starting time of the increase in precipitation may be delayed. In addition, the actual ending time of the precipitation increase

and the exact precipitation between 0.5 and 1.0 mm could not be determined due to the observational resolution of 0.5 mm. However, if delayed, the appearance of precipitation increase at YP was within the Tot-SEED period.

To reduce the above-mentioned uncertainties, a snow depth meter was installed at YP in May 2015. Furthermore, a precipitation collector was developed to build a system for the analysis of the chemical composition of precipitation at

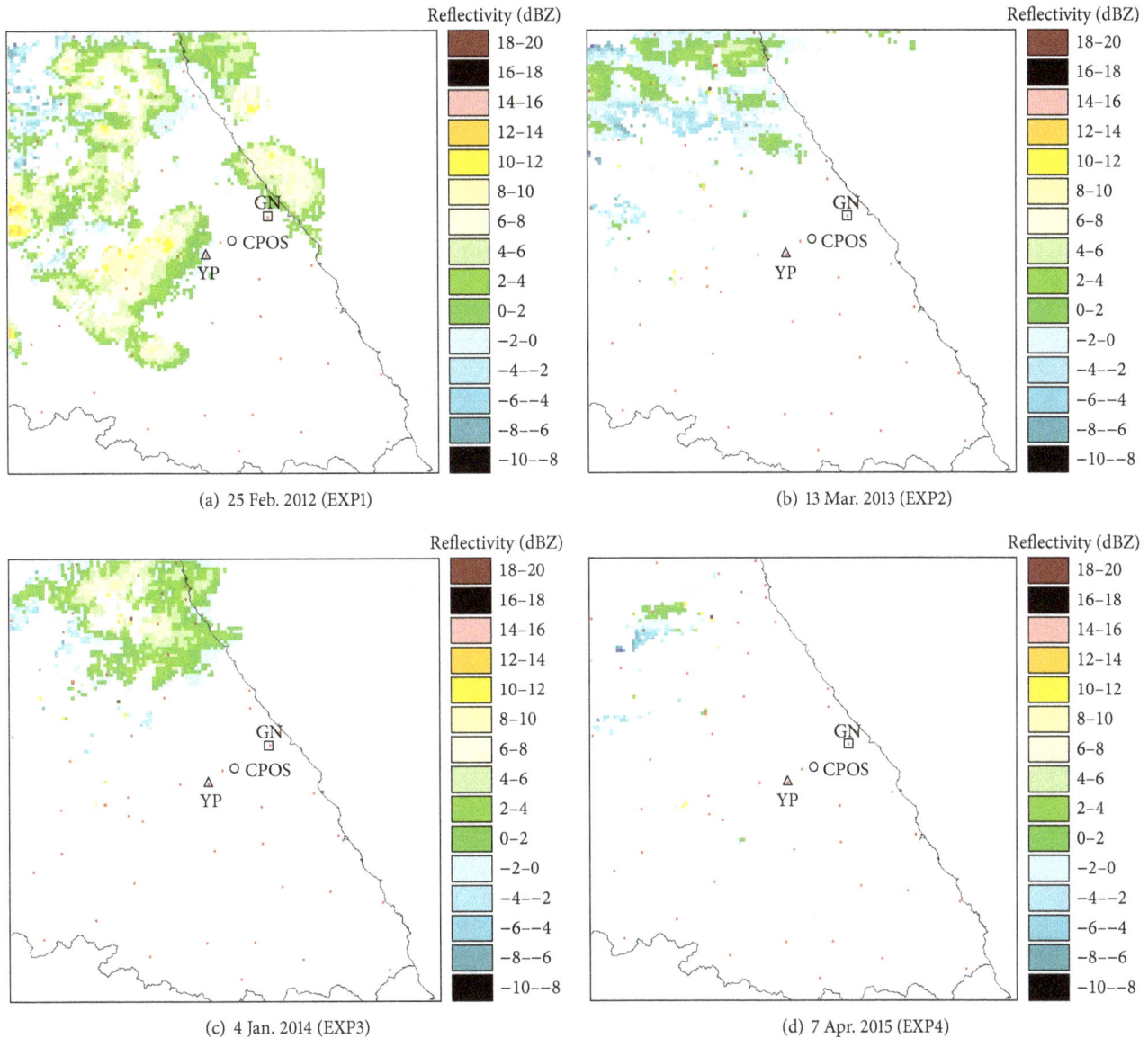

(a) 25 Feb. 2012 (EXP1)

(b) 13 Mar. 2013 (EXP2)

(c) 4 Jan. 2014 (EXP3)

(d) 7 Apr. 2015 (EXP4)

FIGURE 10: The 1 h averaged composite reflectivity of PPI0 after the start of seeding from the Korea Meteorological Administration ground-based operating radars for (a) EXP1, (b) EXP2, (c) EXP3, and (d) EXP4. The red dot indicates the automatic weather station (AWS).

YP in March 2017. Many studies (e.g., [30–32]) have already addressed chemical verification for snow accumulation observed at the target region after seeding. In future experiments, more reliable verifications will be made by analyzing chemical changes in snow and precipitation. To complement these results, an intensive observation and seeding study will be carried out between November 2017 and January 2018 before the Winter Olympics in Pyeongchang.

Disclosure

An earlier version of this work was presented as a poster at Geophysical Research Abstracts Vol. 20, EGU2018-5930,

2018, EGU General Assembly 2018, and Korean Meteorological Society Fall Meeting, 2017.

Conflicts of Interest

The authors declare that they have no conflicts of interest.

Acknowledgments

This work was funded by the Korea Meteorological Administration Research and Development program "Research and Development for KMA Weather, Climate, and Earth System Services-Support to Use of Meteorological Information and

(a)

(b)

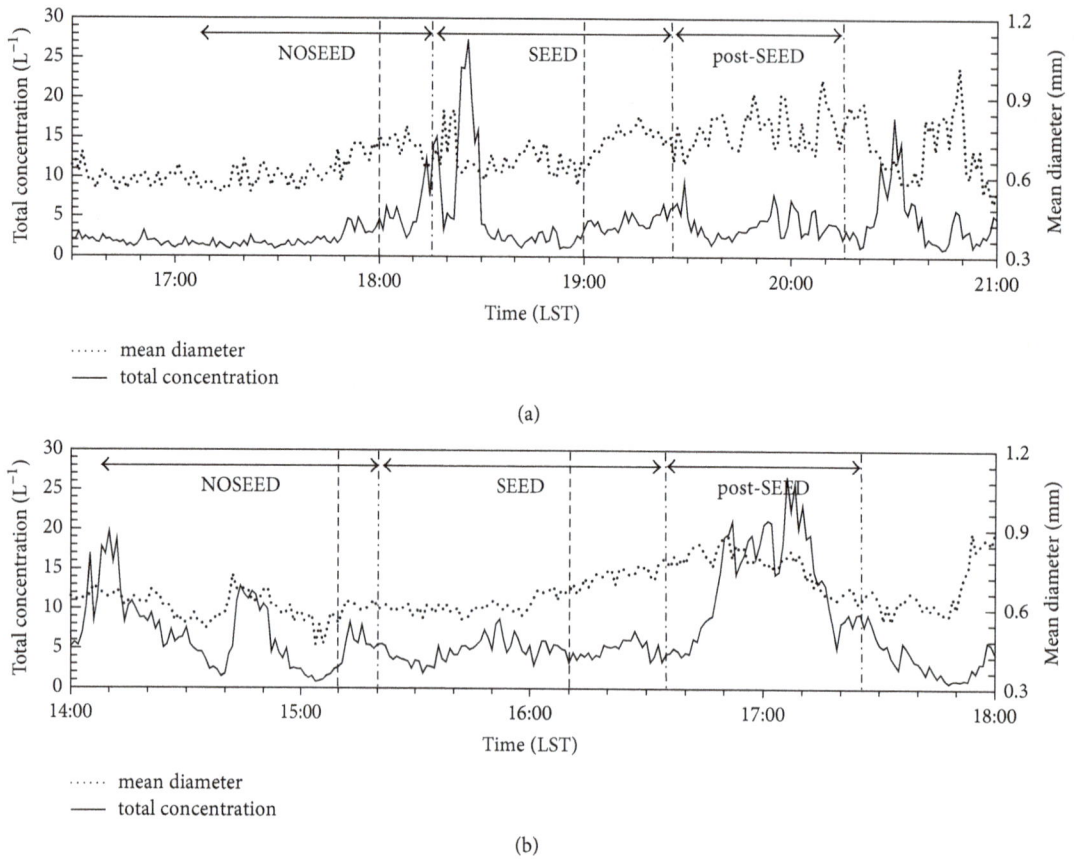

FIGURE 11: Time series of snow size distribution obtained from the PARSIVEL at YP for (a) EXP2 and (b) EXP4. The dotted and solid lines indicate the mean diameter and total particle number concentrations of snow particles, respectively.

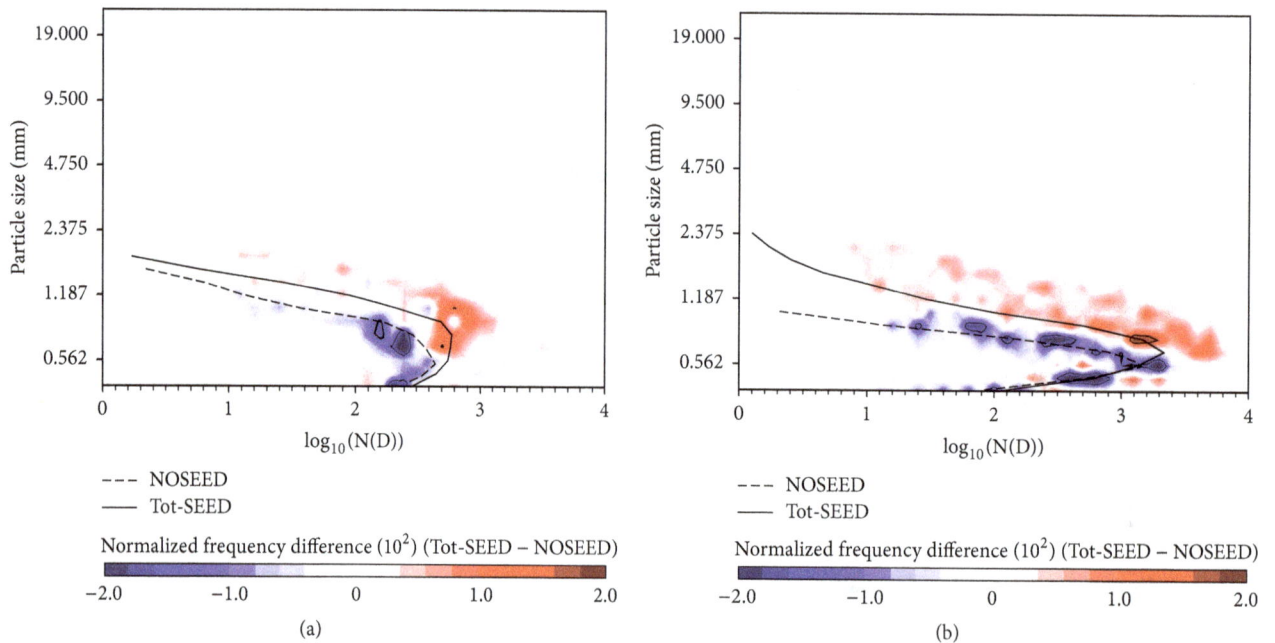

(a)

(b)

FIGURE 12: Normalized frequency difference from the PARSIVEL between NOSEED (dashed) and Tot-SEED (solid) at YP for (a) EXP2 and (b) EXP4.

Value Creation" under Grant KMA2018-00222. The synoptic weather charts of the four cases were downloaded from the Korea Meteorological Administration website. The authors thank Professor Seong Soo Yum's team of Yonsei University for developing the WRF model with the modified Morrison scheme in microphysics.

References

[1] H. R. Pruppacher and J. D. Klett, *Microphysics of Clouds And Precipitation*, Kluwer Academic Publishers, Dordrecht, the Netherlands, 2010.

[2] V. J. Schaefer, "Final report project cirrus. Part I. Laboratory, field and flight experiments," Tech. Rep., General Electric Research Laboratory, Schenectady, NY, USA, 1953.

[3] R. T. Bruintjes, "A review of cloud seeding experiments to enhance precipitation and some new prospects," *Bulletin of the American Meteorological Society*, vol. 80, no. 5, pp. 805–820, 1999.

[4] I. Geresdi, L. Xue, and R. Rasmussen, "Evaluation of orographic cloud seeding using a bin microphysics scheme: two-dimensional approach," *Journal of Applied Meteorology and Climatology*, vol. 56, no. 5, pp. 1443–1462, 2017.

[5] A. B. Super and J. A. Heimbach, "Microphysical effects of wintertime cloud seeding with silver iodide over the rocky mountains. Part II: observations over the bridger range, Montana," *Journal of Applied Meteorology and Climatology*, vol. 27, no. 10, pp. 1152–1165, 1988.

[6] A. B. Super and B. A. Boe, "Microphysical effects of wintertime cloud seeding with silver iodide over the rocky mountains. Part III: observations over the Grand Mesa, Colorado," *Journal of Applied Meteorology and Climatology*, vol. 27, no. 10, pp. 1166–1182, 1988.

[7] B. Geerts, Q. Miao, Y. Yang, R. Rasmussen, and D. Breed, "An airborne profiling radar study of the impact of glaciogenic cloud seeding on snowfall from winter orographic clouds," *Journal of the Atmospheric Sciences*, vol. 67, no. 10, pp. 3286–3302, 2010.

[8] B. Pokharel, B. Geerts, X. Jing, K. Friedrich, K. Ikeda, and R. Rasmussen, "A multi-sensor study of the impact of ground-based glaciogenic seeding on clouds and precipitation over mountains in Wyoming. Part II: seeding impact analysis," *Atmospheric Research*, vol. 183, pp. 42–57, 2017.

[9] M.-J. Lee, K.-H. Chang, G.-M. Park et al., "Preliminary results of the ground-based orographic snow enhancement experiment for the easterly cold fog (cloud) at Daegwallyeong during the 2006 winter," *Advances in Atmospheric Sciences*, vol. 26, no. 2, pp. 222–228, 2009.

[10] P. J. DeMott, "Quantitative descriptions of ice formation mechanisms of silver iodide-type aerosols," *Atmospheric Research*, vol. 38, no. 1-4, pp. 63–99, 1995.

[11] D. N. Blair, "Flame temperature effects on AgI nuclei produced from acetone generators," *The Journal of Weather Modification*, vol. 6, no. 1, pp. 238–245, 1974.

[12] W. C. Skamarock, J. B. Klemp, J. Dudhia et al., "A description of the advanced research WRF version 3," NCAR Technical Note NCAR/TN-475+STR, UCAR Staff Directory, New York, NY, USA, 2008.

[13] C. K. Kim, S. S. Yum, and Y.-S. Park, "A numerical study of winter orographic seeding experiments in Korea using the Weather Research and Forecasting model," *Meteorology and Atmospheric Physics*, vol. 128, no. 1, pp. 23–38, 2016.

[14] S. Chae, K. Chang, S. Seo et al., "Numerical Simulations of airborne glaciogenic cloud seeding using the WRF model with the modified morrison scheme over the Pyeongchang Region in the winter of 2016," *Advances in Meteorology*, vol. 2018, Article ID 8453460, pp. 1–15, 2018.

[15] B.-H. Heo, S. Jacoby-Koaly, K.-E. Kim, B. Campistron, B. Benech, and E.-S. Jung, "Use of the Doppler spectral width to improve the estimation of the convective boundary layer height from UHF wind profiler observations," *Journal of Atmospheric and Oceanic Technology*, vol. 20, no. 3, pp. 408–424, 2003.

[16] G. Ohring, S. Lord, J. Derber, K. Mitchell, and M. Ji, "Applications of satellite remote sensing in numerical weather and climate prediction," *Advances in Space Research*, vol. 30, no. 11, pp. 2433–2439, 2002.

[17] M. Löffler-Mang, "A laser-optical device for measuring cloud and drizzle drop size distributions," *Meteorologische Zeitschrift*, vol. 7, no. 2, pp. 53–62, 1998.

[18] M. Löffler-Mang and U. Blahak, "Estimation of the equivalent radar reflectivity factor from measured snow size spectra," *Journal of Applied Meteorology and Climatology*, vol. 40, no. 4, pp. 843–849, 2001.

[19] M. Maahn and P. Kollias, "Improved Micro Rain Radar snow measurements using Doppler spectra post-processing," *Atmospheric Measurement Techniques*, vol. 5, no. 11, pp. 2661–2673, 2012.

[20] H. Yang, K. Chang, J. Cha, Y. Choi, and C. Ryu, "Characteristics of precipitable water vapor and liquid water path by microwave radiometer," *Journal of the Korean Earth Science Society*, vol. 33, no. 3, pp. 233–241, 2012 (Korean).

[21] S. Jung, Y. Lim, K. Kim, S. Han, and T. Kwon, "Characteristics of precipitation over the East Coast of Korea based on the special observation during the winter season of 2012," *Journal of the Korean Earth Science Society*, vol. 35, no. 1, pp. 41–53, 2014 (Korean).

[22] C.-K. Kim, S.-S. Yum, S.-N. Oh et al., "A feasibility study of winter orographic cloud seeding experiments in the Korean Peninsula," *Asia-Pacific Journal of Atmospheric Sciences*, vol. 41, no. 6, pp. 997–1014, 2005 (Korean).

[23] B.-C. Kwak and I.-H. Yoon, "Synoptic analysis on snowstorm occurred along the east coast of the Korean Peninsula during 5–7 January, 1997," *Journal of the Korean Earth Science Society*, vol. 21, no. 3, pp. 258–275, 2000 (Korean).

[24] J.-G. Lee, "Synoptic structure causing the difference in observed snowfall amount at Daegwallyeong and Gangnung: case study," *Asia-Pacific Journal of Atmospheric Sciences*, vol. 35, no. 3, pp. 321–334, 1999.

[25] G. Vali, D. Rogers, G. Gordon et al., "Aerosol and nucleation research in support of NASA cloud physics experiments in space," Final Report NAS8–32067, 1978.

[26] L. Xue, S. A. Tessendorf, E. Nelson et al., "Implementation of a silver iodide cloud-seeding parameterization in WRF. Part II: 3D simulations of actual seeding events and sensitivity tests," *Journal of Applied Meteorology and Climatology*, vol. 52, no. 6, pp. 1458–1476, 2013.

[27] S. E. Yuter and R. A. Houze, "Three-dimensional kinematic and microphysical evolution of florida cumulonimbus. Part II: Frequency distributions of vertical velocity, reflectivity, and differential reflectivity," *Monthly Weather Review*, vol. 123, no. 7, pp. 1941–1963, 1995.

[28] B. Pokharel, B. Geerts, and X. Jing, "The impact of ground-based glaciogenic seeding on orographic clouds and precipitation:

a multisensor case study," *Journal of Applied Meteorology and Climatology*, vol. 53, no. 4, pp. 890–909, 2014.

[29] B. Pokharel, B. Geerts, X. Jing et al., "The impact of ground-based glaciogenic seeding on clouds and precipitation over mountains: a multi-sensor case study of shallow precipitating orographic cumuli," *Atmospheric Research*, vol. 147-148, pp. 162–182, 2014.

[30] B. Pokharel, B. Geerts, and X. Jing, "The impact of ground-based glaciogenic seeding on clouds and precipitation over mountains: A case study of a shallow orographic cloud with large supercooled droplets," *Journal of Geophysical Research: Atmospheres*, vol. 120, no. 12, pp. 6056–6079, 2015.

[31] J. M. Fisher, M. L. Lytle, M. L. Kunkel et al., "Evaluation of glaciogenic cloud seeding using trace chemistry," *The Journal of Weather Modification*, vol. 48, pp. 24–42, 2016.

[32] A. Zipori, D. Rosenfeld, J. Shpund, D. M. Steinberg, and Y. Erel, "Targeting and impacts of AgI cloud seeding based on rain chemical composition and cloud top phase characterization," *Atmospheric Research*, vol. 114-115, pp. 119–130, 2012.

13

Detecting Anomalies in Meteorological Data using Support Vector Regression

Min-Ki Lee ⓘ,[1] **Seung-Hyun Moon,**[1] **Yourim Yoon ⓘ,**[2] **Yong-Hyuk Kim ⓘ,**[3] **and Byung-Ro Moon ⓘ**[1]

[1]*School of Computer Science and Engineering, Seoul National University, 1 Gwanak-ro, Gwanak-gu, Seoul 08826, Republic of Korea*
[2]*Department of Computer Engineering, College of Information Technology, Gachon University, 1342 Seongnam-daero, Sujeong-gu, Seongnam-si, Gyeonggi-do 13120, Republic of Korea*
[3]*School of Software, Kwangwoon University, 20 Kwangwoon-ro, Nowon-gu, Seoul 01897, Republic of Korea*

Correspondence should be addressed to Byung-Ro Moon; moon@snu.ac.kr

Academic Editor: Alastair Williams

Significant errors exist in automated meteorological data, and identifying them is very important. In this paper, we present a novel method for determining abnormal values in meteorological observations based on support vector regression (SVR). SVR is used to predict the observation value from a spatial perspective. The difference between the estimated value and the actual observed value determines if the observed value is abnormal or not. In addition, SVR input variables are deliberately selected to improve SVR performance and shorten computing time. In the selection process, a multiobjective genetic algorithm is used to optimize the two objective functions. In experiments using real-world data sets collected from accredited agencies, the proposed estimation method using SVR reduced the RMSE by an average of 45.44% whilst maintaining competitive computing times compared to baseline estimators.

1. Introduction

Meteorological observations play an important role in weather forecasting, disaster warning, and policy formulation in agriculture and various industries [1–4]. In addition, meteorological observations are used for efficient operation of alternative energy sources such as solar power, hydropower, and wind power [5–7]. In recent years, as climate change due to global warming has accelerated, the extent of damage due to abnormal weather phenomena is increasing and becoming more difficult to predict. Consequently, there is a greater need for accurate and quantitative climate data based on meteorological observations. The collection of meteorological information, which was previously done manually, has been automated in line with computational advances. An automatic weather station (AWS) is an automated system that allows a computer to observe and collect numerical values of multiple meteorological elements. The development of AWS has enabled (i) real-time information retrieval, (ii) reduced maintenance costs, (iii) increased accuracy of observations, (iv) a larger amount of data, and (v) easier weather observations in poorly accessible regions.

However, meteorological data gathered by AWS often includes errors, and unusual metrics can be observed for a variety of reasons. Causes of unusual values include sensor malfunction, hardware error, power supply error, ambient environment change, and, in some rare circumstances, abnormal weather phenomena. Therefore, a quality control procedure is required for the collected data. Abnormal data identified during the quality control process are examined thoroughly by an expert and may become the subject of further research. If an abnormality is detected due to an error in the measurement process, it is necessary to replace the observed value with a corrected value [8–11]. Quality control of meteorological observations can also be regarded as anomaly detection [12] because anomalous values are of

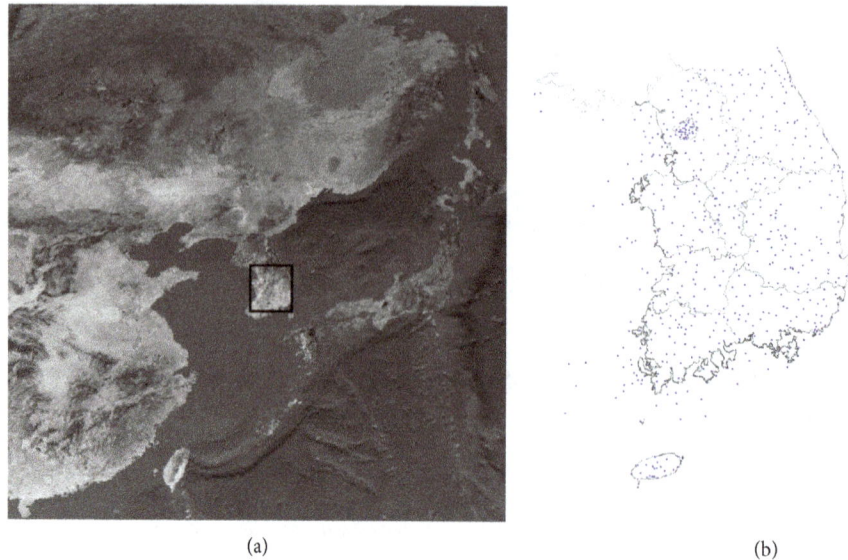

FIGURE 1: Locations of the 572 automatic weather stations (AWSs) in South Korea [18].

substantial interest to researchers. As the installation of AWS is expanding, and the amount and types of collected data are increasing, a fast and reliable quality control algorithm must be developed.

Quality control is achieved by several methods, ranging from simple discrimination using criteria related to physical limits to relatively complex discrimination related to spatiotemporal relationships with other observations [13]. Daly et al. [14] performed quality control of meteorological data metrics using climate statistics and spatial interpolation, and Sciuto et al. [15] proposed a spatial quality control procedure for daily rainfall data using a neural network. We propose a spatial quality control method, which uses values obtained from observational stations surrounding the target observational station to determine spatial compatibility and estimate the value of the observation station. It is possible to determine if an observed value is abnormal or not, based on differences with the estimated value. The developed spatial quality control method uses support vector regression (SVR) and a genetic algorithm. It can be applied to a wide range of meteorological elements to reflect the geographic and climatic characteristics of observation stations by studying past data through SVR. During preprocessing of the SVR, input variables, that is, the surrounding observation stations are selected according to two objective functions: similarity and spatial dispersion. Multiobjective optimization is required to simultaneously optimize the objective function that could be incompatible with these two functions. This is effectively performed by the genetic algorithm, which improves SVR performance and reduces execution time in this study. To verify the performance of the proposed method, we applied it to observational data measured by the Korea Meteorological Administration (KMA) for one year in 2014. Experiments on real-world data sets show that the performance of the proposed method is superior to previous methods such as the Cressman method [16] and the Barnes method [17], which have previously been used for spatial quality control.

TABLE 1: Meteorological elements in automatic weather station (AWS) data.

Meteorological element	Unit
Wind direction	°
Wind speed	m/s
Temperature	°C
Humidity	%
Atmospheric pressure	hPa
Hourly precipitation	
Rainfall occurrence	0 or 1

The remainder of this paper is organized as follows. Section 2 describes the problem that we attempt to solve in this study. Spatial quality control is defined, and we describe real-world data sets used to test the performance of the proposed methods. Section 3 introduces the methods previously used in spatial quality control, Section 4 describes an estimation model using SVR, Section 5 describes the algorithm for selecting SVR input variables, and Section 6 presents the experimental results using the real-world dataset. Finally, Section 7 discusses our conclusions.

2. Problem Description

2.1. Data Sets. This study covers data from 572 AWSs operated by KMA in South Korea. Figure 1 shows the locations of the target AWSs.

Target data include meteorological information measured every 1 minute from January 1, 2014 at 00:00 to December 31, 2014, at 23:59. In one year, 525,600 pieces of observational data are collected for each meteorological element at each station. We selected seven major meteorological elements for analysis: 10-minute average wind direction, 10-minute average wind speed, 1-minute average temperature, 1-minute average humidity, 1-minute average pressure, 1-hour cumulative amount of precipitation, and precipitation. Table 1 shows the types and units of meteorological elements used in this study.

The wind direction is expressed in degrees; however, this leads to a large error in the algorithm internal operation. For example, 1° and 359° differ only by 2°, but arithmetically, they differ by 358°. In addition, the average of the two wind directions should be regarded as 0°, but arithmetically it is 180°. Therefore, to avoid these problems, wind direction was converted into a two-dimensional unit vector, as expressed in [8].

$$\mathbf{v} = \left(\cos\left(\theta \cdot \frac{\pi}{180}\right), \sin\left(\theta \cdot \frac{\pi}{180}\right) \right), \tag{1}$$

where $\mathbf{v}$ is the transformed two-dimensional vector and θ is the original wind direction in degrees. In the quality control process, the components of the two vectors are processed separately. When a quantitative comparison of wind direction is required, the wind direction represented by the vector must be converted back to degrees:

$$\theta = atan2\,(v, u) \cdot \frac{180}{\pi}, \tag{2}$$

where u and v are $\mathbf{v}$'s first and second components, respectively, and $atan2$ is defined as follows:

$$atan2\,(y, x) = \begin{cases} \arctan\left(\dfrac{y}{x}\right), & x > 0 \\[2mm] \arctan\left(\dfrac{y}{x}\right) + \pi, & x < 0, y \geq 0 \\[2mm] \arctan\left(\dfrac{y}{x}\right) - \pi, & x < 0, y < 0 \\[2mm] +\dfrac{\pi}{2}, & x = 0, y > 0 \\[2mm] -\dfrac{\pi}{2}, & x = 0, y < 0 \\[2mm] \text{undefined}, & x = 0, y = 0. \end{cases} \tag{3}$$

2.2. Quality Control. During quality control, when the observed value is determined as not normal, it is then classified as "suspect" or "error" according to its level of abnormality. "Suspect" means that the value is likely to be abnormal, and "error" indicates that it is definitely anomalous.

2.2.1. Basic Quality Control. Basic quality control is a relatively simple quality control procedure performed in real-time. Abnormalities are determined using only the observed value itself or observations of a short time before and after. The data used in this study were first filtered through the following four basic quality control procedures. Each test was performed sequentially. If any test failed, the data were classified as an error, and subsequent tests were not performed. Each test and the numerical criteria are the same as those used by KMA.

(i) *Physical limit test*: If the observed value is higher or lower than the physically possible upper or lower

TABLE 2: Limits for physical limit test.

Meteorological element	Lower limit	Upper limit
Wind direction	0°	360°
Wind speed	0 m/s	75 m/s
Temperature	−80°C	60°C
Humidity	1	100
Atmospheric pressure	500 hPa	1080 hPa
Precipitation	0	400
Rainfall occurrence	0	1

TABLE 3: Maximum amount of change for step test.

Meteorological element	Maximum amount of change
Wind speed	10 m/s
Temperature	1°C
Humidity	10
Atmospheric pressure	2 hPa

TABLE 4: Minimum amount of change for persistence test.

Meteorological element	Minimum amount of change
Wind speed	0.5 m/s
Temperature	0.1°C
Humidity	1.0
Atmospheric pressure	0.1 hPa

limit, respectively, it is classed as an error. The physical limit test is performed on all meteorological elements. Table 2 shows the physical limits of each meteorological element, which are based on World Meteorological Organization (WMO) standards [19].

(ii) *Step test*: The step test is performed for wind speed, temperature, humidity, and atmospheric pressure. If the difference between the current observation value and the value one minute prior is more than a certain value, it is classed as an error. Table 3 shows the maximum variation of each meteorological element.

(iii) *Persistence test*: The persistence test is performed for wind speed, temperature, humidity, and atmospheric pressure. A value is classified as an error when the accumulated change in the observed value within 60 minutes is smaller than a certain value. Table 4 shows the minimum variation within 60 minutes for each meteorological element.

(iv) *Internal consistency test*: The internal consistency test is performed for pairs of wind direction and wind speed data and pairs of precipitation and rainfall occurrence data. If any one of the factors in each pair is determined to be an error in another test, the other factor is also perceived as an error. Also, if the rainfall occurrence value is 0 but the precipitation value is not 0, both values are classed as suspects.

Table 5 shows the percentages of normal, error, and suspect values, respectively, after performing each test on the KMA dataset. If the observed meteorological element is not available due to an absence of observational equipment, or if

TABLE 5: Results of basic quality control.

Meteorological element	Normal (%)	Limit Error (%)	Step Error (%)	Persistence Error (%)	Consistency		Uninspected (%)
					Suspect (%)	Error (%)	
Wind direction	81.97	$1.68e^{-3}$	N/A	N/A	0.00	2.80	15.23
Wind speed	81.90	$2.85e^{-4}$	$1.36e^{-3}$	$8.47e^{-1}$	0.00	2.80	14.45
Temperature	96.41	$3.22e^{-3}$	$7.73e^{-3}$	$1.02e^{-1}$	N/A	N/A	3.48
Humidity	54.47	$2.23e^{-2}$	$3.38e^{-3}$	5.43	N/A	N/A	40.07
Atmospheric pressure	38.30	$1.83e^{-1}$	$5.72e^{-5}$	$3.69e^{-2}$	N/A	N/A	61.48
Hourly precipitation	93.08	0.00	N/A	N/A	1.94	0.00	4.98
Rainfall occurrence	93.08	$2.34e^{-4}$	N/A	N/A	1.94	0.00	4.98

the observed value is missing, it is classified as uninspected. All subsequent experiments were performed only on data determined as normal after basic quality control.

2.2.2. Spatial Quality Control.

2.2.2. Spatial Quality Control. Spatial quality control determines whether the observation data of the target station are abnormal based on the values of other observation stations around the target station. It is sometimes referred to as a spatial consistency test [13]. Because this test is based on a large amount of data, it involves more time and resources than basic quality control. Therefore, spatial quality control is often performed in quasi-real-time. Typical spatial quality control process is as follows:

(i) Estimate the value of the target station using the values of surrounding observation stations.

(ii) If the difference between the observed and the predicted value of target station is greater than the critical value, the observation is considered as abnormal.

The meteorological elements of the KMA dataset, excluding rainfall occurrence, consist of continuous values; therefore, the predicted value can be estimated naturally via the interpolation or regression model. In the case of rainfall occurrence, it has a value of 0 or 1, so the value should be taken as 0 if the estimated value is less than 0.5, and 1 if the estimated value is 0.5 or more. The acceptable range for the difference between the observed value and the predicted value is generally determined using the standard deviation of the surrounding stations, which we set to the observation stations within 30 km of the target station. If the standard deviation is 0, because the observation values of all neighboring stations are the same, it is difficult to determine the acceptable range; therefore, the test is not performed. In the KMA dataset, this was often the case for elements such as precipitation and rainfall occurrence, which are always zero during periods of nonrain. Moreover, if there are less than three stations within 30 km, spatial quality control does not proceed because reliable standard deviations cannot be calculated. Also, observations that are missing or identified as abnormal during basic quality control are not considered for spatial quality control.

If the tolerance of the difference between the observed and predicted value is the same, the accuracy of the predicted value estimation will determine the reliability of spatial quality control. In this study, we aim for more accurate spatial prediction and thus improved spatial quality control performance. Traditional spatial prediction methods include spatial interpolation methods such as the Cressman method [16] and the Barnes method [17]. However, these methods do not reflect the geographical features of each region because they depend only on relative position to estimate the predicted value. Here, we propose a method to improve the accuracy of estimates by overcoming the shortcomings of existing methods by using supervised learning techniques.

3. Previous Methods

This section describes the spatial interpolation methods used in this study: the Cressman method and the Barnes method. The two methods have been slightly modified by the KMA to detect meteorological anomalies in South Korea. Actual observations are compared with estimates generated by the spatial interpolation methods. If there is a significant difference between observed and predicted values, the observation is classed as "suspect" or "error" according to the degree of difference.

3.1. Cressman Method. The Cressman method performs spatial interpolation on a two-dimensional distribution of meteorological elements. Meteorological elements at each station are irregularly distributed in two dimensions and converted into estimated values of the grid points at regular intervals. In this study, the grid interval is 0.2° for both longitude and latitude. The estimated values of the grid points are called the background field and are calculated with respect to the effective radius r. The effective radius is the control parameter describing the maximum station distance when estimating each grid point. Let z_i be the observed value of the station i, and d_{ei} denote the distance between the grid point e and the station i. Then, $Z_r(e)$, the estimated value of the grid point e, is the weighted average of the observations within the effective radius r (Figure 2):

$$Z_r(e) = \frac{\sum w_r(i) \cdot z_i}{\sum w_r(i)}, \qquad (4)$$

where $w_r(i)$, the weight of the station i, depends only on the distance:

$$w_r(i) = \begin{cases} \dfrac{\left(r^2 - d_{ei}^2\right)}{\left(r^2 + d_{ei}^2\right)} & \text{if } d_{ei} \leq r \\ \\ 0 & \text{otherwise.} \end{cases} \qquad (5)$$

To obtain $Z(i)$, the estimated value of a station i, the estimates of the four closest grid points from the station are

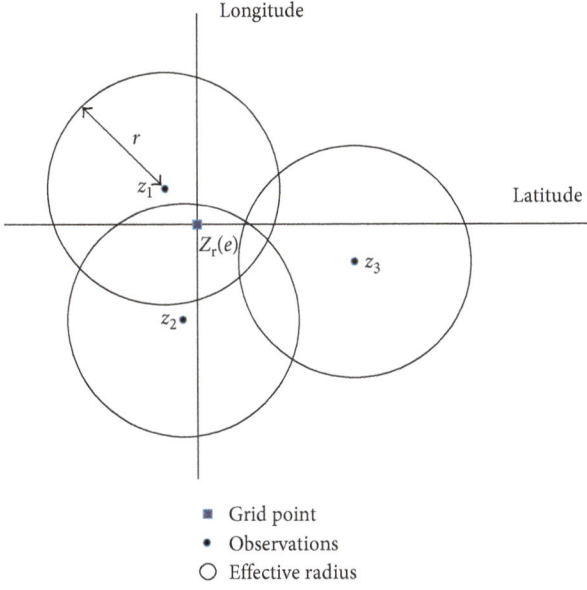

FIGURE 2: Calculation of the $Z_r(e)$, the estimated value of the grid point e in the Cressman method. Only observations of stations located within the effective radius r are used. In this example, z_1 and z_2 are used to calculate $Z_r(e)$, but z_3 is not used.

averaged. After calculating the estimates of all the stations, the background field can be recalculated using the estimates instead of the observations. The estimates of the stations can also be recalculated over the new background field. This process can be repeated as many times as desired. We set the effective radius to 50 km, 30 km, and 10 km and updated the background field and the estimates of the stations.

Let σ_i be the standard deviation of the observations at all stations located within the final effective radius of the station i. If $|z_i - Z(i)|$ is greater than $3 \cdot \sigma_i$, z_i is classified as an error. If $|z_i - Z(i)|$ is greater than $2 \cdot \sigma_i$, z_i is classified as a suspect.

3.2. Barnes Method. The Barnes method is a statistical technique that can derive accurate two-dimensional distribution from randomly distributed data in space. It is similar, in many respects, to the Cressman method, but instead uses a Gaussian function in the weight function:

$$w_r(i) = \begin{cases} \exp\left(\dfrac{-d_{ij}^2}{2r^2}\right) & \text{if } d_{ij} \leq r \\ \\ 0 & \text{otherwise,} \end{cases} \tag{6}$$

where d_{ij} is the distance between stations i and j. The KMA uses observations without using grid points when calculating the estimates by the Barnes method:

$$Z(i) = \frac{\sum w_r(i) \cdot z_i}{\sum w_r(i)}, \tag{7}$$

where r is set to 30 km. The process of determining whether or not observations are normal is almost identical to that of the Cressman method. Let σ_i be the standard deviation of the observations at all stations located within 30 km of the station i.

If $|z_i - Z(i)|$ is greater than $3 \cdot \sigma_i$, z_i is classified as an error. If $(z_i - Z(i))$ is greater than $2 \cdot \sigma_i$, z_i is classified as a suspect.

4. SVR-Based Approach

In this section, we propose a method using support vector regression (SVR) to overcome the spatial prediction limitations of the Cressman and Barnes methods for a target observation station from a spatial quality control perspective. The support vector machine (SVM) is a supervised machine learning method developed by Vapnik and Lerner in 1963 [20]. In the 1990s, nonlinear classification using SVM became popular as an alternative to artificial neural networks, and SVM generally has less overfitting problems than artificial neural networks. In the SVM, learning proceeds in the direction of maximizing the margin of the support vector, which is a hyperplane that divides each class of the given data. During early research, only linear classification was possible; however, nonlinear classification became feasible by mapping data to a higher dimensional space using a kernel function. A typical nonlinear kernel function is a radial basis function (RBF). The RBF function transforms the original space into an infinite dimension Hilbert space. In this study, the RBF function was used because the RBF function is better, on average, than the linear or polynomial function. Support vector regression (SVR) was proposed, which uses SVM for regression with continuous values as the output [21]. Preliminary study on meteorological elements has shown that the estimation capability of SVR is superior to other machine learning techniques [8]. In this study, the SVMlight [22] library was used for C language implementation.

The input and output of the SVR model for spatial quality control are as follows:

(i) *Input*: observations of stations surrounding the target station

(ii) *Output*: observation value of the target station.

In the input, values that are missing or classified as errors during basic quality control are replaced by the temporally closest values. The wind speed converted into a 2D vector representation was learned and tested by two separate models for each dimension. Once the model is learned from the values of the target station and the surrounding observation stations in the past, the predicted value of the target station can be estimated for the input that has not been learned. Because past observation values are not labeled as normal or abnormal with respect to spatial quality control, they are learned regardless of normal and abnormal values. Therefore, this approach assumes that most observations are normal and abnormalities are few.

Once the predicted value of the target station is estimated, the process of determining whether the observed value of the target station is normal is the same as the Cressman method or Barnes method. Let z_i and $Z(i)$ be the observations value and the estimated value by SVR of station i, respectively, and let σ_i be the standard deviation of the observations from all stations within a radius of 30 km of station i. If $|z_i - Z(i)|$ is greater than $3 \cdot \sigma_i$, z_i is classified as an error. If $|z_i - Z(i)|$ is greater than $2 \cdot \sigma_i$, z_i is classified as a suspect.

The SVR model can implicitly capture the geographic characteristics of the target station while learning past data.

FIGURE 3: Examples of neighbor selection. (a) Neighbor stations with high spatial dispersion. (b) Neighbor stations with low spatial dispersion.

Through this process, the combination of each station and each meteorological element has its own specific model. This is an advantage of SVR over non-ML approaches. However, an approach based on machine learning also has its drawbacks; specifically, that it takes a long time to learn. A method to overcome this is introduced in Section 5.

5. Selecting Neighboring Stations

The input of the SVR model uses the observations of neighboring AWSs within a certain radius of the target AWS. However, if there are too many neighbors, the learning time of SVR becomes too long. Also, some neighboring stations act as noise instead of helping to estimate the value of the target station. Therefore, it is necessary to select the best core neighbors to estimate the value of the target station while reducing the number of neighbors used in the input.

5.1. Similarity and Spatial Dispersion. Two criteria were applied to select key neighbors. The first considered the similarity of the observations between the target station and the neighbor station. Observations at locations with similar meteorological phenomena are helpful in deriving observations at the target site. The second considered how widespread the neighboring stations were in space. If one constructs a core neighborhood at stations concentrated in a narrow area, the model cannot be flexible to various situations. For example, if there is a peculiar meteorological phenomenon within a narrow area (e.g., a local storm), the estimate will be misled. Spatial dispersion ensures statistical robustness of the model. Figure 3 shows two different choices of neighboring stations. When the amount of rainfall in target station is estimated, the amount of rainfall in neighboring stations is used. If localized heavy rain happens in an area including neighboring stations with low spatial dispersion, the estimated amount of rainfall will be inclined to be very high even though target station is out of influence of localized heavy rain. On the other hand, it reflects overall surrounding circumstances of rainfall when spatial dispersion of neighboring stations is high."

To measure the similarity of stations according to their meteorological elements, the time series values of the elements are expressed as vectors, and the distance between them is measured in various ways. We used the L_1 distance, L_2 distance, Pearson correlation coefficient, and mutual information to measure the similarity between two vectors. After the distance of all the station pairs was calculated, the smallest value was zeroed and the largest value was normalized to one. The L_1 distance, known as the Manhattan distance or taxicab distance, between two vectors $\mathbf{x}$ and $\mathbf{y}$ was calculated as follows:

$$\|\mathbf{x} - \mathbf{y}\|_1 = \sum_{i=1}^{n} |x_i - y_i|, \qquad (8)$$

where x_i is the i-th element of $\mathbf{x}$. We used $(1-L_1$ distance) as a similarity measure to ensure that larger measurement was given to two vectors which were more similar The L_2 distance, known as the Euclidean distance, between two vectors $\mathbf{x}$ and $\mathbf{y}$ was calculated as follows:

$$\|\mathbf{x} - \mathbf{y}\|_2 = \sqrt{\sum_{i=1}^{n} (x_i - y_i)^2}. \qquad (9)$$

We used $(1-L_2$ distance) as a similarity measure to ensure that larger measurement was given to two vectors which were more similar. The Pearson correlation coefficient is used to measure the degree of the linear relationship between two variables. It has a value 1 when there is a perfect positive linear correlation and -1 when there is a perfect negative linear correlation. The Pearson correlation coefficient is calculated as follows, where $\bar{x} = \sum_{i=1}^{n} x_i / n$:

$$r_{\mathbf{xy}} = \frac{\sum_{i=1}^{n} (x_i - \bar{x}) (y_i - \bar{y})}{\sqrt{\sum_{i=1}^{n} (x_i - \bar{x})^2} \sqrt{\sum_{i=1}^{n} (y_i - \bar{y})^2}}. \qquad (10)$$

Mutual information measures the mutual dependence between two random variables X and Y. It quantifies the reduction in uncertainty of one of the variables due to knowing the other. Mutual information is calculated as

```
non-dominated set E = ∅;
initialize population P;
repeat
      assign a fitness value to each solution in P;
      select 2N parents from P;
      create N offspring applying crossover on the parents;
      mutate offspring;
      repair offspring;
      local-optimize offspring;
      P ← offspring;
      update E;
      remove n_E solutions from P;
      add n_E solutions from E to P;
until stopping condition;
return E;
```

FIGURE 4: The framework of our hybrid multiobjective genetic algorithm.

follows, where $p(x, y)$ is the joint probability function of x and y and $p(x)$ and $p(y)$ are the marginal probability density functions of x and y, respectively:

$$I(X; Y) = \sum_{y \in Y} \sum_{x \in X} p(x, y) \log\left(\frac{p(x, y)}{p(x)p(y)}\right). \quad (11)$$

We computed the mutual information from the observed frequency of two vectors, **x** and **y**, assuming that these vectors constitute an independent and identically distributed sample of (X, Y).

As a measure of spatial dispersion, we used the average of the geographical distance from the nearest station [23]. If the set of target stations and selected neighbors is x, and $d_{x_i x_j}$ is the normalized geographic distance between the two stations x_i and x_j, then the spatial dispersion is calculated as

$$\text{dispersion}(x) = \frac{\sum_{x_i \in x} \min_{x_j \in x, x_i \neq x_j} d_{x_i x_j}}{\sum_{x_i \in x} 1}. \quad (12)$$

The larger the spatial dispersion is, the better the neighborhood selection is. The two criteria of similarity and spatial dispersion often conflict. In general, similarities in climatic characteristics are often due to geographic proximity. Therefore, the key neighborhood screening problem is a multiobjective optimization problem that simultaneously optimizes two or more objectives that are not independent of each other. In this study, we solve the multiobjective optimization problem using genetic algorithms.

5.2. Multiobjective Genetic Algorithm. The genetic algorithm (GA) is a global optimization technique developed by Holland, which mimics the natural evolution of biological selection [24]. It is used to find a solution with high (or low) fitness while repeating a genetic operation that imitates processes such as selection, crossover, and mutation, which are important elements of evolution. GA is a type of metaheuristic that does not depend substantially on the nature of the problem. It can search all ranges and is less likely to fall into a local optimum. Pure GA is disadvantageous in that it takes a long time to

converge. The hybrid genetic algorithm solves this problem by combining the local optimization algorithm with the GA. Several successful attempts have been made to solve multiobjective problems using GA [25–29]. Among them, NSGA-II by Deb [30] is the most well-known. To maximize the function $f_1, f_2, \ldots, f_n$ with n number of objects, if solution x and solution y satisfy the following condition, then it can be said that solution y dominates solution x:

$$f_i(x) \leq f_i(y) \, \forall i \text{ and } \exists j : f_j(x) < f_j(y). \quad (13)$$

When a solution is not dominated by any other solution, the solution is called Paretooptimal. To improve an objective function in a Paretooptimal solution, one has to sacrifice another objective function. The multiobjective genetic algorithm (MOGA) does not output one solution but several Paretooptimal solutions. The final solution selection is performed by the decision-maker.

5.3. Our GA Framework. In this study, we tested the SVR with several Paretooptimal solutions for each meteorological element, and selected the best solution on average. Figure 4 shows the structure of the GA used in this study.

(i) *Encoding*: In a genetic algorithm, one solution is expressed as a set of genes, or a chromosome. Here, one chromosome is represented by a one-dimensional binary string. Each gene corresponds to one station. If the value of the gene is "0," the observation value of the corresponding station is not used as the input of the SVR. If it is "1," it is selected as an input of the SVR.

(ii) *Fitness*: This indicates the validity of the solution for a given problem. When the individual objective function is $f_1, f_2, \ldots, f_n$, the fitness value of solution x is calculated as

$$f(x) = w_1 f_1(x) + w_2 f_2 + \cdots + w_n f_n(x), \quad (14)$$

where $w_1, \ldots, w_n$ are nonnegative and $\sum w_i = 1$, each weight w_i is randomly set for every generation, not as a fixed value. This allows the algorithm to search for various Paretooptimal solutions [31]. This method is more intuitive than the algorithm that uses Pareto ranking-based fitness evaluation [32] and easier to be combined with a local optimization algorithm [33]. In this problem, $n = 2$, and $f_1(x)$ and $f_2(x)$ correspond to similarity and spatial dispersion, respectively, as described in Section 5.

(iii) *Population*: Population is a set of chromosomes. Chromosomes in the population interact each other to generate new solutions and cull existing solutions. In this study, the size of the population was set to 50. The initial population consisted of 50 randomly generated chromosomes.

(iv) *Selection*: This is the operator used to select the parent chromosome for the crossover. To mimic the principle of survival of the fittest in nature, chromosomes with high fitness are selected with high probability. In this study, roulette-wheel selection, one of the most widely used selection operators, was used. The probability that the best fitness solution will be selected is four times the probability that the lowest fitness solution will be selected.

(v) *Crossover*: A key operator of the genetic algorithm. In inheriting the features of the parents, we expect that the different advantageous traits combine to produce an offspring chromosome that is superior to the parents. In this study, we used a two-point crossover with two cut points.

(vi) *Mutation*: This statistically modifies a portion of the offspring chromosome to increase solution diversity and prevent premature convergence. In this study, each gene was flipped with a probability of 10%.

(vii) *Repair*: After crossover and mutation, offspring still may not meet the constraints of the problem. That is, the number of genes with a value of "1" in the chromosome may be different from the number of stations to be selected. If the number of genes with a value of "1" is insufficient, we repeat the process of changing the value of the randomly selected gene among genes with a value of "0" to "1." On the other hand, if the number of genes with a value of "0" is insufficient, we repeat the process of changing randomly selected genes to "0" among genes with a value of "1."

(viii) *Local optimization*: This exchanges the values of two genes whose fitness value increases when exchanged. This process is repeated until the exchange of any two gene values can no longer increase the fitness value.

(ix) *Replacement and elitism*: We used a generational GA to generate offspring as large as the size of the population and replace the entire population. Among the solutions found so far, nondominant solutions that are closest to the Paretooptimal are stored in an external archive. This nondominant solution archive updates every time a new solution is created. In other words, the solution that is dominated by the new solution is removed from the existing nondominant solution archive, and the new solution is stored in the archive when it is a nondominant solution. As survival of good solutions within a population can result in a good solution for the next generation, some of the population are replaced by solutions in nondominant solution archive. In this algorithm, 20% of the entire population was randomly replaced with solutions in nondominant solution archive.

(x) *Stopping condition*: The genetic algorithm stops when 1,000 generations have passed.

Figure 5 describes the whole spatial quality control process including neighbor selection.

6. Experimental Results

In this section, (i) detailed good parameters are selected, (ii) the performances of the estimation methods are compared, and (iii) the results of the proposed spatial quality control procedure are presented using meteorological data collected by the KMA for a year in 2014. The root mean square error (RMSE) was used as a measure to compare the accuracy of each estimation method. RMSE is a standard metrics for dealing with errors between model-estimated values and observed values in a real environment, including meteorology [34]. If θ is the observed vector and $\widehat{\theta}$ is the estimated vector, then the RMSE of $\widehat{\theta}$ is calculated as

$$\mathrm{RMSE}\left(\widehat{\theta}\right) = \sqrt{\mathrm{E}\left((\widehat{\theta} - \theta)^2\right)}. \qquad (15)$$

The lower the RMSE value, the better the model estimate. As the accuracy of estimates should be based on normal observations, only observations classed as normal by the model are used to calculate RMSE. When comparing the RMSE of two or more models, only those observations determined as normal by all models were used.

Performance evaluation of SVR estimation models was achieved through 10-fold cross-validation. All data were divided into 10 folds, of which 9 were used as the training set, and the other was used as the test set. Learning and test are performed 10 times so that each fold can be used as a test set. Due to there being 7 meteorological elements in 572 AWSs, and 10 models must be learned each time, a total of 40,040 models were created for each experiment. The entire training set consists of 473,040 data sets. Because of the large number of models and the overly long total execution time, we sampled 5,000 data sets and used them as training sets for the parameter optimization experiments. We then describe the change in performance and time caused by increasing the size of the training set once the final parameter is determined.

The experiment was performed on an Intel i7 quad-core 2.93 GHz CPU. Each experiment used only one core. Experiments with a long execution time were performed by dividing each of the seven machines by observatories, and the total execution time included the execution time of each machine.

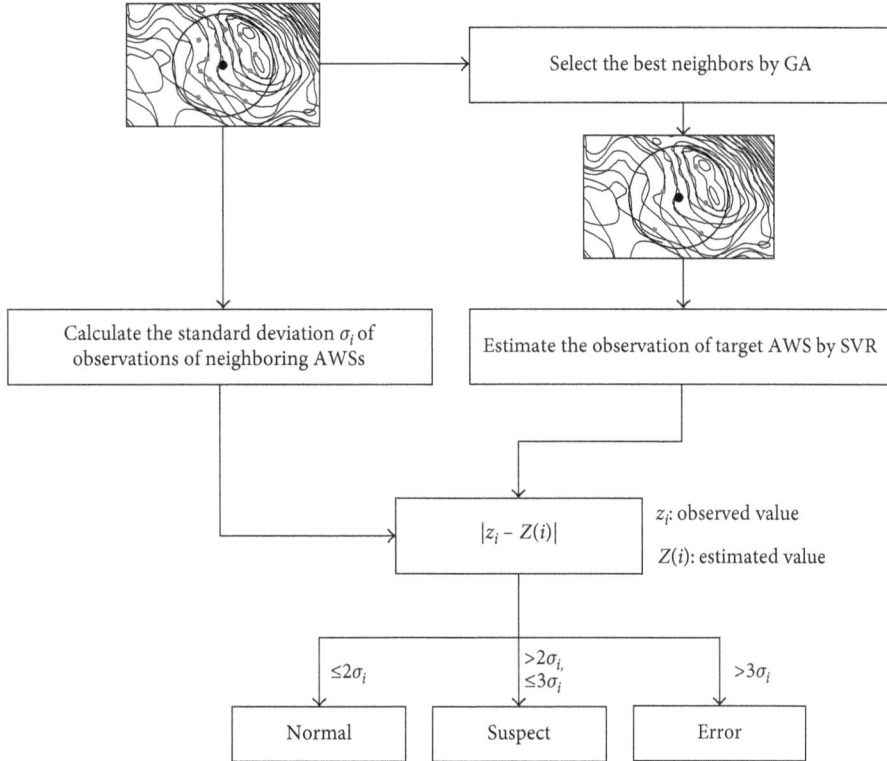

FIGURE 5: The proposed spatial quality control process.

TABLE 6: Accuracy of estimates according to wind direction representation (RMSE).

Representation	RMSE
Degree	92.17
Vector	**68.28**

TABLE 7: Accuracy of estimates for each similarity measure.

Meteorological element	L_1	L_2	PCC[1]	MI[2]
Wind direction	104.574	105.624	102.786	**101.424**
Wind speed	1.228	**1.224**	1.306	1.317
Temperature	**1.241**	**1.241**	1.327	1.319
Humidity	**8.085**	8.086	8.757	8.829
Atmospheric pressure	**6.497**	**6.497**	8.134	7.256
Hourly precipitation	1.074	**1.065**	1.066	1.155
Rainfall occurrence	**0.151**	**0.151**	0.152	0.157

[1]Pearson correlation coefficient. [2]Mutual information.

6.1. Representation of Wind Direction. Section 2.1 describes the process of converting wind direction expressions from degrees to 2D vectors. Table 6 compares the accuracy of SVR estimates for each wind direction representation. The accuracy of the estimate is much higher when expressed in terms of vector expression than degrees. Thus, all subsequent experiments used a vector expression to represent wind direction.

6.2. Similarity Measure. Section 5 describes four measures used to calculate the similarity between two observation stations. To compare the usefulness of each measure, the accuracy of the estimates predicted by the Madsen-Allerup method [35] is examined. The Madsen-Allerup technique selects the stations similar to the target station and then uses the observed values of selected stations to obtain the estimate of the target station; therefore, the higher the quality of the similarity measure, the more accurate the estimate. Table 7 shows the estimation accuracy of the Madsen-Allerup method for each similarity measure. In all subsequent experiments, we used the highest quality similarity measure for each meteorological element. Figure 6 shows the connected station pairs with a similarity greater than 0.5.

6.3. Selecting Neighboring Stations. In Section 5, we proposed MOGA to select input variables to improve SVR performance and speed. Figure 7 shows the accuracy of estimates based on the number of neighboring stations selected by MOGA. The greater the number of parameters (over a certain amount), the worse the performance of the SVR and the longer it takes to train. The optimal number of neighboring stations with the best performance differs with the meteorological element. Table 8 shows the optimal number of neighboring stations according to each meteorological element. All subsequent experiments were fixed using the optimal number of neighbors. Table 9 compares the estimation accuracy of SVR when neighboring stations were selected randomly, with the accuracy of SVR when neighboring stations were selected using MOGA. We confirm that selection of neighbors using MOGA improves the estimation performance of SVR.

6.4. Comparison of Estimation Models. Table 10 shows the accuracy of estimates for each estimation model. Estimation

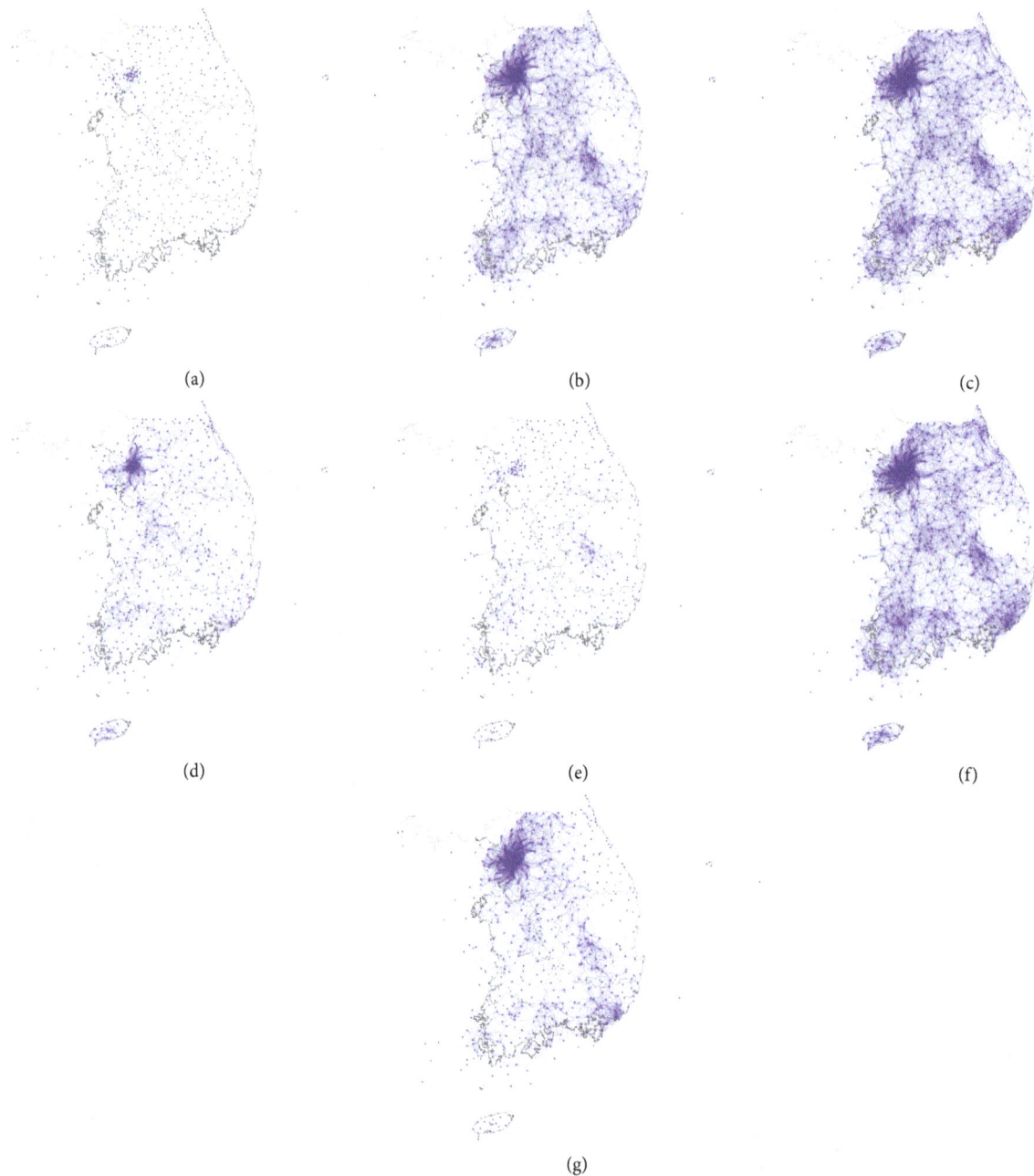

FIGURE 6: Similarity map for different meteorological elements. (a) Wind direction. (b) Wind speed. (c) Temperature. (d) Humidity. (e) Atmospheric pressure. (f) Precipitation. (g) Rainfall occurrence.

using SVR model is better than that using Cressman or Barnes algorithms. Hourly precipitation does not show much improvement compared to other meteorological elements. Because there are many more days without rain than with rain, there is rather sparse data distribution for rainy days, which results in learning difficulties.

Table 11 shows the execution time of spatial quality control according to each estimation model. The execution time might be considered of little importance as a single process of spatial quality control can be executed in a very small time. But if a quality contol process should be performed in a centralized single facility, a large number of meteorological data from every observational station need to

be inspected in real time. For example, in our test data, there are 572 stations, and they collect 7 kinds of meteorological observation data. It takes about 5.77 seconds to inspect every data from every station using the Cressman method, and it should be executed in every minute. Moreover, the execution time becomes more important as the number of stations and the kind of data get bigger and the time interval for collecting data gets shortened.

Spatial quality control is fastest using the Barnes algorithm, but the accuracy of the estimation is very poor. Spatial quality control using SVR is approximately 6 times faster than that using the Cressman algorithm, but more time is required to learn the SVR model. However, as it does not

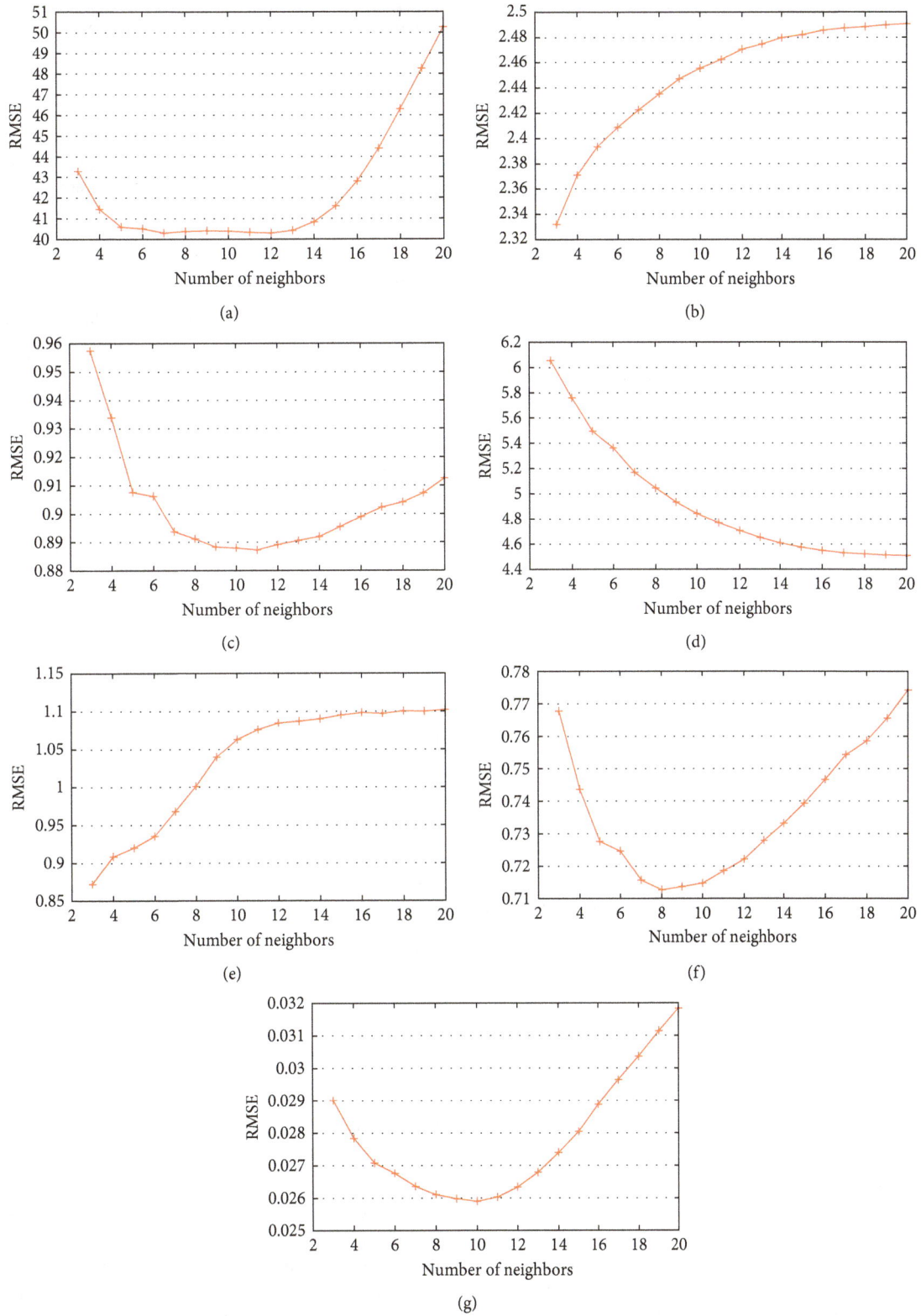

FIGURE 7: Accuracy of estimates according to the number of selected neighboring stations (RMSE). (a) Wind direction. (b) Wind speed. (c) Temperature. (d) Humidity. (e) Atmospheric pressure. (f) Precipitation. (g) Rainfall occurrence.

give weight to more recent data in the learning process, there is no need to learn the model every time the spatial quality control is performed. If the model uses sufficient previous data, the performance of spatial quality control is not adversely affected, even if the learning cycle for model updates is only once a week or a month.

TABLE 8: Optimal number of neighboring stations per meteorological element.

Meteorological element	Optimal number of neighbors
Wind direction	7
Wind speed	3
Temperature	11
Humidity	20
Atmospheric pressure	3
Hourly precipitation	8
Rainfall occurrence	10

TABLE 9: Comparison of SVR estimation accuracy with neighboring stations selected randomly or by MOGA.

Weather element	Random	MOGA
Wind direction	50.390	**48.499**
Wind speed	2.523	**2.513**
Temperature	0.970	**0.902**
Humidity	5.216	**5.038**
Atmospheric pressure	1.066	**1.063**
Hourly precipitation	0.847	**0.762**
Rainfall occurrence	0.028	**0.026**

TABLE 10: Comparison of estimation accuracy based on estimation model (RMSE).

Meteorological element	Cressman	Barnes	SVR
Wind direction	53.568	75.470	**48.341**
Wind speed	2.347	2.315	**2.179**
Temperature	1.180	2.583	**0.880**
Humidity	6.755	12.767	**4.582**
Atmospheric pressure	5.663	11.601	**0.847**
Hourly precipitation	**0.583**	0.833	**0.583**
Rainfall occurrence	0.071	0.137	**0.021**

TABLE 11: Execution time for spatial quality control based on estimation model.

Estimator	Average time spent in learning one model (second)	Average time spent in determining one observation (second)
Cressman	—	$1.442e^{-3}$
Barnes	—	$8.427e^{-5}$
SVR	6.839	$2.303e^{-4}$

6.5. Size of Training Set. In general, the higher the number of training samples in the SVR, the higher the accuracy of the estimate but the longer the learning time. Table 12 shows the accuracy of estimates based on the number of training samples. Exceptionally, in the case of wind speed, the performance tends to decrease as the number of training samples increases. Figure 7 also shows that the fewer input variables of SVR, the better the performance with regards to wind speed. In the present model structure, it is difficult to learn wind speed; thus, overfitting seems to occur if the model becomes overly complicated.

Figure 8 shows the learning time according to the number of training samples, and Figure 9 shows the time taken purely for spatial quality control, excluding learning

TABLE 12: Accuracy of estimated values based on the size of the training set (RMSE).

Meteorological element	5000	10000	15000	20000	25000	30000
Wind direction	43.820	42.831	42.298	41.948	41.691	**41.481**
Wind speed	**2.363**	2.365	2.367	2.369	2.369	2.370
Temperature	0.902	0.879	0.870	0.863	0.860	**0.857**
Humidity	4.710	4.330	4.130	3.998	3.904	**3.831**
Atmospheric pressure	0.871	0.837	0.817	0.807	0.797	**0.785**
Hourly precipitation	0.763	0.746	0.736	0.732	0.727	**0.724**
Rainfall occurrence	0.026	0.025	**0.024**	**0.024**	**0.024**	0.024

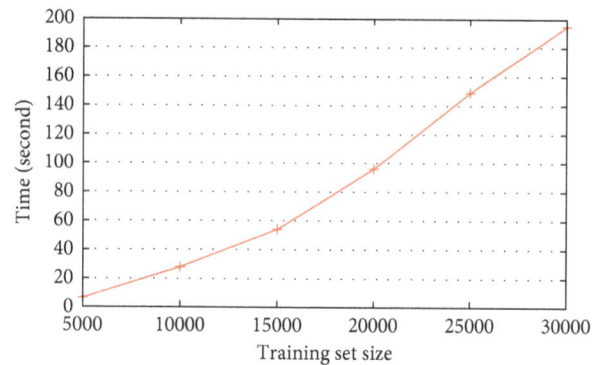

FIGURE 8: Average time spent on learning one model depending on the size of the training set.

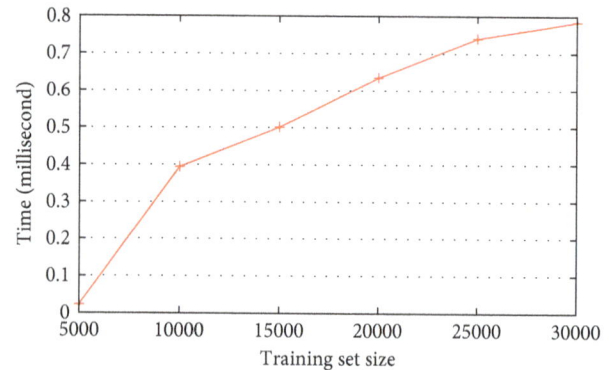

FIGURE 9: Time spent on determining one value depending on the size of the training set.

time. Theoretically, the time taken to test the SVR model is not affected by the size of the training set, but as the training set grows, the complexity of the model becomes larger (e.g., the number of support vectors increases), and the time required for the test also increases. However, as the number of samples increases, the increase in test time gradually decreases. The test time is expected not to increase after the number of samples reaches a certain point. Experiments on all observation stations using 30,000 samples took approximately 15 days on seven machines. Due to time limitations, we could not experiment with more samples, but there seems to be room for further performance improvement. In this

TABLE 13: Results of the proposed spatial quality control method.

Meteorological element	Normal (%)	Suspect (%)	Error (%)	Uninspected (%)
Wind direction	72.9	6.32	$6.31e^{-1}$	20.2
Wind speed	75.7	3.49	$6.56e^{-1}$	20.2
Temperature	93.8	$3.98e^{-1}$	$9.08e^{-2}$	5.67
Humidity	52.8	$2.08e^{-1}$	$5.33e^{-2}$	47.0
Atmospheric pressure	36.4	$4.75e^{-4}$	$1.60e^{-4}$	63.6
Hourly precipitation	87.1	$8.73e^{-1}$	2.99	9.04
Rainfall occurrence	89.2	1.38	$4.16e^{-1}$	9.04

study, all the stations were analyzed together, but the burden of the learning time would not be as great if each test were conducted separately for each observation station.

6.6. Result of Spatial Quality Control. Table 13 shows the results of applying the proposed spatial quality control procedure to actual data. As described in Section 2, the spatial quality control applies only to observations that are determined as normal during basic quality control. Therefore, values that did not pass the basic quality control are classed as uninspected during spatial quality control. The high ratio of uninspected observations of humidity and atmospheric pressure is due to the lack of measuring instruments for those elements in many observation stations.

7. Conclusion

In this study, we proposed a method to detect the abnormality of meteorological data using SVR. First, the value of the corresponding station was predicted using observations made in the surrounding area, and then any abnormality was detected by checking whether the observation differs from the predicted value outside of a predetermined range. SVR was used to create a model to predict the value of observation stations. In addition, we used MOGA to select SVR input variables to improve model performance and to reduce computation time.

Experiments on actual weather data collected by KMA show that using SVR is more accurate than the existing Cressman or Barnes methods for estimating the value of an observation station. Therefore, more accurate anomaly detection is possible through more accurate predictions. If the model can be learned in advance for a fixed cycle rather than learning the model every time, the proposed method has an acceptable execution time. A limitation of the method is that preaccumulated data are required, but it was confirmed through experiments that data collected over approximately one year provide sufficiently high performance.

In future study, the proposed method can be applied to additional meteorological elements such as sunshine duration and cloud height. Other valuable research could examine whether state-of-the-art learning techniques such as deep learning can yield more accurate predictions than SVR, which was not attempted here due to limitations of the system environment. In addition to accurate predictions, additional studies are required on the acceptable difference between the observation and the estimate which we set using the standard deviation. Furthermore, it will be interesting to compare the anomaly detection technique with unsupervised learning technology as opposed to that based on prediction using supervised learning.

Conflicts of Interest

The authors declare that they have no conflicts of interest.

Acknowledgments

This work was supported by Research for the Forecasting Technology and Its Application, through the National Institute of Meteorological Sciences of Korea, in 2015 (NIMR-2012-B-1). This work was also partly supported by BK21 Plus for Pioneers in Innovative Computing (Department of Computer Science and Engineering, SNU) funded by the National Research Foundation of Korea (NRF) (21A20151113068), by the MSIT (Ministry of Science and ICT), Korea, under the ITRC (Information Technology Research Center) support program (IITP-2018-2017-0-01630) supervised by the IITP (Institute for Information & Communications Technology Promotion) and by a grant (KCG-01-2017-05) through the Disaster and Safety Management Institute funded by Korea Coast Guard of the Korean government. The Institute of Computer Technology (ICT) at the Seoul National University provides research facilities for this study.

References

[1] J. H. Seo and Y. H. Kim, "Genetic feature selection for very short-term heavy rainfall prediction," in *Convergence and Hybrid Information Technology*, pp. 312–322, Springer, Berlin, Germany, 2012.

[2] Y. H. Kim and Y. Yoon, "Spatiotemporal pattern networks of heavy rain among automatic weather stations and very-short-term heavy-rain prediction," *Advances in Meteorology*, vol. 2016, Article ID 4063632, 13 pages, 2016.

[3] P. Cortez and A. D. J. R. Morais, "A data mining approach to predict forest fires using meteorological data," in *Proceedings of the 13th Portuguese Conference of Artificila Intelligence*, pp. 512–523, Guimarães, Portugal, December 2007.

[4] A. Stoppa and U. Hess, "Design and use of weather derivatives in agricultural policies: the case of rainfall index insurance in Morocco," in *Proceedings of the International Conference on Agricultural Policy Reform and the WTO: Where are We Heading*, Capri, Italy, June 2003.

[5] M. Kubik, P. J. Coker, J. F. Barlow, and C. Hunt, "A study into the accuracy of using meteorological wind data to estimate turbine generation output," *Renewable Energy*, vol. 51, pp. 153–158, 2013.

[6] H. Yang, L. Lu, and J. Burnett, "Weather data and probability analysis of hybrid photovoltaic–wind power generation systems in hong kong," *Renewable Energy*, vol. 28, no. 11, pp. 1813–1824, 2003.

[7] S. A. Kalogirou, "Applications of artificial neural-networks for energy systems," *Applied Energy*, vol. 67, no. 1, pp. 17–35, 2000.

[8] M. K. Lee, S. H. Moon, Y. H. Kim, and B. R. Moon, "Correcting abnormalities in meteorological data by machine learning," in *Proceedings of the 2014 IEEE International Conference on Systems, Man and Cybernetics (SMC)*, pp. 888–893, San Diego, CA, USA, April 2014.

[9] N. Y. Kim, Y. H. Kim, Y. Yoon, H. H. Im, R. K. Choi, and Y. H. Lee, "Correcting air-pressure data collected by mems sensors in smartphones," *Journal of Sensors*, vol. 2015, Article ID 245498, 10 pages, 2015.

[10] Y.-H. Kim, J.-H. Ha, Y. Yoon et al., "Improved correction of atmospheric pressure data obtained by smartphones through machine learning," *Computational Intelligence and Neuroscience*, vol. 2016, Article ID 9467878, 12 pages, 2016.

[11] J.-H. Ha, Y.-H. Kim, H.-H. Im, N.-Y. Kim, S. Sim, and Y. Yoon, "Error correction of meteorological data obtained with mini-AWSs based on machine learning," *Advances in Meteorology*, vol. 2018, Article ID 7210137, 8 pages, 2018.

[12] V. Chandola, A. Banerjee, and V. Kumar, "Anomaly detection: a survey," *ACM Computing Surveys (CSUR)*, vol. 41, no. 3, pp. 1–58, 2009.

[13] J. Estévez, P. Gavilán, and J. V. Giráldez, "Guidelines on validation procedures for meteorological data from automatic weather stations," *Journal of Hydrology*, vol. 402, no. 1, pp. 144–154, 2011.

[14] C. Daly, W. Gibson, M. Doggett, J. Smith, and G. Taylor, "A probabilistic-spatial approach to the quality control of climate observations," in *Proceedings of the 14th AMS Conference on Applied Climatology*, pp. 13–16, Seattle, WA, USA, January 2004.

[15] G. Sciuto, B. Bonaccorso, A. Cancelliere, and G. Rossi, "Quality control of daily rainfall data with neural networks," *Journal of Hydrology*, vol. 364, no. 1, pp. 13–22, 2009.

[16] G. P. Cressman, "An operational objective analysis system," *Monthly Weather Review*, vol. 87, no. 10, pp. 367–374, 1959.

[17] S. L. Barnes, "A technique for maximizing details in numerical weather map analysis," *Journal of Applied Meteorology*, vol. 3, no. 4, pp. 396–409, 1964.

[18] J.-H. Seo, Y. H. Lee, and Y.-H. Kim, "Feature selection for very short-term heavy rainfall prediction using evolutionary computation," *Advances in Meteorology*, vol. 2014, Article ID 203545, 15 pages, 2014.

[19] M. Jarraud, *Guide to Meteorological Instruments and Methods of Observation (wmo-no. 8)*, World Meteorological Organisation, Geneva, Switzerland, 2008.

[20] V. Vapnik and A. Lerner, "Pattern recognition using generalized portrait method," *Automation and Remote Control*, vol. 24, pp. 774–780, 1963.

[21] H. Drucker, C. J. Burges, L. Kaufman et al., "Support vector regression machines," *Advances in Neural Information Processing Systems*, vol. 9, pp. 155–161, 1997.

[22] T. Joachims, "SVMlight: support vector machine," 1999, http://svmlight.joachims.org.

[23] P. J. Clark and F. C. Evans, "Distance to nearest neighbor as a measure of spatial relationships in populations," *Ecology*, vol. 35, no. 4, pp. 445–453, 1954.

[24] J. H. Holland, *Adaptation in Natural and Artificial Systems: an Introductory Analysis with Applications to Biology, Control, and Artificial Intelligence*, MIT Press, Cambridge, MA, USA, 1992.

[25] E. Zitzler and L. Thiele, "Multiobjective evolutionary algorithms: a comparative case study and the strength Pareto approach," *IEEE Transactions on Evolutionary Computation*, vol. 3, no. 4, pp. 257–271, 1999.

[26] C. M. Fonseca and P. J. Fleming, "Multiobjective optimization and multiple constraint handling with evolutionary algorithms. I. A unified formulation," *IEEE Transactions on Systems, Man, and Cybernetics-Part A: Systems and Humans*, vol. 28, no. 1, pp. 26–37, 1998.

[27] C. A. Coello, "An updated survey of GA-based multiobjective optimization techniques," *ACM Computing Surveys (CSUR)*, vol. 32, no. 2, pp. 109–143, 2000.

[28] L. Xiujuan and S. Zhongke, "Overview of multi-objective optimization methods," *Journal of Systems Engineering and Electronics*, vol. 15, no. 2, pp. 142–146, 2004.

[29] A. Konak, D. W. Coit, and A. E. Smith, "Multi-objective optimization using genetic algorithms: a tutorial," *Reliability Engineering and System Safety*, vol. 91, no. 9, pp. 992–1007, 2006.

[30] K. Deb, A. Pratap, S. Agarwal, and T. Meyarivan, "A fast and elitist multiobjective genetic algorithm: NSGA-II," *IEEE Transactions on Evolutionary Computation*, vol. 6, no. 2, pp. 182–197, 2002.

[31] T. Murata and H. Ishibuchi, "MOGA: Multi-objective genetic algorithms," in *Proceedings of the 1995 IEEE International Conference on Evolutionary Computation*, vol. 1, p. 289, Perth, Australia, 1995.

[32] C. M. Fonseca and P. J. Fleming, "Multiobjective genetic algorithms," in *Proceedings of the IEE Colloquium on Genetic Algorithms for Control Systems Engineering (Digest No. 1993/130)*, London, UK, 1993.

[33] H. Ishibuchi and T. Murata, "Multi-objective genetic local search algorithm," in *Proceedings of the IEEE International Conference on Evolutionary Computation*, pp. 119–124, Anchorage, AK, USA, 1996.

[34] T. Chai and R. R. Draxler, "Root mean square error (RMSE) or mean absolute error (MAE)?—arguments against avoiding RMSE in the literature," *Geoscientific Model Development*, vol. 7, no. 3, pp. 1247–1250, 2014.

[35] P. Allerup, H. Madsen, and F. Vejen, "A comprehensive model for correcting point precipitation," *Hydrology Research*, vol. 28, no. 1, pp. 1–20, 1997.

14

Predictive Contributions of Snowmelt and Rainfall to Streamflow Variations in the Western United States

Xiaohui Zheng [ID], [1] **Qiguang Wang** [ID], [2] **Lihua Zhou,** [1] **Qing Sun,** [3] **and Qi Li** [4]

[1]*College of Global Change and Earth System Science, Beijing Normal University, Beijing 100875, China*
[2]*China Meteorological Administration Training Center (CMA), Beijing 100081, China*
[3]*College of Applied Meteorology, Nanjing University of Information Science and Technology, Nanjing 210044, China*
[4]*School of Atmospheric Physics and Atmospheric Environment, Nanjing University of Information Science and Technology, Nanjing 210044, China*

Correspondence should be addressed to Qiguang Wang; wangqg@cma.gov.cn

Academic Editor: Brian R. Nelson

This study used long-term in situ rainfall, snow, and streamflow data to explore the predictive contributions of snowmelt and rainfall to streamflow in six watersheds over the Western United States. Analysis showed that peak snow accumulation, snow-free day, and snowmelt slope all had strong correlation with peak streamflow, particularly in inland basins. Further analysis revealed that the variation of snow accumulation anomaly had strong lead correlation with the variation of streamflow anomaly. Over the entire Western United States, inner mountain areas had lead times of 4–10 pentads. However, in coastal areas, nearly all sites had lead times of less than one pentad. The relative contributions of rainfall and snowmelt to streamflow in different watersheds were calculated based on the snow lead time. The geographic distribution of annual relative contributions revealed that interior areas were dominated by snowmelt contribution, whereas the rainfall contribution dominated coastal areas. In the wet season, the snowmelt contribution increased in the western Pacific Northwest, whereas the rainfall contribution increased in the southeastern Pacific Northwest, southern Upper Colorado, and northern Rio Grande regions. The derived results demonstrated the predictive contributions of snowmelt and rainfall to streamflow. These findings could be considered a reference both for seasonal predictions of streamflow and for prevention of hydrological disasters. Furthermore, they will be helpful in the evaluation and improvement of hydrological and climate models.

1. Introduction

The Western United States (WUS) is a semiarid region that covers more than half the land area of the U.S. [1]. This area, which receives little precipitation during summer, relies considerably on the wintertime precipitation phase and snowpack accumulation to sustain a multitude of ecosystem goods and services [2]. Thus, the regional ecology and economy are both vulnerable to water resource anomalies caused by seasonal hydroclimatic variations [3]. Recent hydrological disasters in the WUS have been closely related to climate change [4–6]. The occurrence probabilities of hydrological anomalies and of hydrological extremes have both increased [7]; these conditions have affected the

regional agricultural production, the occurrence of forest fires, and the national socioeconomic development [8, 9]. Therefore, it is of great importance to conduct accurate and effective hydrological forecasting for the WUS within the context of climate change.

Streamflow is a hydroclimatic variable that directly influences drought- and flood-related disasters. It is affected both by natural factors, such as the precipitation system, soil state, and land surface, and by human factors, including land use changes and water use efficiency. Because of the considerable reliance of the WUS on snow as a water resource, snow accumulation represents a factor of first-order importance regarding regional water supply [10]. Snowmelt contributes approximately 50%–80% of the total streamflow and is highly seasonal in nature, that is, the majority of streamflow occurs because of snowmelt during the late spring–summer (April–July) [11]. In western coastal areas, streamflow is derived mainly from rainfall. Therefore, streamflow variation in the WUS has strong correlation with changes in snowpack accumulation and rainfall and has obvious seasonal variation [11–14].

A considerable amount of research has been undertaken regarding the relationships between runoff and both precipitation systems and snowmelt. In addition, several studies have documented the correlation between oceanic climatic phenomena and streamflow. For example, Hunter et al. [15] reported a strong relationship of streamflow with ENSO and relatively weak correlations with the PDO, AMO, and NAO anomalies. Except for those that researched atmospheric-oceanic circulations, most of the studies on the seasonal variation of streamflow focused on the relationship between runoff and snowmelt during the melting season. Snowmelt is related closely to the snowpack amount and temperature in winter and spring. The spring temperature can affect streamflow by influencing both the time of the onset of snowmelt and the snowmelt slope [16]. Recently, spring temperatures have demonstrated a warming tendency coincident with an earlier onset of snowmelt. Many researchers have started to focus on the relationship between early snowmelt and spring and summer runoff [17–20]. Some studies have suggested that warmer spring temperatures would lead to earlier onset of snowmelt and greater associated runoff, thereby resulting in increased spring streamflow and decreased summer streamflow [19, 21, 22]. Jeton et al. [23] derived the same conclusion, believing that early snowmelt occurred when vegetation was less active. They claimed that such inactivity disrupts the synchronicity between the water availability and demand, resulting in greater springtime streamflow. Conversely, Bosson et al. [24] thought that earlier snowmelt might contribute more to evapotranspiration (ET) than to streamflow because of the increased vapor pressure deficit due to atmospheric warming, which would result in a lower-than-usual springtime streamflow. Recent findings by Trujillo and Molotch [25] indicated that reduced levels of solar radiation available for driving snowmelt earlier in the year could produce slower rates of snowmelt and decrease the generation of streamflow during the spring.

Based on the relationship between snowmelt and streamflow, snowmelt could be used as a significant predictor of streamflow. However, previous studies simply revealed the possibility of this phenomenon and consequently outlined its potential mechanisms without achieving a uniform conclusion. In addition, earlier work focused primarily on changes in the timing of snowmelt without considering the effects of other snow metrics or of the memory of snowmelt on streamflow. Furthermore, the research cited in this study was based either on the hydrologic situation of the entire WUS or on that of a specific watershed. Therefore, this study separated the WUS into six watersheds and selected eight snow metrics to both investigate the relationships among the snow metrics and streamflow and to determine the lead correlation between

the snowpack and streamflow in each watershed. The findings of this research demonstrate the potential for the use of snow variations to predict streamflow changes in advance.

In addition to snowpack, rainfall represents another important predictor of streamflow in the WUS. However, because the contributions of rain-derived runoff (also called rainfall contribution, referred to as f_{rain} hereafter) and of snow-derived runoff to the total runoff (also called snowpack contribution, referred to as f_{snow} hereafter) are different in each area; both f_{rain} and f_{snow} have different contributions to streamflow. Previous studies focused primarily on f_{snow} [10]. In addition, in most of those studies, f_{snow} was calculated based on metrics such as the total snowfall as a fraction of total precipitation, the total snowfall as a fraction of total runoff, or melt season runoff as a fraction of total annual runoff [26]. Partly because of differences among the methods by which f_{snow} was approximated, large variations in the estimates have been reported. For example, most research has reported that snow contribution ratio was 75% [27–31], whereas other studies have produced values in the ranges of 40%–60%, 50%–80%, or 60%–90% [17, 32–34]. Compared with such ratio metrics, the following methods are considered more reasonable. Li et al. [10] quantified f_{snow} by tracking the fate of snowmelt in modeled hydrologic fluxes and obtained gridded model results. Through calculations of the long-term probability of snowmelt pulse occurrence (i.e., if the snowmelt pulse occurred during the period between the 150th day and the 250th day of the water year), Fritze et al. [35] divided streamflow sites into four categories: clearly rain dominated, mostly rain dominated, mostly snowmelt dominated, and clearly snowmelt dominated. This study used a new method to calculate f_{rain} and f_{snow} at each streamflow site both during an entire year and during the wet season to provide a more accurate basis for a seasonal prediction of streamflow.

To achieve qualitative and quantitative seasonal predictions of streamflow, this study considered three issues based on long-term observational data. First, among the eight snow metrics used widely throughout the literature, those that have significant correlations with the peak streamflow were selected. Second, the lead correlation between the SWE and streamflow at each stream site was calculated. Third, the relative contributions of rain and snowmelt to streamflow were quantified. These findings will provide a basis for seasonal streamflow predictions and benefit the evaluation of hydroclimatic models. The remainder of this paper is organized as follows. Descriptions of the study area and of the observational data are provided in Section 2. The methods adopted for the analysis of the data are presented in Section 3. Section 4 presents the results with respect to the three issues listed above. Finally, the conclusions are summarized in Section 5.

2. Research Area and Data

This study focused on six hydrologic regions of the WUS which correspond to USGS Regions 13–18 (https://water.usgs.gov/GIS/regions.html): Rio Grande (RG), Upper

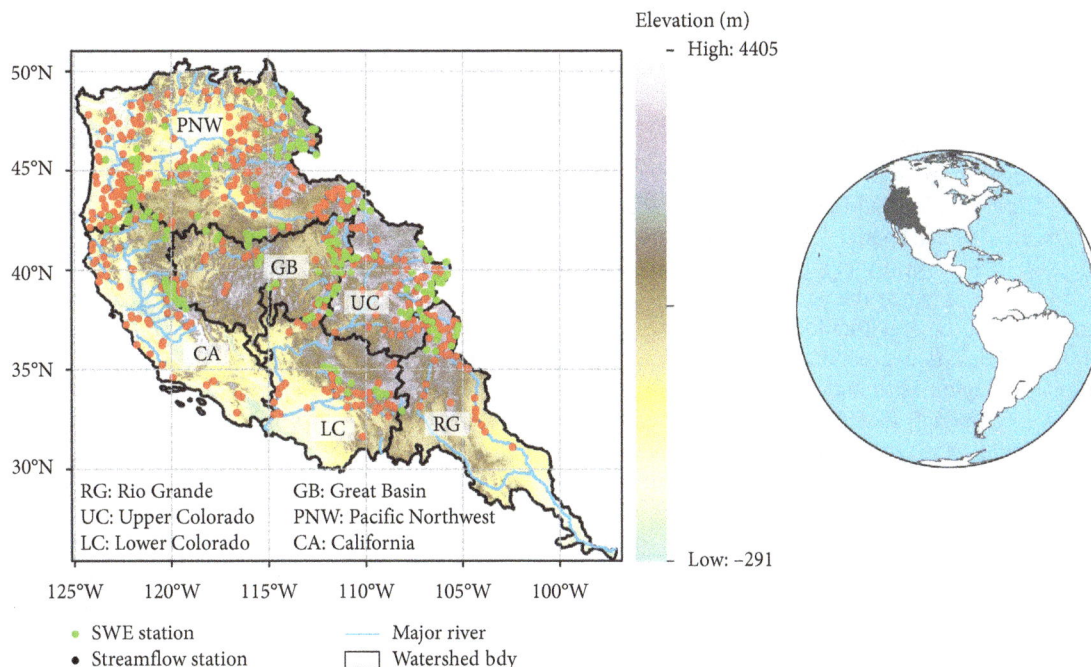

FIGURE 1: Research area, station distribution, and the location of the study area in global (gray area in western hemisphere).

Colorado (UC), Lower Colorado (LC), Great Basin (GB), Pacific Northwest (PNW), and California (CA). The distribution of the watersheds and details of their elevations and major rivers are illustrated in Figure 1.

2.1. Snow Water Equivalent (SWE). The primary daily SWE record was obtained from the Natural Resource Conservation Service, which has operated the SNOTEL automated network of snowpack monitoring sites in the WUS since 1978 (http://www.wcc.nrcs.usda.gov/snow/). At each SNOTEL site, the weight of snow on a liquid-filled pillow is measured hourly by a pressure sensor and converted to SWE [16, 36, 37]. The high temporal resolution and long time series of the SNOTEL record (since the late 1970s) make it uniquely suited to the primary goals of this study. Data were selected over a 35-year period (1 October 1981 through 30 September 2016) from SNOTEL sites with continuous daily records during the snow accumulation period (1 September through 30 April the following year). Overall, data from 222 SNOTEL sites located within the 6 watersheds were considered suitable. The numbers of selected SNOTEL sites in the six watersheds are listed in Table 1, and the locations of the stations are marked in Figure 1. The selected stations covered all the mountains of each watershed; thus, the snow metrics could be calculated reliably without undue interference from sampling errors and/or seasonal effects.

2.2. Streamflow. The streamflow data are available on the Internet from the National Water Information System of the USGS (http://waterdata.usgs.gov/nwis/). The stations were categorized according to the hydrologic drainage basins in the US known as hydrologic unit codes developed by the USGS [12]. Because human interventions such as reservoirs

TABLE 1: Numbers of SNOTEL and streamflow stations used in this study.

	RG	UC	LC	GB	PNW	CA	Total
SNOTEL	12	39	15	54	92	10	222
Streamflow	43	63	51	54	209	41	461

and other diversions could potentially alter the routes of river systems, the USGS maintains unimpaired streamflow stations within the Hydro-Climatic Data Network 2009 (HCDN-2009) [38]. In the site selection process of this study, these sites were chosen first. However, the number of HCDN-2009 stations is limited and their spatial distribution is uneven. Therefore, other USGS sites (added stations) were incorporated based on the following quality assurance/quality control procedure. First, manual screening was performed to remove those with incomplete daily data. Second, the watershed mean pentad anomaly streamflow of all the HCDN-2009 stations in each watershed was calculated. Then, sites that were obviously inconsistent with the HCDN-2009 stations, based on a comparison of the correlation coefficient of the pentad anomaly streamflow between each added station and the watershed mean, were discarded. This step was performed to remove any station that had been obviously influenced by anthropogenic activities. The data period was the same as for the SWE sites. The numbers of selected stations in each of the six watersheds is listed in Table 1, and the locations of the stations are marked in Figure 1. The selected streamflow stations covered nearly all the major rivers in each watershed, providing strong support for the subsequent data analysis.

2.3. Precipitation and 2-m Temperature. Daily precipitation and daily mean 2-m temperature data were obtained from the PRISM dataset named AN81d [39, 40]. This daily dataset, covering the conterminous US, began on 1 January 1981, and it continues to the present day, and the spatial resolution was 4 km. To derive best estimates, this dataset uses all the station networks ingested by the PRISM Climate Group (13 station networks for temperature and 20 station networks for precipitation). Climatologically aided interpolation (detailed in Descriptions of PRISM Spatial Climate Datasets, 2015 [41]), using 1981–2010 monthly climatologies as predictor grids, was adopted to enhance the accuracy of the climate analysis. Furthermore, the precipitation dataset was created by computing an elevation-precipitation relation for all stations within a subregion, with individual stations weighted differently, and by applying that relation to a topographic basemap. In this study, the data period was the same as for SWE and streamflow. The precipitation data with 4 km high spatial resolution and correction for elevation were considered suitable for the requirements of precipitation system analysis over high mountain areas and for determining effective rainfall areas. The 2-m temperature data were used to classify precipitation as either rain or snow for analysis of the rainfall contribution.

3. Methodology

3.1. Snow Metrics. This study adopted eight snow metrics following the work of Trujillo and Molotch [25]. These metrics comprise the initial snow accumulation day and snow-free days, the peak of snow accumulation and date of the peak, the length of the snow accumulation season and the length of the snowmelt season, and the snow accumulation slope and snowmelt slope. This research focused on the relationship between long-term snow accumulation and streamflow. Therefore, the calculations for all of the metrics were based on the longest snow cover period (>20 d) in each water year, and short-term snow periods were neglected. The eight metrics were derived from the daily SWE data at each station. The daily SWE curve (Figure 2) for accumulation and melt seasons shows the basic metrics that can be used to characterize the snowpack dynamics at a particular station. Metrics (1) and (2) mark the days of initial snow accumulation and snow disappearance, respectively. Metrics (3) and (4) mark the peak SWE date and peak SWE value, respectively. Metrics (5) and (6) indicate the lengths of the accumulation and melt seasons, respectively. The snow accumulation slope and snow snowmelt slope are represented by the peak snow accumulation divided by the length of the snow accumulation season and by the length of the snowmelt season, respectively.

In the correlation analysis with peak streamflow, the snow state in each watershed is considered to have consistent characteristics; thus, the snow metrics are the mean values of all the SWE stations in each individual watershed.

3.2. Lead Correlation of Snow Accumulation to Streamflow. Because streamflow has a long-term snowpack memory, that memory could be calculated as the predictive time of the influence of snowpack on streamflow. The lead time

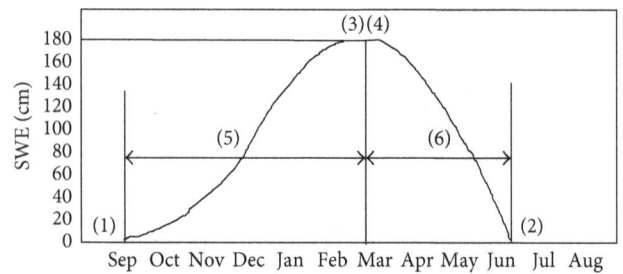

FIGURE 2: Sketch of the basic snow metrics based on the daily SWE curve. Metrics (1) and (2) mark the days of the initial snow accumulation and snow disappearance, respectively. Metrics (3) and (4) mark the peak SWE date and peak SWE value, respectively. Metrics (5) and (6) indicate the length of the accumulation and melt seasons, respectively.

(memory) was derived from the lead correlation between the long-term SWE pentad anomaly and the streamflow pentad anomaly. Here, because the study area was divided into six watersheds, it was assumed that the snow characteristics of each individual basin were uniform. The lead correlation was calculated between the basin mean long-term (i.e., 35 years) SWE pentad anomaly and the long-term streamflow pentad anomaly at each stream site in the corresponding basin. As the response time of streamflow to snowpack is different at each site, a group of correlation coefficients was calculated with lead times ranging from 0 to 73 pentads for every site. Then, the maximum correlation coefficient corresponding to the lead time was considered the predicting period for that site.

3.3. Relative Contributions of Rainfall and Snowpack. In this study, rainfall and snowpack were chosen as the principal streamflow contributors. For snowpack, snow accumulation was adopted rather than snowfall. This is because the influence of snowpack on streamflow is realized when the snow accumulated in mountainous areas is released during the spring, discounting the negligible amounts of snow that fall directly into rivers. Because it is difficult to obtain observed rainfall data, PRISM precipitation data were classified as snowfall or rainfall based on PRISM 2-m-resolution temperature data. To verify the feasibility of this method, the classified snowfall data were evaluated against SNOTEL in situ SWE data. Subsequent to data preparation, the contributions of both rainfall and snowpack to streamflow were calculated according to the following procedure.

First, it was necessary to determine the rainfall area with a direct impact on each streamflow station. As rainfall has a direct influence on streamflow, it was assumed that a change in the streamflow anomaly would have the same pattern as a change in the contributed rainfall anomaly. The rainfall area that affected each streamflow site was chosen based on the criterion that the pentad rainfall anomaly had a significant correlation with the runoff anomaly of that site during the wet season. The wet season was chosen instead of the entire year to avoid seasonal variations. The wet season, which differs among each basin, was selected based upon

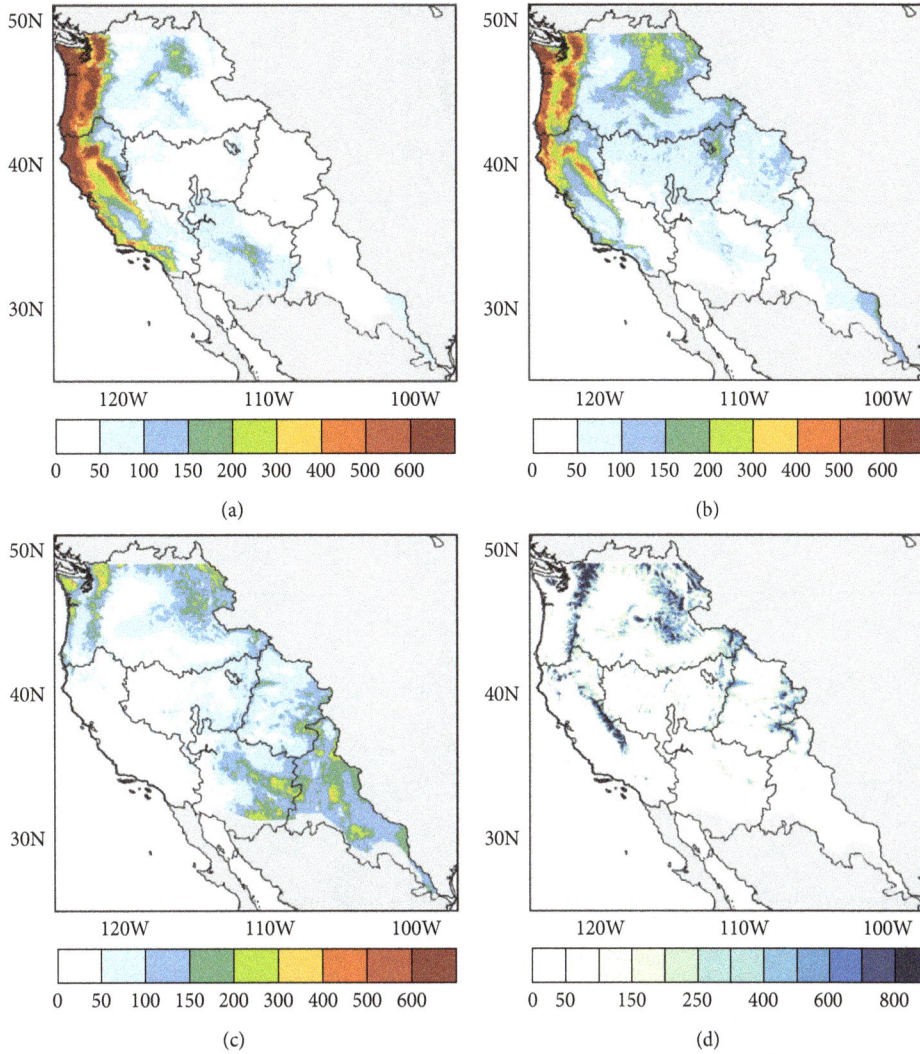

FIGURE 3: Mean precipitation during 1981–2016 (unit: mm): (a) winter mean rainfall, (b) spring mean rainfall, (c) summer mean rainfall, and (d) annual mean snowfall.

the long-term averaged seasonal variation in the rainfall (Figure 3, RG: Jul–Sep; UC: Jul–Sep; LC: Jul–Sep; GB: Mar–May; PNW: Oct–Dec; CA: Dec–Mar). The regional mean rainfall anomaly of the selected area was the effective rainfall anomaly of each station.

Second, the long-term pented rainfall and SWE anomalies at each streamflow site were normalized, and then use (1) to fit the two normalized variables into a new variable, which is called $comb_a$.

$$comb_a(i) = a(i) \cdot Rain_a + (100 - a(i))(SWE_a + \text{lead time}),$$
$$a(i) = i - 1, i = 1, 2, \ldots, 101. \tag{1}$$

where SWE_a is the normalized basin mean long-term pentad SWE anomaly, $Rain_a$ is the normalized long-term pentad rainfall anomaly, lead time is the highest probability of all the lead times in each basin, and a is the relative contribution ratio of rainfall from 0 through 100. The snow accumulation contribution ratio is represented as $100 - a$ because only these two contributors were considered in this study.

Third, using the group of the combined array to perform the correlation with streamflow at each site in each basin, the peak correlation among the group of correlations was selected. The peak correlation corresponding to a and $100 - a$ represents the contribution ratio to streamflow at each station.

4. Results

4.1. Rain, Snow, and Streamflow in the WUS. Because streamflow variation depends largely on the precipitation system, the variation and distribution of precipitation should first be established. Two principal precipitation mechanisms affect the WUS. The first is associated with eastward-moving winter Pacific storms, which bring heavy precipitation to coastal areas and the western highlands, thereby reflecting the considerable orographic effects generated by upslope motions [42–44]. The second precipitation system is associated with subtropical summer monsoon rainfall.

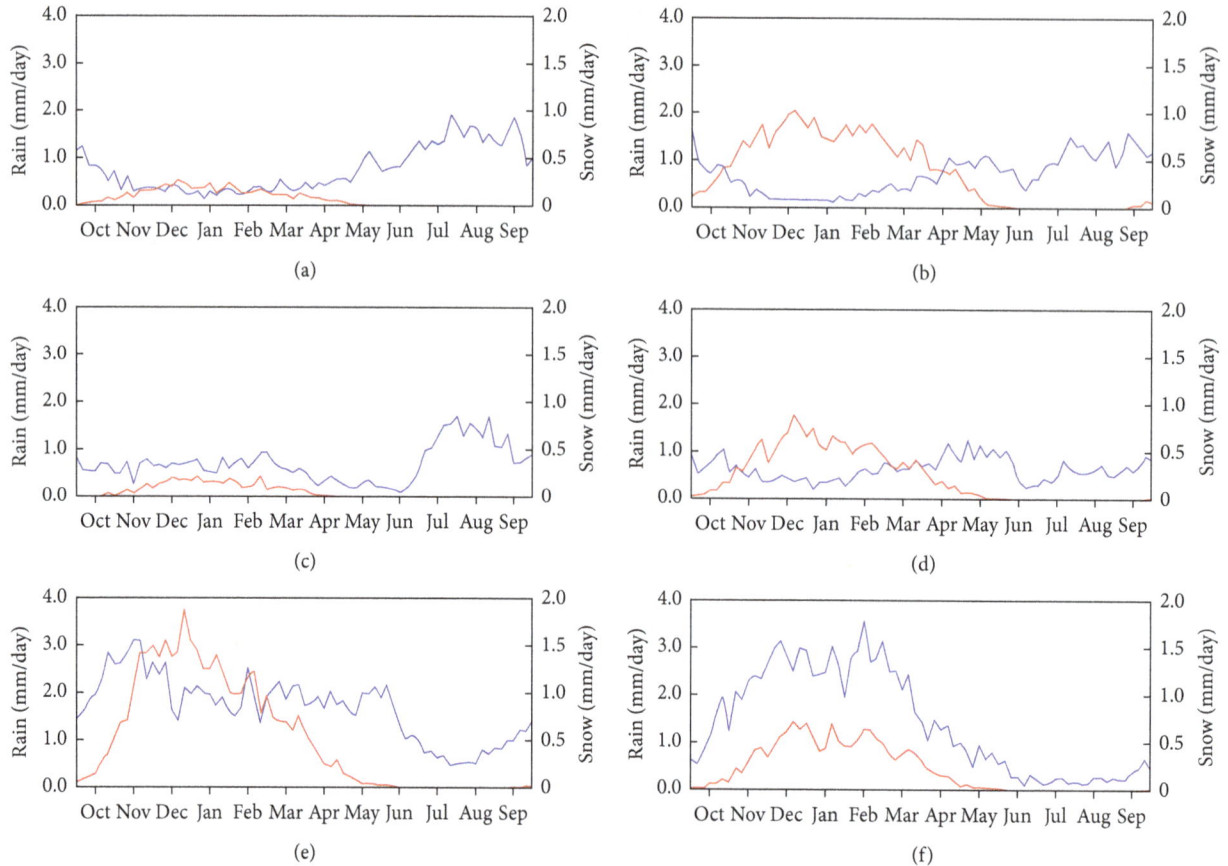

FIGURE 4: Annual time series (1981–2016) of the regional mean precipitation (blue line: rain; red line: snow). Panels (a)–(f) successively represent watersheds (a) RG, (b) UC, (c) LC, (d) GB, (e) PNW, and (f) CA.

The temporal variations and distributions of seasonal rainfall and annual snow amounts over the WUS are illustrated in Figures 3 and 4. It is evident that considerable rainfall occurs during winter, particularly in the Cascade, Klamath, and Sierra Nevada mountains and over coastal areas. Compared with winter, both the amount and the area of rainfall in spring are generally diminished, but the locations of the rainfall centers remain similar. However, in the Rocky Mountains, the rainfall amount increases in spring because snowfall transfers to rainfall with rising temperatures. Summer precipitation is located primarily within the RG and LC watersheds (i.e., summer monsoon rainfall), whereas precipitation is reduced considerably in the PNW because Pacific storm systems are weaker in this season. In the WUS, snowfall generally commences at the end of autumn and it ends in early spring (Figure 4). The area of snow cover is generally concentrated within inland areas of the Cascade, Sierra Nevada, and Rocky Mountains, and the amount of snowfall increases from south to north.

Because of its low latitude, the RG basin receives the least amount of snowfall; however, its large runoff in spring suggests that snowmelt in the southern Rocky Mountains is the main source of its streamflow. The summer monsoon constitutes the primary precipitation system affecting the RG basin throughout the entire year, and it provides a major source of summer runoff (Figures 4(a) and 5(a)). Most of the UC basin lies within the southern Rocky Mountains. The considerable accumulation of wintertime snow therein indicates that subsequent snowmelt provides abundant water for spring runoff; thus, the period of high flow, during which time the flow is substantially higher than during the other months, is concentrated during May–July (Figures 4(b) and 5(b)). Streamflow in the LC basin is similar to that in the RG basin. The spring streamflow in both basins is fed primarily by snowmelt from the southern Rocky Mountains. However, because the Rocky Mountains are located further to the south than in the RG basin, the period of snowfall is short and the amount of accumulation is low; therefore, the spring streamflow contributed by snowmelt is less prominent compared with the other seasons. Furthermore, the contributions to runoff from both winter and summer rainfall produce a curve of seasonal runoff variations with reasonably gentle features (Figures 4(c) and 5(c)). The GB watershed exhibits a midlatitude desert climate; the amount of rainfall in this basin is the least among all of the studied watersheds. The GB is bordered by the Sierra Nevada range to the west and the Wasatch and Uinta Mountains to the east. Therefore, its streamflow is supplied primarily by snowmelt from these two mountain ranges (Figures 4(d) and 5(d)). The PNW has a largely marine oceanic climate with considerable rainfall throughout the entire year, except during the summer. Further to the east, it has a semiarid-steppe

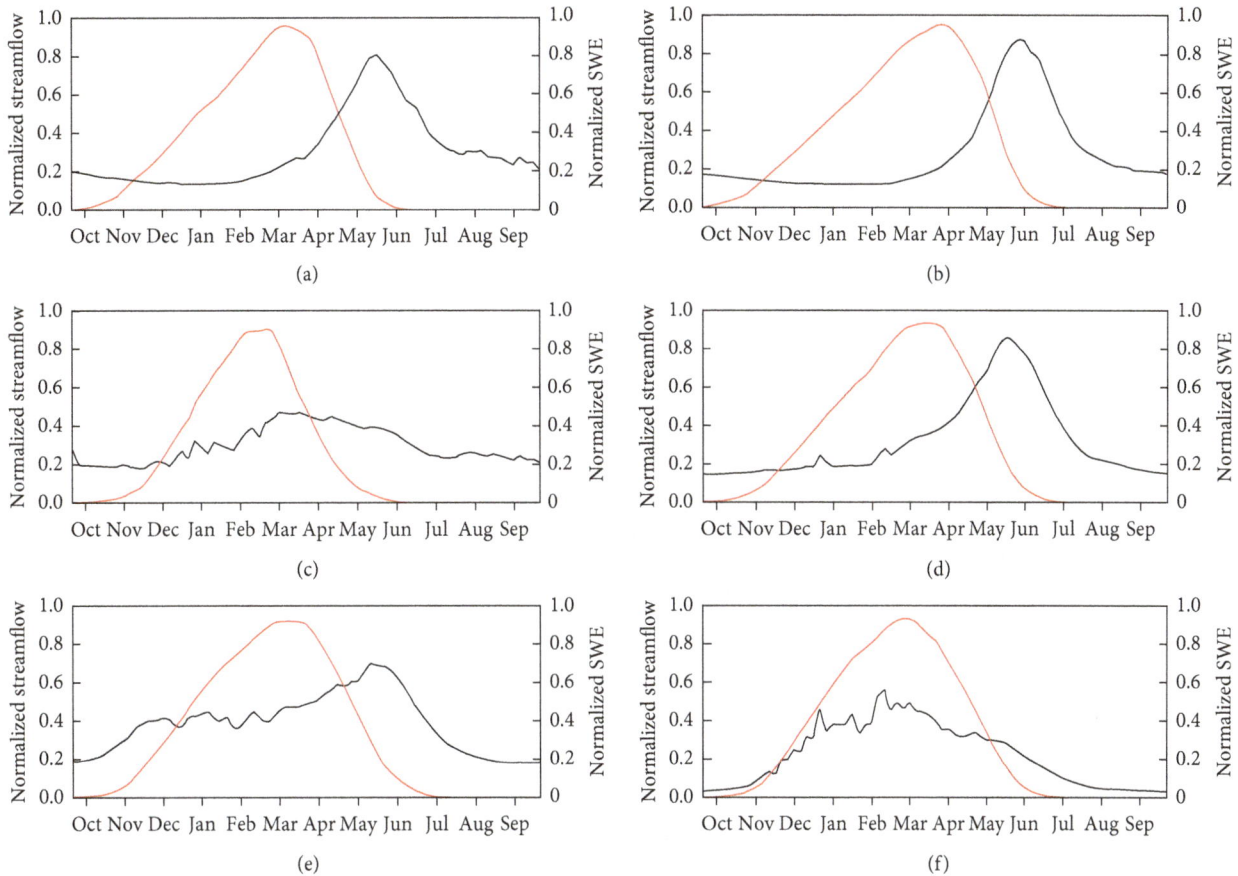

FIGURE 5: Annual time series (1981–2016) of the normalized SWE and streamflow (black line: streamflow; red line: SWE). Panels (a)–(f) successively represent watersheds (a) RG, (b) UC, (c) LC, (d) GB, (e) PNW, and (f) CA.

climate. Snow accumulation over the northern and central parts of the Rocky Mountains stores vast quantities of water for the PNW. Thus, based on winter rainfall and spring snowmelt, the annual streamflow variations display a characteristically sustained increase from the early winter to the end of the spring (Figures 4(e) and 5(e)). The Mediterranean climate of the CA basin is reflected by cool, wet winters and hot, dry summers. This area has more precipitation in winter and spring, which is the main source of streamflow; thus, runoff and precipitation show synchronized seasonal variations (Figures 4(f) and 5(f)).

Although the basins of the WUS exhibit a certain degree of consistency in terms of their hydroclimatic situations, the seasonal variations in both precipitation and streamflow reveal that each basin does have its own characteristics. Therefore, to enhance an understanding of the hydroclimatic situation of the WUS, the relationships between the precipitation systems and streamflow must be explored for each individual basin.

4.2. Correlation between Snow Metrics and Streamflow Peak. Because streamflow has a short response time to rainfall, only snowfall was considered in this analysis. To obtain meaningful seasonal predictors of runoff, the relationships to runoff of eight snow metrics (detailed in Section 3) were

explored. Based on the work of Trujillo and Molotch [25], the selected metrics have different nonlinear relationships in the different regions of WUS (i.e., maritime climate, intermountain climate, and continental climate); thus, the relationships between the snow metrics and the runoff peak can be discussed independently. The focus of this study was on the relationships between the snow metrics and the runoff in each of the six individual watersheds. Therefore, the snow metrics were taken as the average values from all of the SWE sites in each watershed. The 95th percentile of the annual streamflow, which is strongly related to the wet season streamflow amount and flood disasters, was used to perform the correlation for the streamflow metric.

Figure 6 shows that each snow metric has different characteristics in each region. Except for the initial snow accumulation day, which primarily has a negative correlation with the peak streamflow, all of the other metrics have positive correlations. The negative correlations displayed in Figure 6(a) indicate that an earlier onset of snow accumulation corresponds to a higher peak streamflow. This is intuitively correct because an early occurrence of snow would generally correspond to deeper snowpack, which would cause greater streamflow during the snowmelt season. However, this correlation is not significant for all of the six basins, that is, this metric is relevant only in the northern RG, UC, and GB watersheds. Among the remaining metrics,

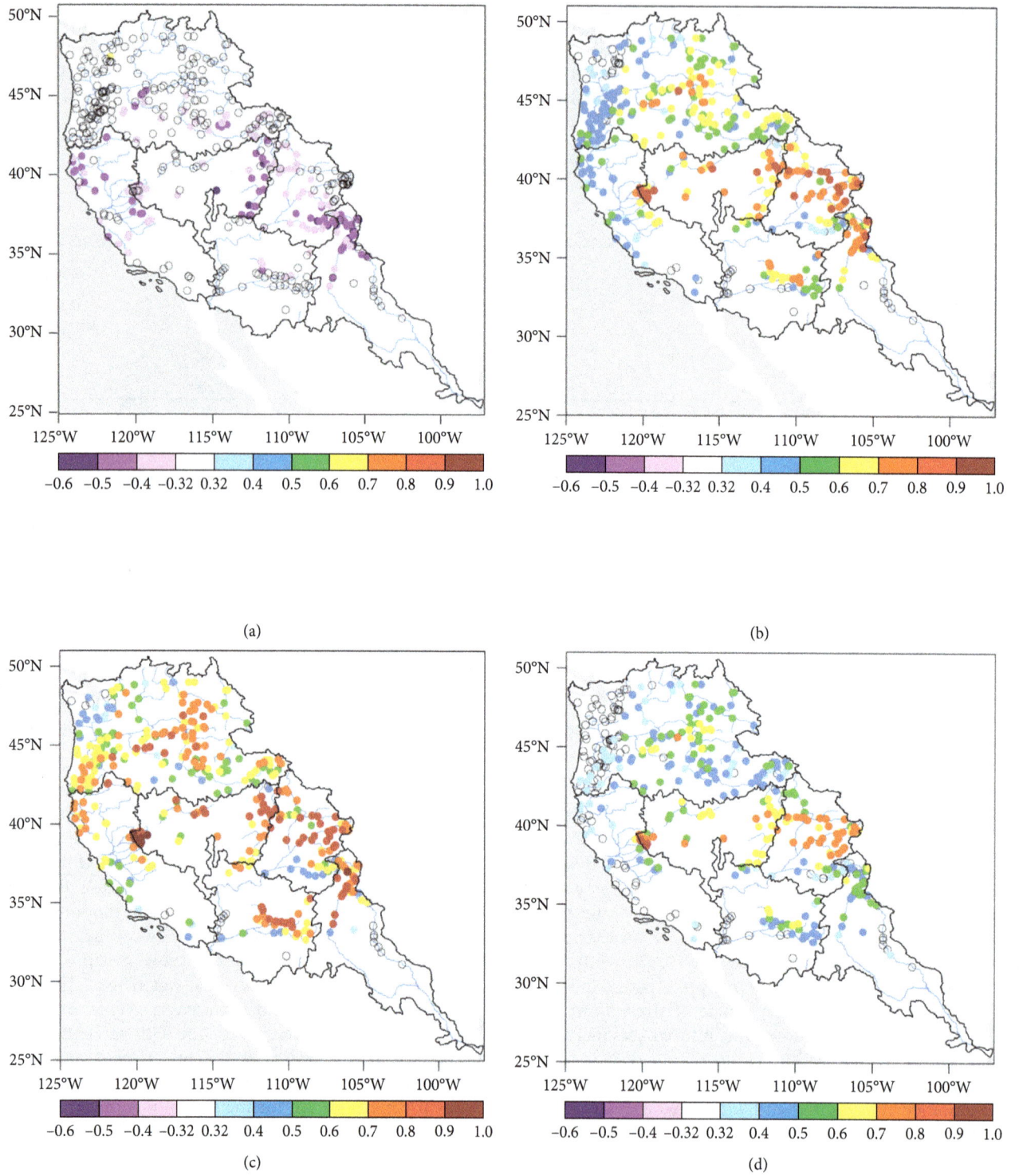

FIGURE 6: Continued.

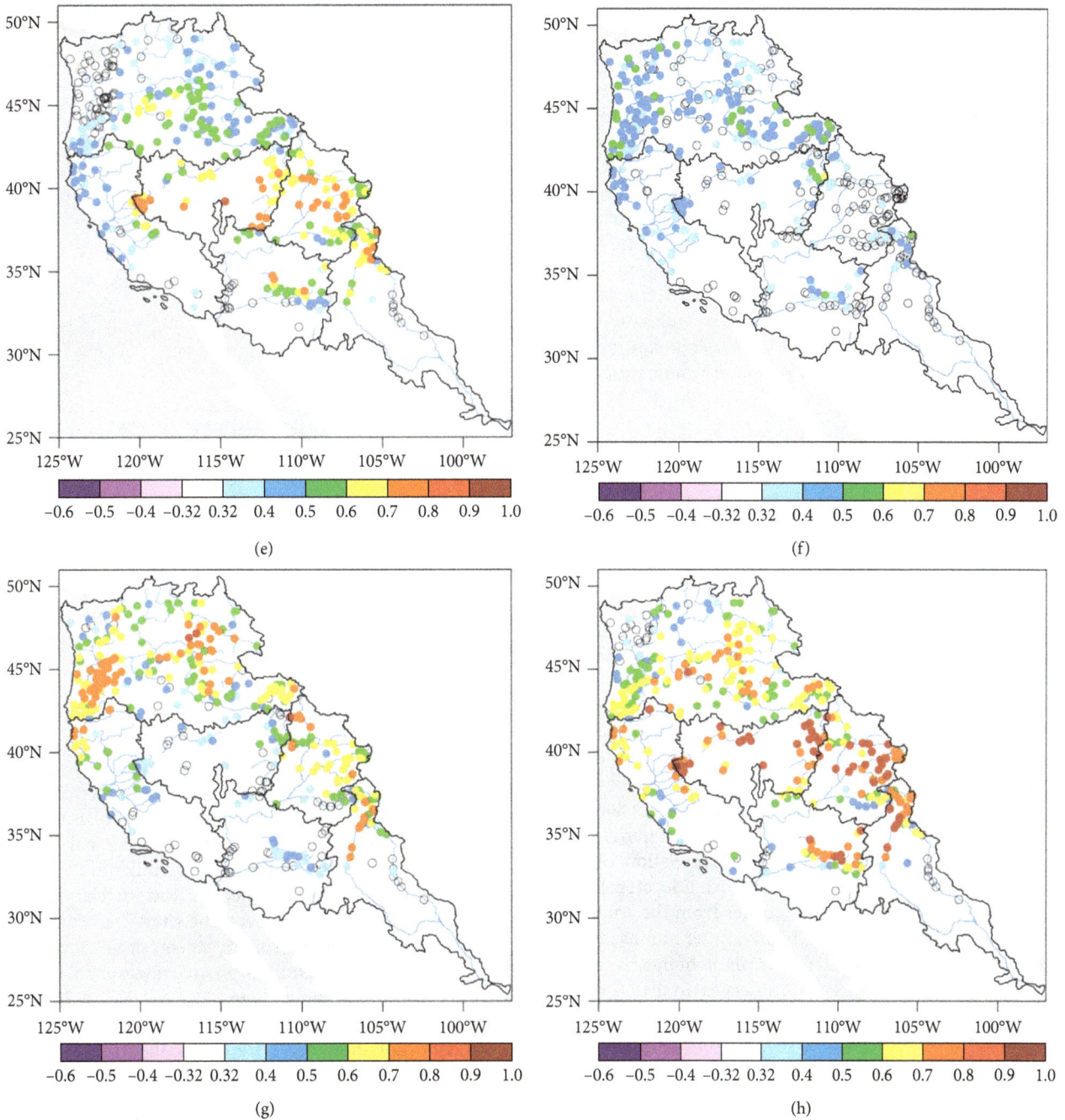

FIGURE 6: Distributions of the correlation coefficients between the snow metrics and peak streamflow in each watershed (1981–2016). Panels (a)–(h) represent different snow metrics: (a) the initial snow accumulation day, (b) snow-free day, (c) peak SWE value, (d) peak SWE day, (e) length of snow accumulation season, (f) length of snow melting season, (g) snow accumulation slope, and (h) snowmelt slope. Circles represent stations that did not pass the 95% significance test.

the peak of snow accumulation, the snow-free days, and the snowmelt slope all demonstrate strong positive correlations with the peak runoff (i.e., considerable snow accumulation, late snow disappearance, and rapid snowmelt all correspond to a high peak streamflow), followed by the date of peak accumulation, the length of snow accumulation, and the snow accumulation slope. The distribution of these correlations (Figure 6) shows that most of the metrics are strongest in the UC and GB watersheds. This is because

water storage is dominated by snowmelt in these areas; thus, the snow metrics are more informative. In addition, both the RG and the LC basins are located in the lower latitudes of the WUS; therefore, they receive less snow than the other basins (Figures 3 and 4), and their downstream runoff is affected principally by rainfall. Those runoff sites that have strong correlations with the snow metrics are concentrated in the upstream areas near the mountains. It is noteworthy that in the coastal areas of the PNW and CA basins, although runoff is dependent primarily on winter rainfall, these results do show certain correlations with the snow metrics. This is because both rainfall and snowfall in these two basins are caused by the same weather systems; thus, they appear almost synchronized, which means that these snow metrics appear to have certain correlations that represent the regional rainfall characteristics. In conclusion, the peak of snowfall accumulation, the snow-free days, and the snow snowmelt slope could be used as the principal metrics for predicting the peak value of the runoff throughout the WUS, especially in the inland mountain areas. Among the three most significant metrics, the peak of snowfall accumulation demonstrated the strongest correlation with the peak streamflow. This result shows that snow accumulation could be used to quantify the forecast period of runoff variations.

4.3. Lead Correlation of Snow Accumulation and Streamflow.
This analysis explores the forecasting effect of snow accumulation on runoff based on the long-term pentad anomaly of SWE and runoff (detailed in Section 3). Figure 7 shows that almost all runoff sites passed the 95% significance test and most have strong correlation. The lead correlation between SWE and streamflow was examined from 0 through 14 pentads. Overall, in the inland mountains, the lead time is more than 4 pentads, while in coastal areas, the lead time is 0-1 pentads.

Among the six watersheds, the UC and GB basin show the longest snow memory in terms of the runoff, that is, generally 4–10 pentads. Snowfall in the RG basin is concentrated largely over the southern Rocky Mountains, which is where the source of the Rio Grande River (the westernmost of the two main rivers in the RG basin) is located. Therefore, the runoff into the river is derived primarily from snowmelt, and the river shows a lead correlation of 5–7 pentads. In contrast, the Pecos River, which is located on the eastern side of the RG basin, originates from the southern edge of the Rocky Mountains. Thus, summer rain has much greater impact than snowmelt on the runoff of this river, and no strong correlation with SWE is observed. In the LC basin, the response time of snow to runoff is short, and most runoff sites do not have a strong correlation with snowmelt. Because the snow storage in the LC basin is small, the snowmelt period is short compared with those of other watersheds, and most of the water resources in this basin are derived from summer monsoon rainfall. The occurrences of rainfall and snowfall during the winter are highly consistent, indicating that they are both caused by the same weather systems, that is, mountainous areas receive snow while areas at lower elevations receive rain. The response of the runoff to rainfall is very short, that is why the lead correlation in the basin is just 0-1 pentads showed in Figure 7. The coastal area of the PNW shows a short lead time and a low correlation coefficient. This is because eastward-moving Pacific storm systems during the winter generate considerable rainfall in front of the mountainous areas and abundant snowfall at high elevations, that is, the rainfall and snowfall in this basin are generated by the same weather systems, as discussed for the LC basin. The amount of precipitation over the eastern area of the PNW is much smaller than that over the coastal

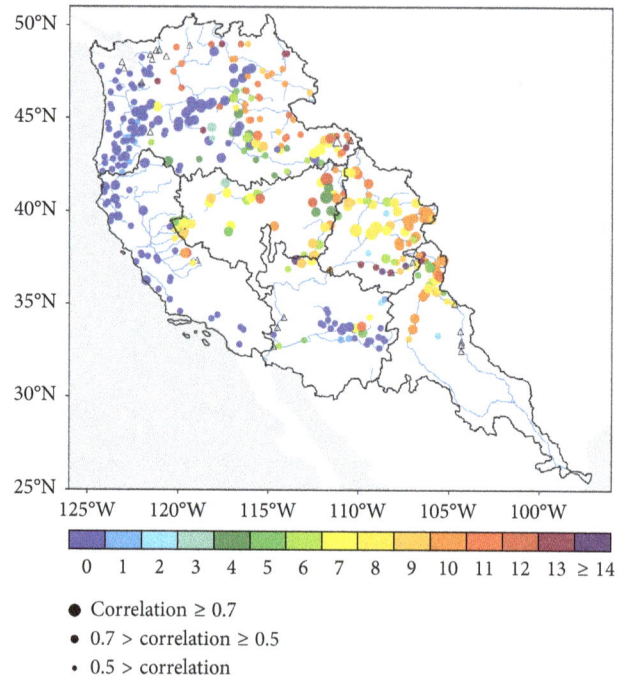

FIGURE 7: Distribution of the peak lead time and correlation coefficient between the snow anomaly and streamflow anomaly (unit: pentad). Hollow triangles represent stations that did not pass the 95% significance test.

areas. Runoff in the eastern PNW is caused predominantly by snowmelt; thus, Figure 7 shows that this region has a significantly longer lead time correlation than the western part. In the CA basin, except for the eastern Sierra Nevada range, where snow contributes considerably to runoff and causes a long lead correlation, most of the regional runoff is affected by rainfall with characteristics similar to the rainfall in the western PNW. This results in a lead correlation of 0-1 pentads with the SWE; however, the short lead correlation simply reflects the relationship between rainfall and runoff.

In summary, these findings provide a reasonably reliable basis for streamflow predictions over the WUS. It is also established that the lead times calculated using the pentad mean data (not show here) are longer than those obtained using the pentad anomaly data. This is because the lead time of the former one represents the period between the snow peak and the runoff peak, whereas the pentad anomaly reflects the relationship between changes in snow accumulation during the snowmelt period and the runoff anomaly. Thus, the lead time of the latter is shorter than the former. Furthermore, the above analysis highlights some problems. In some areas where rainfall dominates over runoff, the lead correlation reflects the relationship of streamflow with rainfall rather than with snow. Therefore, the relative contributions from both rainfall and snow accumulation at each runoff site must be combined to perform runoff anomaly predictions.

4.4. Relative Contributions of Rainfall and Snowmelt.
To obtain predictions with greater accuracy at each streamflow site, the relative contributions from both rainfall and snowpack to each streamflow station were calculated. The

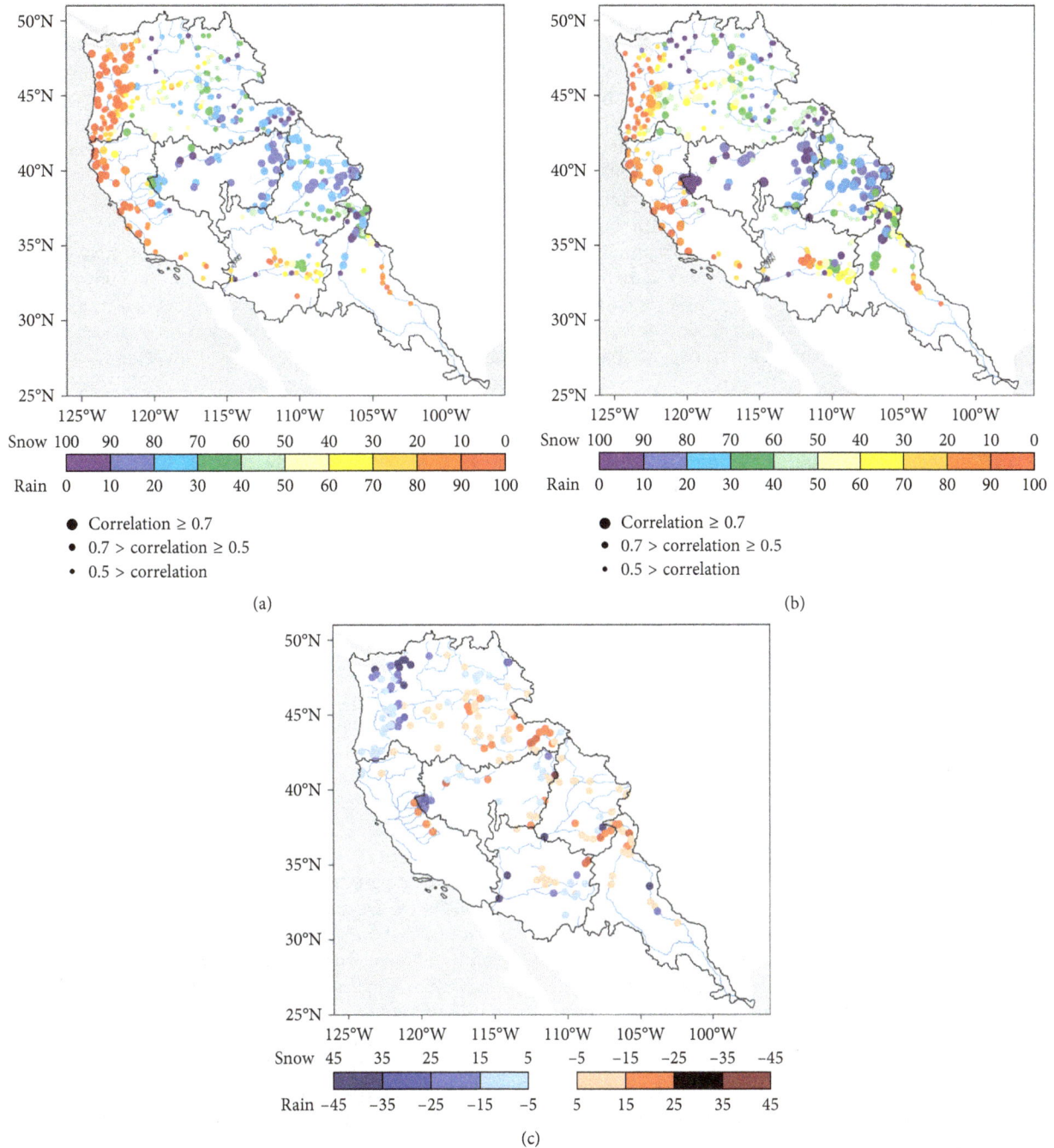

FIGURE 8: Distributions of the relative contributions of rainfall and snowfall to each streamflow station (unit: %): (a) annual contribution, (b) high-flow season contribution, and (c) difference between the high-flow season and annual contributions (i.e., high-flow season-annual). Hollow triangles represent stations that did not pass the 95% significance test.

rainfall and snowfall areas and their ratio were not used to directly distinguish between f_{rain} and f_{snow} for two reasons. First, the impact of the snowmelt area is greater than that of the snow cover area, that is, a site located in a rainfall area might receive a greater contribution from snowmelt than from rainfall. Second, the response of runoff to rainfall is short, while snowmelt has a longer runoff memory with a contribution that is mainly in the spring. Therefore, for a particular runoff site, if the rainfall season and snowmelt seasons occur at different times, the site will receive different

water resources during different seasons. Conversely, if the wet and snowmelt seasons occur at the same time, the site will receive contributions from both resources. Therefore, this study adopted a new method (detailed in Section 3) to calculate the annual f_{rain} and f_{snow} values and to determine the relative contribution rates for the entire year and the high-flow season (i.e., the wet season).

The geographic distribution of the annual relative contributions of rainfall and snowmelt presented in Figure 8 (a) shows that values of $f_{snow} > 50\%$ are distributed primarily in the UC, GB, eastern PNW, and northern RG basins. In these areas, the upper streams in the mountains receive greater contributions from snow accumulation, that is, up to 70% or more. Based on the terrain elevation and distribution of precipitation systems (Figures 1 and 3), it is evident that these areas are high mountainous regions that receive large amounts of snowfall. Snowpack provides abundant water storage in these large areas. In the coastal areas of the PNW and CA basins, the value of f_{rain} is >50%. Despite the large amounts of snowfall received by the Cascade Mountains and the Sierra Nevada, the contributions from rainfall still dominate these areas. For the LC basin, monsoon rainfall provides a greater contribution to streamflow than snowmelt; thus, the streamflow situation in the LC basin is dominated by rainfall.

Runoff variations are an important indicator for local hydrological disasters and agricultural irrigation. Of the entire year, the high-flow season is the most crucial period for early hydrological warnings. Thus, based on Figure 5, three consecutive months with high flow were selected as the wet season for each of the six basins (RG: Apr–Jun, UC: May–Jul, LC: Mar–May, GB: Apr–Jun, PNW: Apr–Jun, and CA: Feb–Apr). Figure 8(b) shows that the distribution of relative contributions in the high-flow season is similar to the annual distribution (Figure 8(a)), that is, rainfall dominates the coastal areas and snowmelt dominates the inner mountain areas. However, some differences between the annual and high-flow season f_{rain} are shown in Figure 8(c). For example, f_{rain} is reduced in the area of the Cascade Mountains in the PNW, while it is increased in the southeastern PNW. This is because the amount of rainfall over the Cascade Mountains is diminished during the high-flow season of this basin (Figure 3); thus, snowmelt becomes comparatively more prominent. The southeastern PNW experiences its heaviest rainfall during the high-flow period (Figure 3); thus, the value of f_{rain} is increased to 5%–25%. For the southern UC and northern RG basins, the contribution from rainfall is increased; alternatively, it could be considered a reduction of the snowmelt contribution. This is because the high-flow period also corresponds to the beginning of the wet season; therefore, the contribution rate of rainfall becomes more prominent. In summary, the relative contributions of rainfall and snowmelt to the runoff could support runoff predictions and could help produce seasonal runoff forecasts that would be more accurate in combination with the lead time results from the previous section.

5. Conclusions

Runoff is affected by many factors, such as the precipitation, soil state, land surface state, and water use. However, snow accumulation and rainfall both constitute important factors with direct impacts on the regional water supply of the WUS. Based on statistical analyses of long-term hydroclimatic data from each watershed, the predictive contributions from both rainfall and snowmelt to streamflow variations were investigated. First, the hydroclimatic characteristics of each basin were summarized in terms of the seasonal rainfall, snow, and runoff variations. Determining the basin-independent features

of runoff that affect the rainfall and snow contributions made it possible to establish numerous runoff forecasting factors. Second, eight snow metrics were selected based on variations in the snow accumulation. Among all of the variables, the peak of snow accumulation, snow-free days, and snow snowmelt slope all have strong correlations with the peak runoff. Over the entire WUS, the correlations among the metrics were stronger in inland regions compared with coastal areas. Thus, these strong significant factors could be used as primary predictors for the runoff amounts during the wet season since the peak streamflow also represents the state of the high-flow season. Meanwhile, it was found that the variations in the snow accumulation anomaly exhibited a leading correlation with the runoff anomaly. This means that the streamflow prediction time based on the snow accumulation could be quantified in advance. Third, the leading relationship between snow and streamflow was analyzed, the results of which showed a lead time ranging from 0 to 14 pentads. Overall, the lead time exceeded 4 pentads in the inland mountain regions, while it was 0-1 pentads in the coastal areas. Because of the considerable impact of snowmelt on the runoff in the UC, GB, and eastern PNW basins, the snow pentad anomaly in these areas had a 4–10 pentad lead period. In comparison, the LC, western PNW, and CA basins had short lead times largely representative of the rainfall effect. This is because their streamflows were affected primarily by rainfall and because the rainfall and snowfall therein generally occurred simultaneously. Finally, based on the pentad anomaly lead time findings, the relative contributions from both snowmelt and rainfall to the runoff at each site in each basin were analyzed. It was found that the UC and GB basins are snow-dominated areas, in which the f_{snow} was greater than f_{rain}. The f_{snow} was the greatest in the northern RG and eastern PNW watersheds, whereas f_{rain} was the greatest in the other parts of these two basins. The greatest contributions in the LC and CA basins were largely from rainfall. It was found that the contribution rate during the wet period was slightly different from the annual rate, that is, the snowmelt contribution increased in the western PWN while the rainfall contributions increased in the southeastern PWN, southern UC, and northern RG basins. In conclusion, quantitative and qualitative analyses regarding runoff predictions for the studied watersheds of the WUS constitute an important reference for hydrological management and will be useful in the evaluation and improvement of hydrological and climate models.

6. Discussion

This research explored the predictability of runoff in the WUS; however, scope remains for further study. (1) In this study, for the discussion of the relationship at each runoff station in each basin, the snow state in each basin was based on the average at all snow sites. However, for large watersheds, for example, the PNW, the in situ data were distributed in different mountain areas with different snow factor characteristics; thus, the method of averaging would cause some bias. In future work, subbasins will be defined to reduce this deviation. (2) When considering the contribution rates to runoff, data limitations restricted the discussion to just rainfall and snow. In fact, runoff is affected by many other factors such as evaporation, soil state, and groundwater. In future research, these factors will be taken into consideration. However, the findings of this study have value regarding the seasonal prediction of runoff, and they could serve as a basis for future research and model testing.

Conflicts of Interest

The authors declare that they have no conflicts of interest.

Acknowledgments

The authors thank James Buxton, MSc, from Liwen Bianji, Edanz Group China (http://www.liwenbianji.cn./ac), for editing the English text of this manuscript. This research was supported by National construction of high-quality University projects of graduates and National Natural Science Foundation of China (41590874, 41675107, and 41775081).

References

[1] D. W. Pierce, T. P. Barnett, H. G. Hidalgo, T. Das, C. Bonfils, and B. D. Santer, "Attribution of declining Western U.S. Snowpack to human effects," *Journal of Climate*, vol. 21, no. 23, pp. 6425–6444, 2008.

[2] P. Z. Klos, T. E. Link, and J. T. Abatzoglou, "Extent of the rain-snow transition zone in the Western U.S. under historic and projected climate," *Geophysical Research Letters*, vol. 41, no. 13, pp. 4560–4568, 2014.

[3] J. A. Johnstone, "A quasi-biennial signal in Western US hydroclimate and its global teleconnections," *Climate Dynamics*, vol. 36, no. 3-4, pp. 663–680, 2011.

[4] J. T. Abatzoglou, "Influence of the PNA on declining mountain snowpack in the Western United States," *International Journal of Climatology*, vol. 31, no. 8, pp. 1135–1142, 2011.

[5] H. Chang and I. W. Jung, "Spatial and temporal changes in runoff caused by climate change in a complex large river basin in Oregon," *Journal of Hydrology*, vol. 388, no. 3-4, pp. 186–207, 2010.

[6] D. W. Clow, "Changes in the timing of snowmelt and streamflow in Colorado: a response to recent warming," *Journal of Climate*, vol. 23, no. 9, pp. 2293–2306, 2010.

[7] M. Ashfaq, S. Ghosh, S. C. Kao et al., "Near-term acceleration of hydroclimatic change in the Western U.S.," *Journal of Geophysical Research Atmospheres*, vol. 118, no. 19, pp. 10676–10693, 2013.

[8] N. Water, N. Resources, and C. Service, "A recent increase in Western US streamflow variability and persistence," *Journal of Hydrometeorology*, vol. 6, no. 2, pp. 173–179, 2005.

[9] J. T. Abatzoglou and C. Kolden, "Climate change in Western US deserts: potential for increased wildfire and invasive annual grasses," *Rangeland Ecology & Management*, vol. 64, no. 5, pp. 471–478, 2011.

[10] D. Li, M. L. Wrzesien, M. Durand, J. Adam, and D. P. Lettenmaier, "How much runoff originates as snow in the Western United States, and how will that change in the future?," *Geophysical Research Letters*, vol. 44, no. 12, pp. 6163–6172, 2017.

[11] A. Kalra, S. Ahmad, and A. Nayak, "Increasing streamflow forecast lead time for snowmelt-driven catchment based on large-scale climate patterns," *Advances in Water Resources*, vol. 53, pp. 150–162, 2013.

[12] A. Kalra, T. C. Piechota, R. Davies, and G. A. Tootle, "Changes in U.S. streamflow and Western U.S. snowpack," *Journal of Hydrologic Engineering*, vol. 13, no. 3, pp. 156–163, 2008.

[13] A. Kalra and S. Ahmad, "Estimating annual precipitation for the Colorado River Basin using oceanic–atmospheric oscillations," *Water Resources Research*, vol. 48, no. 6, article W06527, 2012.

[14] A. Nayak, D. Marks, D. G. Chandler, and A. Winstral, "Modeling inter-annual variability in snowcover development and melt for semi-arid mountain catchment," *Journal of Hydrologic Engineering*, vol. 17, no. 1, pp. 74–84, 2012.

[15] T. Hunter, G. A. Tootle, and T. C. Piechota, "Oceanic-atmospheric variability and western US snowfall," *Geophysical Research Letters*, vol. 33, no. 13, article L13706, 2006.

[16] N. Knowles, M. D. Dettinger, and D. R. Cayan, "Trends in snowfall versus rainfall in the western United States," *Journal of Climate*, vol. 19, no. 18, pp. 4545–4559, 2006.

[17] D. W. Clow, L. Nanus, K. L. Verdin, and J. Schmidt, "Evaluation of SNODAS snow depth and snow water equivalent estimates for the Colorado Rocky Mountains, USA," *Hydrological Processes*, vol. 26, no. 17, pp. 2583–2591, 2012.

[18] A. Harpold, P. Brooks, S. Rajagopal, I. Heidbuchel, A. Jardine, and C. Stielstra, "Changes in snowpack accumulation and ablation in the intermountain west," *Water Resources Research*, vol. 48, no. 11, article W11501, 2012.

[19] A. Harpold and N. P. Molotch, "Sensitivity of soil water availability to changing snowmelt timing in the western U.S.," *Geophysical Research Letters*, vol. 42, no. 19, pp. 8011–8020, 2015.

[20] T. B. Barnhart, N. P. Molotch, B. Livneh, A. A. Harpold, J. F. Knowles, and D. Schneider, "Snowmelt rate dictates streamflow," *Geophysical Research Letters*, vol. 43, no. 15, pp. 8006–8016, 2016.

[21] A. F. Hamlet, P. W. Mote, and M. P. Clark, "Twentieth-century trends in runoff, evapotranspiration, and soil moisture in the western United States," *Journal of Climate*, vol. 20, no. 8, pp. 1468–1486, 2007.

[22] I. T. Stewart, D. R. Cayan, and M. D. Dettinger, "Changes toward earlier streamflow timing across Western North America," *Journal of Climate*, vol. 18, no. 8, pp. 1136–1155, 2005.

[23] A. E. Jeton, M. D. Dettinger, and J. Smith, *Potential Effects of Climate Change on Streamflow, Eastern and Western Slopes of the Sierra Nevada, California and Nevada*, United States Geological Survey, Sacramento, CA, USA, 1996.

[24] E. Bosson, U. Sabel, L.-G. Gustafsson, M. Sassner, and G. Destouni, "Influences of shifts in climate, landscape, and permafrost on terrestrial hydrology," *Journal of Geophysical Research: Atmospheres*, vol. 117, no. 5, article D05120, 2012.

[25] E. Trujillo and N. P. Molotch, "Snowpack regimes of the Western United States," *Water Resources Research*, vol. 50, no. 7, pp. 5611–5623, 2014.

[26] M. C. Serreze, M. P. Clark, R. L. Armstrong, D. A. McGinnis, and R. S. Pulwarty, "Characteristics of the western United States snowpack from snowpack telemetry (SNOTEL) data," *Water Resources Research*, vol. 35, no. 7, pp. 2145–2160, 1999.

[27] K. T. Chang and Z. Li, "Modelling snow accumulation with a geographic information system," *International Journal of Geographical Information Science*, vol. 14, no. 7, pp. 693–707, 2000.

[28] J. J. Simpson and T. J. McIntire, "A recurrent neural network

classifier for improved retrievals of areal extent of snow cover," *IEEE Transactions on Geoscience and Remote Sensing*, vol. 39, no. 10, pp. 2135–2147, 2001.

[29] M. K. Cowles, D. L. Zimmerman, A. Christ, and D. L. McGinnis, "Combining snow water equivalent data from multiple sources to estimate spatiotemporal trends and compare measurement systems," *Journal of Agricultural, Biological, and Environmental Statistics*, vol. 7, no. 4, pp. 536–557, 2002.

[30] C. Powell, L. Blesius, J. Davis, and F. Schuetzenmeister, "Using MODIS snow cover and precipitation data to model water runoff for the Mokelumne River Basin in the Sierra Nevada, California (2000–2009)," *Global and Planetary Change*, vol. 77, no. 1-2, pp. 77–84, 2011.

[31] C. M. Welch, P. C. Stoy, F. A. Rains, A. V. Johnson, and B. L. McGlynn, "The impacts of mountain pine beetle disturbance on the energy balance of snow during the melt period," *Hydrological Processes*, vol. 30, no. 4, pp. 588–602, 2016.

[32] I. T. Stewart, D. R. Cayan, and M. D. Dettinger, "Changes in snowmelt runoff timing in western North America under 'a business as usual' climate change scenario," *Climatic Change*, vol. 62, no. 1–3, pp. 217–232, 2004.

[33] T. Barnett, J. Adam, and D. Lettenmaier, "Potential impacts of a warming climate on water availability in snow-dominated regions," *Nature*, vol. 438, no. 7066, pp. 303–309, 2005.

[34] B. J. Gillan, J. T. Harper, and J. N. Moore, "Timing of present and future snowmelt from high elevations in northwest Montana," *Water Resources Research*, vol. 46, no. 1, article W01507, 2010.

[35] H. Fritze, I. T. Stewart, and E. Pebesma, "Shifts in Western North American snowmelt runoff regimes for the recent warm decades," *Journal of Hydrometeorology*, vol. 12, no. 5, pp. 989–1006, 2011.

[36] K. A. Dressler, S. R. Fassnacht, and R. C. Bales, "A comparison of snow telemetry and snow course measurements in the Colorado River basin," *Journal of Hydrometeorology*, vol. 7, no. 4, pp. 705–712, 2006.

[37] G. E. Maurer and D. R. Bowling, "Seasonal snowpack characteristics influence soil temperature and water content at multiple scales in interior western U.S. mountain ecosystems," *Water Resources Research*, vol. 50, no. 6, pp. 5216–5234, 2014.

[38] H. F. Lins, *USGS Hydro-Climatic Data Network 2009 (HCDN-2009), U.S. Geological Survey Fact Sheet 2012-3047*, United States Geological Survey, Reston, VA, USA, 2012, http://pubs.usgs.gov/fs/2012/3047/.

[39] C. Daly, R. P. Neilson, and D. L. Phillips, "A statistical-topographic model for mapping climatological precipitation over mountainous terrain," *Journal of Applied Meteorology*, vol. 33, no. 2, pp. 140–158, 1994.

[40] C. Daly, M. D. Halbleib, J. I. Smith et al., "Physiographically sensitive mapping of climatological temperature and precipitation across the conterminous United States," *International Journal of Climatology*, vol. 28, no. 15, pp. 2031–2064, 2008.

[41] *Descriptions of PRISM Spatial Climate Datasets for the Conterminous United States*, 2015.

[42] J. D. Goodridge, "A study of 1000 year storms in California," in *Preprints, Predicting Heavy Rainfall Events in California: A Symposium to Share Weather Pattern Knowledge*, C. Dailey, Ed., pp. 3–72, Sierra College, Rocklin, CA, USA, 1994.

[43] A. Kalra and S. Ahmad, "Evaluating changes and estimating seasonal precipitation for the Colorado River Basin using a stochastic nonparametric disaggregation technique," *Water Resources Research*, vol. 47, no. 5, article W05555, 2011.

[44] D. J. Seidel, Q. Fu, W. J. Randel, and T. J. Reichler, "Widening of the tropical belt in a changing climate," *Nature Geoscience*, vol. 1, no. 1, pp. 21–24, 2007.

Permissions

All chapters in this book were first published in AM, by Hindawi Publishing Corporation; hereby published with permission under the Creative Commons Attribution License or equivalent. Every chapter published in this book has been scrutinized by our experts. Their significance has been extensively debated. The topics covered herein carry significant findings which will fuel the growth of the discipline. They may even be implemented as practical applications or may be referred to as a beginning point for another development.

The contributors of this book come from diverse backgrounds, making this book a truly international effort. This book will bring forth new frontiers with its revolutionizing research information and detailed analysis of the nascent developments around the world.

We would like to thank all the contributing authors for lending their expertise to make the book truly unique. They have played a crucial role in the development of this book. Without their invaluable contributions this book wouldn't have been possible. They have made vital efforts to compile up to date information on the varied aspects of this subject to make this book a valuable addition to the collection of many professionals and students.

This book was conceptualized with the vision of imparting up-to-date information and advanced data in this field. To ensure the same, a matchless editorial board was set up. Every individual on the board went through rigorous rounds of assessment to prove their worth. After which they invested a large part of their time researching and compiling the most relevant data for our readers.

The editorial board has been involved in producing this book since its inception. They have spent rigorous hours researching and exploring the diverse topics which have resulted in the successful publishing of this book. They have passed on their knowledge of decades through this book. To expedite this challenging task, the publisher supported the team at every step. A small team of assistant editors was also appointed to further simplify the editing procedure and attain best results for the readers.

Apart from the editorial board, the designing team has also invested a significant amount of their time in understanding the subject and creating the most relevant covers. They scrutinized every image to scout for the most suitable representation of the subject and create an appropriate cover for the book.

The publishing team has been an ardent support to the editorial, designing and production team. Their endless efforts to recruit the best for this project, has resulted in the accomplishment of this book. They are a veteran in the field of academics and their pool of knowledge is as vast as their experience in printing. Their expertise and guidance has proved useful at every step. Their uncompromising quality standards have made this book an exceptional effort. Their encouragement from time to time has been an inspiration for everyone.

The publisher and the editorial board hope that this book will prove to be a valuable piece of knowledge for researchers, students, practitioners and scholars across the globe.

List of Contributors

Aldo S. Moya-Álvarez, José L. Flores, Daniel Martínez-Castro and Yamina Silva
Instituto Geofísico del Perú, Lima, Peru

Daniel Martínez-Castro
Instituto de Meteorología de Cuba, La Habana, Cuba

Iman Rousta
Department of Geography, Yazd University, Yazd 8915818411, Iran

Iman Rousta
Senior Researcher, Institute for Atmospheric Sciences, University of Iceland and Icelandic Meteororological Office (IMO), Bustadavegur 7, IS-108 Reykjavik, Iceland

Farshad Javadizadeh
Department of Environment, Collage of Natural Resource, Bandar Abbas Branch, Islamic Azad University, Bandar Abbas, Iran

Fatemeh Dargahian
Desert Research Division, Research Institute of Forests and Rangelands, Agricultural Research Education and Extension Organization (AREEO), Tehran, Iran

Haraldur Ólafsson
Department of Physics, University of Iceland, Institute for Atmospheric Sciences and Icelandic Meteororological Office (IMO), Bustadavegur 7, IS-108 ReykjaviK, Iceland

Amin Shiri-Karimvandi, Mehdi Doostkamian and Anayat Asadolahi
Department of Geography, University of Zanjan, Zanjan 3879145371, Iran

Sayed Hossein Vahedinejad
Department of Geography, University of Kharazmi, Tehran 1491115719, Iran

Edgar Ricardo Monroy Vargas
Department of Civil Engineering, Universidad Catolica de Colombia, Bogotá, Colombia

Wei Sun, Yonghua Sun, Xiaojuan Li, Tao Wang, Yanbing Wang, Qi Qiu and Zhitian Deng
College of Resource Environment and Tourism, Capital Normal University, Beijing 100048, China
College of Geospatial Information Science and Technology, Capital Normal University, Beijing 100048, China

Beijing Laboratory of Water Resource Security, Capital Normal University, Beijing 100048, China

Shaobo Zhong, Chaolin Wang, Zhichen Yu, Yongsheng Yang and Quanyi Huang
Department of Engineering Physics, Institute of Public Safety Research, Tsinghua University, Beijing 100084, China

Fabiani Denise Bender and Paulo Cesar Sentelhas
Department of Biosystems Engineering, ESALQ, University of São Paulo, 13418-900 Piracicaba, SP, Brazil

Stephen M. Jessup and Amanda L. Burke
Department of the Earth Sciences, The College at Brockport, State University of New York, Brockport, NY, USA

Chenkai Cai, Jianqun Wang and Zhijia Li
College of Hydrology and Water Resources, Hohai University, Nanjing 210098, China

Tomi Afrizal
Interdisciplinary Graduate School of Earth System Science and Andaman Natural Disaster Management, Prince of Songkla University, Phuket Campus, Phuket 83120, Thailand

Chinnawat Surussavadee
Telecommunications Engineering Department, Faculty of Engineering, King Mongkut's Institute of Technology Ladkrabang, Bangkok 10520, Thailand
Research Laboratory of Electronics, Massachusetts Institute of Technology, Cambridge, MA 02139, USA

Shibo Gao and Jinzhong Min
Key Laboratory of Meteorological Disaster, Ministry of Education (KLME), Joint International Research Laboratory of Climate and Environment Change (ILCEC), Collaborative Innovation Center on Forecast and Evaluation of Meteorological Disaster (CICFEMD), Nanjing University of Information Science and Technology, Nanjing 210044, China

Qingping Cheng, Lu Gao, Ying Chen, Meibing Liu, Haijun Deng and Xingwei Chen
College of Geographical Science, Fujian Normal University, Fuzhou 350007, China

Qingping Cheng
Northwest Institute of Eco-Environmental and Resources Research, Chinese Academy of Sciences, Lanzhou 730000, China

Qingping Cheng
University of Chinese Academy of Sciences, Beijing 100049, China

Lu Gao, Ying Chen, Meibing Liu, Haijun Deng and Xingwei Chen
Institute of Geography, Fujian Normal University, Fuzhou 350007, China
Fujian Provincial Engineering Research Center for Monitoring and Assessing Terrestrial Disasters, Fujian Normal University, Fuzhou 350007, China
State Key Laboratory of Subtropical Mountain Ecology (Funded by Ministry of Science and Technology and Fujian Province), Fujian Normal University, Fuzhou 350007, China

Matthew J. Lauridsen and Brian C. Ancell
Texas Tech University, Lubbock, TX, USA

Ha-Young Yang, Ki-Ho Chang, Sanghee Chae, Seongkyu Seo, Jin-Yim Jeong, Jung-Ho Lee, Yonghun Ro and Baek-Jo Kim
Applied Meteorology Research Division, National Institute of Meteorological Sciences, Seogwipo 63568, Republic of Korea

Eunsil Jung
School of Constructional Disaster Prevention and Environmental Engineering, Kyungpook National University, Sangju 37224, Republic of Korea

Min-Ki Lee, Seung-Hyun Moon and Byung-Ro Moon
School of Computer Science and Engineering, Seoul National University, 1 Gwanak-ro, Gwanak-gu, Seoul 08826, Republic of Korea

Yourim Yoon
Department of Computer Engineering, College of Information Technology, Gachon University, 1342 Seongnam-daero, Sujeong-gu, Seongnam-si, Gyeonggi-do 13120, Republic of Korea

Yong-Hyuk Kim
School of Software, Kwangwoon University, 20 Kwangwoon-ro, Nowon-gu, Seoul 01897, Republic of Korea

Xiaohui Zheng and Lihua Zhou
College of Global Change and Earth System Science, Beijing Normal University, Beijing 100875, China

Qiguang Wang
China Meteorological Administration Training Center (CMA), Beijing 100081, China

Qing Sun
College of Applied Meteorology, Nanjing University of Information Science and Technology, Nanjing 210044, China

Qi Li
School of Atmospheric Physics and Atmospheric Environment, Nanjing University of Information Science and Technology, Nanjing 210044, China

Index

www.ingramcontent.com/pod-product-compliance
Lightning Source LLC
Chambersburg PA
CBHW080527200326
41458CB00012B/4354